Authority and Expertise in Ancient Scientific Culture

How did ancient scientific and knowledge-ordering writers make their work authoritative? This book answers that question for a wide range of ancient disciplines, from mathematics, medicine, architecture and agriculture, through to law, historiography and philosophy (focusing mainly but not exclusively on the literature of the Roman Empire). It draws attention to habits that these different fields had in common, while also showing how individual texts and authors manipulated standard techniques of self-authorisation in distinctive ways. It stresses the importance of competitive and assertive styles of self-presentation, and also examines some of the pressures that pulled in the opposite direction by looking at authors who chose to acknowledge the limitations of their own knowledge or resisted close identification with narrow versions of expert identity. A final chapter by Sir Geoffrey Lloyd offers a comparative account of scientific authority and expertise in ancient Chinese, Indian and Mesopotamian culture.

JASON KÖNIG is Professor of Greek at the University of St Andrews. This is the third in a trilogy of volumes arising from a Leverhulme-funded research project, 'Science and Empire in the Roman World', which ran from 2007 to 2010 in St Andrews; the other two volumes, *Ancient Libraries* and *Encyclopaedism from Antiquity to the Renaissance*, were both published by Cambridge University Press in 2013.

GREG WOOLF is Professor of Classics and Director of the Institute of Classical Studies in London. He co-directed the project, 'Science and Empire in the Roman World', at St Andrews and co-edited the two previous books resulting from it.

Authority and Expertise in Ancient Scientific Culture

Edited by JASON KÖNIG AND GREG WOOLF

CAMBRIDGE
UNIVERSITY PRESS

CAMBRIDGE
UNIVERSITY PRESS

University Printing House, Cambridge CB2 8BS, United Kingdom

One Liberty Plaza, 20th Floor, New York, NY 10006, USA

477 Williamstown Road, Port Melbourne, VIC 3207, Australia

314-321, 3rd Floor, Plot 3, Splendor Forum, Jasola District Centre, New Delhi - 110025, India

79 Anson Road, #06-04/06, Singapore 079906

Cambridge University Press is part of the University of Cambridge.

It furthers the University's mission by disseminating knowledge in the pursuit of education, learning and research at the highest international levels of excellence.

www.cambridge.org
Information on this title: www.cambridge.org/9781107629646
10.1017/9781107446724

© Cambridge University Press 2017

This publication is in copyright. Subject to statutory exception and to the provisions of relevant collective licensing agreements, no reproduction of any part may take place without the written permission of Cambridge University Press.

First published 2017
First paperback edition 2021

A catalogue record for this publication is available from the British Library

Library of Congress Cataloging in Publication data
Names: König, Jason. | Woolf, Greg.
Title: Authority and expertise in ancient scientific culture / edited by Jason König and Greg Woolf.
Description: Cambridge : Cambridge University Press, [2017] | Includes bibliographical references.
Identifiers: LCCN 2016023672 | ISBN 9781107060067
Subjects: LCSH: Science, Ancient. | Science – Greece – History.
Classification: LCC Q124.95.A98 2017 | DDC 509.38 – dc23
LC record available at https://lccn.loc.gov/2016023672

ISBN 978-1-107-06006-7 Hardback
ISBN 978-1-107-62964-6 Paperback

Cambridge University Press has no responsibility for the persistence or accuracy of URLs for external or third-party internet websites referred to in this publication, and does not guarantee that any content on such websites is, or will remain, accurate or appropriate.

Contents

List of Contributors [*page* vii]
Preface [x]
List of Abbreviations [xi]

1 Introduction: Self-Assertion and Its Alternatives in Ancient Scientific and Technical Writing [1]
JASON KÖNIG

2 Philosophical Authority in the Imperial Period [27]
MICHAEL TRAPP

3 Philosophical Authority in the Younger Seneca [58]
HARRY HINE

4 *Iurisperiti*: 'Men Skilled in Law' [83]
JILL HARRIES

5 Making and Defending Claims to Authority in Vitruvius' *De architectura* [107]
DANIEL HARRIS-MCCOY

6 Fragile Expertise and the Authority of the Past: The 'Roman Art of War' [129]
MARCO FORMISANO

7 Conflicting Models of Authority and Expertise in Frontinus' *Strategemata* [153]
ALICE KÖNIG

8 The Authority of Writing in Varro's *De re rustica* [182]
AUDE DOODY

9 The Limits of Enquiry in Imperial Greek Didactic Poetry [203]
EMILY KNEEBONE

10 Expertise, 'Character' and the 'Authority Effect' in the *Early Roman History* of Dionysius of Halicarnassus [231]
NICOLAS WIATER

11 The Authority of Galen's Witnesses   [260]
DARYN LEHOUX

12 Anatomy and *Aporia* in Galen's *On the Construction of Fetuses*   [283]
RALPH M. ROSEN

13 Varro the Roman Cynic: The Destruction of Religious Authority in the *Antiquitates rerum divinarum*   [306]
LEAH KRONENBERG

14 Signs, Seers and Senators: Divinatory Expertise in Cicero and Nigidius Figulus   [329]
KATHARINA VOLK

15 The Public Face of Expertise: Utility, Zeal and Collaboration in Ptolemy's *Syntaxis*   [348]
JOHANNES WIETZKE

16 The Authority of Mathematical Expertise and the Question of Ancient Writing *More Geometrico*   [374]
REVIEL NETZ

17 Authority and Expertise: Some Cross-Cultural Comparisons   [409]
G. E. R. LLOYD

Bibliography   [424]
Index   [463]

Contributors

JASON KÖNIG is Professor of Greek at the University of St Andrews. He works broadly on the Greek literature and culture of the Roman Empire. His books include *Athletics and Literature in the Roman Empire* (2005), *Saints and Symposiasts: The Literature of Food and the Symposium in Greco-Roman and Early Christian Culture* (2012) and *Ordering Knowledge in the Roman Empire* (2007) (edited jointly with Tim Whitmarsh).

GREG WOOLF is Professor of Classics and Director of the Institute of Classical Studies in London. His books include *Becoming Roman: The Origins of Provincial Civilization in Gaul* (1998), *Et tu Brute: The Murder of Julius Caesar and Political Assassination* (2006), *Tales of the Barbarians: Ethnography and Empire in the Roman West* (2011) and *Rome: An Empire's Story* (2012). He has also edited volumes on literacy, on the city of Rome and on Roman religion and has published widely on ancient history and Roman archaeology.

AUDE DOODY is Lecturer in Classics at University College Dublin. Her research focuses on ancient scholarship and technical writing, especially Pliny the Elder; her publications include *Pliny's Encyclopedia: The Reception of the Natural History* (2010).

MARCO FORMISANO is Professor (*docent*) of Latin Literature at Ghent University. He has published extensively on ancient technical and scientific writing, on late Latin literature, on martyr acts and on the reception of antiquity. Relevant publications include *Tecnica e scrittura. Le letterature tecnico-scientifiche nello spazio letterario tardolatino* (2001), *Flavio Vegezio Renato, L'arte della guerra romana* (2003), *War in Words: Transformations of War from Antiquity to Clausewitz* (edited with Hartmut Böhme) (2011), *Knowledge, Text and Practice in Ancient Technical Writing* (2017) and *Vitruvius: Text, Architecture, Reception* (2016).

JILL HARRIES is Emeritus Professor of Ancient History in the School of Classics at the University of St Andrews. She is the author of *Sidonius Apollinaris and the Fall of Rome* (1994), *Law and Empire in Late Antiquity* (1999),

Cicero and the Jurists (2006), *Law and Crime in the Roman World* (2007) and *Imperial Rome, AD 284–363: The New Empire* (2012).

DANIEL HARRIS-MCCOY is Assistant Professor of Classics at the University of Hawai'i at Mānoa. His edition of Artemidorus' *Oneirocritica* was published in 2012.

HARRY HINE is Emeritus Professor in the School of Classics at the University of St Andrews. He has edited Seneca's *Natural Questions*, and has published widely on Seneca and on Latin technical and scientific literature.

EMILY KNEEBONE is College Lecturer in Classics at Newnham College, Cambridge, and Co-Investigator of the AHRC research project 'Greek Epic of the Roman Empire: A Cultural History' (University of Cambridge). She has published articles on imperial Greek prose and poetry, including Josephus and Quintus of Smyrna, and is currently completing a monograph on Oppian's *Halieutica*.

ALICE KÖNIG is Lecturer in Latin at the University of St Andrews. She works on Domitianic, Nervan and Trajanic literature, ancient technical/scientific writing (especially Vitruvius and Frontinus) and military writing. She is currently working on a monograph which will bring Frontinus' surviving works into dialogue with each other. She is also running a collaborative research project on 'Literary Interactions under Nerva, Trajan and Hadrian' based at the University of St Andrews.

LEAH KRONENBERG is Associate Professor of Classics at Rutgers University. She is the author of *Allegories of Farming from Greece and Rome: Philosophical Satire in Xenophon, Varro, and Virgil* (2009), as well as articles on Catullus, Virgil and Ovid.

DARYN LEHOUX is Professor of Classics at Queen's University, Ontario. He is the author of *Creatures Born of Mud and Slime* (forthcoming), *What Did the Romans Know?* (2012), and *Astronomy, Weather, and Calendars in the Ancient World* (2007), as well as co-editor (with A. D. Morrison and A. Sharrock) of *Lucretius: Poetry, Philosophy, Science* (2013).

SIR GEOFFREY LLOYD is Emeritus Professor of Ancient Philosophy and Science at the University of Cambridge and Senior Scholar at the Needham Research Institute. His most recent books are *The Ideals of Inquiry: An Ancient History* (2014) and *Analogical Investigations: Historical and Cross-Cultural Perspectives on Human Reasoning* (2015).

REVIEL NETZ is Professor of Classics at Stanford University. His main field is the history of pre-modern mathematics. His research involves the wider issues of the history of cognitive practices, e.g. visual culture, the history of the book, literacy and numeracy. His books include *The Shaping of Deduction in Greek Mathematics: A Study in Cognitive History* (1999), *The Transformation of Early Mediterranean Mathematics: From Problems to Equations* (2004) and *Ludic Proof: Greek Mathematics and the Alexandrian Aesthetic* (2009).

RALPH M. ROSEN is Vartan Gregorian Professor of Humanities and Classical Studies at the University of Pennsylvania. He has published widely in various areas of Greek and Roman literature, intellectual history and ancient medicine. His most recent book is *Making Mockery: The Poetics of Ancient Satire* (2007). He is also co-editor (with Lesley Dean-Jones) of *Ancient Concepts of the Hippocratic: Papers Presented at the XIIIth International Hippocrates Colloquium* (2015).

MICHAEL TRAPP is Professor of Greek Literature and Thought at King's College London. His books include *Maximus of Tyre: The Philosophical Orations* (1997), *Philosophy in the Roman Empire: Ethics, Politics and Society* (2007) and (as editor) *Socrates from Antiquity to the Enlightenment* and *Socrates in the Nineteenth and Twentieth Centuries* (both 2007). He is now working on a revision and completion of the Loeb Classical Library edition of Aelius Aristides.

KATHARINA VOLK is Professor of Classics at Columbia University and the author of monographs on Latin didactic poetry, the astrological poet Manilius and Ovid. She is currently working on a book on the intellectual history of the late Roman Republic.

NICOLAS WIATER is Lecturer in Classics at the University of St Andrews. His major publications include *The Ideology of Classicism: Language, History, and Identity in Dionysius of Halicarnassus* (2011), a new German translation with commentary of Dionysius' *Roman Antiquities* (volume 1 published in 2014) and *The Struggle for Identity: Greeks and Their Past in the First Century BCE* (2011) (jointly edited with Thomas A. Schmitz). His current projects include a new commentary on the third book of Polybius' *Histories*.

JOHANNES WIETZKE received his PhD in Classics from Stanford University and is currently Visiting Assistant Professor in the Department of Classics at Carleton College, Texas. He is preparing a monograph on authorship and self-canonisation in ancient scientific, technical and expository writing.

Preface

This book is the third in a trilogy of edited volumes arising from a project on 'Science and Empire in the Roman World' which ran in St Andrews from 2007 to 2010, funded by the Leverhulme Trust; the others are *Ancient Libraries* and *Encyclopaedism from Antiquity to the Renaissance*, both published by Cambridge University Press in 2013. The project as a whole was a joint enterprise, but Jason König took the lead in assembling this final collection and bringing its editing to completion. This third volume arose from a series of workshops and conferences on scientific writing and on scientists and professionals in the ancient world which took place during the second half of the project. Very few of the chapters below were delivered as papers – most were commissioned and designed afterwards – so this is not in any sense a conference volume, but we are very grateful to all who attended those events to give papers and to contribute to our discussions, which had a formative influence over our plans for the volume as a whole. Among other things, they helped to convince us even further of the value of examining a wide range of different disciplines together in order to see better the cross-fertilisation between them. In that sense the word 'science' in the title is intended in the most capacious sense possible, to encompass the whole industry of ancient knowledge-ordering, not just what we would refer to now as 'scientific' and technical topics, but also fields like law, historiography and generalship, which are not often studied as part of the history of ancient science. We are grateful to the Leverhulme Trust for funding that project, to Katerina Oikonomopoulou, who collaborated with us in organising all of these events as postdoctoral fellow for the project, and to Michael Sharp and all at Cambridge University Press. In what follows we have generally preferred Latinate spellings of Greek names, except where the Greek version seems more familiar or more appropriate.

Abbreviations

Journal titles are cited in full in the bibliography. Abbreviations in chapters follow the *Oxford Classical Dictionary* (or for works not listed there, the *Greek-English Lexicon* (LSJ) or the *Oxford Latin Dictionary*). Title abbreviations for the works of Galen follow Hankinson 2008b: 391–7. The following abbreviations are not listed in any of those sources:

EAE	*ṭupšarru Enūma Anu Enlil*
EANS	Keyser, P. T. and Irby-Massie, G. L. (eds) (2008) *The Encyclopedia of Ancient Natural Scientists*, London.
K	Kühn, K. G. (ed.) (1821–33) *Opera omnia Claudii Galeni*, Leipzig.
N	Nickel, D. (2001) *Galen: Über die Ausformung der Keimlinge* (*CMG* 5.3.3), Berlin.
Varro, *ARD*	Varro, *Antiquitates rerum divinarum*

1 | Introduction: Self-Assertion and Its Alternatives in Ancient Scientific and Technical Writing

JASON KÖNIG

Interdisciplinary Approaches to Authority and Expertise

One of the most distinctive developments in the history of science as an academic discipline within the last few decades has been a new attention to rhetoric.[1] The personas of modern scientific writing, even in their most objective and dispassionate forms, are the product of culturally specific institutional pressures and educational histories; they are shaped by culturally specific assumptions about what makes an argument authoritative, or what makes an individual or an institution trustworthy. Scientific writing seeks not only to be truthful but also to persuade, and while the two may often go hand in hand, ideas about what is convincing and deserving of attention, what carries weight, inevitably vary from decade to decade and even from discipline to discipline.

Hand-in-hand with that work on modern scientific rhetoric – though not always in close communication with it – there has been an increasing volume of publications on related issues in ancient science. A major strand of recent ancient-science scholarship has given prominence to questions about how ancient scientists represented themselves and their disciplines, and particularly how they made their writings authoritative. Geoffrey Lloyd's work has had a pioneering influence in that respect, not least his comparative research on the relations between Greek and Chinese science. A central contribution of his work has been to elucidate the competitive quality of ancient Greek and Roman scientific discourse, which was formed in a world without formal scientific or educational qualifications. That situation meant that ancient experts had to work much harder than their modern counterparts to convince their audiences and potential clients and students, by

---

[1] Among many other works, see Latour 1987: esp. 21–62 on the authority-claiming rhetoric of modern scientific publications; Gross 1996, heavily revised as Gross 2006; Pera 1994; Pera and Shea 1991; Ziman 1984: esp. 58–80 on communication and authority in modern science, and Ziman 2000; also Asper 2013b: 3–4 for brief reflections on the relevance of those approaches to ancient science.

rhetorical means, of their competence,² and so tended to reach for self-assertive and ostentatiously innovative first-person personas.³ In that sense attention to persona and authority is perhaps an even more obvious priority for the ancient world than for modern science, given that the scientific 'I' was often so much more prominent.

There are now many publications which pay serious attention to these issues for specific texts and authors. The work of Galen, the great medical writer of the second century CE, is an obvious example. The extraordinary range of his surviving works and the prominence of his own personality in many of his writings make him an ideal candidate for viewing ancient scientific self-assertion in action. And in many respects his work is typical of ancient scientific writing more broadly. He gives a prominent role to his own persona.⁴ He is consistently competitive: he regularly debunks rival practitioners and rival disciplines which do not measure up to the philosophically inspired medical knowledge he himself espouses.⁵ He draws attention to his own moral virtue in ways which bring an impression of reliability.⁶ He also draws attention to his remarkably wide learning. His authority rests in part on his intricate knowledge of the work of his predecessors, and his alignment of himself with that tradition, especially in his opportunistic appropriation of the writings of the Hippocratic corpus so that they come to match his own medical views.⁷ At the same time he repeatedly challenges received wisdom. In some cases, he does that through a claim to personal experience and observation,⁸ for example in his frequent narration of incidents

---

² E.g. Lloyd 1979: 86–98, 1987a: 50–108 and 1996b: 20–46 (although stressing here the importance of resisting over-generalisation about the agonistic quality of Greek science). See also Jouanna 1999: 75–111 on the competitive context of the Hippocratic corpus.
³ See Lloyd 1987a: 56–78; cf. Thomas 2000: 235–47 on Herodotus; Goldhill 2002 for an account of the way in which the development of prose in classical Greece made available new models of authority, based on analytical argument, which in some respects gave new prominence to the figure of the researcher, in contrast with the language of divine inspiration traditionally applied to the poet figure.
⁴ E.g. see Barton 1994b: esp. 143–7; Nutton 2009; von Staden 1997; also van der Eijk 2013, who draws attention to what he refers to as Galen's 'rhetoric of confidence'.
⁵ See Barton 1994b, esp. 147–9.
⁶ See Barton 1994b, esp. 145–7; Boudon-Millot 2009; cf. Fögen 2009 on similar effects in a range of Latin technical writers, e.g. 189–96 on Columella.
⁷ See Lloyd 1988 and 1991a; cf. van der Eijk 2013; von Staden 2009; and more generally Sluiter 2013 on the use of commentary to project control over the texts of the past in rivalry with other commentators.
⁸ Cf. Lloyd 1979: 126–225 for an overview of the importance of empirical research for ancient Greek science; but also Lloyd 1983: 135–49 for the point that uncritical dependence on written authority can sometimes mean that the commitment of ancient scientific authors to personal observation is relatively superficial, with particular reference to Pliny the Elder.

from his wide clinical experience and in his frequent accounts of experimentation on the bodies of animals (in some cases conducted in front of an audience, in a way which allows him to indulge in public refutation and humiliation of his poorly informed rivals). He also repeatedly draws attention to the complexity of the medical expertise which he espouses, writing at length on the various subdivisions of the art of medicine, and then in turn subdividing, in enormously complicated ways, the various subdisciplines of medical knowledge, so as to leave an impressive sense of his command over a very sophisticated body of knowledge.[9] Many other authors too have begun to be analysed for their use of these and other related techniques of self-presentation, even if there are few other authors who use them anything like so richly and forcefully as Galen.

One of our arguments in this volume, however, is that even more needs to be done to understand the connections between different bodies of expertise and different authors in their techniques of self-authorisation.[10] The word 'scientific' in our title is thus intended in the most capacious terms possible to encompass (provocatively) the whole industry of ancient knowledge ordering – including even areas like law, historiography, philosophy and generalship, which are still rarely read together with ancient writing on what we would more naturally refer to as 'scientific' or 'technical' topics (mathematics, medicine, architecture). These different fields have in the past sometimes been kept separate from each other, in part because of the anachronistic application of the idea of 'history of science' as a discrete academic discipline on to the ancient world, which never saw an absolute dividing line between 'scientific' and 'non-scientific' knowledge.[11] Ancient writers clearly thought about all of these bodies of expertise as part of a spectrum of different fields of knowledge. From Plato onwards it is commonplace to list a wide

---

[9] Barton 1994b: 152–66.

[10] For recent parallels, which share some of that ambition for wide coverage of many disciplines, see Asper 2007; Barton 1994b; Taub and Doody 2009. König and Whitmarsh 2007b similarly deal with a range of different knowledge-ordering authors and fields, although without the sustained focus on authority and self-definition which is our main priority in what follows; the same goes for Lehoux 2012, who offers a broad survey of the procedures and priorities that underlay ancient enquiry into the natural world in a wide range of disciplines.

[11] The alternative category of 'technical' literature or 'Fachtexte' has been used as a fruitful working category in some recent scholarship, allowing us a glimpse of precisely the kinds of cross-fertilisation between ancient intellectual disciplines that we are most interested in here: e.g., see Fögen 2005 and 2009. Nevertheless, even that category risks narrowness and anachronism if we try to circumscribe it too rigidly: e.g. see Asper 2007: 35–53 on the lack of any self-conscious generic markers for what he refers to as 'Wissenschaftstexte' (but cf. Fögen 2009: 9–66, where he argues that technical literature is distinguished at least in some respects by its use of practical, non-literary language).

range of different disciplines in comparison with each other. For example, in the opening sentence of Philostratus' *Gymnasticus* to take one of many similar examples, athletic training (the subject of Philostratus' treatise) is compared with philosophy, rhetoric, poetry, music, geometry, astronomy, generalship, medicine, painting, sculpting and piloting. But modern classical scholarship has sometimes been reluctant to take such a capacious view of the interconnection between different bodies of ancient knowledge.

The juxtaposition between many kinds of expertise within this volume will, we hope, bring to light some surprising points of contact between texts which are usually kept separate from each other. The same challenges of self-representation and the same self-authorising gestures outlined above for Galen recur over and over for other disciplines as they are discussed within other chapters later in the volume.[12] Michael Trapp's chapter in this volume demonstrates the importance of many of these same techniques of self-authorisation for the philosophical culture of the Roman Empire, in which Galen participated: he emphasises among other things the importance of rivalry between philosophers, and between philosophers and others, the value of moral self-presentation (also discussed at length by Nicolas Wiater for Dionysius of Halicarnassus), the value of technical mastery, and the importance of alignment with philosophical tradition. Jill Harries makes many of the same points for legal expertise, showing how the jurists of the Roman Empire buttressed their own authority, in rivalry with other strands of the legal profession, by emphasising their educational accomplishments, and by representing themselves as intellectual descendants of earlier legal experts.[13] For Vitruvius too, intellectual tradition is crucial, although he takes a rather different approach, as Daniel Harris-McCoy shows: Vitruvius appropriates a host of earlier authorities, but tends not to name them, foregrounding instead his own sole and comprehensive control over the whole discipline of architecture. Geoffrey Lloyd, in his closing chapter, shows how widespread many of these same approaches are – and especially the appropriation of canonical authority – in the scientific writing of other ancient cultures too.

At the same time, we hope that this juxtaposition will also help to make clearer some of the differences, the things which make individual disciplines

---

[12] For one pioneering attempt to sketch out the contours of a 'grammar of scientific discourse', in other words to map out the range of different possibilities and conventions for scientific argument and scientific self-presentation, across a wide range of different genres of knowledge-ordering, see van der Eijk 1997.

[13] Cf. Eshleman 2012 on the construction of intellectual genealogies as a technique of self-authorisation in both pagan and early Christian culture.

(and even individual texts, bearing in mind the perils of over-generalisation in any study of ancient science)[14] distinctive and unusual in relation to wider trends of 'scientific' self-assertion. For example, Reviel Netz makes clear the oddity, by modern standards, of ancient conceptions of mathematical authority, which is not borrowed by other disciplines.

Not only that, but an interdisciplinary approach to scientific authority also needs to pay attention to recent scholarship on personas in ancient literary writing (and perhaps also vice versa).[15] It is striking, for example, that one important recent volume on precisely that topic – de Jong, Nünlist and Bowie 2004 – sticks very closely to canonical texts, with prose literature represented by historiography, philosophy, oratory, biography and the novel, but with no chapters at all on scientific, technical or miscellanistic writing in ancient Greek literature. Clearly, there are differences between literary verse and scientific prose in their techniques of persona construction, but we should surely think of them more as two ends of a spectrum, with a surprising amount of cross-fertilisation, rather than entirely different genres.[16] For example, the ancient scientific preface has its own conventions and motifs which are reused and varied in ways which are just as complex and sophisticated as anything we find in ancient verse prefaces, and which in some cases even borrow from them.[17]

Finally, we will see in at least some of what follows that an interdisciplinary approach to ancient intellectual authority needs to pay attention to the way in which ancient scientific writing borrows from and in turn even influences ideas about social and political authority in ancient culture. Some of the chapters below look back in passing to classical Greek culture, but the majority focus on Late Republican/Hellenistic texts and especially on the Roman imperial period. That is a deliberate choice. Our hypothesis is that the experimentation and cross-fertilisation in techniques of self-authorisation that we are interested in is at its richest and most variable in the globalised intellectual culture of the Roman Empire, i.e. in the vast corpus of knowledge-ordering writing that survives from the late Republic, in the first century BCE, through to the beginnings of what we refer to as 'late antiquity', in the fourth century CE. This is also where we can see most clearly the intersection between knowledge-ordering and politics. It is fairly clear that the models of intersection between science and empire which have

---

[14] See Lloyd 1996b: 4–5 and *passim*. [15] See Asper 2013b: 3 for the same point.
[16] Roby 2016, on ekphrasis of mechanical objects in ancient scientific and technical writing, makes that point vividly.
[17] See König 2009 on Galen's inventive reshaping of the standard prefatory claim to have written in response to the request of a friend.

been developed so fruitfully for modern European history cannot be transferred straightforwardly to the ancient world. It is much harder for classical culture to find examples of scientific investigation and knowledge-ordering writing which is actively enabled by or in the service of political power,[18] or to find evidence for ancient science and other kinds of expertise embedded within the politically charged struggles between rival institutions and rival groups within the life of ancient cities.[19] The key factor once again is the lack of any sustained institutionalisation of ancient expertise:[20] the dominant pose in ancient knowledge-ordering writing is of the intellectual as free agent, working within an imagined virtual community of experts, willing perhaps to dedicate his work to a powerful patron, but without following the agenda of any professional or political body (whereas for us, certainly in modern English usage, the word 'expert' is standardly used for knowledge in the service of political or legal judgements:[21] 'expert witness', 'expert report', 'expert opinion').

And yet despite all of those caveats, it is clear that changing habits of scientific self-presentation in the Roman Empire were often responses to political developments. Katharina Volk, for example, shows in her chapter how divinatory expertise became a valuable and highly contested commodity among the Late Republican senatorial elite in Rome, who used it in some cases to further their political goals. Michael Trapp examines the idea that philosophical authority could stand in opposition to political power (although he cautions against overstating the oppositional character of philosophical expertise per se). The encounter between knowledge-ordering expert and the political world was often envisaged specifically as an encounter with the emperor. Emperors – usually as dedicatees in prefaces – were repeatedly used as powerful images against whom expert writers could measure up their own control over their material, as well as showing their expertise in the service of politics: Vitruvius is an obvious example.[22] Nor was it only the image of the emperor that ancient scientific writers borrowed: Johannes Wietzke shows in his chapter how the language of benefaction, which was so familiar to the inhabitants of the Greek cities of the

---

[18] Cf. König and Whitmarsh 2007a: esp. 4–6; Woolf 2011: 59–88.
[19] For recent accounts of the capital cities of early modern Europe as contexts for institutionalised scientific activity and scientific rivalry, see Harkness 2007 and Rabier 2007a.
[20] However, see Cuomo 2007b for exceptions, drawing among other things on evidence for ancient guilds.
[21] See Rabier 2007b: 1–2.
[22] E.g. on Vitruvius, see A. König 2009 and McEwen 2003; on Frontinus see Fögen 2009: 278–85; and A. König 2007.

east in the imperial period, gave expert writers a powerful image to use in describing their own achievements.

Self-Effacing, Anti-Competitive and Anti-Expert Authority

The image of ancient experts basing their authority on competitive self-promotion is thus an important one for this volume. As we have already seen, Geoffrey Lloyd's work has shaped our understanding of the importance of that phenomenon for ancient scientific culture. He has also, however, made crucial contributions to our understanding of the way in which that model needs to be qualified and nuanced. He has shown, for example, how frequently ancient authors chose to assert their own authority precisely by avoiding prominent uses of the first person, so as to stress their own objectivity and their own distance from excessively rhetorical modes of self-presentation (sometimes a difficult thing to achieve, given that the claim to speak the truth was itself a recognised technique of rhetorical persuasion).[23] He has also shown how Galen, like others, stresses his own suspicion of excessive *philotimia* (competitiveness) by directing that accusation instead against his rivals, and by foregrounding his own conformity with the work of his predecessors (although that rhetorical pose is of course not incompatible with innovation).[24] And he has pointed to the way in which we see a move towards more tradition-centred models of scientific discourse in the Hellenistic and Roman worlds, and a shift (albeit not an entirely uniform one) away from some of the more aggressively innovative and self-promoting techniques of self-advertisement which were so widespread within the classical Greece of the fifth and fourth centuries BCE.[25]

Many of the chapters in this volume extend those insights further, showing how the avoidance of rivalry,[26] along with various other kinds of

---

[23] See Lloyd 1996b: 74–92, esp. 90–2.
[24] See Barton 1994b: 150; Lloyd 1991a: 400, esp. n. 8. Cf. König 2005: 254–300 on the way in which Galen's attack on the incompetence and competitiveness of athletic trainers allows him to articulate the moderate quality of his own indulgence in rivalry; König 2009: 50–8 on Galen's pose of reluctant self-promoter, publishing his work only at the repeated request of friends, in *On the Order of My Own Books* and elsewhere; König 2011: 185–7 on the alternation of an intrusive authorial persona with more dispassionate, self-effacing language in Galen's *On the Natural Faculties*.
[25] E.g. Lloyd 1987a: 104–8.
[26] Cf. König 2010: esp. 279–83 for more extensive discussion of the way in which the stereotypes of competitiveness and winning at all costs need to be qualified for both the athletic and intellectual culture of the ancient world; Tarrant 2003 for discussion of the way in which the sophists appropriate athletic language to describe intellectual competitiveness, and 355–8 on

self-deprecation, constitute one very prominent strand in the knowledge-ordering culture of the Roman Empire in particular. I want to look in turn at two (closely interrelated) strands: first, self-effacement; second, resistance to narrow professional affiliation.

Of course, self-effacement could carry authority in itself, just as it does for us in the objective, dispassionate language of much modern scientific discourse. Different authors and even different disciplines made use of that kind of pose to varying degrees.[27] In ancient mathematics in particular the author is very often absent: mathematical authority is founded not on rhetorical self-presentation, but on logical demonstration to a degree which is unusual in ancient science,[28] and mathematical authors tend to take the avoidance of competitiveness further than their counterparts in other fields, as Johannes Wietzke shows in Chapter 15 below (although he also stresses that claims about collaboration in ancient mathematical writing tend to be relatively superficial). The self-effacement we characteristically find in dialogue is of a different type, but in its own way equally authoritative: there the author's position is concealed beneath the range of views which are in dialogue with each other, and responsibility is transferred at least in part to the reader, who is led to find his or her own solution in partnership with the author, as Katharina Volk shows in discussing the difficulty of extracting clear messages from the dialogue form of Cicero's *De divinatione*. The same goes for many exempla texts, which present collections of anecdotes without necessarily guiding us about how to read them: here again this kind

---

the way in which Plato and Socrates view that model of sophistic competitiveness with suspicion, overlaying it with the ideals of cooperative excellence. For related discussion outside the field of scientific and technical writing, see among many others Scodel 2008 on Homer; Graziosi 2001 on anti-competitive pressures in ancient wisdom literature; Hesk 2007 on Aristophanes' comical exploration of the problems of excessively combative rhetorical practices within fifth-century Athenian political culture; Crowther 1992 and 2000 on boasts about second-place finishes and drawn contests among athletes, which challenge the still widespread winning-is-everything model of Greek sport; Brown 1978: 38–9 on avoidance of naked competitiveness in the elite culture of the Roman world generally.

[27] See König 2011: 180–7 for broad reflections on self-effacement in ancient scientific and technical writing in the ancient world. For other work on the range of different possibilities for first-person usage, with reference to various degrees of prominence or self-effacement and their implications for authorial authority, see von Staden 1994, whose methodology is followed in adapted form by Hine 2009 and Nutton 2009; van der Eijk 1997: esp. 115–20 and 2005: esp. 40 on the alternation between rhetoric of confidence and rhetoric of modesty in scientific and philosophical writing; also Clarke 1997: esp. 94–8 for debate over the appropriate degree of explicit self-characterisation in ancient geographical and historiographical writing, with special reference to Strabo; Goldhill 2002: 28 on the hesitancy of Herodotus' pronouncements as a ploy which in itself enhances his authority and expertise.

[28] Cf. Lloyd and Sivin 2002: 132–3.

of self-effacement does not seem to have been incompatible with authority, as Alice König shows for Frontinus' *Strategemata*.

Important also is the fact that authority in ancient knowledge-ordering writing is often envisaged as a two-way process: it is rarely a simple matter of top-down assertion. Nicolas Wiater makes that point for ancient historiographical writing, especially for the *Roman Antiquities* of Dionysius of Halicarnassus, emphasising the importance of the author's ongoing relationship with his readers. Daryn Lehoux shows how Galen's authority in his anecdotes of encounters with rival experts often relies on the presence of witnesses, who in many cases are themselves his addressees, as a way of papering over possible weaknesses.

In some cases, we even find imperial authors opting not just for self-effacement, but for various kinds of self-deprecation or self-doubt designed precisely as authorising gestures. The most extreme examples of authorial self-abasement are in later Christian authors, for whom humility, which is equated with piety, is a necessary starting point for religious authority.[29] For an extreme example, one might look at the *Letters* of Ignatius of Antioch to a series of different congregations in Asia Minor (usually dated to the reign of Trajan, although some scholars take it to be a late second-century forgery). Ignatius is vehement in his enforcement of orthodox doctrine: he repeatedly denounces heresy and those who spread it: 'For some carry about the name of Jesus Christ with terrible deceit, while at the same time doing things that are unworthy of God. You must flee from these people as from wild beasts. For they are mad dogs, who bite secretly. You must guard against them, as people who are hard to cure' (Ignatius, *Ephesians* 7.1). And yet at the same time he is also (following the example of Paul) intensely self-abasing: 'But I am ashamed to be spoken of as one of them [i.e. as one of the Syrian bishops]; for I am not worthy, being the least of them, and born out of due time; but I have found mercy to be someone, if I should reach God' (Ignatius, *Romans* 9.2).[30]

Scientific and technical self-deprecation is usually much more muted, but it is nevertheless widespread. It may be explained partly by the convention that self-praise was more acceptable and more effective if it included a degree of self-criticism. Plutarch makes that point memorably in his work *On Self-Praise* 543f: 'some choose not to introduce praise of themselves in an entirely glittering and undiluted form, but instead throw in certain slips

---

[29] See Krueger 2004.
[30] This passage closely echoes 1 Cor 15.8–9. On the influence of Pauline models of authority over Ignatius, see Lindemann 2005; Mitchell 2006: esp. 35–6 on this passage; Reis 2005; Smith 2011.

and failures and minor faults, and so guard against offensiveness and disapproval …'. Scientific self-deprecation may be motivated in some cases by the same principle. It is often also intended to show that the author is fully aware of the complexity and difficulty of his subject: in that sense acknowledgement of the ultimate inadequacy or incompleteness of the author's attempt at the subject under discussion may paradoxically enhance his authority.[31] For example, Ralph Rosen makes that point for Galen's emphasis on the impossibility of adequate knowledge of the soul, in his work *On the Construction of Fetuses* (although he also stresses that this is in itself a means of attacking Galen's rivals, whose failure to acknowledge the limitations and problems of human knowledge is a sign of over-confidence and incompetence).[32]

Second: linked with that ambivalence about competitive self-promotion and foregrounding of the self is a tendency to be wary about identifying too readily with narrow, clearly definable areas of expert knowledge. It is surprisingly common to find members of the ancient elite laying claim to intellectual authority while at the same time espousing a rather hesitant or stand-offish relationship with certain kinds of expertise. In some cases, that involves the avoidance of technical language: Reviel Netz discusses in Chapter 16 the surprising absence of specialist mathematical argumentation, which we might expect to be an authority-enhancing feature, in non-mathematical treatises. In other cases, it involves the deliberate avoidance of any close link with practical skills.

That latter effect is partly a response to the deep-rooted assumption, which had its origins in classical Athenian culture if not before, that some kinds of skill – particularly skills which were manual or which were primarily concerned with earning money – were not admirable.[33] There was a range of possible responses to that problem. One common response was to go out of one's way to dissociate oneself from these inferior kinds of expertise, for example by presenting one's own skills as complex and sophisticated, in the manner outlined above for Galen. In some cases, that involved drawing an explicit contrast between high-status and low-status

---

[31] See Lloyd 1987b; van der Eijk 1997: 120.
[32] Cf. Harris-McCoy 2013 for the way in which Artemidorus' manual of dream-interpretation, the *Oneirocritica*, combines an aspiration to exhaustiveness with an awareness of the impossibility of covering all that needs to be covered: that helps to advertise both the sophistication of the art of dream interpretation, and his own sensitivity to local cultural difference, which is one of the factors which increases the range of possible dreams.
[33] See (among many others) Cuomo 2007b: 7–40 for exhaustive discussion of the range of different opinions about *technê* in classical Athenian culture, esp. 9 for the widespread distinction between *technê* and banausic or base *technê* (e.g. at Aristotle, *Politics* 1258b26–35 and 1337b8–18); also Whitney 1990: 23–55.

skills. Galen's *Protrepticus* is a case in point:[34] there he opposes the category of *technê* to the category of *kakotechnia* – false arts like juggling and athletic training. Within the overarching category of *technê* he then postulates a set of further subdivisions with the most divine, rational arts (including medicine) at the top, and the banausic craft activities at the bottom:

> Given that there is a distinction between two different types of art (*technê*) – some of them are rational and highly respected, whereas others are contemptible, and centred around bodily labour, in other words the ones we refer to as banausic or manual – it is better to take up one of the first category… In the first category are medicine, rhetoric, music, geometry, arithmetic, logic, astronomy, grammar and law; and you can also add sculpting and drawing if you wish. (Galen, *Protrepticus* 14 [K1.38–9])

Similarly, Pollux, in his *Onomasticon* (dating from the second century CE), whose importance for our understanding of the way in which imperial Greeks valued different kinds of activity as part of Hellenic tradition has only recently begun to be recognised,[35] offers an unusually extensive account of the distinction between banausic *technai*, the province of workers or craftsmen (in Book 7), and liberal *technai*, which belong to the highly educated (in Book 4) (although even in the case of the latter he is not universally complimentary, offering in particular a rather ambivalent image of sophists, who are linked with money-earning).

In response to those distinctions, there are large numbers of surviving texts dedicated to debating the technicity of particular types of expertise (in other words their eligibility for the prestigious category of *technê*).[36] Many ancient technical treatises open with claims about the technicity of the expertise under discussion, as a way of differentiating their own subject from inferior kinds of expertise. In some cases, we see authors on practical subjects going out of their way to supplement the utility of their work with evidence for its intersection with the other liberal arts: the famous opening to Vitruvius' *De architectura* (discussed in Chapter 5 by Daniel Harris-McCoy) is a case in point. 'The architect's knowledge', Vitruvius suggests, 'should be adorned with many different branches of study' (1.1.1), and he goes on to name history, philosophy, mathematics, music and medicine, among others (1.1.4–11).[37] In other cases again, we find ancient intellectuals advertising their own distance from the more mundane, introductory

---

[34] See König 2005: 291–300. [35] For one attempt, see König 2016.
[36] See Blank 1998: xvii–xxxiv for an overview.
[37] Cf. Riggsby 2007: 105–6, pointing out that excessive specialisation in architecture 'would be inappropriate to the "omnicompetent" citizens of the political class' (105).

facets of the areas of expertise they profess, as we shall see further for Philostratus' *Lives of the Sophists* below.

Some of these authors (Vitruvius included) represent their own elevation above narrow concepts of expert or professional authority as a consequence of their philosophical identity. That approach of course looks back to Plato's Socratic dialogues, where Socrates repeatedly sets out to demolish his interlocutors' certainties about the value of specific *technai*, and famously opposes an alternative model of authority based on knowledge of his own ignorance.[38] Many of the attacks on technicity mentioned above are conducted from a philosophical perspective: texts like Philodemus' *On Rhetoric* or Sextus Empiricus' *Against the Grammarians* express scepticism about all arts which are not subordinated to philosophy as the art of living.[39] Harry Hine shows below that Seneca even chooses to avoid identifying himself as a professional philosopher, even as he insists on the importance of philosophical thinking for living a good life.[40]

Of course, those manoeuvres are sometimes more about rejecting (once again) an association with inferior kinds of expertise than about rejecting expertise per se. Galen criticises his rivals for not being philosophical enough, but that is perfectly compatible with his construction of an idealised vision of medicine as a philosophical art, for example in his treatise *That the Best Doctor Is Also a Philosopher*. Others, however, make their own non-expert identity absolutely clear, setting themselves up as literary chroniclers of practical skills and offering an overview of particular branches of knowledge without claiming any specialist command over the varieties of expertise they are writing about, almost as though their distance from their subject matter makes them more qualified to articulate its significance. That motivation goes some way towards explaining the fact that ancient scientific and technical texts often make it difficult to apply the knowledge they contain.[41]

Admittedly, we need to be wary of the assumption that ancient thinkers worked with a clear division between high-status theoretical knowledge and low-status practical knowledge, even if we do find versions of that division in Plato and Aristotle and the later philosophical tradition.[42] It is clear, in

---

[38] See esp. Plato, *Apology* 20e–3c.   [39] See Blank 1998: xxviii.
[40] And see Trapp, section entitled 'The Extension of "Philosopher"', p. 41, below, for parallels.
[41] See Formisano and van der Eijk 2017 for wide-ranging discussion of that phenomenon, although many of the chapters in that volume focus on successful ancient attempts to bridge the gap between theory and practice.
[42] For useful overview, see Parry 2007; Rihll 1999: 13–16.

fact, that the gap between theoretical knowledge on paper and practical application is not always intended to imply the superiority of the former. In some cases, it seems to be intended precisely to advertise the practical skills of the author, implying that full transmission of knowledge is not possible without personal teaching and real-life experience, and so suggesting a positive valuation for hands-on expertise and one-to-one transmission of knowledge. Dio Chrysostom, for example, at the end of his treatise offering advice on training for public speaking, writes as follows:

> Just as it is not enough to say to painters and sculptors that their colours should be like this or their lines like this, but instead they get the greatest benefit if someone sees them painting or sculpting; and just as it is not enough for athletic trainers to talk about wrestling holds, but instead they must also show them to their pupils; in just the same way, in consultations like this the benefit will be greater if one sees the person who has given advice in action himself. (18.21)

In that case the advice offered in the text is a close substitute for intimate personal communication – Dio repeatedly stresses his closeness to the addressee – but it is still no substitute for teaching in person.[43]

Nevertheless, it is clear that in many cases authors present themselves as theorists of knowledge capable of re-envisaging a particular branch of expertise within a literary framework, but without taking any great interest in the prospect of practical application. Some authors represent themselves, or their addressees, as members of the political elite, who may need only partial understanding of a particular body of knowledge.[44] At the same time, that is often accompanied by an interest in throwing doubt on the validity of precisely that kind of purely literary, theoretical authority, as if advertising one's self-consciousness about the problem is in itself a way of assuring the reader of the intelligence and subtlety of the treatise he or she is reading. Emily Kneebone argues in Chapter 9 below that Greek didactic verse authors tend to be primarily and unashamedly interested in representing literary accomplishment as the basis of their authority, rather than practical experience (although with occasional exceptions, where practical knowledge is more highly valued). At the same time, however, she shows that they frequently draw attention to the limitations of their knowledge and authority. Aude Doody similarly shows that Varro downgrades and occludes the practical knowledge of the slave in favour of his own theoretical knowledge

---

[43] Cf. van der Eijk 1997: 96, with reference to a range of passages from Aristotle (esp. *EN* 1181b2–6) and Galen.
[44] See Meissner 1999: esp. 189, 255.

of agriculture. At the same time, however, he hints repeatedly at the inadequacy of writing as a source of knowledge. Marco Formisano shows how writing about military strategy too tends to value literary authority over practical experience, although there are also plenty of passages in Latin historiography where we see expressions of contempt for those whose knowledge of warfare is purely theoretical. Alice König discusses many of the same themes in her treatment of Frontinus' *Strategemata*: Frontinus draws attention to his own authoritative systemisation and theorisation of military strategy in a lost earlier work, the *De re militari*, and leads us to expect that the *Strategemata* will give us more of the same, although it also undermines those claims as we read by hinting at the possibility that practical, improvisational intelligence may be more valuable after all.

Others in turn go much further in sceptical or satirical attitudes to all forms of established knowledge and tradition. In some cases, they take the form of self-parody, where we are presented with a narrating voice whose claims to authority are implicitly debunked. Leah Kronenberg argues in Chapter 13 below that Varro's antiquarian writing on religious history should be interpreted in precisely those terms, as a parody of scholarly pedantry – taking on the pose of an expert but debunking it – in a world dominated by increasingly narrow kinds of expertise.[45] Lucian is another obvious example. Repeatedly in his work he satirises the claimants to intellectual authority and expertise that he saw all around him in the elite culture of the mid to late second century CE – for example in texts like *On the Parasite*, where a parasite attempts to defend the claim that his own profession should be respected as a serious *technê*, with fantastically ingenious reshaping of many of the standard motifs of ancient expert self-definition.[46] Lucian turns his criticisms against philosophers too, mocking those whose allegiance to particular philosophical schools is superficial and hypocritical.[47] The pose he exemplifies so well, of a figure whose outsider status in relation to mainstream intellectual culture and in relation to any narrow intellectual and professional affiliation gives him special authority to comment on

---

[45] See Wallace-Hadrill 1997 for the claim that the late Republic saw a transfer of control over knowledge from members of the elite to experts; however, see also Chapter 14, n. 62, this volume, for possible problems with that picture.

[46] See Nesselrath 1985: 123–239 for ancient writing about *technai* and Lucian's engagement with it in *On the Parasite*.

[47] For an obvious example, see Lucian's *Symposium*, with König 2012: 248–51, where each of the philosophers at a dinner party misbehaves in a way which is ironically appropriate to his own philosophical allegiances.

it and to criticise it, is widespread (albeit usually in less extravagant form) throughout ancient writing on knowledge and wisdom.[48]

Plutarch

The poses of avoiding excessively strident self-promotion and deprecating narrow professionalism and expertise in favour of higher and less restricting claims to knowledge are thus very widespread in the ancient world, and in Roman imperial culture in particular. I want to illustrate some of these points now in relation to two imperial texts which at first sight appear to give us some of the most undiluted representations of intellectual competitiveness in surviving ancient literature, but which on closer inspection turn out to be imbued with precisely the kinds of hesitancy I have been discussing.

The first of those is Plutarch's *Sympotic Questions* (*Quaest. conv.*), a collection of learned mini symposium dialogues between Plutarch and his fellow guests which purports to be an accurate record of events stretching across several decades of Plutarch's life.[49] The range of topics the text covers is enormous. Many of the dialogues are on scientific subjects; some are on questions connected with the traditions of the Greek symposium; others involve questions of local history or analysis of famous texts from earlier Greek literature. One reason why they are so fascinating is because of the way in which they show us competitive self-presentation in action not just on paper, but also within more or less realistic social situations.[50] The conversations have a strikingly competitive quality, in line with the competitive traditions of sympotic conversation which date back at least to Athens in the fifth century BCE. Most often, one of the guests proposes a problem or puzzle for solution, and the others take it in turn to offer an explanation for the phenomenon in question. Some of these responses cite explanations from earlier Greek texts, but others are strikingly original and ingenious (sometimes playfully and implausibly so), and speculative ingenuity seems to be highly prized by Plutarch's sympotic communities. What we are

---

[48] Morgan 2013 uses the term 'xenological authority' to describe that kind of outsider perspective, and sees it as central to the influence of ancient wise man sayings.
[49] See König 2012: 60–89 for detailed analysis, with further bibliography.
[50] Not surprisingly, many scholars have doubted whether these are accurate reports of real conversations, but even if that is right it clearly is the case that they are meant to have some connection to actual practice, giving readers a set of images against which to measure their own learning and their own conversational practice.

seeing, then, is a playful experimentation with skills of competitive argumentation which could be reused in other more serious situations outside the symposium, for example in contexts of rhetorical, philosophical or scientific debate.[51] The setting of many of these conversations at agonistic festivals, combined with repeated mentions of the victors in both musical and athletic contests, contributes to a sense that the intellectual competitiveness of Plutarch and his fellow guests is an elevated version of the competitive virtues on display in the festival contexts themselves.[52]

Not only that, but Plutarch himself takes a very prominent role in many of the dialogues.[53] Frequently he speaks last and most convincingly.[54] In the *quaestiones* that date from late in his life he is often shown giving authoritative examples of sympotic argumentation as a model for the younger men who are present. In those that date from his early adulthood – many of them featuring his famous teacher Ammonius – he is often shown as the most promising of his fellow students.

When we look more closely, however, it becomes clear that there are pressures in the other direction. For one thing, Plutarch's prominent first-person presence as a character in the dialogues sometimes disappears from view, and especially so in the prefaces, which in other knowledge-ordering writing from the imperial world is often the place where an author is most prominent of all, but which see Plutarch hiding behind a vague first-person plural and talking in generalising terms about the value of sympotic speech and sympotic friendship.[55] As I have argued elsewhere,[56] that self-effacement seems to be intended partly as an acknowledgement that Plutarch's authority must be understood as part of a cooperative endeavour: cooperation with his fellow guests, with the authors of the past, with whom they often seem almost to be entering into dialogue, and even cooperation with his readers, given that one of the main functions of the dialogue form as Plutarch uses it is to engage his readers in puzzling and problem-solving and thinking philosophically for themselves.

---

[51] For related discussion of the links between sympotic debate and the wider intellectual culture of the Roman Empire, see Lim 1995: 1–4 and Schmitz 1997: 127–33.

[52] See König 2012: 81–8.

[53] See also Klotz 2007 (reprinted in slightly adjusted form as Klotz 2011) for similar discussion of Plutarch's self-presentation in the QC, stressing especially the way in which he presents himself at a range of different ages.

[54] For good examples see QC 1.9, 5.2, 5.4, 6.4, 6.5, 6.6, 7.5.

[55] See König 2011: 190–4, with lots of examples. There is an obvious difference here from his essays on practical ethics, where he portrays himself more consistently as the sole authority on the subjects he discusses: see Van Hoof 2010: 66–80 for that point.

[56] König 2011.

The work also offers a fascinating image of expertise and professional affiliation, and at first sight we might expect it to be a celebration of expert knowledge. Across the nine books we meet a remarkable range of different types of expert: grammarians, rhetors, doctors, military writers, geographers, mathematicians, philosophers of many different philosophical schools, even a farmer and an athletic trainer.[57] There is a sense that each of these professions can meet on common ground, and that each of them can contribute something distinctive from his own expertise to the discussions. At the same time, however, the text implies that it is the sum of these different approaches, in collaboration with each other, rather than any of them individually, that leads to the richest solutions to the problems under discussion, and that narrow devotion to a particular field of knowledge can be limiting.

Grammarians, for example, are frequently portrayed as pedantic and inflexible, by contrast with Plutarch's own more wide-ranging styles of argumentation.[58] *Quaest. conv.* 1.9 is a good example. There the Stoic philosopher Themistocles asks a question which he represents as particularly appropriate to the literary skills of his interlocutor, the grammarian Theon: 'tell us why Homer has described Nausicaa doing her washing in the river rather than in the sea, even though the sea was nearby, and even though the sea is likely to have been warmer and clearer and more cleansing' (*Quaest. conv.* 1.9, 627a). Theon's (rather cursory) solution involves using the typical grammarian's approach of quoting from his reading, in this case from Aristotle, arguing that sea water has impurities which make it wash less effectively. Plutarch, however, refutes that Aristotelian suggestion, and instead comes up with two alternative solutions of his own, quoting in the process from an alternative passage of Aristotle, but also making a set of arguments from his own observation: 'For I see that people often thicken their water with ash or with soda, or with dust or if they do not have these other substances available, as if the earthy material is more able to wash out the dirt because of its roughness, whereas the water on its own, because of its lightness and weakness, does not achieve this in the same way' (1.9, 627b–c). Plutarch there uses the common scientific gesture – discussed in the opening section of this chapter – of refuting written authority, in the process foregrounding his own authoritative ability to challenge received wisdom by independent thought and observation (although it is also important to stress that there are other passages in the work, and even in 1.9 itself, where excessive ingenuity is criticised in favour of book learning: Plutarch dramatises

---

[57] See Hardie 1992: 4754–6. [58] See Horster 2008: esp. 618–22.

the centuries-old tension in ancient knowledge-ordering between innovation and tradition rather than favouring one consistently over the other).[59] It also seems likely, however, given the work's repeated suspicion of grammatical expertise, that Plutarch intends this as a demonstration of the value of thinking ingeniously without being bound by particular disciplinary habits. That suspicion of experts who do not raise their heads from their own areas of expertise is of course a version of the Socratic stand-offishness towards *technê* in the Platonic dialogues: like so many of his philosophical contemporaries and predecessors, Plutarch recommends a broad, overarching devotion to philosophy above specialist knowledge.

Moreover, within the ebb and flow of sympotic conversation excessive competitiveness is repeatedly greeted with suspicion by the other characters, and there are even times when friendly rivalry threatens to become seriously divisive. In some cases, rivalry is even represented precisely as a consequence of the clash between experts of different types. Those problems become particularly prominent in Book 9, which is unusual in having fifteen *quaestiones* instead of ten, all of them from a single occasion, a dinner given by Ammonius in Athens. The dinner follows a set of literary and musical contests in the Diogeneion between young men studying within the *ephebeia*. Ammonius invites the successful teachers to dinner, in the hope that dining together will lead them to discard any ill-feeling, just as Achilles gives dinner to the contestants after the funeral games of Patroclus, but things do not go according to plan: 'the competitiveness and rivalry between the teachers took on a sharper edge over drinking, and it was not long before they began to pose challenges and questions for discussion, all mixed together and confused' (*Quaest. conv.* 9.1, 736e). Ammonius therefore asks one of the guests, Erato, to sing the passage from the opening of the *Works and Days* where Hesiod distinguishes productive rivalry, between people of the same profession, from its destructive equivalents.

At the beginning of the next *quaestio*, 9.2, we hear that it is traditional at the festival of the Muses for lots to be drawn so that different guests take it in turns to pose learned problems to each other. Ammonius, however, is anxious that there may be trouble if guests from the same profession are drawn together (he does not seem confident in Hesiod's judgement about the productive quality of professional rivalry), and instead he decides that all the pairs should be made up of representatives from different professions,

---

[59] For a good example, see *Quaest. conv.* 8.4, 723f–724a, where one of the speakers criticises his fellow guest, a rhetor, for his excessively speculative explanation and then proceeds to give an alternative explanation he has found in his reading.

without a ballot: 'a geometer should pose problems to a grammarian, a musician to a rhetor, and then vice versa' (9.2, 737e). Even that goes wrong, however. The geometer Hermeias attempts a complex numerological explanation for the number of the letters of the alphabet; the grammarian Zopyrion laughs at him and dismisses his explanation as nonsense (9.3); Hermeias then has to be stopped by the other guests from posing a problem to Zopyrion in turn (9.4, 739b); and another grammarian, Hylas, is teased by the others for his bad temper, caused partly by his lack of success in the contests earlier in the day. Eventually the conversation settles down into a more harmonious vein, but this is nevertheless a striking example of the way in which even a work like this, which in some respects celebrates intellectual competitiveness and expert identity, is also at the same time acutely aware of the problems they bring with them.

Philostratus

My second text is Philostratus' *Lives of the Sophists* (VS). No one could accuse Philostratus' sophists of being self-effacing: they are among the most spectacular self-dramatists in the whole of ancient literature. They are also among the most competitive of all ancient intellectuals.[60] The famous rivalry between Polemo and Favorinus[61] is replicated in a less extravagant form in the many other rivalries which are spread right through the text. As in the *Sympotic Questions*, the great agonistic festivals of Greece are represented as important venues for the display of the sophists' expertise[62] – in line with the long-standing tradition of Greek orators and other intellectuals giving speeches at the Olympic festival – and the competitiveness of the sophists themselves may be one reason why Philostratus is so fascinated by that theme, and why he sees those venues as so significant.

Nevertheless, it is also clear on closer inspection that the text at the same time has a deeply rooted ambivalence about the value of intellectual rivalry.[63] Many of Philostratus' authorial interventions suggest something close to admiration of contest. The following is typical:

> The quarrel (*diaphora*) that arose between Favorinus and Polemo began in Ionia, the Ephesians siding with Favorinus, from the moment when Smyrna began to express their admiration for Polemo; but it escalated in Rome; for there consuls and sons of consuls by praising one or the other

---

[60] Cf. Whitmarsh 2005: 37–40.   [61] See Gleason 1995.   [62] See König 2014.
[63] See König 2010 for a more extended version of that argument.

> of them started between them feelings of *philotimia* ('love of glory', 'competitiveness'), which even among wise men kindles great envy. They may be forgiven for their *philotimia*, since human nature views *to philotimon* ('love of glory', 'competitiveness') as something which never grows old; but they are to be blamed for the speeches which they composed against each other, for abuse (*loidoria*) is brutal, and even if it is true, that does not absolve from shame even the person who has spoken about such things. For those who called Favorinus a sophist precisely this fact – that he had quarrelled with a sophist – was sufficient proof, for that spirit of *to philotimon* of which I have spoken here is usually directed against rivals from the same profession (*tous antitechnous*). (VS 1.8, 490–1)

Here Philostratus represents competitiveness almost as a source of glory and even immortality ('something which never grows old'), in a way which recalls Homeric characterisation of the heroes of the *Iliad*,[64] and even as a kind of behaviour which can define the sophists' professional identity. The point about *philotimia* directed against rivals recalls *Works and Days* 23–6 (discussed briefly above in relation to Plutarch), where Hesiod talks about the kind of strife which inspires competition between neighbours as 'good for mortals: potter is angry with potter, carpenter with carpenter...' However, Philostratus also acknowledges some of the potential problems of competitiveness, when it is taken too far, in stressing the shameful nature of personal abuse.

Elsewhere too he seems to be working with a distinction between good and bad competitiveness. For example, in VS 1.21, 514–15, he rebukes the critics of Scopelian as follows:

> I shall speak now of the sophist Scopelian, once I have grappled with those who try to reproach (*kakizein*) him. For they deem him unworthy of the category of sophists, calling him bombastic and undisciplined and bloated in his style of speech. This is what quibblers and slow-witted people say of him, and those who do not themselves have the skill of vigorous improvised speech; for humans are by nature envious.

In this case the problem seems to be the kind of empty, hypocritical denigration of rivals which is not underpinned by talent. In VS 2.33, 627–8, by contrast, Philostratus is more open to the idea that mutual criticism and rivalry can have positive consequences, when it is between equals (as seems to be the case also for Polemo and Favorinus in the passage quoted above from 1.8):

---

[64] See Bowie 2006 on Homeric overtones in this text.

> The quarrel (*diaphora*) between Aspasius and Philostratus of Lemnos began in Rome, but it escalated in Ionia, encouraged by the sophists Cassianus and Aurelius... The saying that it is possible to learn good things even from an enemy has been illustrated often in human affairs, and most of all in the case of Philostratus and Aspasius. For during the course of their quarrel, Aspasius acquired an ability in fluent extempore speaking, from the moment when Philostratus too came to be held in renown in this field. Conversely Philostratus pruned back his own style of speech, which up to that time had been running riot, taking Aspasius' precision as his model.

That range of judgements – many of which seem rather at odds with each other on first reading – makes it hard to generalise about the value of intellectual rivalry, and hard to be sure how we should judge any one individual case. It is clear that competition can have very positive effects, but the text also insistently portrays it as something which needs to be treated with caution.

Philostratus' own authorial self-presentation shares some of those hesitations. His voice is certainly an authoritative one. It has a didactic quality – like the voice of a sophist, perhaps, talking to his students.[65] And yet at the same time there are some respects in which he stands apart from the sophist figures he describes, despite the fact that he was one of them. That is obvious most of all in the preface to the work, where he reminds the future emperor Gordian of a conversation they once had about sophistry in the temple of Daphnaean Apollo in Antioch:

> I have written up for you in two books a record of those who were philosophers but enjoyed the fame of sophists, and also of those who were rightly called sophists; partly because I realise that you trace back your ancestry to that profession (*technên*), to the sophist Herodes; but also because I remember the discussions we once had about the sophists at Antioch, in the temple of Daphnaean Apollo. (*VS* preface, 479)

That involvement in non-confrontational dialogue immediately sets him apart from the very different, more public, more self-promoting speech-acts of the sophists: there are very few examples of the sophists involved in dialogue in the rest of the work.[66] In the very final lines of the work we hear the following (about Philostratus' relative, Philostratus of Lemnos):

---

[65] Anderson 1986: 77–96, Campanile 2005, Schmitz 2009 and Whitmarsh 2004b all stress the similarities between Philostratus and his subjects.
[66] See König 2014 for more extensive justification of that claim.

> Concerning Philostratus of Lemnos, his skill in the law-courts, in speeches in the assembly, in writing treatises, in declamations, concerning his stature as an expert in extempore speech, and concerning Nicagoras the Athenian, who was even crowned as herald at the temple of Eleusis, and Apsines the Phoenician and the extent of his abilities in memorisation and precision, it is not for me to write, for I would certainly be mistrusted as having shown favour, since I was joined to them by ties of friendship. (VS 2.33, 627–8)

There Philostratus acknowledges his own involvement in the networks of friendship and intellectual patronage which structure the sophists' world, but he also declares his determination to stand apart from them, and that makes him very different from the very partial figures he has been representing to us through the rest of the work.

Not only that, but the text also shows some ambivalence about the value of easily codifiable rhetorical expertise. Many of the sophists seem to stand above the more mundane aspects of their own discipline. Some context may be helpful. Rhetoric was an area of expertise which was particularly committed to demonstrating its own technicity. Admittedly, some of the most memorable discussions of that subject are critical ones, denying the status of a *technê* to rhetoric,[67] but in some cases even they can help to reveal the weight of the rhetoricians' arguments for their own profession. Philodemus' *On Rhetoric* is a case in point: it denies technicity to two of the three branches of ancient rhetoric – the deliberative and the forensic – but also in the process gives us access to many of the arguments used by those who disagreed.[68] Much of the surviving writing on ancient rhetorical theory represents rhetorical skill as something which requires an intricate knowledge of the different kinds of rhetorical speech, and categorisation and codification of different methods.[69] Rhetorical handbooks and works of rhetorical theory also tend to represent rhetorical training as a painstaking activity, heavily dependent on repetitive training.[70]

In *Lives of the Sophists* those facets of rhetorical practice are often sidelined. In some contexts, *technê* is a label that seems to raise a discipline above

---

[67] Probably the most influential examples are Plato, *Phaedrus*, esp. 260e and *Gorgias*, esp. 462b–c.
[68] See Chandler 2006: 26–31 for translation and 70–80 for discussion.
[69] See Heath 2004: 254–76; Steel 2009. That is not to say that ancient engagement with theory was mechanical; rather, theory gave ancient orators a set of starting points from which to proceed, which could be varied creatively: see Quintilian *Institutio Oratoria* 2.11 and 10.3.15, quoted by Heath 2004: 17–18.
[70] See Heath 2004: 217–54, esp. 245 and 253–4 on the importance of lifelong practice (including discussion of Galen's claims – *PHP* 2.3.16 and 9.2.31 – about the importance of daily practice in rhetoric, as in medicine).

more mundane types of skill, but here the opposite seems to be true: Philostratus seems to want to raise his sophists above *technê*. There are admittedly quite a few passages where Philostratus uses the word *technê* in a relatively neutral fashion simply to describe the skills and the profession of the men he is talking about: examples include the very opening sentence of the preface, quoted already above: 'because I realise that you trace back your ancestry to that profession (*technên*)'.[71] Some sophists are praised quite explicitly for their command over *technê*. For example, Lollianus at 1.23, 527 is described as 'very proficient in his *techne* (*technikôtatos*) and very clever at effective exposition of the kind of reasoning that depends on artful inventiveness (ἐπινοίᾳ τεχνικῇ)'.[72] Their pupils are also regularly mentioned. And Philostratus himself draws on standard literary-critical language – albeit in rather unsystematic and idiosyncratic ways – in his assessments of the styles of his subjects.

However, it is striking that Philostratus seems on the whole rather uninterested in rhetorical theory, and tends not to mention the technical publications of his subjects, perhaps because he views theoretical teaching as something which should be confined to the early stages of rhetorical education.[73] That is perhaps one reason why Philostratus' text often gets little if any attention in modern introductions to rhetoric, which tend to be more interested in those ancient texts which are self-reflexive about rhetorical expertise.[74] Lollianus is again an exception: he is said to have given classes which were 'not only about the practice of declamation (μελετηράς) but also involved theoretical instruction (διδασκαλικάς)' (1.23, 527).[75] In this case, however, it is striking that Lollianus' publications on rhetorical theory go unmentioned. It may even be that the apparently complimentary reference to Lollianus as *technikôtatos* earlier in 1.23 is intended not so much to praise him as a model example of what all sophists should aspire to, but rather to point out his oddity in relation to the criteria Philostratus generally uses for judging sophistic

---

[71] Other examples include VS 1.9, 492; 1.24, 528; 2.27, 617.
[72] Other examples include VS 1.15, 500; 1.25, 542; 2.24, 607.
[73] See Heath 2004: 27 and 227: 'the implication is that the groundwork of theory would be laid at relatively elementary stages, and many, if not most, of the top-level sophists in their advanced classes would concentrate exclusively on the development of style and performance through practical classes' (227); also 23 and 228 on evidence of disdain for theory among other imperial writers on rhetoric, with particular reference to Phrynichus and Quintilian: 'disdain for theory is a pose most easily adopted by those who have absorbed the theory and achieved distinction' (228).
[74] E.g. the indices to Gunderson 2009 list Philostratus, VS only four times, all of them passing mentions in the main text.
[75] See Civiletti 2002: 473.

ability. One might view this as an attempt by Philostratus to privilege practice over theory, but that is not quite the right way of putting it: he is interested above all in the contrast between rhetoric-as-theory and the virtuoso performances of rhetoric-as-improvisation (rather than a contrast between rhetoric-as-theory and rhetoric as mundane day-to-day experience). In the majority of cases where teaching relations are mentioned, it seems to be primarily because of Philostratus' fascination with the way in which authority is passed on from teacher to pupil,[76] and with the loyalties and rivalries of the sophistic community, rather than because of any detailed interest in the systematic quality of day-to-day rhetorical training.[77] Philostratus' sophists tend to be represented instead by a very different strand of imagery linked with oracularity and divine inspiration, which makes them appear far removed from the routine business of day-to-day expertise, or indeed the kinds of formal theorisation often linked with professional rhetorical identity.

Not only that, but there are several striking passages in the work which seem to express doubt about these more technical aspects of rhetoric, as if a concern with rules and conventions is incompatible with the kinds of inspired, flashy display he seems to value most highly. In 2.9, 585, for example, we hear that Aristides was 'of all sophists most proficient in his *techne* (*technikôtatos*), and was much preoccupied by theoretical reflections (*theoremasi*), for which reason he refrained from extempore speaking. For the desire to speak always according to theory (*kata theorian*) keeps the mind busy and robs it of boldness.'[78] This is the only other use of the superlative *technikôtatos* in the text apart from 1.23, 527, and backs up the impression that the characterisation of Lollianus in those terms may not be straightforwardly complimentary. In 1.22, 523, Philostratus refutes a tradition that the sophist Dionysius used to train his pupils in the art of memory as follows: 'The arts (*technai*) do not exist nor could they, for memory gives us the arts (*technas*), but it is itself unteachable and not attainable by any *technê*, for it is a gift of nature or a part of the immortal soul.'

Elsewhere we hear of sophists who are praised despite or even because of their preference for not following the rules of *technê*. For example, at

---

[76] Cf. Eshleman 2012: 125–48. One of the arguments of her book as a whole is that we need to take more account of the overlaps between Greco-Roman and early Christian groups in their strategies of self-authorisation.

[77] Cf. Heath 2004: xvi, pointing out that Philostratus' lack of interest in day-to-day training makes him very unreliable as a witness to mainstream rhetorical culture in the Roman Empire.

[78] For discussion of these terms, see Civiletti 2002: 576–7, with ref. also to *VS* 2.7 and Philostratus' lack of interest in Hermogenes' theoretical writings there.

2.10, 590, we hear that Hadrian of Tyre 'was abundant in his ideas and brilliant... but his speech was not orderly, nor did he follow the rules of *technê*...'. And at 2.12, 592: 'Pollux had been sufficiently well trained in the art of criticism, having been the pupil of his father, who was an expert in criticism; but he composed his sophistic speeches with the aid of audacity rather than *technê*, trusting in his own natural abilities, for he was indeed very talented by nature.' Here again Philostratus seems to value styles of speech which are quite separate from a narrow concern with theory or *technê*.

Conclusion

These two texts, for all their idiosyncrasies, are typical of a set of wider phenomena within ancient scientific and knowledge-ordering culture. The figure of the expert was often very prominent, much more so than in modern scientific writing. This was a world without any formal educational qualifications: young men would study not at formal institutions of higher education, but instead with individuals – philosophers, like Plutarch's teacher Ammonius or Plutarch himself, or orators, like Philostratus' sophists, who are often described surrounded by large groups of admiring students. Virtuoso displays of knowledge could bring social as well as intellectual prominence. Both of these texts are valuable among other things for the way in which they bring home the link between knowledge and elite status: in both we see great intellectual figures mingling in the very highest social circles. Most other ancient scientific and technical texts are not quite so revealing of their social and educational contexts. Nevertheless, most of them are similarly interested in self-assertive styles of self-representation. We see ancient writers competing for attention, laying claim to the intellectual and even moral high ground, stressing their own mastery and control over their sources and the disciplines they are working in.

That is not the full story, however. We also see at the same time – and often in the same texts – frequent attempts to avoid excessively strident forms of self-assertion. In some cases, that involves stressing one's dependence on past authorities in order to avoid the impression of innovation for its own sake. In other cases, we see scientific writers exercising what they represent as proper caution about the limitations of human knowledge or about the complexity of the material they are dealing with, or else acknowledging the difficulty of applying theoretical knowledge in practice. Self-effacement can also be a response to the worries about excessive competitiveness that stretched back into classical Greek culture and beyond. Hesiod's *Works and Days* makes the claim that some kinds of rivalry, between people of the same profession, can be productive. That sentiment is echoed repeatedly in later Greek and Latin literature too, and competition was undeniably important in many different areas of ancient cultural activity. But side by side with that view we also find worries about the destructive power of excessive competitiveness.

Attitudes to disciplinary affiliation were also often complex. The label *technê* ('skill', 'art', 'expertise') was a highly valued one; indeed, many experts fought hard to claim that label for their own disciplines (or indeed to deny it to areas of expertise they wished to downgrade). Others, however, went out of their way to avoid identifying themselves with narrowly defined bodies of expertise. One reason for that was the way in which association with partisan intellectual groupings was itself often linked with problematic kinds of competitiveness, for example in the disputes between different medical sects that Galen is so keen to dissociate himself from. In other cases, the ability to step back from narrow varieties of expert knowledge was linked with philosophical identity, and with the kind of high-status educational breadth which proclaimed its ability to master many different disciplines rather than just a single one (as in Vitruvius' famous claim that the architect must have a knowledge of a wide range of disciplines, including philosophy). Expertise and authority did not always go together straightforwardly in ancient knowledge-ordering culture.

2 | Philosophical Authority in the Imperial Period

MICHAEL TRAPP

Introduction: Prehistory

Seen in its full chronological span, the story of the evolution of the notion of philosophical authority is long and involved. At its outset, in the sixth and earlier fifth centuries BC, stands a landscape in which there are individuals who have no generic name (they are only subsequently and retrospectively to be dubbed 'philosophers'), but are nonetheless keen to establish that they possess a distinctive expertise and form of intellectual status, and to find ways of articulating it. This is the experimental period of the new-style masters of insight, later to be called 'Presocratics', with their diverging essays in self-definition and promotion, from the sardonic Delphic riddling of Heraclitus, to the varied blendings of mystagogue, shamanistic traveller and rhapsode contrived by Parmenides and Empedocles.[1] The successors of these pioneers through the middle years and second half of the fifth century observed keenly as other nascent professional groups – doctors, civic educators (sophists) – set about the business of defining and advertising their group identities and brands of expertise, creating and problematising the very notion of an art/skill/expertise (τέχνη) as they did so; to some degree they even participated in the process on their own account.[2] But it was really an achievement of the fourth century to create the contours of something called 'philosophy', and thus also to establish the basis for articulating and debating a distinctively philosophical expertise and authority (as well as providing the conceptual tool with which a coherent back-history for the newly defined calling could be retrospectively constructed from the untidy diversity of earlier intellectual experimentation). Giant steps are taken by Plato: in his construction of Socrates as an emblematic figure for a kind of knowing and acting pointedly different from the ostentatiously professional professionals with whom he is shown interacting; and in his essays in imagining what might result if whole communities are structured in deference

---

[1] This is a topic still not directly confronted in the scholarly literature, though the contributions of Detienne 1996, Humphreys 1975 and Most 1999 provide some essential materials.
[2] Heinimann 1961; Roochnik 1996: 17–88.

to philosophical aims and insights. This Platonic project is then rephrased and consolidated by Aristotle, and the next wave of innovators and system-builders that breaks with Zeno, Xenocrates and Epicurus at the end of the century. For present purposes, however, we must jump forward from here to the early centuries AD and the world of the Empire.[3]

Two Vignettes

I begin with two episodes from writing of the 160s and 170s AD which together paint a suggestive even if incomplete picture of contemporary expectations of the philosopher, and the ways in which his authority could be both perceived and challenged.

At the beginning of Lucian's *Nigrinus*, the dialogue's principal interlocutor reports on the visit he has just paid to the (perhaps fictitious) Platonist philosopher in Rome,[4] and on the effect the sage's words have had on him. On admission to his house, he found Nigrinus 'with a book in his hands, encircled by numerous images of philosophers of old; set out on view there were also a blackboard (πινάκιον) with some geometrical shapes drawn on it, and a wicker sphere apparently representing the Universe' (*Nigr.* 2). Upon asking the philosopher what he was engaged on, and whether he had it in mind to return to Greece, he was, he reports, treated to what he describes as a display of such ambrosial eloquence as to put the Sirens and Homer's lotus into the shade:

> Led on by his theme, he spoke the praises of philosophy, and of the freedom which philosophy confers, and expressed his contempt for the vulgar error which sets such a high value on wealth and fame and dominion and power, on gold and purple, and all that dazzles the eyes of the world, and once attracted my own! I drank this in with rapt and receptive (ἀναπεπταμένη) soul. In the heat of the moment I could not even guess what had come over me, and was awash with conflicting emotions. My dearest idols, riches and renown, lay shattered; one moment I was ready to shed bitter tears over the disillusionment, the next, I could have laughed

---

[3] The question of how best to disentangle the intermediate phases and processes of development between these two points is a complex one, made more difficult by the relatively patchy nature of our evidence for the Hellenistic period, and the need so often to rely on later summaries and quotations rather than strictly contemporary material. Many features of the picture to be outlined in what follows must already have been in evidence in, say, the third or second century BC, but it is hard always to be certain exactly which.
[4] Clay 1992 dismisses Nigrinus as a fictional construction; Tarrant 1985 proposes that he is meant as an ironic caricature of the Platonist Albinus.

for scorn of these very things, and was exulting in my escape from the murky atmosphere of my past life into the brightness of the upper air. The result was curious: I forgot all about my ophthalmic troubles,[5] in the gradual improvement of my spiritual vision; for until that day I had been grovelling in spiritual blindness without realizing it. (*Nigr.* 4, tr. Fowler [adapted])

The second episode is one recorded by Aulus Gellius in *Attic Nights* 19.1 as an experience of his own, and perhaps belongs to the late 140s or 150s AD. Sailing once across the Adriatic from Cassiopa to Brundisium, his ship was caught in a prolonged and severe storm, which put the passengers in fear of their lives. Among these passengers was a celebrated Stoic philosopher, whom Gellius had got to know in Athens, and knew as a man of considerable authority (*non parua ... auctoritate*) with a good record of ensuring respectable behaviour in his pupils (*satisque attente discipulos iuuenes continentem*). In the storm, this individual was observed showing similar signs of fear to everybody else: he may not have wailed out loud as they did, but his pallor and expressions of distress were very much the same (*coloris et uoltus turbatione non multum a ceteris differentem*). Afterwards, he was challenged for this by a fellow passenger, an over-dressed and conceited plutocrat from somewhere in Asia Minor, who asked mockingly why he had been so afraid, when he himself (the plutocrat) had not. After some hesitation over whether to answer at all, the philosopher replied that, though there was a good reason, his questioner did not deserve to hear it. He would have to make do with the same answer as the Socratic Aristippus gave in similar circumstances: he had the life of Aristippus to be afraid for, whereas his challenger had only the life of a worthless scoundrel, and so could afford to stay calm. Safely in port, and with the storm abated, Gellius questioned the Stoic further, and was granted the real explanation, which had been denied to the bumptious tycoon. According to the founders of Stoicism, Zeno and Chrysippus, and the faithful record of their thinking given in Book 5 of Arrian's *Discourses of Epictetus*,[6] even the most virtuous person is permitted – indeed cannot avoid – an initial, automatic and purely corporeal reaction to sudden external events. What matters, and what establishes his true moral character, is whether or not he then mentally 'assents' (συνκατατίθεται) to this initial external appearance (φαντασία) of something bad. Initial physical reactions like turning pale and grimacing are non-rational reflexes, not signs that the

---

[5] The speaker had originally gone to Rome to seek treatment for an eye complaint.
[6] Now lost; Gellius' translation of Arrian's (Epictetus') words in 19.1.15–20 is fr. 9 in Schenkl's 1916 edition.

person manifesting them is really afraid, or really believes he is faced with something bad.[7]

Taken together, these two passages speak to the issue of philosophical authority on a number of different levels. On the one hand, they suggest some of the main areas in which that authority is exercised, embracing both moral character and conduct and technical (doctrinal) knowledge, and knowledge of external, macrocosmic, and human nature. They suggest expectations that the authority of these paragons of knowledge and conduct will be manifested in a distinctive appearance, accompanied by a charismatic impact on those who come into contact with them. They hint that their authority is not a purely personal matter, but derives in large part from the great figures of the past, whose thought they are qualified to expound. And finally, and very importantly, they also suggest that philosophical authority, rather than being a secure given, was open to various kinds of contestation. This last point is easily visible in Gellius' story. His Stoic may indeed be shown as vindicating himself in the end, but he has to overcome several layers of challenge in order to do so: he is from the start under observation by his fellow passengers, who regard it as something remarkable and potentially discreditable that his facial features should behave as they do, and he then has to meet the directly sarcastic scepticism of the tycoon. But sceptical or at least ironic stirrings can be detected in Lucian's evidence too. Nigrinus' interlocutor's reaction to the philosopher's fine words and high thoughts, both as sketched in §§ 3–4 and even more as developed later on in the dialogue (§§ 35–7), seem to tip over from level-headed acknowledgement of a philosopher's impact into the kind of rhapsodising that throws the speaker's good sense, and grip on what the process is really meant to achieve, into serious doubt. Nigrinus is not directly criticised or challenged, and no sceptic confronts him within the world of the dialogue, but the fact that he has been given such a 'pupil' as this arouses strong suspicions of authorial, Lucianic irony at his expense.[8]

All such thoughts about the implications of these opening passages will need further development and supplementing in the light of other evidence from the first and second centuries AD. But before that, we need to take a step back and remind ourselves of the essential framework within which both of the two episodes described make sense. What was

---

[7] For discussion of this point of Stoic doctrine, with further texts, see Sorabji 2000: 66–132, 375–84.

[8] For other views of the operation of irony in this dialogue, see Hunter 2012: 15–19; Schlapbach 2010: 261–75; Tarrant 1985. Schmitz 2010: 304 bluntly declares that attempts to detect irony have so far proved unsuccessful.

philosophy (*philosophia*) at this time, and what was the standard sense of what counted as being a philosopher (*philosophus*)?

Definitions and Understandings

A handy starting point is provided by the second-century *Manual of Platonic Doctrine* (*Didaskalikos*) attributed to the otherwise unknown Alcinous.[9]

> 'Philosophy' is an urge for wisdom (ὄρεξις σοφίας), or a releasing and reorientation of the soul away from the body, as we turn towards intelligibles and what truly exists; 'wisdom' is knowledge of divine and human matters. A 'philosopher' is someone who bears the name derived from 'philosophy', as a 'musician' does from 'music'; such a person must first of all have a natural bent for learning... and secondly he must have a passion for truth, and never accept falsehood; in addition to this, he must have a certain innate self-control (πως σώφρονα εἶναι) and a natural restraint in respect of the emotional component of his soul. The prospective philosopher must also have a nobility of spirit... [10]

Although the formulation here has an identifiable Platonist inflection in the sharp distinction of soul from body and the privileging of intelligibles as the true reality, the remainder is common property for the period, for all but a minority of styles of philosophy. Philosophy and philosophers are defined by the combination of mastery of a body of knowledge, and a set of abilities and dispositions that embrace both the intellectual and the moral. Philosophy both requires and promotes not just the unsullied and single-minded pursuit of truth, but also excellence of moral character; at the same time, the truth or truths towards which it is oriented are of the broadest and deepest conceivable kind.

This last point, about the breadth of philosophical knowledge, is already contained in the formula 'divine and human matters', but can be expanded and clarified by referring to the standard map of the territory of philosophy, first drawn up by Plato and Aristotle in the fourth century BC, but long common ground (again, for *almost* all philosophical persuasions) by the period with which we are concerned.[11] According to this, the proper

---

[9] The διδασκαλικὸς τῶν Πλάτωνος λόγων: ed. Whitaker 1990, tr. Dillon 1993 (as *The Handbook of Platonism*).
[10] *Didaskalikos* 1.2–3.
[11] Cf. e.g. Apuleius, *De Platone* 2.6.228; Philo, *De congressu* 79; Quintilian, *Inst.* 12.2.8; SVF 2.35–6, 1017; Plato, *Resp.* 486a, 593e, *Symp.* 186b, *Laws* 631b.

concerns of the philosopher embrace physics (ἡ φυσικὴ τέχνη, the science of nature), ethics (ἡ ἠθικὴ τέχνη, the science of moral character) and logic (ἡ λογικὴ τέχνη, the science of words). This already looks broad enough, but the aim for complete comprehensiveness only properly emerges when it is appreciated that, under the banner of 'nature', physics includes theology, metaphysics and meteorology as well as what a modern reader instinctively understands as physics, while logic, the science of words, embraces etymology, grammar and rhetoric as well as the skills of logical definition and argumentation. There is, in short, no aspect of the world and human experience worthy of any kind of serious attention that philosophy does not claim to be able to give a privileged account of.

At the same time, as also emphasised by Alcinous, the pursuit of deep and broad knowledge is supposed to support and be supported by the pursuit of a morally virtuous character, which is in turn seen as the highest kind of human realisation – fulfilling the blueprint of the human as no other activity or achievement can. In the terms of another image, originating from a Stoic source, but expressing a more general conviction, once more in the 'Socratic' tradition from Plato and Aristotle, if the whole of philosophy is a walled garden, then logic is the garden's walls, physics its soil and ethics its fruit.[12] Philosophy can thus be characterised as the τέχνη τοῦ βίου, the *ars vitae*,[13] claiming ultimate authority over character and life quite as much as over the pursuit of truth about the external world, from its lowliest to its most exalted manifestations.

In its full form, the understanding of philosophy just sketched is, as already suggested, the understanding of the mainstream of the 'Socratic' tradition running from Plato through Aristotle to the Stoics. At its heart lies a privileging of reason as the highest human faculty, enabling both the intellectual apprehension and synthesis of truths about the world and the cultivation of a virtuous moral character, and thereby the realisation of nature's blueprint for the human being. Other philosophical sects, though not sharing that central preoccupation with rationality, nevertheless formulate their own models of the philosophic in reaction to it, and to the overall shape of the Platonic-Stoic configuration of ideas. The Cynics, renegade Socratics, laid exclusive emphasis on the cultivation of moral virtue through stern self-discipline in the pursuit of the simplest viable mode of existence, and discard the pursuit of knowledge through intellectual speculation. The Epicureans

---

[12] Diogenes Laertius, *Lives* 7.40; Sextus Empiricus, *Against the Professors* 7.17–19.
[13] Among others, see Cicero, *De finibus* 3.2.4; Sextus Empiricus, *Against the Professors* 11.168–257; Maximus, *Oration* 1.3.

break the tight bond that for Platonists, Stoics and Peripatetics makes the exercise of reason constitutive of moral virtue and the full realisation of the human, but are in their own way equally insistent that the insight into the nature of reality that they offer is also the only path, for all but a tiny fraction of humanity, to a truly happy and morally virtuous life.[14] The Sceptics, like the Epicureans, hold that the exercise of reason is only contingently required for the achievement of a fulfilled and virtuous existence: thought and argument are only needed by someone aware of and tempted by the positive doctrines of other sects. But since all but a tiny fraction of those they take themselves to be addressing are so aware, rational exertion cannot be avoided in their eyes either. Thus, in effect all the major philosophies bar the Cynics buy into one version or another of the understanding of the philosophical as the combined pursuit of the intellectual and the ethical.

Philosophical Mastery

If, then, philosophers are those who profess a dedication to the pursuit of such goals as these, and in virtue of this profession are assumed to be (at the very least) closer to achieving them than their ordinary fellow citizens, it seems easy enough to see what kind of standing they ought to enjoy. Approximate modern parallels, illuminating if not pressed too hard, suggest a combination of the respect paid to scientists (or the abstract 'Science') with that paid to spiritual leaders, at whatever level of attainment: so to speak, from the lofty eminence represented by a combination of Stephen Hawking with the Dalai Lama, down to the more workaday level of a parish priest or imam who comments on recent developments in particle physics or the study of genetics in sermons or the parish magazine. The imagery commonly used to acknowledge this standing tends to concentrate on two leading ideas, that of philosophers as supervisors and guides to the rest of humanity (or at least, that part of humanity that recognises its need of supervision and guidance), and that of philosophers as pioneers or champions, who use their experience to help others towards the same goals. Maximus of Tyre, in his introductory and protreptic *Oration 1*, may have an eye to his own position as a philosophising orator when he compares philosophers to actors and athletes, riveting the attention of their audiences,[15] but

---

[14] The exception: people naturally constituted not to worry, so not in need of therapeutic argument.
[15] Maximus, *Oration* 1.1, 4–6, 10.

he blends this with pictures of the philosopher as a doctor, managing the physical vagaries of his patient's body, a herdsman tending and preserving his flock, and a musical director training a choir.[16] Each of this trio of images of benevolent and skilled control forms part of a larger system, which can be exploited to express thoughts about the nature of philosophy itself, and the effects of philosophical concepts or discourse on their audience, as well as about the authority of the philosopher: philosophy as music (either in the sense of harmony, or in the sense of the greatest of the activities patronised by Apollo and the Muses); philosophy as the tending and shaping of natural growths (animal or vegetable); philosophy as operating on the complex of dispositions, emotions and moral character (the soul) as medicine does on the analogously complex physical body. The expressive possibilities of medical imagery are particularly commonly resorted to across the range of first- and second-century philosophical writing, with varying emphases, but the expertise and controlling ability of the philosophical 'doctor' is never far from the surface. Musonius Rufus is reported by Aulus Gellius as imagining the philosopher's audience necessarily shuddering, with silent shame or joy, as he probes 'the sound and the diseased parts of their souls' like a doctor palpating his patient (*proinde ut eum conscientiamque eius adfecerit utrarumque animi partium aut sincerarum aut aegrarum philosophi pertractatio*).[17] Musonius' pupil Epictetus, as reported by Arrian, characteristically paints the picture with still more acerbic precision, though also with more than one eye on the possibility of the philosopher himself failing to live up to his responsibilities:

> The philosopher's school, gentlemen, is a surgery: you ought not to go out of it with pleasure, but with pain. For you are not in sound health when you enter: one has dislocated his shoulder, another has an abscess, a third a fistula, and a fourth a headache. Then am I to sit and utter to you little thoughts and witty little comments so that you can praise me and go away, one with his shoulder in the same condition in which he entered, another with his head still aching, and a third with his fistula or his abscess just as they were? Is it for this then that young men shall leave home, and abandon their parents and their friends and kinsmen and property, so that they can say to you, 'Bravo!' when you are uttering your witty little comments. Did Socrates do this, or Zeno, or Cleanthes?[18]

Images of the philosopher as pioneer and champion, making his own gains available to assist those less advanced, are most commonly cast in terms of

---

[16] Maximus, *Oration* 1.2–3, 7. [17] Gellius, *Noctes Atticae* 5.1 = Musonius fr. 49 (Hense 1905).
[18] Arrian, *Discourses of Epictetus* 3.23.30–2 (tr. G. Long, slightly adapted).

journeying and ascent. The most important single precedent for this field of imagery and its application is undoubtedly Plato's allegory of the Cave in Book 7 of the *Republic*, which images the acquisition of knowledge and insight as the difficult upward path from the shadows and inanimate models of the cavern to the sunlit originals in the world outside and above, but also pictures the successful philosophical pilgrim returning to try to impart some of his enlightenment to the still bound prisoners below.[19] But *Nigrinus* 4, cited above, reminds us also of the importance of the flight imagery of *Phaedrus*, and the *Phaedo*'s contrast between the murky, obscuring atmosphere of the human realm and the bright clarity of the real world above.[20] And we might well also recall Lucretius' depiction of Epicurus in the prologue to Book 1 of the *De rerum natura*, as a cosmic voyager returning to the Earth like a triumphing general to report on the true nature of reality, as discerned by his penetrating intellect.[21] For the Imperial period probably the richest development of the imagery of journeying and ascent comes in the pseudepigraphic *Tablet of Cebes*. In this elaborate allegorical development of the Cave, together with Hesiod's two roads and Prodicus' Choice of Heracles, the attainment of the knowledge, virtue and happiness that go with true Culture (*Paideia*) is figured as arrival at a lofty citadel at the top of a steep and rocky ascent.[22] Once arrived and welcomed, the successful pilgrim is then sent back down to the zone from which he ascended, back down among the unenlightened, where he can now wander at will immune from harm, while 'all will welcome him gladly, as sick people do a doctor'.[23] Didactic authority in this text is additionally embodied in the parallel figures of the Daimon (Genius, tutelary spirit) who instructs the newborn at the gates of Life within the allegory, and the mysterious old man, complete with his lecturer's staff (ῥάβδος), who expounds the allegory itself to his young enquirers.[24]

A number of significant claims and assumptions about the standing of philosophy and philosophers underlie and are underscored by these two sets of imagery. To figure philosophical activity as a path or an ascent, which all sensible human beings aim to follow, but philosophers cover faster or further than the rest, is to present it not as some optional pursuit, one choice of occupation among others, but as the one and only right course for a human being. On this strongly objectivist and normative view, the facts of human nature dictate a single way of developing that nature to its fullest degree;

---

[19] Plato, *Republic* 7.514a–17b, noting particularly 516e–17a; 7.519d–20d.
[20] Plato, *Phaedrus* 246a–50c; *Phaedo* 110b–11c.
[21] Lucretius, *De rerum natura* 1.62–79. [22] *Tabula Cebetis* 17–21.
[23] *Tabula Cebetis* 24–6. [24] *Tabula Cebetis* 2–4.

and it is philosophy, the 'art of life', which holds the key to it. At the same time, the imagery of ascent to and return from an eminence that normal humanity will struggle desperately to reach, unless supported, assisted and inspired by a more talented few, dramatises a claim to transcendence: philosophy speaks to everyday attainment and comprehension from a point beyond the everyday, to which the class of philosophers have a privileged access. Third, the repeated and insistent co-option of models of authority, leadership and control from other areas of activity to provide images of philosophical direction and control, besides clarifying the role that is being claimed for philosophy in life, also carries a competitive charge: by being comparable to a whole series of reputed forms of expertise or skilled attainment, philosophy emerges as better than any. In addition to the images of doctor, herdsman and musical conductor already mentioned, the philosopher is also commonly figured as helmsman and general (in the voyage and the campaign of life),[25] as well as seer and legislator.[26] Just as comparison with the seer links assertion of a precious skill with another version of the claim to transcendent authority (the discerning of truths out of the normal scope of human wit), so comparison with the legislator, and to some extent with the general, links technical expertise with the entitlement to control humanity *en masse*, in military and political conglomerates. It is also significant that a large proportion of this range of imagery of leadership, guidance and skilled supervision is regularly applied, in the appropriate Platonic-Stoic tradition, to the supervising role and activity of god, or the gods:[27] philosophers are to mundane life what the divine is to the cosmos as a whole, and enjoy that status precisely because they have a privileged contact with that higher level.

Philosophical and Political Authority

The potential for a challenge to conventional political authority in this kind of imagery is clear, and easy to substantiate on the level of formal doctrine. Plato had after all famously declared that salvation for politics and human affairs in general could only come 'if philosophers rule in their states, or those who are now called kings and rulers genuinely and adequately practice philosophy',[28] and shown in the surrounding argumentation of the *Republic*

---

[25] E.g. Maximus, *Orations* 29.7, 30.1–2.   [26] E.g. Maximus, *Orations* 21.7, 29.7, 37.2.
[27] E.g. Maximus, *Oration* 4.9; Philo, *De sacrificiis Abeli et Caini* 131, *De opificio mundi* 46, *De plantatione* 2; Dio Chrysostom, *Or.* 12.34, 36.32.
[28] Plato, *Resp.* 473c–d.

how the resulting political community might look. Stoic paradox declared the sage (the perfect Stoic) to be the only true king, just as he alone was also truly wise, brave, beautiful, wealthy and free, though this was admittedly as much a way of declaring political values and political status irrelevant to true virtue and happiness as of inserting philosophy into the political arena.[29] And philosophy certainly claimed rights of comprehension and authoritative exegesis over key political concepts, above all justice, temperance and courage.

In their own eyes, and those of their admirers, philosophers certainly felt able and entitled to speak truth to power, and to receive a respectful hearing as spokesmen for a higher level of insight and understanding. In his essay *On the Proposition that the Philosopher Ought to Converse Above All with Political Leaders*,[30] Plutarch asserts in sound Platonic fashion that philosophical enlightenment is required to make a truly good and beneficent ruler, and that candid and respected philosophical advisers are needed to inculcate it:

> If the philosopher's discourse gets hold of a private individual, someone who is happy to play no role in public business and confines himself to his bodily requirements as if drawing a circle in geometry, it is not diffused to others, but withers away and vanishes once it has created peace and calm in that one person. But if it takes possession of a political leader and a statesman and a man of action, it brings benefits to many through that one person, as Anaxagoras did by associating with Pericles, and Plato with Dion, and Pythagoras with the leading citizens of Magna Graecia. Cato himself left his army behind and sailed off to find Athenodorus, and Scipio sent for Panaetius when he himself was dispatched by the Senate… [31]

And in the adjacent essay in the *Moralia*, traditionally given the slightly misleading title *To an Uneducated Ruler*,[32] Plutarch argues, in similarly Platonising vein, that it is only through philosophical teaching that rulers can fulfil their proper function of being God's likeness on earth, spreading justice and virtue in the human world as the sun spreads warmth and light in the natural world.[33]

---

[29] Cicero, *Acad. Pr.* (*Lucullus*) 136; Lucian, *Fisherman* 20; cf. *SVF* 3.619 and Clement, *Strom.* 2.5.20.1.

[30] Plutarch, *Moralia* 776b–9c (*Maxime cum principibus*, περὶ τοῦ ὅτι μάλιστα τοῖς ἡγεμόσι δεῖ τὸν φιλόσοφον διαλέγεσθαι).

[31] *Maxime cum principibus* 1.776f–7a.

[32] *Ad principem ineruditum*, πρὸς ἡγέμονα ἀπαίδευτον, *Moralia* 779d–82f (incomplete). 'Misleading' because the essay is about the need for rulers in general to be philosophically enlightened, rather than the failings of any specific individual.

[33] *Ad principem ineruditum* 3.780f–1a, 5.781f–2b; the debt to the Sun simile of Pl. *Resp.* 6.508a–9c is obvious.

There is no strong sense, however, in either of Plutarch's essays of any kind of tension or struggle in the encounter of philosophical with political authority. Rather, it is suggested, in a way that implicitly flatters any ruler who happens to be reading, that as thinking beings, right-minded leaders will of course acknowledge the value and truth of philosophical discourse when exposed to it, and act accordingly. This sort of insinuation by flattery – advice cast as encouragement to continue on a course already begun, or begin a course to which the addressee is already naturally inclined, rather than as reproof or challenge to change course for the better – can be seen also in Seneca's *De clementia*, directly addressed to Emperor Nero.[34] Challenge and confrontation tend to be envisaged mainly in the safety of the past, as when Dio Chrysostom in his *Fourth Kingship Oration* conjures up Diogenes needling and disconcerting the young Alexander,[35] or still more when Philostratus in his *Apollonius of Tyana* depicts his sage in angry, and triumphant, confrontation with the Emperor Domitian.[36]

Dio himself might just be thought an exception to the general pattern. If the four *Kingship Orations* were genuinely delivered before the emperor,[37] they would be a written record of four occasions on which a self-identified philosophical adviser dispensed firm moral advice to his supreme ruler, and at least twice (in *Orations* 2 and 4) depicted a great historical model for the ruler (Alexander) on the defensive in the face of the same advice, as dispensed by individuals with authority over him, in Philip and Diogenes. But it is debatable whether this analysis can stand. In the first place, even if the orations, or some versions of them, were indeed really delivered before the emperor, the way Dio modestly conceals himself in them behind a series of *personae* from the nearer and further past (Diogenes, Socrates, Philip, his own past self) dulls any real sense of confrontation and suggests instead the kind of flattering insinuation (that the emperor is of course already following the advice being tendered to him) identified above. In addition, actual delivery before the emperor can by no means be taken for granted. In publishing the orations, Dio could equally well have been conjuring up an imagined set of occasions, hoped for but never actualised, but serviceable nonetheless as further elements in the continuing project of self-presentation and career creation constituted by his written speeches.[38]

---

[34] Seneca, *Clem.* 1.1.1–7, 2.1–2.
[35] Dio, *Or.* 4, on which see above all Moles 1983 and 1990, but also Brancacci 2000 and Trapp 2000: 225–7.
[36] Philostratus *VA* 7.32–8.10.
[37] Arguably Trajan in all four cases: see Moles 1990.
[38] On Dio's self-creation through publication, see further Trapp 2012: 120–5.

In this case the *Kingships* would count as one more assertion from the philosophical side of the fence of what the authority of philosophy ought to be rather than proof of how it could actually function in relation to high political power. On the level of what can be securely documented, the effect of imperial authority on philosophers, whether in the repeated decrees of banishment that punctuate the first century AD in particular,[39] or in the curtailment of concessions of immunity (ἀτέλεια) in the later second century,[40] remains more striking than any flow in the other direction.

Even the claim that philosophical authority must have been taken seriously by emperors because they repeatedly saw fit to banish philosophers can only be allowed to stand in a severely qualified form. Banishments of specific individuals, as of Musonius Rufus and Cornutus under Nero,[41] and Dio Chrysostom under Domitian,[42] are always likely to have had as much to do with political connections as with any desire to remove a perceived intellectual challenge. Wholesale expulsions of philosophers as a class, as by Vespasian in 71 AD and Domitian in 93 AD,[43] acknowledge the availability of philosophers' vocabulary and ideas, and emblematic figures from the history of philosophy, to articulate and lend extra respectability to political opposition or subversion, but it is again at the political exploitation not the intellectual challenge that the punitive measure is directed.[44]

Philosophical Authority and Social Superiority

Possession of the identity 'philosopher' nevertheless remained, in spite of uncertainties over relations with the high coercive authority of political power, a weighty weapon to be able to wield in confrontations with others on a more everyday level. It was not, to be sure, uniformly or invariably effective, since its power to efface or outflank normal social distinctions depended heavily on the willingness of the socially superior to defer to it, and that willingness was clearly not always forthcoming. The case of the venerable philosopher Thesmopolis, cited (or invented?) by Lucian in his pamphlet *On Salaried Posts*, provides a vivid example of normal social hierarchy taking precedence over mere intellectual distinction: however doted

---

[39] MacMullen 1966: 70–94; Trapp 2007: 226–9.
[40] Trapp 2007: 19–20, 246, 252, with references to earlier literature.
[41] Tac. *Ann.* 15.71, Dio Cassius 62.29.   [42] Dio, *Or.* 13.1–2; cf. Jones 1978: 45–6.
[43] Suetonius, *Vesp.* 13, Dio Cassius 6.12–13, Gellius, *NA* 15.11.3.
[44] On the wider question of 'philosophical opposition' to the principate, see Trapp 2007: 226–30 and the older studies cited there.

on by his patroness, this elderly, heavily bearded Stoic remained a trophy philosopher, and as a hired hand had not only to share a carriage with her favourite eunuch, but also take care of her Maltese lap dog.[45]

Striking cases can nevertheless be found of a more successful deployment of philosophical credentials. One example of how their competitive potential could be realised has already been seen in the case of Gellius' Stoic in the storm. Identification as a philosopher may initially expose him to sceptical scrutiny, but his philosopher's knowledge of both fact and doctrine enables him to turn the tables: as well as being able to face down his boorish critic, he also succeeds in confirming Gellius' own existing good opinion of him, and thus earning his place in the *Attic Nights*. The social dimension to the encounter is striking too: Gellius and the Stoic bond as fellow members of an intellectual elite, the crass and hostile materialist is put firmly in his (inferior) place.

A second example of much the same kind of manoeuvring from much the same period is provided by the *Apology* of Apuleius, dating from the year 158/9 AD.[46] On trial for having allegedly secured the affections of a widowed heiress by magical means, Apuleius represents himself as having come under attack for compounding the felony by cloaking his lust and his dark practices under a hypocritical pretence of philosophy. Whether or not this is how the prosecution actually presented its case, it is exactly what Apuleius needs as the springboard for his defence, a central strand in which is the assertion that they have crassly and ignorantly failed to understand what philosophy is, and that what they have taken as evidence of magical malpractice is in fact proof of learned commitment and expertise. When he had fish bought for him and cut them up, this was not a magical operation, but a piece of Aristotelian biological investigation.[47] When he whispered in a slave-boy's ear and the boy thereupon collapsed, he was investigating what he had diagnosed as a case of epilepsy, not casting a spell.[48] Similarly, what the prosecution have presented as conduct suspiciously unbecoming to a philosopher only serves further to prove their cloddish ignorance of what any cultivated individual knows about philosophy: it is a not unphilosophical vanity but a mark of the physicist's interest in optics to possess a

---

[45] Lucian, *De mercede conductis* 33–4. In this pamphlet, Lucian alleges a special proneness on the part of Romans to behave in this manner in their treatment of their Greek intellectual employees, and treats it as one more indication of their inability to appreciate true civilisation and civilised values (cf. Swain 1996: 315–21, citing also *Nigrinus* and *The Ignorant Book-collector*); the pattern is in fact as likely to have been general across the moneyed elite, without any necessary coordination with the Greek-Roman divide.
[46] Harrison 2000: 39–88; Harrison *et al.* 2001: 11–24; Riess 2008; Schindel 2000.
[47] Apuleius, *Apol.* 29–41.   [48] Apuleius, *Apol.* 42–5.

mirror.⁴⁹ And when they confusedly charge him with sordid poverty, and fling in an insult about his 'wallet and staff', all this shows is their utter inability to tell the difference between a Platonist and a Cynic.⁵⁰ But what gives all this its similarity to the Gellius episode is the fact that, within the dramatic world of the trial that the speech reports, Apuleius' superior scorn against the prosecution is directed not to some general audience, but directly to the court president, the proconsul Claudius Maximus.⁵¹ Apuleius appeals to him, collusively, as one educated and philosophically responsible man to another, over the heads of the prosecution – and displays himself doing so in the written version of the speech that he is now circulating.

The Extension of 'Philosopher'

The case of Apuleius, however, also points to a further complexity for any discussion of philosophical authority in this period, in particular when he is taken together with the somewhat contrasting cases of Thesmopolis and Gellius' Stoic. There can be no mistaking the energy and persistence he puts into his claims to philosophical status, not only as described above in the *Apology*, but also in the fragments of his declamations preserved as the *Florida*,⁵² in the fully surviving declamation *On Socrates' God*, and perhaps in other more technical works as well.⁵³ He parades a wide-ranging knowledge not only of philosophical doctrine, but also of the history of philosophy from the earliest times, and the personal achievements and idiosyncrasies of great philosophers of the past. Yet in the modern historiography of philosophy he does not count as a central case of the genus 'philosopher'. He has evidently had a philosophical education in some form,⁵⁴ but he equally evidently operates outside any formal scholastic environment (unlike Gellius' Stoic and, presumably, Thesmopolis as well, he gives no courses of lectures, expounds no texts and has no pupils), and shows no sign of independent speculation on doctrines and problems; although he can – probably – rise to systematic summary of doctrines in handbook form, his natural habitat is that of the popular declaimer, in lecture hall or theatre, addressing a general audience rather than training the next generation of professional

---

⁴⁹ Apuleius, *Apol.* 13–16.   ⁵⁰ Apuleius, *Apol.* 18–22.
⁵¹ See Apuleius, *Apol.* 1, 13, 19, 25, 36, 38, among other passages.
⁵² Apuleius, *Flor.* 9, 13, 14, 15, 22.
⁵³ Depending on whether or not *On Plato and his doctrines* and *On the cosmos* are accepted as genuine works of his: see Harrison 2000: 174–80 for the case for.
⁵⁴ Cf. Apuleius, *Flor.* 16.26 and 18.5.

thinkers. None of this, however, prevents him from claiming the status and its authority, or – if the inscribed statue-base from Madauros really does refer to him[55] – being allowed it by general consent in his home city.

Apuleius was in no sense an isolated case in this regard. As both contemporary usage and a host of further instances show, neither regular engagement in teaching and intellectual debate, nor a commitment to debating and developing doctrines and arguments was required to qualify for the title of philosopher. On the one hand, the first- and second-century landscape is not short of figures who, like Apuleius, devote time and energy, even large proportions of their lives, to the presentation of philosophical material in elegant literary guise, both for live performance and for written circulation, without ever having 'professional' philosophical careers. Dio Chrysostom, Favorinus, Euphrates and Maximus of Tyre all clearly fall into this category on the Greek side; on the Roman side, there is the declaimer Fabianus. All of these are looked back on only a few generations after their deaths as philosophers, and there is every reason to suppose that this reflects the perception of their contemporaries as well. On the other hand, both inscriptional and textual evidence points to the widespread use of 'philosopher' as an honorific, indicating a sympathy for higher values and the superiority of the life of the mind and the pursuit of virtue over crass materialism and self-indulgence, rather than any particular sectarian allegiance or commitment to scholastic activity.[56]

It might at first sight seem that a very loose and tolerant extension of what ought to be an exclusive label is in evidence here. It is in fact better seen as an easy and unexceptionable consequence of the name and nature of philosophy in this period. Etymologically, as Alcinous' treatment emphasised, the name denotes an attitude and an ambition – a love for wisdom – rather than an achieved competence; the more weight is placed on this, the more reasonable it can seem to allow in those who have themselves more of the attitude than the competence. But in substance, too, philosophy is meant to be as much about the cultivation of good character and virtuous action as about knowledge, argument and insight; in this way as well the door is opened to those who can be praised for their moral conduct and attitudes without reference to extensive formal study. The point can also be made by

---

[55] *ILAlg.* 2115: ... *[ph]ilosopho [Pl]atonico | [Ma]daurenses ciues | ornament[o] suo. D(ecreto) d(ecurionum), [p(ecunia) p(ublica)]*.

[56] See for instance Walz, *Rhetores graeci* 1, p. 38 (Hermogenes); p. 274 (Nicolaus); p. 466 (Nicephorus); *PMil. Vogl.* 11 and *POxy.* 3069 (letters between private individuals addressing each other as 'philosopher'); and honorific and funerary inscriptions such as *Inschriften griechischer Städte aus Kleinasien* 17.2, Ephesos 8.2.3901 and 4340; *ILAlg* 1, 2115; *CIL* 13.8159; cf. Trapp 2007: 246–9.

reference to another standard understanding of philosophy: if philosophy is indeed 'the art of life', identifying patterns of existence and states of the self that are uniquely right for all human beings as human beings, then it is a path for all, irrespective of social or professional identity, and there should be nothing surprising about encountering a philosopher in any walk of life. As Maximus of Tyre insists in his introductory oration, in his own version of a widespread assertion, almost everyone has the basic endowment needed to begin:

> Physical skills and the exertions needed to acquire them cannot be sustained by every bodily constitution… The contests of the soul, on the other hand, have quite the opposite character. It is only a tiny proportion of the human race, one almost never encountered, that is not naturally endowed for them.[57]

In theory, then, the task of tracing the operations of philosophical authority becomes even more diffuse and complex than at first appeared. If this view of 'philosophy' as the name for a set of aims and a disposition as well as a professional identity is pressed, it follows that philosophical achievement and authority could be encountered – asserted, deferred to or resisted – in a whole range of contexts and interactions, not only (so to speak) within groups of full-time professionals, or between full-time professionals and laypeople, but also among the laity. In practice, given the nature of the available evidence, concrete instances of the last of the three can be hard to detect in any very useful form. The private letters referred to above reveal individuals asserting or awarding the title 'philosopher' in private interactions, inscriptions show individuals claiming it or having it claimed for them in public space; but further details of the circumstances in each instance, and the ways in which the claims were received and reacted to are lacking. The focus of this chapter will therefore remain where the evidence (mainly textual) is richer and more atmospheric, with the play of philosophical authority between professionals, and even more, between professionals and a wider public.

Modes of Challenge: Philosopher to Philosopher

For both domains, a revealing question to ask is how claims to philosophical authority, whether explicit or implied, could be challenged or

---

[57] Maximus, *Oration* 1.5; cf. e.g. Seneca, *Ep.* 44.1–2, 90.1, 90.46, Apuleius, *De Platone* 2.3.222, Galen, *Quod animi mores* 11; and see Trapp 2007: 49–51.

impugned, and what was at stake when such a challenge was made. Where interactions among professionals are concerned, the reasonable default expectation might seem to be that authority will normally have been challenged or denied on straightforwardly intellectual grounds: demonstrable falsehood of doctrines, technical flaws in argumentation, infidelity to observed facts, ignorance or misunderstanding of other philosophers' views and arguments. And one might further expect readiness to challenge and impugn to run along sectarian lines, aimed outwards at all traditions of thought bar the critic's own. These expectations are in the main answered, but with some qualifications and extra features.

Perhaps most strikingly, the disposition to work along sectarian lines is not absolute. Instances of strong polemical hostility to rival sects are certainly found, for instance in Platonist Plutarch's anti-Stoic and anti-Epicurean treatises,[58] and in Epictetus' fulminations against Peripatetics, Academics (i.e. Academic sceptics), Pyrrhonists and Epicureans;[59] the fervour with which Sceptics and Epicureans returned the compliment can be seen in the treatises of Sextus Empiricus and the fragments of Diogenes of Oenoanda.[60] Sceptics of course on principle deny authority to all dogmatic thinkers categorically; other polemicists aim their fire at the falsehood of individual doctrines, but with the regular implication that a thinker who can err in this way is more generally unworthy of a respectful hearing. But what look like more tolerant and welcoming attitudes can also be found. Maximus of Tyre may be as vehement as anybody in his refusal to take Epicurus seriously as a thinker – he is the traitor who sells out to the very thing (pleasure) that true philosophy is there to resist, the maimed intellect as incapable of appreciating the divine as a Cimmerian is of understanding the sun[61] – but with that one exclusion Maximus is happy to maintain that all philosophers deserve an admiring and receptive audience, whatever the differences in their appearance, social status or teaching style.[62] From a different angle, Stoic Seneca ostentatiously uses thoughts and maxims of Epicurus in the early stages of his *Moral Epistles* to his friend and trainee Lucilius.[63]

---

[58] Anti-Epicurean: *Non posse* (*Mor.* 1086c–107c), *Adversus Colotem* (*Mor.* 1107d–27e), *Latenter vivendum* (*Mor.* 1128b–30e). Anti-Stoic: *Stoicorum repugnantia* (*Mor.* 1033a–57c), *Absurdiora poetis* (*Mor.* 1057c–8d), *Communes notitiae* (*Mor.* 1058e–86b) – but there are anti-Epicurean and anti-Stoic moments and emphases in many other works; see Russell 1973: 65–9.
[59] Arrian, *Discourses of Epictetus* 1.5, 1.22, 2.19.20–4, 2.20, among other passages.
[60] Diogenes of Oenoanda, fragments 5–7, 11 and 34–5 (Chilton 1971).
[61] Maximus, *Orations* 33.3 and 25.4; cf. 4.4, 11.5, 15.8, 19.3.
[62] Maximus, *Oration* 1.10; cf. Trapp 1997: xxii–xxv.
[63] Seneca, *Ep.* 2.5, 8.11, 8.7–8, 11.8–9, 12.10–11, 13.16–17, 16.7, 17.11, 18.14, among other passages. Cf. Griffin 2007: 90–3.

Seneca's declared rationale for his borrowing from Epicurus is that well-perceived and well-expressed truths, regardless of who formulated them, are common property.[64] Quoting from Epicurus is thus not exactly an appeal to the support of his personal authority (as the act of quoting might standardly be taken to be), but something more back-handed. The shock effect of finding an Epicurean asserting what a good Stoic ought to believe as well lends extra emphasis to the thought in question, but without at the same time giving the Epicurean strong credit for it. It is also noteworthy that Seneca ceases to quote Epicurean maxims as in his sequence of letters he moves on from preliminaries (*Letters* 1–29) and into the real business of helping Lucilius to advance along a Stoic path:[65] borrowed formulations are fine for protreptic and consciousness-raising preliminaries, but have ultimately a limited usefulness that reflects the limited value of their originator.

In Seneca's view, then, truth may incidentally circulate via non-Stoics, but real authority over central issues remains within the school. Maximus' apparently relaxed pluralism also rests on ideas about the location and accessibility of truth, but with a different focus. As is most directly set out in the complementary *Orations* 4 and 26, he is working with a model of philosophical insight in which all the most important elements of a unitary truth were already known in the far distant past of Greece, when they were enshrined in the works of the old poet-sages (Musaeus, Orpheus, Hesiod and Homer). Subsequent philosophy has simply re-expressed the same truths in a more pedantically literal (less poetic and allegorical) form;[66] at the same time, a tendency on the part of individual philosophers to separate off and concentrate on just one part of the original conglomerate has produced a spurious and undesirable impression of diversity and disagreement.[67] In theory, this could have been a basis for denying real authority to all but the original, poetic philosophers; in practice, Maximus avoids any such severity, and instead – as *Oration* 1 demonstrates – allows a large degree of credit to rub off on the successors as well, for their ability both to understand the original insights, and to re-express them effectively. Plato may be presented as one of Homer's many derivatives and debtors in

---

[64] Seneca, *Ep.* 8.7–8, 12.10–11, 16.7, 33.2; in the second of these – *quid est tamen quare tu istas Epicuri uoces putes esse, non publicas?*, 'why though should you think that these are Epicurus' sayings, not common property?' – there is a covert play on the contrast between the title of Epicurus' maxim collection, κυρίαι δόξαι ('authoritative' or 'valid doctrines', but with overtones of 'properly his own'), and the Stoic term κοιναὶ ἔννοιαι, 'common concepts'.

[65] Note *Ep.* 33 as a commentary on this shift in emphasis: Griffin 2007: 90, citing Wilson 2001: 183–5.

[66] Maximus, *Oration* 4.3, 4.5–7.   [67] Maximus, *Oration* 26.2; cf. 29.7.

Oration 26,⁶⁸ yet at the start of *Oration* 11 a strong contrast can still be drawn between the power of his insight and expository authority on the one hand, and mere interpreters on the other.⁶⁹

The sheer variety of attitudes available and in play is, however, underlined by the distinctive case of Galen. Describing his own intellectual formation in his treatise on *The Affections and Errors of the Soul*, he records how his father encouraged him not to declare an allegiance too quickly to any particular philosophical sect, but to scrutinise them all carefully, and how he has faithfully carried this precept into adult life, maintaining his critical independence.⁷⁰ This is not scepticism, as Galen is happy to accept that there are truths, and that (some) philosophers can (sometimes) get to them; but the philosophers are to be tested against Galen's own independent sense of the truth, rather than relied on to be telling the truth because they are philosophers. In practice, the greatest authority for him is the doctor (but also master philosopher) Hippocrates,⁷¹ who in his eyes assembled the greatest collection of truths available, though Plato, thanks to his perceived closeness to Hippocrates, also scores high marks.⁷²

Further space for reciprocal criticism and denials of authority within the circle of professionals is created by other elements in their shared sense of the name and nature of their calling. The fact that philosophy is understood to span the theoretical and the practical, holding together intellectual insight, expository competence and active moral striving opens the way to reciprocal accusations of betraying one part of the complex by concentrating too hard or too narrowly on another. In one perspective, too great a concentration on technicalities of theory, and what is seen as the willfully complicated terminology in which they are expressed, can be seen as a betrayal of the essentially moral and practical point of the whole venture:

> If you think that philosophy is simply a matter of nouns and verbs, or skill with mere words, and refutation and argument and sophistry... then there is no problem in finding a teacher. The world is full of that kind of

---

⁶⁸ Maximus, *Oration* 26.3: the striking claim that Plato is closer to Homer than he is to Socrates, for all the superficial appearance he gives of admiring the latter and distancing himself from the former.

⁶⁹ Maximus, *Oration* 11.1–2: Plato's words are as the all-illuminating sun, compared to the feeble bonfire lit by his expositors, or a rich gold-mine, necessarily superior to the assayers who sort and classify its yield.

⁷⁰ Galen, *Affections and Errors* 8.

⁷¹ E.g. Galen, *That the Best Doctor is Also a Philosopher* and *Doctrines of Hippocrates and Plato* (*PHP*).

⁷² E.g. *PHP* 3.4, *Protrepticus* 5, *Usefulness of the Parts of the Body* (*De usu partium*) 17.1; see in general De Lacy (1972).

sophist... But if all that is only a small part of philosophy, and such a part that ignorance of it is disgraceful and knowledge no cause for pride, by all means let us escape reproach by knowing it, but let us not therefore give ourselves airs... The summit of philosophy and the road that leads there demands a teacher who can rouse young men's souls and guide their ambitions.

So Maximus in his introductory oration;[73] his use of the word 'sophist' here (σοφιστής) to designate a *soi-disant* philosopher who misses the real point of the exercise through dishonest or pointlessly over-complicated argument is characteristic both of his own usage, and that of imperial period intellectual discourse more generally.[74] From a different part of the landscape, the Stoic Epictetus might at some moments seem to be in agreement. In an imagined confrontation with a lecturer keen to give a reading from his commentary on Chrysippus and to show how well he can clarify the philosopher's diction (τὴν λέξιν διαλύσω καθαρώτατα), he retorts indignantly: 'Is it for this that young men are to leave their homelands and families, in order to come and hear you expounding trivial turns of phrase?'[75] In another passage, however, he is to be found insisting in good Stoic style that scrupulousness over the technicalities of argument and terminology is in fact essential. Discoursing on the utility of 'changing' and 'hypothetical' arguments, he recalls how, when his teacher Musonius took him to task for not detecting an omission in a syllogism, and he retorted that 'It isn't as if I burned down the Capitol!', he was put down with the crushing 'Slave, in this case the omission *is* the Capitol!'.[76] Epictetus' own chief grouse is against those (among whom he would probably have included Maximus, had he known him) who pay too much attention to elegant style, and gathering plaudits for their own eloquence, and not enough to the real (painful) business of doing moral good to their audience.[77] There is agreement that authority belongs to the philosopher who correctly appreciates and sticks to the real point of the

---

[73] Maximus, *Oration* 1.8; for the coolness towards technical language as a mark of real philosophical authority, compare *Orations* 4.3 and 21.4.

[74] Compare e.g. *Oration* 1.4, τί οὖν, εἰ ἀκόλαστος ὁ σοφιστής ...; Aristides, *Or.* 1.306, *Or.* 2.177; Philo, *Quod Deterius Potiori Insidiari Soleat* 72, *De posteritate Caini* 86; Josephus, *Jewish Antiquities* 17.152 and 155, *Jewish War* 2.433 and 445 (σοφιστής being Josephus' regular Greek equivalent for 'Pharisee'). There is useful discussion of the workings of the sophist/philosopher antithesis in Sidebottom 2009 and Stanton 1973; Winter 1997 is an awful warning of the consequences of misunderstanding the use of the term 'sophist'.

[75] Arrian, *Epict. diss.* 3.21.6–8.

[76] Arrian, *Epict. diss.* 1.7.32–3; cf. Dobbin 1998: 118 *ad loc.*

[77] See above all Arrian, *Epict. diss.* 3.23, 'To those who give readings and lectures for display', Πρὸς τοὺς ἀναγιγνώσκοντας καὶ διαλεγομένους ἐπιδεικτικῶς.

exercise, but disagreement over what this commits him to and how he gives evidence of it.[78]

Modes of Challenge: Philosophers and Others

In the eyes of the wider, lay public, the most obvious grounds for challenge, to particular philosophers individually, or to all of them as a class, is the perception of failure to practise what they preach and to live up to their own standards. We have already seen Apuleius in his *Apology* representing himself as confronted with the charge, and rebutting it with what he hopes will appear contemptuous ease. It is also a recurring element in Lucian's satirical play with the perceived failings of contemporary philosophers (the classic figures of the past being carefully exempted). In the *Fisherman*, for instance, it is the promise of free food that brings would-be philosophers swarming to the Acropolis, and the supposedly hardy and ascetic Cynic who is fished up turns out to have perfume, money, a mirror and dice in his bag;[79] in the *Hermotimus*, Lucian's interlocutor Lycinus vividly proves the inadequacy of the elderly would-be Stoic Hermotimus' tutor by revealing that he has a hangover, following a rowdy party at which he not only got drunk, but also beat up a fellow diner who disagreed with him on a point of argument.[80]

The idea that philosophers can speak with authority to others only if they follow their own high-minded precepts, and are regularly bereft of that authority because they do not (a feeling clearly also underlying the episode of Gellius' Stoic), was evidently a useful one for resisting philosophy's claims to legislate for human life in general. But, as the example of the *Hermotimus* shows, resistance could have broader grounds as well. The anti-protreptic that Lucian depicts Lycinus developing in that dialogue ultimately charges philosophers with concocting a completely fictitious and fantastical model of human aspiration and fulfilment. They image the life lived under their direction as a long and difficult road, leading eventually to the lofty citadel of Virtue and Happiness.[81] But in reality there is no such road and no such citadel;[82] philosophers are deceivers, who can only maintain their hold over

---

[78] Compare also Seneca's ostentatiously Roman treatment of Chrysippus in the *De beneficiis*, discussed by Hine elsewhere in this volume. In this instance, Chrysippus' tendency to blunt the force of his insights with over-subtle analysis and argument (*acumen nimis tenue... pungit, non perforat*, 1.4.1) is stigmatised as typically Greek (*magnum mehercules uirum, sed tamen Graecum*).

[79] Lucian, *Fisherman* 41–2 and 45. [80] Lucian, *Hermot.* 11–12.
[81] Lucian, *Hermot.* 2–8, 24–6. [82] Lucian, *Hermot.* 71–3.

their pupils (dupes) because of the insidious consistency of their teaching, once the first false step is conceded, or because of the power of embarrassment in preventing them from admitting their error.[83]

A more narrowly targeted resistance, motivated by professional rivalry, could also be mounted on behalf of particular groups with reasons for resenting the encroachment of philosophers onto what they saw as their territory. Most obviously, teachers and students of rhetoric, in competition with philosophers for pupils and for the prestige of representing the pinnacle of educated culture, and heirs to the longest-running of all confrontations with philosophy, had strong reasons for denying philosophy's claims to deal with their material better than they themselves could. Like his master and hero Cicero, Quintilian is ready to concede that the trainee orator needs to know a good deal of philosophy, and that all of philosophy's three main divisions (*naturalis, moralis, rationalis*) contain matter relevant to the business of oratorical persuasion,[84] but he does so with an edge of wariness and even indignation. The point is not just that philosophers are theorists, whose over-delicate argumentation, however insightful, is unfit for the purposes of public persuasion and action (they are like little creatures (ants?) that manoeuvre agilely in a confined space, but are easy prey on open ground);[85] it is also that their claim to special rights over moral values (both in theory and in practice) is unjustified. Historically, after an early, happy period in which the study of eloquence and the study of morality were united, the two went their separate ways; and as the study of morality was abandoned by students of eloquence, it 'fell prey to weaker intellects' (*infirmioribus ingeniis uelut praeda fuit*).[86] Subsequently, this act of appropriation (*nomen... sibi insolentissimum adrogauerunt*) has been compounded by a second betrayal, since those who now call themselves philosophers, in addition to illegitimately claiming exclusive rights over moral theory, signally fail to put their own precepts into practice.[87] The topics of philosophy are quite in general not exclusive to philosophers (*haec autem quae uelut propria philosophiae adseruntur, passim tractamus omnes*); but morality in particular belongs to orators, who should fearlessly reclaim what they are entitled to (*nunc necesse est... uelut nostrum reposcere, non ut nos illorum utamur inuentis, sed ut illos*

---

[83] Lucian, *Hermot.* 74–5.
[84] Quintilian, *Education of an Orator* (*Inst.*) 1.4.4–5, 10.1.35, 12.2.
[85] Quintilian, *Inst.* 12.2.14: *itque reperies quosdam in disputando mire callidos, cum ab illa cauillatione discesserint, non magis sufficere in aliquo grauiore actu quam parua animalia, quae in angustis mobilia campo deprehenduntur*. The word *cauillatio* here is the Latin equivalent of σόφισμα (Seneca, *Ep.* 111.2).
[86] Quintilian, *Inst.* 1.pr.13–14.   [87] Quintilian, *Inst.* 1.pr.15.

alienis usos esse doceamus).[88] The very breadth of philosophy's claims here becomes grounds for denying it unique authority, though Quintilian does also appeal to the cliché of the degeneracy and hypocrisy of modern philosophy, and indeed to the old local sense of philosophy as something not quite right for Romans.[89]

That Quintilian's resistance to philosophical authority on behalf of rhetoricians, orators and politicians is not an isolated piece of eccentricity might in any case be assumed, but is helpfully confirmed by Aelius Aristides' essay *Response to Plato, in Defence of Oratory* (*Or.* 2). In this spirited rebuttal of the Platonic attack on oratory in the *Gorgias*, Aristides' central strategy is to demonstrate that Plato is inconsistent, and speaks with two voices: his own words, not only in other dialogues but even in the *Gorgias* itself, show that in fact in his better moments he endorses the same values as are (according to Aristides) central to orators and oratory. Philosophy cannot therefore claim unique responsibility for defining and upholding the core political value of justice, since orators do so too, indeed more effectively.[90]

It is not only in the name of oratory that the intellectual imperialism of philosophers could be resisted. Galen, characteristically, contributes a more measured, less aggressive challenge, on behalf of medical knowledge, and above all of his own prime authority, Hippocrates. His chief objection is to the knowledge claims of the modern representatives of the principal schools of thought, whom he accuses of defending their own doctrines out of partisan loyalty rather than a disinterested concern for truth,[91] and indeed of choosing their sectarian allegiances in the first place for accidental rather than principled reasons.[92] The great minds of the more distant past, on the other hand – Aristotle, Theophrastus, Hipparchus, Archimedes and above all Plato – he finds worthy of genuine respect. Yet even the greatest of them, the 'divine' Plato,[93] must ultimately yield to the authority of Hippocrates. He may sometimes usefully clarify and supplement what Hippocrates' works leave incomplete or obscure, just as his own work is sometimes usefully

---

[88] Quintilian, *Inst.* 1.pr.16–17.
[89] Quintilian, *Inst.* 12.2.7, 12.2.29–31. For further discussion of the context of Quintilian's manoeuvring, see Trapp (2007: 226–57, esp. 249–51), and for the playing of the Roman card, Trapp 2014.
[90] Aristides, *Or.* 2.204–318, 394–437.
[91] Galen, *Doctrines of Hippocrates and Plato* 4.4 and 9.7; cf. De Lacy (1972: 27).
[92] Galen, *The Order of My Own Books* 1; the similarity to the charge brought by Lucian's Lycinus in *Hermotimus* 15–20 is intriguing.
[93] E.g. *That the Faculties of the Soul Follow the Mixtures of the Body* (QAM = *Quod Animi Mores*) 7; further references are given by De Lacy 1972: 31, n. 24.

extended and clarified by Aristotle,⁹⁴ but he is manifestly inferior to Hippocrates in his medical knowledge⁹⁵ and indeed historically dependent on Hippocrates for the content of his principal doctrines.⁹⁶ Though not inconsiderable, and certainly greater than that of his latter-day disciples who often misunderstand him, his is a derivative not an originative authority. Truth is to be established by looking outward from medical knowledge, not from the philosophers.⁹⁷

Authority and Tradition

A large point to emerge from much of the material surveyed in this chapter is the importance of tradition in determining the location of authority. Galen looks back to the foundational authority of Hippocrates, Maximus to that of Homer; Gellius' Stoic vindicates both his own personal honour and his school's doctrines by referring back to the work of Epictetus, who in turn is faithfully reproducing the ideas of the founding fathers Zeno and Chrysippus; Lucian's Nigrinus is discovered 'encircled by numerous images of the philosophers of old (παλαιῶν φιλοσόφων).' This cast of thought, clearly related to the strongly classicising modes of Imperial period literary culture, manifests itself on a number of levels. Contests of authority between rival schools of thought can be fought out in terms of the antiquity of their ancestors, and the relations of dependence between those ancestors and other more recent figures. Just as Galen proposed that Plato depended on Hippocrates, so others could argue that he owed the centre of his thought to Pythagoras.⁹⁸ The turn could be especially useful to those wishing to relativise the great figures of Hellenic wisdom not to earlier authorities within the same tradition, but to another tradition of wisdom entirely, as when Philo, Aristobulus and Numenius on behalf of Judaism connect Plato back not to Homer, but to Moses.⁹⁹ Within individual schools of thought, the

---

⁹⁴ E.g. *Natural Faculties* 2.9, *Commentary on the Aphorisms of Hippocrates* 3.16, *Doctrines of Hippocrates and Plato* 8.4, *Faculties of the Soul* 3; De Lacy 1972: 38–9.
⁹⁵ De Lacy 1972: 33–5.
⁹⁶ E.g. *Usefulness of the Parts of the Body* (*De usu partium*) 1.8; De Lacy 1972: 36–8.
⁹⁷ This is a conviction Galen shares with, and presumably derives from, such Hippocratic treatises as *On Ancient Medicine* and *The Nature of Man*.
⁹⁸ E.g. Apuleius, *Florida* 15.26, *De Platone* 1.3; Maximus, *Oration* 27.5; Cicero, *Tusc.* 4.5.10.
⁹⁹ Philo, *De aeternitate mundi* 18–19, *Legum allegoriarum* 1.108, *Quis rerum divinarum heres sit* 214, among other passages; Aristobulus as cited by Eusebius *Preparation for the Gospel* 9.6; Numenius fr. 8 (from Eusebius, *Preparation* 11.10). For the Christian continuation of this turn of thought, see Ridings 1995.

authority and importance of founding figures is both reflected and sustained in the dominance of textual exposition as the main mode of philosophical teaching and writing (the commentary).[100]

This is a state of affairs reflected in the self-presentation of the modern representatives of the various schools, who regularly characterise themselves as trustworthy middle-men, able to expound 'our' doctrines with authoritative fluency, but at the same time consistently deferential to the superior originative insight of the founders. Maximus of Tyre, Seneca and Epictetus, for all the differences between them in context and target audience, all share this qualified modesty, confident in their ability to help their pupils on from a position of superior grasp and achievement, but careful not to over-claim on their own behalf. In writing his *Moral Epistles* to Lucilius, Seneca inevitably casts himself as an adviser with some degree of superior grasp and achievement over his correspondent, who can speak as one who has already confronted the challenges and problems that the latter now encounters. But he also speaks (writes) as one friend and social equal to another;[101] and he takes care to make clear that he is not claiming yet to have solved all the problems in his own life, or to have arrived long since at their shared destination. Rather, he and Lucilius are fellow travellers, equally to be contrasted with the fully realised sage.[102] It is characteristic that in one celebrated passage, he makes the point by referring to a great figure from the Stoic past, approvingly quoting the words of Panaetius to an enquiring pupil: 'We will see about the Sage later; you and I, who are a long way from being sages ourselves, should not make the mistake of tumbling into a condition that is disordered and uncontrolled' (the pupil had asked whether the sage would fall in love) (*Ep*. 116). Maximus of Tyre in his introductory oration may be ready to sing his own praises as a rhetorical model and teacher 'with all the pride and vanity at my command'; but when it comes to his demeanour towards philosophically interested pupils, 'I furl the sails of my proud speech, and humble myself; I am not the same man. This is a weighty matter, God knows, and demands a champion out of the common run.'[103] Confronted with the request for an exegesis of an area

---

[100] On the importance of the commentary as a philosophical medium in this period, see: Sedley 1997; Sorabji 1990.

[101] See Hine's discussion of the *Epistles* elsewhere in this volume. The sense of the relationship between Seneca and Lucilius as essentially one between like-minded friends is of course embodied in the use of letter form (Trapp 2003: 249–50), and reflected also in the definition of philosophy as *bonum consilium* that Seneca advances in *Ep*. 38.

[102] E.g. *Ep*. 75.8–10 and 13–16.   [103] Maximus, *Oration* 1.8.

of Platonic doctrine (his theology), he consents, on the grounds that even in a gold mine, assayers are needed to distinguish what is really valuable in what the earth has yielded up; but this is only after he has first protested that going to someone else (an exegete) rather than Plato's own work is like turning one's back on the bright sun in favour of a feeble bonfire.[104] Epictetus berates, exhorts and cajoles his various audiences with great vigour and confidence,[105] yet at the same time is reluctant to allow himself to be called a philosopher in so many words,[106] and refers back constantly to great figures of the past as touchstones of both doctrine and good philosophical pedagogy: Zeno and Chrysippus for doctrine, Socrates and Diogenes for style and relationships.[107] The ability to place himself in a tradition both confers authority on the modern thinker and teacher, and at the same time limits its extent.[108]

Authority Made Visible

This chapter has so far concentrated on verbal and textual embodiments and representations of philosophical authority. A strong visual dimension has, however, been in evidence intermittently along the way, and should be given its proper weight too. Lucian's Nigrinus, it will be remembered, was discovered seated over a book, with images (presumably portrait-heads or herms) of great philosophers of the past around him, along with geometrical diagrams and a celestial globe. This is a set of props and attributes easy to parallel from contemporary visual culture: the philosopher portraits from what we know of the presence of philosophers in mosaics, frescos and herm-galleries in private space throughout the Imperial period,[109] the emblematic use of books and globes from particular portrayals like the mosaic group from Pompeii ('Plato's Academy') and the lone philosopher on the floor of the villa at Brading on the Isle of Wight.[110] The messages encoded in this

---

[104] Maximus, Oration 11.1–2; for the imagery employed here, cf. Trapp 1997: 97.
[105] Long 2002: 52–64 for a study of Epictetus' preaching styles.
[106] Arrian, Epict. diss. 1.8.14, 1.9.1, 2.19.23–4.
[107] See e.g. Epict. diss. 1.17.11–19, 1.20.14, 1.4.6–9, 1.10.10, 1.2.33–6, 2.2.8–20, 2.18.22, 1.24.6–10, 2.3.1.
[108] See also Van Hoof 2010: 73–9 on the importance of tradition to Plutarch's construction of his philosophical authority in his treatises of practical moral advice.
[109] Trapp 2007: 247–8, with further references; see in particular Lorenz 1965.
[110] Naples, Museo nazionale archeologico, Inv. 124545 (from the House of T. Siminius Stephanus), showing book-rolls, sundial, celestial globe and chest (book box?). The Brading philosopher is seated beneath a sundial, with a celestial globe and a mortar at his feet

imagery also fit comfortably into well-established patterns: the book and the portraits asserting the basis of the philosopher's authority in his subscription to and mastery of a venerable tradition, the globe the comprehensive reach of his insight and his concern with the highest dimensions of reality, as well as (in combination with the geometrical diagrams) his specifically Platonist loyalties.

Equally clearly, however, Lucian's vignette in this particular case exploits only selected aspects of a larger and richer visual repertoire. As the case of Apuleius, as well as numerous further passages from Lucian, makes clear, costume and coiffure provided what was perhaps the most widely and readily recognised set of tokens of philosophical authority – or, more accurately, a set of tokens through which that authority could be challenged and questioned as well as asserted.[111] Thanks largely to the textual description of Socrates by the first generation of his disciples (Antisthenes and Aeschines as well as Plato and Xenophon), and its uptake in fourth-century portraiture,[112] the philosopher was expected to dress with (at best) deliberate simplicity, to be heavily bearded, and to have a hairstyle similarly betraying a lack of concern for conventional elegance (whether that meant close cropping or untidy exuberance).[113] The addition of the simple traveller's stick and wallet, symbolic of tough self-sufficiency and freedom from attachment to a particular place, strictly speaking was specific to Cynics, but as both Apuleius and Lucian again suggest, popular cliché tended to extend these conveniently visible attributes to philosophers more generally. In his (surely heavily fictionalised) account of how he became a philosopher during his period of exile under Domitian, Dio Chrysostom reports how his simple traveller's garb induced complete strangers to approach and question him on weighty issues in their lives, and to call him 'philosopher' as well as 'vagabond'.[114]

The possibility of challenge and questioning comes in with the reflection that the more there is a readily identifiable philosophical 'uniform', the easier

(Ling 1991); like one of the figures in the Pompeii mosaic, he holds a pointer (staff, ῥάβδος) out towards the globe. On the iconography of the Pompeii mosaic, and its close relative from Sarsina, now in the Villa Albani, see Elderkin 1935.

[111] Apuleius, *Apology* 4 and 22; Lucian, *Hermotimus* 86, *Fisherman* 42, *Runaways* 12–14, among many other passages.
[112] Zanker 1996: 32–9, 57–62.
[113] Cf. e.g. Lucian, *Hermotimus* 86; Philostratus, *VS* 2.3.567.
[114] Dio, *Or.* 13.10–12. In the light of what has been said above about deference to the superior authority of the classics of the past, it is interesting that Dio goes on to record that, when faced with the need to speak at length, rather than respond to individual questions, he took refuge in an imitation of Socrates (*Or.* 13.14ff, the opening of which is modelled on the Platonic *Clitophon*).

it becomes for a manipulative individual to adopt the style without the substance. Since what really matters is internal and invisible – the state of the individual's soul and the working of his rational faculty – external appearance does not strictly matter either way, conventional philosophical aspect being neither a necessary nor a sufficient indication of real philosophical authority, just as even the most degenerate external appearance is neither a necessary nor a sufficient indication of moral turpitude. A classic anecdote that explores the uncertainty thus opened up is the story of Socrates and the physiognomer Zopyrus, perhaps first told by Phaedo in his dialogue *Zopyrus*, and repeated by Cicero, Alexander of Aphrodisias and Eusebius.[115] Examining Socrates in ignorance of his identity and reputation, the physiognomer declared him to be, on the evidence of his grotesque facial features, a deeply vicious person. When his pupils expostulated in outrage at this, Socrates assured them that the physiognomer was right: such was indeed his underlying nature, but he had overcome it by stern moral effort. There is a comparable subtlety about Gellius' story of the Stoic in the storm, which similarly seeks to suggest that inferences from observed external signs to inner character and worth (and thus also title to authority) are more slippery and complicated than is commonly assumed.[116] On a cruder (and more self-serving) level, Maximus is to be found protesting, on behalf of those of elegant appearance, that

> merely carrying a wallet and a staff does not count as imitating Diogenes, as one can surely have these accoutrements and still be more wretched than Sardanapallus – the famous Aristippus, who wore purple and anointed himself with myrrh, was no less continent than Diogenes... The philosopher must be judged not by appearance, nor by age, nor by social status, but by his mind and his words and the disposition of his soul.[117]

Most simple-mindedly of all, Lucian's satirical ribbing of contemporary philosophers depends time and again precisely on pointing to the discrepancy between a hypocritically adopted external aspect and the actual conduct of its wearer.[118]

Pulling in the other direction, however, so as to underline the collective authority of philosophers as a class rather than challenging their credentials individually, was the sheer pervasiveness of the artistic representations

---

[115] Cicero, *Tusculans* 4.80, *On Fate* 5.10; Alexander of Aphrodisias, *On Fate* 6; Eusebius, *Preparation* 6.9. See McLean 2007.
[116] I have attempted to probe some of the issues arising in Trapp (forthcoming).
[117] Maximus, *Oration* 1.9–10.
[118] E.g. Lucian, *Fisherman* 44–5, *Hermotimus* 11–12, *Symposium* (passim).

already referred to. In private space, the great philosophers of the past were depicted in mosaic and fresco on floors and walls, and in stone in both individual portrait-busts and sets of herms.[119] In public spaces – libraries, gymnasia, stoas, odeia and so on – there were likewise statues of both great past figures and contemporaries honoured in their capacity as philosopher. Depictions on coins, characteristically commemorating the classical thinkers born in the issuing cities, moved between public and private space, as (though less visibly) did images on signet rings. Here, too, aspects of the standard iconography for living philosophers – simple clothing, heavy beard and untended hair, along with intent expressions and such gestures of intellectual authority as the fore-thrust hand with the fingers configured for disputation – were prominently on display.[120] The suggestion that the elongated faces and gravely calm expressions characteristic of sculpted portraits of Epicurean philosophers were deliberately designed not just to stress the loyalty and similarity of all subsequent representatives to the great founder, but also to spread Epicurean serenity and win converts, may be an extravagant one;[121] but the cumulative sense of status and authority conveyed by all images of philosophers collectively should surely not be underestimated.

Coda

Any claim to superior expertise on the part of a particular group creates the potential for conflict and contestation: between that group and rival specialists claiming rights over the same territory, between self-selected specialists and the lay public, and between rival members of the same group. In the case of philosophy, a number of aspects of the way it was conceived in the Imperial period give an extra depth and edge to each kind of conflict: the sheer scope of philosophy's claims to superior knowledge and insight; the unusual overlap between the specialist territory of philosophy and the world of the general public (philosophy as 'the art of life'); and the division of philosophy into competing sects, more or less hostile to each other's assertions of mastery. The position that the thinkers of the fourth century BC had constructed for themselves, out of the materials provided by the sixth and fifth centuries, had turned out to be one that offered great rewards, but could also be a precarious one, poised on the margins of both public and private life, even as it claimed centrality to both.[122] Philosophical authority was widely

---

[119] Trapp 2007: 246–8. [120] Zanker 1996: 90–115. [121] Frischer 1982.
[122] More on this in Trapp 2007: 233–56.

in evidence, and widely deferred to, but also widely resisted, and where not directly denied, could also be tamed and deflected in ways that will have frustrated some as much as they relieved and reassured others. It is hard to think of any other ingredient in the high culture of the Imperial period over which such a tangle of perceptions and attitudes could have arisen.

3 | Philosophical Authority in the Younger Seneca

HARRY HINE

Introduction

Unlike many of the writers who feature in this volume, Seneca was not a professional of any sort. He wrote on philosophical subjects, but he was not a *philosophus*, which for him and his contemporaries meant a professional philosopher, someone whose full-time occupation was philosophy, and someone normally of a much lower social rank.[1] Seneca was one of the most prominent political and literary figures of his generation. By the time of what is probably his earliest work, the *Consolatio Ad Marciam* (*Dial.* 6, probably written in AD 39 or 40),[2] he was a successful orator, and had held the quaestorship and become a senator; he spent eight years in political exile, from 41 to 49, and then became tutor, later speech-writer and adviser, to Nero, with a suffect consulship in 55 or 56. In the later years of Nero he became estranged from the emperor, and effectively withdrew from the court, until in 65 he was accused of involvement in the Pisonian conspiracy and ordered to commit suicide. As a literary figure, Quintilian later characterised him as an immensely popular writer across the genres of oratory, poetry, epistolography and dialogues (*Inst.* 10.1.125–31).

Seneca's literary talents are everywhere on display in his writings, and from time to time he enters into controversy and discussion on literary and rhetorical topics. Yet, as is well known, one could read his surviving works from end to end and have hardly any awareness of his political eminence, or of his status as an orator or poet, for, in stark contrast to Cicero's letters and philosophical works, Seneca's contain very little specific information about his political and literary activities.[3] Conversely, as a recent writer has observed: '[The] modern interest in Seneca as philosopher is perhaps not

---

[1] See Hine 2016.   [2] Bellemore 1992 argues for an earlier, Tiberian, date.
[3] One does learn from the *Consolationes Ad Polybium* and *Ad Helviam* (*Dial.* 11 and 12) that he was in exile, but he is reticent about the causes of his exile; and the *De Clementia* presupposes a close relationship to the young Emperor Nero, but there is nothing about the specific nature of Seneca's role in Nero's service. Though the day-to-day particulars of public life rarely surface in Seneca, there is a great deal about broader political issues; see particularly Griffin 1976, and also Ker 2009.

altogether indicative of his reputation in Neronian Rome. Whilst it is clear that Seneca's philosophical interests and writings were known and read by his contemporaries, it seems that he was identified more readily as a politician or orator than as a philosopher.'[4]

All this complicates the question of authority in Seneca's prose works. One can surmise that, once he had established his public and literary reputation, this eminence contributed to the impact of any new work by him. The focus of this chapter is on his creation of philosophical authority in his prose works, and it will be argued that we should not think of all his 'philosophical' works as alike, for in some of them the issue of philosophical authority is at the forefront, but in others it is very muted; and his authority as a Roman, embedded in the shifting network of personal and political relationships that was essential to public life in Rome, also comes into play.[5] The first section of the chapter will consider the question in what philosophical guise he presents himself in his works, if he was not a professional *philosophus*, and will argue that he principally presents himself as a trusted friend (or sometimes a relative) offering advice. The second part of the chapter – which will focus on just a few of his works – will show how the level of philosophical commitment and engagement on display varies greatly from work to work, and how philosophy is sometimes presented within an authorising Roman framework. The third part will compare him to some other philosophical writers of the late Republican and early imperial period, to bring out the distinctive features of Seneca's presentation of his own philosophical authority. A brief conclusion will consider how we should read his philosophical oeuvre as a whole, given the variations in the prominence of philosophical authority between individual works.[6] Little will be said about how Seneca uses his literary talents to enhance his authority, mainly for reasons of space, and also because this aspect of Seneca's writings is well covered in existing scholarship.[7]

---

[4] Bryan 2013: 142. Compare the point made by Levick 2003: 211, that it can be 'legitimate to consider Seneca, as some of his contemporaries may have done, less as a philosopher in politics than as a politician who happened also to be a philosopher ...'.

[5] The diversity of Seneca's philosophical works has also been well emphasised by Wilson 2013: 94–5.

[6] The *Naturales Quaestiones* will not be discussed in this chapter, for reasons of space. The work deals with scientific rather than ethical topics (though there is also an important ethical dimension to it). It is the most doxographic of Seneca's works, and hence is able to create a stronger sense of Seneca's position in the ongoing history of the subject than one finds in the other works (see my remarks in Hine 2006: 53–67).

[7] See particularly Setaioli 2000; Traina 1995; von Albrecht 2008a, 2008b.

Openings – the Wise Friend

If Seneca did not regard himself as a *philosophus*, then in what capacity or role did he write about philosophy? This question will be approached by looking at the ways in which he starts his works. Openings are important in establishing a writer's identity and authority.[8] Some of Seneca's works open with a request or a question that is presented as coming from the person to whom the work is addressed.[9] Thus, the *De Prouidentia* opens with: *Quaesisti a me, Lucili, quid ita, si prouidentia mundus ageretur, multa bonis uiris mala acciderent* ('You asked me, Lucilius, why it was that, if the world was controlled by providence, many bad things happened to good men', *Dial.* 1.1.1). The fact that Lucilius is presented as having asked this question suggests that he has thought to some extent about this philosophical issue, and that he thinks Seneca is capable of giving a worthwhile answer. Similarly, *De Ira* opens with: *Exegisti a me, Nouate, ut scriberem quemadmodum posset ira leniri, nec immerito mihi uideris hunc praecipue affectum pertimuisse maxime ex omnibus taetrum ac rabidum* ('You demanded, Novatus, that I should write on how anger can be soothed, and you seem to me quite justified in being particularly afraid of this emotion, which is the foulest and most violent of them all', *Dial.* 3.1.1). Here, the question is about a pressing issue of everyday behaviour, and shows Novatus' regard for Seneca's ability to answer, as well as Seneca's regard for Novatus, whom he compliments on his perceptiveness in singling out anger for treatment.[10]

Some openings are more complex. Uniquely among the surviving works, the *De tranquillitate animi* (*Dial.* 9) opens with a long speech from the dedicatee Serenus. Serenus (as Seneca portrays him – his very name suggests his affinity with the topic of the dialogue) reveals himself in this speech as well practised in the moral self-examination that Seneca repeatedly advocates. He knows the difference between virtue and specious things like honour and eloquence (9.1.3), but still finds himself pulled in different directions when it comes to deciding on three major issues (all relevant to Roman public life): frugality versus luxury (9.1.5–9), political involvement versus

---

[8] The openings of *De Otio* (*Dial.* 8) and *Ad Polybium de consolatione* (*Dial.* 11) are missing in the manuscript tradition.

[9] For the purposes of this chapter I take all statements about requests and questions from the addressee at face value, without asking whether or to what extent they are fictional, for my concern is with Seneca's self-presentation. Miriam Griffin's discussion of his portrayal of Lucilius' philosophical development is a warning not to assume that his presentation of his addressees is biographically true (Griffin 1976: 350–3).

[10] Novatus' request is again mentioned at the start of the third book, *Dial.* 5.1.1.

otium (9.1.10–12), and simple versus ornate style (9.1.13–14); and he seeks Seneca's advice on handling this malaise (9.1.15–17). He refers to following the instructions of Zeno, Cleanthes and Chrysippus (the founding fathers of Stoicism) on political involvement, though he observes that while they preached political involvement, none of them was himself politically active (9.1.10). So Serenus is characterised as someone who has set out on the Stoic path, and is well acquainted with Stoic teaching, but still finds his footsteps are faltering. Hence his turning to Seneca validates Seneca's authority to give philosophical guidance.

Some other dialogues begin not with a question or request from the recipient, but with a general statement by Seneca about people's moral behaviour or attitudes. Thus, *De vita beata* (*Dial.* 7) begins with the fact that everybody desires happiness, but people are unsure where to find it; *De breuitate vitae* (*Dial.* 10) begins by saying that most people complain about life's shortness; and *De beneficiis* starts by describing the widespread ignorance of how to give and receive benefits properly. In each case, the opening generalisation leads into a closer analysis of the problem and its solution.

De constantia sapientis (*Dial.* 2), however, starts with an assertion of a rather different kind, about the superiority of Stoic wisdom to all other forms of wisdom (*Dial.* 2.1.1–2). It soon emerges that the issue had arisen during a recent conversation in which Serenus, the addressee, had made an indignant intervention: *Nuper cum incidisset mentio M. Catonis, indigne ferebas, sicut es iniquitatis inpatiens, quod Catonem aetas sua parum intellexisset, quod supra Pompeios et Caesares surgentem infra Vatinios posuisset...* ('Recently, when someone happened to mention Marcus Cato, you, with your intolerance of injustice, expressed indignation that Cato's contemporaries hardly understood him, that, though he towered above the likes of Pompey and Caesar, they rated him below the likes of Vatinius...', 2.1.3). Seneca reports how he then responded to Serenus' outburst (2.2.1–3), though what starts as a report in indirect speech about what he said on that occasion soon turns into direct speech that seems to flow from Seneca's pen as he writes (the turning point is at 2.2.1 *Hos enim Stoici nostri sapientes pronuntiauerunt...*, 'For our Stoics declared that wise men are those who...').[11] Then in chapter 3 Seneca imagines Serenus being provoked to respond: *Videor mihi intueri animum tuum incensum et efferuescentem, paras adclamare: 'haec sunt quae auctoritatem praeceptis uestris detrahant* ('I think I can see your mind incensed and incandescent, you are getting

---

[11] Compare the way in which the words attributed to another person sometimes merge into Seneca's own words; see Hine 2010: 217–23.

ready to cry out: "This is what robs your precepts of their authority"', 2.3.1), and we are into a whole paragraph of arguments put into Serenus' mouth in direct speech. The immediacy of conversation, real or imagined, between Seneca and Serenus does not continue, though there is a further reference to Serenus' views at 2.7.1.

This evocation of face-to-face encounter with the addressee is also found in *De beneficiis*. In the first four books Liberalis is addressed a number of times, and near the start of Book 3 (*Ben.* 3.1.2) Seneca lets drop in passing that Liberalis and he have previously engaged in discussion – *disputationem* – on a point of disagreement. But at the start of Book 5 Liberalis comes into sharper focus. Seneca declares that the first four books have covered the central issues, and to continue with further questions will be an unnecessary indulgence; nevertheless, he will continue, because that is what Liberalis wants (*Verum, quia ita uis, perseueremus*, 'But, since you want to, let us press on', 5.1.2). There follows a long tribute to Liberalis' generosity and sensitivity in the matter of benefits (5.1.3–5). Then at the start of Book 6 Seneca says that some questions are intellectually fascinating but of little practical use, so he will follow Liberalis' guidance on how far he should pursue them, and he imagines that he can watch Liberalis' expression (6.1). Later in the same book Seneca twice imagines reading Liberalis' thoughts from his face (6.7.1, 12.1).[12]

The references to recent or imagined exchanges with Serenus in *De constantia* and with Liberalis in *De beneficiis* evoke the world of philosophically informed conversations that can be seen from time to time in Seneca's *Letters*. Occasionally, we glimpse Seneca and others discussing technical philosophical issues: for instance, letter 58, an important exploration of Platonic ontology, opens with a passing reference to conversations about Plato on the previous day, and later in the letter Seneca says that an unnamed 'friend of ours, a very learned man' (*amicus noster, homo eruditissimus, Ep.* 58.8) had explained that according to Plato there are six categories of being; in letter 66 Seneca has recently met Claranus, a fellow student whom he had not seen for many years, and the two of them spent several days talking, Seneca says, beginning with the question of how it can be right to claim that all goods are equal (66.5); in letter 64 there is a dinner party where conversation flits around various topics, before they settle on listening to a reading from a book by the philosopher Q. Sextius the elder; in letter 65, friends visit Seneca when he is unwell, conversation turns to thorny philosophical questions about causes, and Lucilius is invited to arbitrate (65.2, 10, 15). Letter 30

---

[12] On Liberalis' role in the work, see Dalle Vedove 2009; Griffin 2013: 142–8.

describes how Aufidius Bassus, the historian, is facing death and the debilitations of old age: Seneca says that he has frequently been to visit Bassus (30.13), and he reports a number of the things he has heard Bassus say about death and old age. Interestingly, it is only three-quarters of the way through the letter that we learn that Bassus was an Epicurean – though no doubt some of the original readers would have known this already. Other friends are not so committed to philosophy: letter 29 is about a mutual friend, Marcellinus, who, Seneca says, rarely comes to see him now, because he does not want to hear the truth, and has a poor view of philosophers (29.1–6). A later letter (77) reports the death of a mutual friend Tullius Marcellinus – some suspect that this is a different Marcellinus, for if it is the Marcellinus of letter 29, he seems to have undergone something of a transformation since the earlier letter. Seneca describes how he had been suffering from a long illness, and started to think of suicide. He gathered together several friends to ask their advice, and most of them were timid or obsequious, but someone referred to as 'our Stoic friend' (*Amicus noster Stoicus*, 77.6) gave advice that helped Marcellinus to make a decision for suicide.

The dialogues we have so far discussed can be viewed, I suggest, as an extension into writing of these philosophically informed discussions among friends. Seneca's relationship to his addressee or reader can be described in terms of 'vertical' relationships – such as philosopher and student, teacher and pupil, lecturer and audience, doctor and patient – and indeed there are such images in Seneca's work,[13] while at the same time he stresses that he is still stumbling along the road to wisdom and goodness, still a sick patient himself (e.g. *Dial.* 7.17, *Ep.* 27.1, 68.9).[14] Here, I am emphasising the 'horizontal' framework: Seneca and the addressees of his dialogues are social equals, and he presents himself as the kind of person to whom friends turn for philosophically informed advice and insight. Sometimes they are explicitly said to have asked him a question or discussed a topic with him, but in the other cases too it is implied that they will be interested in the issue Seneca raises and will value his contribution. Some of them are more advanced in

---

[13] Though one should note that, besides never calling himself a *philosophus*, he never refers to himself as a teacher, and never refers to his addressees or readers as pupils or students (*discipuli*, *auditores*).

[14] Griffin 1976: 417 argues that Seneca tries, sometimes unconvincingly, to 'unite in his therapy the didactic tone of the stirring lecturer with the supporting one of a teacher who is himself learning'; for her the uncertain tone of the *Letters* is a sign of their fictional character. But perhaps the 'didactic' tone, with its sometimes abrupt commands to Lucilius, is less troubling when seen in the context of the prominence of self-command in Seneca's philosophy, recently analysed by Star 2012, whereby one might say that Seneca, paradoxically, adopts a 'vertical' relationship towards himself.

their own philosophical thinking and moral progress than others, and in some cases the addressee is presented as progressing philosophically in the course of a longer work.[15]

These friends and acquaintances whom we encounter in the *Letters* do their networking in traditional Roman ways. There is the regular call on friends: Seneca visits Aufidius Bassus, and Marcellinus at one time used to be in the habit of visiting Seneca.[16] There are dinner parties with philosophical conversation and reading. There is the invitation to friends to come and give their advice regarding an important decision, as we see in the case of Tullius Marcellinus. These encounters seem to take place in the setting of everyday urban life, in contrast to those philosophical works of Cicero where the participants have retreated to a country villa during a public holiday, a contrast that perhaps reflects Seneca's concern to emphasise the weaving of philosophy into the fabric of everyday life.[17]

The *Letters* to Lucilius fit into the same pattern of an extension of conversation, and the relationship between the two men is a close one from the start. Right from the opening phrase of the first letter (*Ep.* 1.1, *Ita fac, mi Lucili*... , 'Do just that, my dear Lucilius... '), Seneca represents his letters as being one side of a two-way correspondence, and he regularly says or implies that he is responding to a question or remark from Lucilius (cf. particularly 118.1).[18] Once he says that he can imagine Lucilius actually in conversation with him when he reads his latest letter (67.2).

There are other works not yet discussed where Seneca's relationship to the addressee is different, but it can be argued that in them too Seneca demonstrates – or implicitly lays claim to – a sufficiently close relationship to the addressee to justify his writing the work for them, and a relationship that in most cases is not a 'vertical' one. The clear exception to this is *De clementia*, addressed to the young Nero. This work posed a unique challenge, because it would be infelicitous to portray the emperor as in need of advice, or as actively seeking it; so Seneca adopts the strategy of saying that he is holding up a mirror to the emperor, showing him as he already is, and so he

---

[15] On the 'progress' of Lucilius in the *Letters* and of Liberalis in *De beneficiis*, see now Griffin 2013: 142–8.

[16] Compare how Cornelius Senecio, whose sudden death is reported in letter 101, had started his day by calling on Seneca, as was his habit (*ex consuetudine*, 101.3); no philosophical dimension to his visits is recorded.

[17] I here stress the conventional Roman nature of these interactions; but at the same time in other passages Seneca rewrites the traditional value system of Roman friendship: see Wilcox 2012, esp. 115–31.

[18] On apparent allusions to Lucilius' letters, see Griffin 1976: 416–17.

creates a panegyric framework for his advice on clemency;[19] and the mirror imagery implies that Seneca knows the emperor closely enough to portray this reflection, and indeed Seneca offers a detailed account of Nero's character, behaviour and even his thoughts, in the course of the work.

The three consolations are different again. For a start, consolatory works were normally unsolicited, so a request from the recipient was not to be expected; and also two of Seneca's consolations are addressed to women. The start of the consolation to Polybius is missing, so we do not know what sort of relationship with him Seneca claimed or implied; but the dialogue presents Seneca as closely aware of Polybius, his family, his various activities and his relationship with Emperor Claudius. Seneca scarcely needs to justify addressing a consolation to his mother Helvia on the subject of his own exile, and we do find a considerable amount of detail about her and other members of the family. To Marcia he begins by asserting that he knows her character, and is confident that she has the resilience and moral strength that will enable her to accept his advice on how to handle her grief (*Dial.* 6.1). The work is puzzling in that Seneca does not claim any specific connection with Marcia or her family, though he does express admiration for the writings of her father Cremutius Cordus, which were banned by Tiberius (6.1.2–4, 22.4–8) – and hence there has been speculation about Seneca's political affiliations and motives.[20] But certainly Seneca presents himself as someone who knows, or knows about, Marcia.

To return to the works that open with a question from the addressee, we may be reminded of how the discourses of Musonius Rufus and Epictetus sometimes start from a question or remark from a student or visitor (Musonius 3, 14, 16, 17, cf. 8, 9; Epictetus, 1.10.2–6, 1.11, 1.13, 1.14, etc.), though we should remember that these are reports by third parties of live encounters, and that, unlike Seneca, Musonius and Epictetus were teachers of philosophy. There was of course a very widespread literary convention of presenting a work as a response to a request from a patron or friend, and this applied in all sorts of genres, including philosophy and other forms of technical writing. In Latin, the prefaces of two of Cicero's surviving philosophical works refer to general encouragement from the addressee, Brutus, to write about philosophy (see *Tusc.* 1.1, cf. *Fin.* 1.8), but there is never mention of a request to write on a particular topic or discuss a specific question; after all, Cicero had his own clearly conceived programme for his great series of philosophical works. Varro's *De re rustica* is presented as a response to a

---

[19] See Braund 2009, particularly 19–21, 53–7.
[20] See Bellemore 1992; Griffin 1976: 22–3, 47–51.

request from his wife for practical guidance on running her farm (1.1.2). The first book of Columella's *De re rustica* does not mention any request from Publius Silvinus, the addressee, but several of the later books start with a specific question, comment or request from him (2.1.1, 4.1.1, 5.1.1, 10.pr.1, 11.1.1–2 – where Columella also mentions the request of a certain Claudius). In the *Moralia* Plutarch several times says he is responding to a specific enquiry or request (see *Bravery of Women* 242F; *On Tranquillity of Mind* 464E; *Sympotic Questions* 612E; *Precepts of Statecraft* 798A–C; *On the Generation of the Soul in the Timaeus* 1012B; and, though it is not by Plutarch himself, *On Fate* 568B–C). So there is nothing unusual in Seneca's presenting some of his works as responding to requests; but the frequency with which he does so is greater than in Cicero's philosophical works (of which only two mention Brutus' general encouragement) or in Plutarch's treatises (only a handful of the dozens of treatises and dialogues open with a request). For if one leaves aside the works where, as said above, a request is not to be expected (the consolations, *De clementia* and *Letters*), and the incomplete *De otio*, then three of the remaining eight works open with some kind of request or question (*De providentia*, *De ira*, *De tranquillitate animi*), and two more later on mention or evoke discussions with the addressee (*De constantia sapientis*, *De beneficiis*).

Nor is there anything unusual about Seneca referring to recent conversations with the addressee and others: but it throws into relief the fact that he never turned such conversations into dialogues proper such as we find in the Platonic or Aristotelian tradition; the nearest he comes to it is with Serenus' opening speech in *De tranquillitate animi*. By contrast, Cicero presents several of his works as conversations in which he recently or not so recently participated; as does Varro in the *De re rustica* (1.1.7, 2.pr.6, 3.2.1); and Plutarch offers accounts of recent or not so recent conversations in a number of works, including *The E at Delphi*, *On the Delays of Divine Vengeance* and *Sympotic Questions*. Why Seneca did not write such dialogues is a matter for speculation, but earlier Stoics had hardly ever written dialogues,[21] and maybe he would have sympathised with Epictetus' strictures (2.1.32–6) on those who devoted more energy to writing polished literary dialogues than to their own moral self-improvement. When it came to one's reading, Seneca urged Lucilius not to be content with knowing what earlier writers had said, but to think independently (*Ep.* 33), and he said that one should absorb and refashion one's reading from various authors in the same way as a bee makes honey (*Ep.* 84). Analogously, one may surmise that he counted

---

[21] See Hirzel 1895: 1.365–7, 370–3, 415–16.

recording who said what in a dialogue, real or imagined, as less important than framing one's own thoughts about the topic.

Levels of Philosophical Engagement

Traditionally, the prose works of Seneca have often been treated as a homogeneous body of philosophical writing, but there is a growing awareness among scholars of the great diversity among these works. For instance, attention is now paid to generic differences: in particular, the moral letters are no longer treated as philosophical essays, but the epistolarity of the collection is properly recognised.[22] Here I am concerned with diversity in the degree to which philosophy is given an explicit role, and philosophical authority is at stake, in different works, and shall argue that there is a spectrum of degrees of engagement. There is not space here to look at all Seneca's works, so I consider a few that occupy different points on the spectrum.

Michael Trapp in this volume reviews the range of ways in which philosophical authority was conceived and displayed in the Imperial period, including: technical knowledge of philosophical doctrine, in ethics, physics and logic; the ability to defend the doctrines of one's own school against challenges from other schools; emphasis on one's continuity with the great figures of the past to enhance one's own authority; a charismatic character, and one that lives up to the moral principles one teaches; the ability to defend oneself against criticism; the ability to teach, guide and inspire one's followers; and often a distinctive appearance, with one's beard and one's clothes marking one out as a philosopher. A distinctive external appearance was never part of Seneca's philosophical persona: he expressly warns against long beards, unkempt appearance and other superficial markers of the philosophical life (*Ep.* 5.1–5, cf. 103.5); but otherwise most of these aspects of philosophical authority are on display in his works, particularly in the *Moral Letters* written in his final years, with which I start.

In the *Letters* Seneca from time to time refers to himself and Lucilius living a life guided by and devoted to *philosophia* (e.g. *Ep.* 4.2, 8.7, 14.11, 16, 17.5–7, 20.1, 72, 78.3, 82.5, 90.1, 103.4), and he includes a considerable amount of material that shows how he has acquired the authority to write about philosophy: there are reminiscences of the philosophers he admired

---

[22] See, e.g.: Edwards 1997; Henderson 2004; Wilcox 2012; Wilson 1987, 2001.

in his early years (cf. 11.4–5, 49.2, 52.11, 63.5–7, 64, 67.15, 72.8, 73.12–15, 100, 108.3–4, 110.14–20); he comments on things he has been reading, and describes his own reading practices (cf. 2, 33, 84); he sometimes quotes approvingly from earlier philosophers, appropriating their authority for himself (e.g. in the quotations from Epicurus and others at the end of each letter in the first three books);[23] in other cases there are in-depth critiques of leading philosophers of the past, including Stoics, sometimes involving discussions of highly technical philosophical questions (e.g. 58, 65, 66, 83.8–12, 85, 87, 89, 90, 94, 95, 113, 117); the duty to evaluate what philosophy tells us is implicitly present throughout, and sometimes explicit (e.g. 21.9, 80.1). We find descriptions of recent philosophical discussions with friends (see above); he mentions contemporary philosophers, sometimes people whom he or Lucilius has recently encountered, and he apportions criticism or praise as appropriate (cf. 20.9–10, 29.5–7, 40, 52.7–15, 62.3, 67.14, 76.1–4, 91.19, 108.5–8). At the same time, the *Letters* repeatedly make clear that philosophy is a matter of how one lives one's life, of moral progress, not just of academic knowledge: philosophy must govern our deeds, not just our words, and we must not let ourselves be drawn into profitless logical quibbles (e.g. 16.2–3, 20.2, 24.15, 45, 48.4–12, 49.5–12, 82.19–24, 88.42–6, 111). Repeatedly he uses himself as an example of how to behave, or how not to behave: he defends himself against criticisms that are presented as coming from Lucilius, sometimes acknowledging his own imperfections (e.g. 8.1–2, 27.1–3, 57.3, 68.8–9), but he can also use himself as an example of the transformative potential of philosophy (particularly 6.1–3). Recent scholarship has emphasised the importance of the epistolary form, and the epistolary collection, to the enterprise of putting on display Seneca's own philosophical self-fashioning and his encouragement of Lucilius along the same journey of philosophical self-improvement.[24]

So in the *Letters* in these various ways Seneca supports his implicit claim to be a reliable and authoritative philosophical guide. But the *Letters* lie at one end of the above-mentioned spectrum. At the other end are the two consolations Seneca addresses to bereaved people, *Ad Marciam* and *Ad Polybium*.[25] They have traditionally been regarded as works of Stoic philosophy: thus commentators typically talk about the philosophical background

---

[23] See Wilson 2001: 176 on the quotations from Epicurus and others at the end of the *Letters* in the first three books: '... the voice of philosophical authority is displaced from Seneca onto the broad philosophical tradition'.

[24] E.g. Edwards 1997; Wilcox 2012.

[25] The *Consolatio ad Heluiam* is unusual, as Seneca says at the start (*Dial.* 12.1), for it consoles his mother on his own exile.

in their introduction, and philosophical topics feature prominently in the commentaries.[26] But Marcus Wilson has recently stressed that these works contain very little expressly philosophical content.[27] This is not altogether surprising, in that philosophers had no monopoly over consolation, which was found in poetry and letter-writing and other genres too. All the same, it was perfectly possible to write a philosophically oriented consolation: the Plutarchean *Consolatio ad Apollonium* contains references to, and quotations from, a considerable number of philosophers, as well as poets;[28] and the short consolatory letter that Seneca wrote to his friend Marullus, which he forwards to Lucilius in letter 99, discusses in some detail whether the Stoic wise man will ever weep (*Ep.* 99.18–21), quotes in Greek a sentence from the Epicurean Metrodorus, criticises it at length (99.25–31), and refers to the progress that Marullus has made in philosophy (99.14).[29] By contrast, *Ad Marciam* and *Ad Polybium*, as Wilson stresses, contain few references to philosophers or philosophy. It will be instructive to look in some detail at these works, to see how these few references are framed, and also to see how non-philosophical forms of authority are explicitly brought into play.

The consolation to the Imperial freedman Polybius, so far as it survives (for, as stated already, the opening is lost), contains no references to named philosophers, and just a couple of references to unspecified ones. In chapter 9, talking of the sufferings that Polybius' brother may have been spared by his early death, Seneca says: *Si uelis credere altius ueritatem intuentibus, omnis uita supplicium est* ('If you would believe those who examine the truth more deeply, all life is a punishment', *Dial.* 11.9.6). Those 'who examine the truth more deeply' presumably include philosophers, though it could also include poets and other traditional wise figures; but 'if you would believe', leaving open the possibility that we shall not believe it, is not a ringing endorsement of their idea (which is not one with which a Stoic would be comfortable). Later, in the course of a lengthy description of the support that he imagines Emperor Claudius is offering to Polybius, Seneca says: *iam omnium praecepta sapientium assueta sibi facundia explicuit* ('he has already expounded the precepts of all the wise men with his customary eloquence', 11.14.1). Again, 'wise men' could include more than philosophers, but combined with *praecepta*, 'precepts', which is a technical term of Seneca's philosophy (see particularly *Epp.* 94, 95, and *Dial.* 6.2.1 quoted below), it certainly

---

[26] So, in recent commentaries, Kurth 1994: 9–11 and *passim*; Manning 1981: 12–20 and *passim*.
[27] Wilson 2013: 104–8.
[28] Boys-Stones 2013 argues also for a deeper Platonist philosophical programme in the work.
[29] Wilson 2013: 96–7 is, to my mind, over-confident in his assertion that Marullus is an invented character (cf. Wilson 1997: 66–7), but this issue does not really affect the argument above.

includes them. This is the only straightforward reference in the dialogue to what philosophers have said or written about grief, and, significantly, it is attributed to the emperor; a few lines later Seneca stresses the extraordinary authority that his words carry (this is the sole occurrence of the word *auctoritas* in this dialogue): *aliud habebunt hoc dicente pondus uerba uelut ab oraculo missa; omnem uim doloris tui diuina eius contundet auctoritas* ('when he is speaking, the words will have a special power, as if they were uttered by an oracle; his divine authority will crush all the force of your grief', *Dial.* 11.14.2). Later on, now speaking in his own voice, Seneca repudiates the view that the wise man will not grieve at all, a view he attributes to *quosdam durae magis quam fortis prudentiae uiros* ('men of a harsh rather than a courageous wisdom', 11.18.5); he does not spell it out, but he is dismissing the strict Stoic view – as he does elsewhere (e.g. *Dial.* 6.4.1, *Ep.* 63.1, 99.15).

So in the *Ad Polybium* explicit references to philosophy are few, lacking in detail and sometimes non-committal, with the most positive use of philosophical teaching being attributed to the authoritative emperor. One can add that much of the dialogue is devoted to themes unrelated to philosophy – the merits of Polybius' dead brother, Polybius' own merits and literary activities, and his obligations to family, friends, colleagues and emperor, as well as Seneca's pleas for an end to his exile. In short, in this work one can scarcely detect any interest on Seneca's part in claiming or establishing philosophical authority for himself, except that once or twice he can be seen implicitly, and briefly, assuming the ability to judge the appropriateness of the philosophical ideas to which he refers.

The consolation to Marcia contains slightly more in the way of overt philosophical content, and like *Ad Polybium* it explicitly touches on the question of authority. Seneca opens by drawing on his knowledge of Marcia's character and behaviour, and focuses on her loyalty to her father and his historical works (a focus that also serves as a *captatio beneuolentiae* for a wider readership; *Dial.* 6.1.2–4). Aggressive treatment is needed, not gentle handling, because Marcia has proved resistant to previous attempts to alleviate her grief, which is undiminished by the passage of time (6.1.5–8). Those previous attempts are only briefly mentioned, with what is the first occurrence of *auctoritas* in the work: *Quis enim erit finis? Omnia in superuacuum temptata sunt: fatigatae adlocutiones amicorum, auctoritates magnorum et adfinium tibi uirorum* ('For where will it end? Everything has been tried to no avail: the comforting words of friends and the authority of great men who are related to you have been exhausted', 6.1.6). The friends are not specified, nor the influential relatives. But Marcia is impervious to the *auctoritas* that they should exert over her.

After this opening chapter Seneca launches into the consolation proper by offering Marcia a different source of *auctoritas*: *Scio a praeceptis incipere omnis qui monere aliquem uolunt, in exemplis desinere. Mutari hunc interim morem expedit; aliter enim cum alio agendum est: quosdam ratio ducit, quibusdam nomina clara opponenda sunt et auctoritas quae liberum non relinquat animum ad speciosa stupentibus* ('I know that everyone who wants to give advice begins with instructions and ends with examples. Sometimes it is useful to change this pattern. Different people need different treatment: some are guided by reason; some need to be confronted with famous names, with authority that will constrain their thinking when they are captivated by superficial appearances', 6.2.1). The reference to the conventional order of instructions and examples places Seneca firmly in the philosophical and literary tradition of consolation, and credits Marcia and the reader with being aware of this tradition, while at the same time his departure from the traditional order signals his independence and innovativeness (something calculated to appeal to a woman with strong literary enthusiasms of her own, cf. 6.1.3, 6.1.6). Seneca proceeds to offer Marcia two examples, of her own sex and generation (6.2.2 *et sexus et saeculi tui exempla*), Octavia, sister of Augustus, whose mourning for her dead son Marcellus was excessive and unending, and Livia, wife of Augustus, a model of appropriate mourning after the death of her son Drusus. Marcia must have been quite young when Octavia died in 11 BC,[30] but Seneca later refers to Marcia's friendship with Livia (6.4.2 *familiariter coluisti*), which increases the likelihood that she will prefer to follow her example.

We shall return to Livia shortly, but first let us look at the third and last reference to *auctoritas* in the dialogue, at the start of the final chapter, where Seneca puts into the mouth of Marcia's dead father a speech exhorting her to cease grieving for her son Metilius, who now enjoys a blessed existence; and he emphasises the *auctoritas* of her father (*cui tantum apud te auctoritatis erat quantum tibi apud filium tuum*, 'who had as much authority with you as you had with your son', 6.26.1). So after the introductory chapter, where the failure of Marcia to respond to the *auctoritas* of her relatives is noted, the work is flanked by the examples of Octavia and Livia, and by the speech of Cordus, both flagged as carrying *auctoritas*. Seneca's own authority is mediated via the authority of Roman figures who were both important in their own right, and (at least in the cases of Livia and Cordus) important to Marcia personally as friend or relative.

---

[30] See Manning 1981: 2.

We have already seen how Seneca announces his creative engagement with the literary and philosophical tradition at the start of chapter 2, and a different sort of creative engagement is on display at the end of the dialogue, for chapters 23 to 26 develop the theme that Metilius' soul has now escaped his body and enjoys a better existence along with the souls of great men, including his grandfather's. A rich literary and philosophical tradition lies behind this concluding section: Plato is mentioned near the start (6.23.2), and throughout there are points of contact with the conclusion of Cicero's *De republica*, the *Somnium Scipionis* (the Scipios are named at 6.25.2).[31]

The work is framed by important Roman figures, and this Roman emphasis is found throughout the work. These *exempla* of Octavia and Livia are followed by a number of others later in the consolation, and they are preponderantly Roman: Sulla, Pulvillus, Paulus, Bibulus, Julius Caesar, Augustus, Tiberius and two Cornelias, are examples of how to bear the loss of a child (chapters 12 to 16); Pompey, Cicero and the younger Cato are examples of people for whom it would have been better to have died earlier than they did (chapter 20); and the work ends with the picture of Marcia's son being welcomed into the afterlife by Scipios and Catos, as well as Marcia's father Cordus (6.25.2). At one point Seneca expressly passes over a possible Greek example and gives a Roman one instead: *Ne nimis admiretur Graecia illum patrem qui in ipso sacrificio nuntiata filii morte tibicinem tantum tacere iussit et coronam capiti detraxit, cetera rite perfecit, Puluillus effecit pontifex, cui postem tenenti et Capitolium dedicanti mors filii nuntiata est. Quam ille exaudisse dissimuluuit…* ('Pulvillus the pontifex ensured that Greece should not be over-proud of the father who, when news of the death of his son reached him as he was in the middle of a sacrifice, simply told the piper to stop playing, removed the garland from his head, and then completed the rest of the ritual;[32] for Pulvillus, when he received news of the death of his son as he was holding onto the doorpost during the dedication of the Capitol, pretended he had not heard … ', 6.13.1). The one place where Greek history is prominent is when Seneca imagines a prospective visitor to Syracuse being given a frank account of the positive and negative features of the city: if, after that, they still visit the city, they have only themselves to blame for the consequences (6.17.2–6). Syracuse, and Sicily, are evoked as places with literary resonance,[33] and as sites of famous historical events

---

[31] On the background to the chapter, see the commentary of Manning 1981.
[32] This refers to Xenophon, when he heard of the death of his son Gryllos in 362.
[33] There is a quotation from Vergil, *Aen.* 3.418, on the separation of Sicily from Italy; Charybdis is *illam fabulosam Charybdin*, 'the Charybdis of mythology'; the spring of Arethusa is *celebratissimum carminibus*, 'famed in poetry' (6.17.2–3).

(the defeat of the Athenian expedition in the Peloponnesian War, the Syracusan quarries, the tyrant Dionysius). The argument of the passage requires a city whose good points and bad points are evenly balanced, and Syracuse fits the bill well.

And what of Greek philosophy in the *Ad Marciam*? There is very little explicit mention of it. Only three Greek philosophers are named. Plato gets a passing reference to the ineffectiveness of his philosophy in influencing Dionysius the tyrant (*Dial.* 6.17.5), and, near the start of the final chapters on Metilius and the afterlife, he is cited for the view that the wise man's soul is constantly directed towards death (6.23.2). Socrates' behaviour during imprisonment is briefly mentioned, sandwiched between two Roman examples, Rutilius' behaviour in exile, and the younger Cato's behaviour when facing suicide (6.22.3). The only philosopher to get more extended treatment – and the only one given the label *philosophus* (perhaps needed because he was a less familiar figure than Plato or Socrates[34]) – is Areus, and he is firmly placed in a Roman frame. After Seneca has compared Octavia's and Livia's handling of the loss of a son, and stated his confidence that Marcia will prefer to follow the example of her friend Livia, he continues, referring to the latter: *illa te ad suum consilium uocat* ('she is summoning you to join her discussion', 6.4.2), and there follows an account of how the philosopher Areus offered her advice on coping with her grief. Areus is given a double imperial endorsement: he is introduced as 'her husband's [i.e. Augustus'] philosopher' (*philosopho uiri sui*, 6.4.2), which gives him Augustus' backing; and Seneca tells us that Livia later said that in her grief Areus influenced her more than the Roman people, or Augustus, or her surviving son Tiberius (6.4.2). After putting direct speech into Areus' mouth, Seneca continues: *Tuum illic, Marcia, negotium actum, tibi Areus adsedit; muta personam – te consolatus est* ('Your own circumstances were addressed on that occasion, Marcia, it was you that Areus sat beside; change the cast list, and he offered comfort to you', 6.6.1). Not only does Areus' advice apply to Marcia, but by implication Seneca himself is carrying out the same role with Marcia as Areus did with Livia. Yet we should observe that, although Areus is introduced as a philosopher, his speech does not contain specifically philosophical content: his arguments are about her consistent concern to maintain her good reputation, and about how she should be willing to let her friends talk about Drusus, willing to enjoy positive memories of Metilius, and ready to show she is equal to adverse circumstances (6.4.3–5.6).

---

[34] On the circumstances in which Roman writers label individuals as *philosophus*, see Hine 2016: 24.8.

In the work as a whole the more philosophical ideas are blended in seamlessly with other ideas. In fact, the modern reader may be interested in distinguishing between 'popular' consolatory themes that are found in a wide range of literature (including letters and poetry) and specifically philosophical arguments (e.g. debate about whether or not the soul survives death, and argument that on either view death does not involve suffering; the Stoic argument that life is not inherently good or bad, but morally indifferent); Seneca, however, makes no such distinction, either here or in his other consolations, and he does not on the whole use technical philosophical language. Stoic ideas appear, but with a light touch, and they are not labelled as Stoic or as philosophical. Thus, fairly early on Seneca says he will not appeal to the harsh view that the bereaved person should not grieve at all, but he does not identify it as the Stoic view (6.4.1, cf. 11.18.5, quoted above). The dialogue ends with the Stoic doctrines of ecpyrosis and the absorption of individual souls at the ecpyrosis – again not labelled as Stoic, and this time validated by being put into the mouth of the dead Cordus (6.26.6–7). Much of the argument uses conventional moral terms rather than Stoic ones. At 6.19.5, *Mors nec bonum nec malum est* ('Death is neither a good nor an evil') looks as though it could be an instance of the Stoic doctrine of moral indifferents, but in the context Seneca is rather arguing that death is nothingness, non-existence, and so cannot be either good or evil. Here and throughout the dialogue Seneca uses 'good' and 'bad' in their everyday senses, not in the Stoic sense: *mala* are always 'sufferings' in the normal sense, from their first appearance (*antiqua mala in memoriam reduxi*, 'I have reminded you [Marcia] of sufferings that are long past', 6.1.5) to their last (*Regesne tibi nominem felicissimos futuros si maturius illos mors instantibus subtraxisset malis?*, 'Should I give you the names of kings who would have been supremely happy if death had rescued them sooner from the ills that threatened them?', 6.26.2).[35]

So the *Ad Marciam* is a work in which Seneca offers advice as a friend or at least an acquaintance, drawing on the prestige of other Romans, particularly those known to Marcia during her lifetime, but also people of past generations. Philosophy colours the argument at various points, but where it is most prominent it is in a firmly Roman frame, as though to assure us that, before Seneca writes, philosophy has already secured a place in the Roman world, and at the highest level, in the imperial court; and there is no serious demarcation between philosophical or Stoic arguments and more conventional thinking.

---

[35] Cf. 6.2.2, 3.4, 9.5, 12.6, 16.8, 19.1, 19.4–5, 20.1, 20.6.

In neither of these consolations, we may note before moving on, does Seneca use himself as an example of how to respond, or how not to respond, to bereavement, though in *Ad Polybium* he does talk about the tears he has shed because of his exile (*Dial*. 11.2.1, cf. 11.18.9). One may contrast his reference to his own excessive grief for his brother in the consolatory *Letter* 63.14–15.[36]

Finally in this section, let us compare a very different work, the *De beneficiis*, whose seven books make it the longest of Seneca's surviving moral treatises. It is a work firmly embedded in Roman social and political practice, but equally a work deeply indebted to Greek philosophical thinking. As Griffin has shown, the work is structured to take the addressee Aebutius Liberalis, and the reader, into gradually deeper levels of discussion. The first three books cover the central practical issues of giving and receiving benefits (i.e. gifts and services), then the central fourth book explores the Stoic ethical basis for Seneca's view of benefits. At the start of book 5 he says that the central questions have been covered, but he will continue with discussion of subsidiary issues, though he will follow Liberalis' guidance on how far to pursue what are often rather theoretical discussions. Scattered throughout the work there are references to named philosophers, including Socrates, Plato and the Stoics Zeno, Cleanthes, Chrysippus and Hecaton. So the explicit engagement with philosophy is more extensive than in the consolations, and integral to the work.[37]

At the same time, Greek philosophy is clearly placed in a Roman framework right at the start, in a way that does not happen in the *Letters*. The opening chapter of book 1 is a prolonged indictment of failings regarding benefits that Seneca observes in contemporary society. Chapter 2 begins with two lines from an unknown Roman comedy, whose sentiments Seneca proceeds partly to criticise, partly to applaud. His central points are that benefits should not be treated like loans, and we should continue to give even when we encounter ingratitude. In chapter 3, Seneca, under the guise of a digressive *praeteritio*,[38] launches into a dismissive discussion of allegorisations of the three Graces that were intended to bring out lessons on giving benefits. Among other objections, he finds such approaches frivolous, and they show poor understanding of how poets operate. Only when we are

---

[36] See Wilcox 2012: 40–63, 157–74, on eristic elements and use of self as an example in Cicero's and Seneca's consolatory letters.

[37] Griffin 2013 now provides a fundamental guide to the *De beneficiis*; on Liberalis' 'progress', see 125–48; on philosophy in the work, particularly 15–29.

[38] Note 1.3.2 *transilire* ('to pass over, put aside'). On the allegory of the Graces, see Griffin 2013: 99–110, 178–80, who discusses important positive aspects that are not touched on here.

halfway through the chapter does it emerge that Chrysippus is the prime culprit: *Chrysippus quoque, penes quem subtile illud acumen est et in imam penetrans ueritatem, qui rei agendae causa loquitur et uerbis non ultra quam ad intellectum satis est utitur, totum librum suum his ineptiis replet, ita ut de ipso officio dandi accipiendi reddendi beneficii pauca admodum dicat; nec his fabulas, sed haec fabulis inserit* ('Chrysippus too, the possessor of that acute shrewdness that gets right to the heart of the truth, who speaks to get things done and uses no more words than are sufficient for understanding, he fills the whole of his book with this nonsense, so that about giving, receiving and returning benefits he says very little, and what he says is not interrupted by myths, but the myths are interrupted by it', 1.3.8). Here Seneca acknowledges Chrysippus' merits, but makes no bones about his defects, speaking forthrightly about his 'nonsense'. There is a similar but even more pointed contrast in chapter 4: *Tu modo nos tuere, si quis mihi obiciet quod Chrysippum in ordinem coegerim, magnum mehercules uirum, sed tamen Graecum, cuius acumen nimis tenue retunditur et in se saepe replicatur; etiam cum agere aliquid uidetur, pungit, non perforat* ('Will you please mind my back if anyone accuses me of cutting Chrysippus down to size, a great man, by Hercules, but still a Greek, whose excessively refined acuteness gets blunted and often turns back on itself; even when he appears to be getting somewhere, he just pricks the surface, he does not run through', 1.4.1). The lesson is clear, and further spelt out in the rest of the chapter: Seneca is a Roman, not a Greek, he will not be deflected by the self-indulgent fripperies that mar Greek philosophy, but will keep his eye firmly on the task of laying down the *lex uitae*, the proper handling of benefits, which 'are the chief bond of human society' ((*res*) *quae maxime humanam societatem adligat*, 1.4.2). Now Seneca could have chosen simply to omit any mention of what Chrysippus said about the Graces, but by introducing Chrysippus in this way Seneca puts down a firm marker implying that Greek philosophy, for all its undoubted strengths, must when necessary be subject to critique from Roman common sense and practicality.

The sharply critical tone does not continue throughout the entire work, for Chrysippus and others are regularly cited approvingly, but it surfaces occasionally, as at 2.21.4, where an example given by Hecaton is dismissed as 'foolish and frivolous' (*ineptum et friuolum*), and a sound Roman example is offered instead; and in the last three books there is the recurrent question of how far (Greek) theoretical discussion may profitably be pursued – and by this stage Liberalis has advanced far enough to be invited to judge the question for himself (see above). To return briefly to the *Letters*, in them Seneca seems fairly relaxed about moving between the Greek and Roman spheres; from time to time there is the contrast of Greek and Roman, with insistence that the Romans are equal and sometimes superior to the Greeks (e.g. *Ep.* 40.9–12, 59.7, 64.10, 82.8–9, cf. 113.1), but there is not the sort of anxiety that is arguably present at the start of *De beneficiis*.[39]

The Status of Philosophy

Even this limited cross-section of Seneca's works shows that philosophy and philosophical expertise are on display to different degrees and in different ways in his various writings. It is possible to explain these differences in terms of the various recipients of his works – this is the emphasis of Wilson, who stresses that Marcia and Polybius are ordinary people; and he also argues that Seneca's general dislike of 'the desire to attribute ideas to specific philosophers or philosophical schools' supplies another reason for the avoidance of philosophical specifics in the consolations.[40]

This is undoubtedly true of the consolations, but if one considers also the different treatment of philosophy in *De beneficiis* and the *Letters*, one may suggest that there is another factor to take into account, namely the widespread antipathy towards philosophy in Seneca's society: this may help to explain the minimal presence of philosophy in the consolations, and the aggressively critical stance of the opening of *De beneficiis*. But we need to examine Seneca's context, after briefly reviewing the background.

Ever since the trials of Anaxagoras (perhaps 437/6 BC) and Socrates (399 BC), philosophers in the Greek world had regularly encountered hostility, and the suspicion that philosophy was subversive and threatened traditional values; and even when there was no active hostility towards philosophers and their claims, there were always some of their contemporaries who were sceptical or indifferent. When philosophers and philosophy arrived in Rome, there was a further ground for suspicion, because philosophy was widely seen as part of a foreign intellectual culture that was alien and potentially damaging to Roman traditions. But societal attitudes to philosophy varied with place and time, so each philosophical writer found himself in his own situation and handled it in his own way. Cicero remained interested in philosophy throughout his career, but that interest is more evident in some genres of his writing than in others. In his public speeches he hardly ever reveals his knowledge of philosophy, and occasionally adopts a

---

[39] There is a comparable absence of anxiety in the *Natural Questions*; cf. Hine 2006: 53–60.
[40] Wilson 2013: 104–8; quotation from 107.

hostile stance towards philosophically minded Roman contemporaries (not just Piso in *In Pisonem*, but also Cato in *Pro Murena* 60–6). In his *Letters* he generally touches on philosophical matters only when writing to philosophically minded friends.[41] Near the end of his life he embarked on the great series of philosophical works designed to present the case for philosophy to his Roman contemporaries, and in the process to create a philosophical literature in Latin that (unlike the dismal earlier efforts of Amafinius and other Epicureans, see *Acad.* 1.5–6, *Tusc.* 1.6, 2.7–8, 4.6–7) could bear comparison with the Greeks. The series began with the lost protreptic, the *Hortensius*, in which Cicero says he sought to answer philosophy's detractors (*uituperatores*, *Fin.* 1.2, *Tusc.* 2.4). As far as worthwhile philosophical writing in Latin was concerned, Cicero presented himself as a pioneer, but at the same time he could look back to a small number of Romans of previous generations who had taken an interest in Greek philosophy; though in reality he was creating a semi-fictional history of philosophy at Rome in the dialogues he set in the past – and the hazards of the exercise were thrown into relief by the fate of the *Academica*, where he came to feel that the original setting, in the age of Lucullus, was historically too implausible, and so he produced a second version with contemporary speakers.[42]

For Seneca, a century later, the Roman philosophical landscape was in some ways very different:[43] there was now a significant body of Latin philosophical literature, not just the works of Cicero, but works by writers of the stature of Livy and Asinius Pollio, and worthwhile contributions by lesser writers such as Papirius Fabianus (see *Ep.* 100.1–9); and not only were there individuals interested in philosophy in the generations after Cicero, but a Roman philosophical school had been established by the Sextii (*QNat.* 7.32.2). But the story was far from entirely positive. The school of the Sextii had faded away almost as soon as it sprang up, and philosophy and other intellectual pursuits were increasingly neglected (ibid.). There was still widespread antipathy towards philosophy at Rome, and at times Seneca was personally affected by it. His own father disliked philosophy: Seneca's early vegetarianism, adopted for philosophical reasons, was ostensibly abandoned because of its sinister religious associations (*Ep.* 108.22 *inter argumenta superstitionis*), but Seneca says his father's real motive for urging

---

[41] Cf. Griffin 1995.
[42] Only Book 2 of the *Academica priora* survives (also known as the *Lucullus*), and only Book 1 of the *Academica posteriora*.
[43] For a fuller discussion of Seneca's philosophical context, with some emphases different from mine, see Inwood 1995.

its abandonment was his dislike of philosophy (ibid., *qui... philosophiam oderat*).⁴⁴ Seneca was recalled from exile at Agrippina's instigation to be teacher to her son Nero, but philosophy was not on the syllabus: according to Suetonius (*Ner.* 52) she banned her son from studying philosophy, as not fit for a future ruler (*imperaturo contrariam*). After Seneca's death, several philosophers were banished or forced to suicide by Nero, and later there were more banishments under Vespasian and Domitian. There were generally political motives for these sentences of exile, but philosophy was a prominent component of the public stance the victims adopted.⁴⁵

It is against this sort of climate that in the *Letters* Seneca advises Lucilius not to flaunt his philosophy and court adverse publicity. He pithily sums up his advice in an early letter in which he warns Lucilius against attracting unwelcome attention to himself: *Satis ipsum nomen philosophiae, etiam si modeste tractetur, inuidiosum est: quid si nos hominum consuetudini coeperimus excerpere?* ('The mere word "philosophy" is enough to provoke ill feeling, even if it is used unostentatiously: so what will happen if we start to deviate from normal standards of behaviour?', *Ep.* 5.2). This could be seen as the motto of many of Seneca's earlier philosophical works, for there is a striking difference in the frequency with which Seneca uses the Greek loan-word *philosophia* in various works: in the *Dialogues*, *De clementia* and *De beneficiis* it occurs only three times, all in *De beata vita* (which is of Neronian date), whereas in the *Letters* it occurs about 140 times.⁴⁶ Add to this that some works, namely the *Ad Polybium*, *De clementia*, *De otio* (*Dial.* 8) and *De prouidentia* (*Dial.* 1), do not use the related Greek loan-words *philosophus* and *philosophari* either. Seneca does regularly use the native Latin words *sapientia* and *prudentia* and their cognates to refer to philosophy and philosophers, but such words can always have a more general meaning (as we have seen in *Dial.* 11.14.1, discussed above). The treatments of philosophy and philosophers in *Ad Marciam* and *Ad Polybium*, and in *De beneficiis*, can be seen as different strategies with the same aim of reducing the risk of alienating a readership some of whom were suspicious of philosophy.

As has already been said, each philosophical writer lived in his own sociopolitical context, and each adapted to it differently. We have already seen

---

⁴⁴ Compare Musonius Rufus 16, advising a son whose father has forbidden him to study philosophy. On the Seneca passage, see Griffin 1976: 40.

⁴⁵ On philosophers under Nero, see Bryan 2013. On philosophers and politics at Rome, see Griffin 1989; Trapp 2007: 226–33, and in this volume.

⁴⁶ Also six times in the *Natural Questions*.

some of the differences between Seneca and Cicero. If we turn to the Stoics of the early empire who used Greek, then we find that in different ways Musonius Rufus and Epictetus acknowledge that elements of their society are hostile to philosophy: for example, Musonius (16) advises a young man whose father has forbidden him to study philosophy,[47] and Epictetus refers to philosophers who fell foul of Nero (1.25.22, fr. 21; cf. 1.1.28–32) or Vespasian (1.2.19–24). But the audiences they are addressing are self-selecting individuals or groups who have sought them out; that is, they already have some degree of interest in philosophy, hence Musonius and Epictetus have no reservations about speaking openly and incisively about philosophy, even if the individual concerned is a Syrian king (Musonius 8) or a government official (Epictetus 1.11).

For a closer parallel to Seneca's situation, we can turn to Plutarch. The range of his literary output is more diverse than Seneca's in form and subject matter, but within his philosophical works there is a contrast between on the one hand the seriously philosophical – particularly the works on Platonic philosophy, and the attacks on Stoicism and Epicureanism – and on the other a group of works that have commonly been labelled 'popular philosophy', such as *On Feeling Good*, *On Exile* and *On Curiosity*. These and other dialogues have recently been studied by Van Hoof, who shows that the label is somewhat misleading, in that these works are not addressed to a 'popular' audience, but to members of the social and political elite.[48] They are people who have no particular interest in philosophy, and certainly no inclination to devote themselves to philosophy – people, if you like, who are in that respect comparable to Marcia or Polybius. But Plutarch's strategy is rather different from Seneca's: where Seneca, we have seen, keeps his philosophical cards almost invisibly close to his chest, Plutarch's strategy, as Van Hoof has shown, is to argue that philosophy will benefit even the person who is committed to pursuing the traditional goals of political success and honour; he makes no attempt to persuade his addressees that the philosophical life would be superior to the political life. Plutarch, it seems, is addressing people who may be indifferent to philosophy, but there is no sign of the hostility that Seneca encountered and allowed for. A similar impression emerges from comparison of Plutarch's *Consolation to his Wife* with the Senecan consolations we have examined. There are clear similarities: Plutarch names no philosophers, though there is brief reference to the fact

---

[47] Cf. Epictetus 1.26.5–7.   [48] Van Hoof 2010.

that his wife does not believe the views on death of the (unnamed) Epicureans (611d). Yet far from this silence suggesting that his wife has any dislike of philosophy, he says at one point that 'none of the philosophers who has enjoyed our company and acquaintance has failed to be struck by the simplicity of your appearance and the unaffectedness of your lifestyle' (609c), so she is clearly used to the company of philosophers. But in what was originally a personal letter to his wife, it was natural enough that Plutarch did not go into detail about philosophers and their doctrines.

Conclusion

I began this chapter by saying that Seneca was not a professional philosopher. But he presents himself as a model of how to bring philosophy into the warp and weft of Roman life. The differences between the consolations and the *Letters* in the amount of philosophical engagement and knowledge on display are striking, and *De beneficiis* seem to be somewhere in between. At first sight this might suggest a chronological story, starting from the early consolations where the philosophical content is relatively small, and culminating in the final years of effective retirement from Nero's court, which allowed Seneca to become more deeply engaged in philosophy, or at least to feel freer to reveal his deep engagement more openly and honestly in his writings. There may be something in this, but one must be cautious, both because we have looked at only a few of the surviving works, and because we do not know enough about the lost works, particularly the protreptic *Exhortationes*, whose date is very uncertain, and whose remains are meagre;[49] and even if there was some chronological development, differences of genre and addressee were also important factors. Furthermore, Seneca may well have been concerned for the impression made by the entire corpus of his prose works.[50] One aspect of the corpus as a whole is that it illustrates the breadth of his interests – including ethics (political, social and individual), physical science and even geography, ethnography and biography.[51] Another aspect is that in their different ways of handling philosophical issues, his works reveal a man who does not, as it were, declaim from his philosophical soapbox to every passer-by in the same tone, but is sensitive to the wide spectrum

---

[49] On the date of the *Exhortationes* (F76–89 Vottero), see Vottero 1998: 63–4.
[50] For a different perspective on the corpus as a whole, see Ker 2006.
[51] On geography and ethnography: T19–21 and 56 Vottero; the biography of his father: F97 Vottero.

of attitudes to philosophy in Roman society. Addressing those at one end of the spectrum, he can handle philosophy with a light touch, gently indicating that philosophy has been accepted in the highest echelons of Roman society. To those at the other end he can emphasise his own engagement with both the theoretical side of philosophy and his struggle with the practicalities of seeking to live one's life philosophically.[52]

[52] I am grateful to the editors for their comments on an earlier version of this chapter.

4 | *Iurisperiti*: 'Men Skilled in Law'

JILL HARRIES

Lawyers, Emperors and Authority

Early in the third century CE, Julius Paulus, legal adviser to Septimius Severus and his successors, set out the inside story of several legal hearings, showing the reasoning behind a selection of the emperor's decisions on petitions and appeals.[1] Paul's motives for publishing his collection would have been shared by Roman elite writers in general:[2] to create a permanent literary memorial to himself; to educate lawyers, judges, litigants and other interested parties in the law; and to advertise his own proximity to the heart of power. Like his contemporary, the senatorial historian Cassius Dio, Paul laid claim to the authority of autopsy;[3] he was personally present at the emperor's legal hearings and, as he was careful to emphasise, his contribution to the discussions helped to formulate the legal rationale for the emperor's verdict. By his own account, therefore, Paul's authority as a *iuris peritus*, a 'man skilled in law', derived from his access to and influence on imperial decision-making. He was a privileged member of a team of lawyers chosen for their expertise by an emperor who, as Paul's records show, was keen to involve himself in the minutiae of legal argument, and contradict his advisers, if he saw fit.[4]

---

[1] Originally in six books, Paul's *Decreta* (Decrees) or *Sententiae* (Legal Decisions) were reworked as two collections of two and three books. Dig. 10.2.41 (*Decreta*) is the same case as Dig. 37.14.24 (*Sententiae*); the latter supplies the names of those involved, while the former has been reworked to create a 'general' case.

[2] Cf. Peachin 2002: 11: 'Those Romans who chose to write about law belonged to a class of people for whom writing and reading were an integral part of life... Thus... we should probably regard legal literature much as we do the other types of writing potentially engaged in by a Roman aristocrat'.

[3] Dio, *Roman History* 73.4.2 ('I state these and other facts later not, as up to now, based on the authority of the reports of others but from my own observation'); 73.7.1 (Dio present at the interrogation of an imposter); 73.18.2–4 (on excesses of Commodus: Dio was present and 'participated in everything seen, heard and spoken'; more detail offered on events of Dio's lifetime 'because I was present in person when they happened and know no-one else... who knows about them as accurately as I do').

[4] Dig. 4.4.38, where Paul does not approve the decision: see Honoré 1994: 21–2.

The terms of the legal debates in council, however, conformed to a totally different construction of authority, which was independent of imperial power and endorsement. This was based, not on social or political status, but a long and rigorous intellectual tradition. Using the long-established conventions of legal discourse, Paul exploited the setting of imperial conciliar deliberations to examine how legal questions arose and the methods used by the experts to resolve them. In so doing, he also explored what critics of lawyers consistently complained of as legal 'ambiguity'.[5] Laymen, who included many emperors, wanted to know 'what the law was' and were impatient with lawyers for their failure to provide straight answers. For lawyers, however, the contradictions inherent in human affairs were essential to their discipline. As Fergus Millar said, of the supposedly definitive collection of juristic writings in Justinian's Digest, issued in 533, 'far from being the monolith which Justinian wanted, the Digest is in fact a repertoire of varied and mutually contradictory opinions and approaches.'[6]

Paul assumed, as lawyers always did, that each question had potentially contradictory answers, both (or all) of which could be justified in law. His purpose was to lay out what the options were in any given case and the rationale for adopting one solution in preference to the other(s). In some respects, this reflects the conditioning offered by the rhetorical schools, where the children of the elite were trained in arguments for either or both sides of *controversiae*, but the lawyers' handling of 'controversy' was, as we shall see, based on a canon of authorities and a legal discourse that were distinctively their own.

The citation of authorities, on one side or the other, and assessment of their value were the means by which ambiguities were resolved. 'Authorities' is itself a difficult term; what the emperors said as supreme legislators in edicts, letters or even rescripts created precedents and points of reference and by the late second century CE, the jurists had evolved their own self-perpetuating canon of *iuris periti*, who analysed not only written laws, such as statutes (*leges*), senatorial resolutions (*senatus consulta*) and the Edicts of the Praetor and the Aediles, but also unwritten custom and equity. But the discussions recorded by Aulus Gellius in the mid second century CE show that 'authority' could, at least in informal settings, be more broadly defined. In Gellius' Roman Forum, which was crammed with jurists 'teaching the *ius civile* or issuing *responsa*',[7] the question arose as to whether a quaestor could

---

[5] Hence the late Roman enthusiasm for legal codification as a means of removing legal ambiguities, e.g. at *De Rebus Bellicis* 21 (c. 369); *Nov. Theod.* 1.1 (Feb. 15, 438).
[6] Millar 1986: 274.   [7] Gell. *NA* 13.13, discussed by Howley 2013: 24–5.

be summoned into the praetor's court. The case discussed is not hypothetical, but, as Gellius concedes, it could have been. The discussants at the legal seminar thought that the quaestor was protected by the *maiestas*, dignity or privilege, of his office. Gellius, however, produced two separate citations from Varro's *Human Antiquities*, which made the opposing case: 'when both citations from the book had been read out, everyone came round to agreeing with Varro's opinion (*sententia*)'.[8] For students of Roman law, conditioned by the exclusive use of jurists in Justinian's Digest some four centuries later, it could come as a surprise that a debate about a question of legal procedure could be 'settled' by the citation of an antiquarian (although Varro was also the author of a now lost treatise on the *ius civile*). In fact, the discourse of law was far from static; even allowing for Gellius' preference for interdisciplinarity, jurisprudence in the mid second century was clearly more flexible and versatile in its use of authorities than it was later to become.

If authorities contradicted each other, how were decisions to be reached? One of Paul's cases concerned a dispute brought before Severus and his *consilium*, which turned on the reading of a phrase in a will and entailed setting the authority of one emperor against that of another.[9] The testator, Fabius Antoninus, had bequeathed 'in trust' his entire estate to his under-age son, to take effect 'when he reached his twentieth year' (*cum ad annum vicensimum pervenisset*).[10] His daughter Honorata would inherit the whole if the son died before he had reached the stated age. Fabius died and his widow, the trustee, later died intestate, leaving her children as joint default heirs. Then the young son also died, at the age of 19, but before reaching his twentieth birthday, leaving his little daughter, Fabia, as his heir. Honorata, the aunt, promptly claimed the whole estate under the trust, on the grounds that her brother had died while still under-age, and her claim was upheld by the court of the governor (*praeses*). When little Fabia's guardians appealed to the emperor (and Paul takes up the story), both opposing sides adduced imperial rescripts to justify their reading of the meaning of the crucial words 'when he had reached his twentieth year'. A rescript of Hadrian was cited for little Fabia, adducing the analogy of liability for *munera*,[11] and, for Honorata, the more recent opinion of Marcus Aurelius on liability for

---

[8] Gell. NA 13.13.3: *utraque igitur libri parte recitata in Varronis omnes sententiam concusserunt.*

[9] Dig. 36.1.76.

[10] Cf. Dig. 36.1.48 (Iavolenus Priscus, *Epistulae* 11) on an inheritance held in trust for a son to be restored *cum ad annos sedecim pervenisset*. Here the relevance of the age of death (which invalidates the bequest) is not at issue.

[11] Hadrian's ruling was that on the age of liability to undertake council services, the year on which 'he had entered should be reckoned as if completed'.

guardianships.[12] Severus, backed by Paul, ruled for Fabia on two grounds. One was equity (*aequitas*): Fabia's guardians had pleaded her poverty (*egestas*), as part of a wider argument for natural justice. The second, on a more technical level, came from Paul, who, inter alia, adduced parallel usages for the age limits for manumissions under Augustus' Lex Aelia Sentia; it helped that he was also the author of a commentary on that statute.[13]

Paul's explanation of how a contradiction of authorities was resolved demonstrates the importance of access to and control of knowledge, and of understanding of how that knowledge is to be deployed. As narrator, he controls the parts played by the advocates of the contending parties, the emperor and himself. The advocates play a key role. Both sides come armed with rescripts, bearing the authority both of the emperors as lawgivers and, tacitly, of their legal draftsmen. The winner, however, adds the argument, based on legal *aequitas*, that Fabia will lose out disproportionately if the vote goes against her, as she, unlike Honorata, will 'unfairly' suffer 'poverty'.[14] Paul's job, in this case, is to strengthen the verdict, which will also act as a precedent for future cases, by adducing parallels from his own legal learning to justify the preferred reading of the disputed text.

In the world of Paul's *Decreta*, the authority of jurists and emperors is complementary. The emperor has power as the lawgiver and ultimate adjudicator; the men skilled in law have the knowledge to ensure he gets it right. The ideal relationship between emperors and their legal advisers was therefore one of cooperation, each deferring to the separate and distinct authority of the other, and drawing on the decisions reached by past emperors and past jurists, as conversations between the two were renewed through the generations. Thus, Paul's colleague, Ulpian, writing in privileged retirement after 212 CE, cited a rescript, concerning a property dispute, issued by the joint emperors Marcus Aurelius and Lucius Verus (161–7 CE), which illustrates the interdependence between emperor, expert advisers and their expertise.[15] The text opens with a statement that the emperor is indebted

---

[12] Marcus ruled that exemptions from guardianships which applied to 70-year-olds should not extend to men aged 69 (i.e. who had not completed their 70th year). Although technically inconsistent, the policy effect of both was to extend outwards the age limits for liability for services.

[13] *Nobis et legis Aeliae Sentiae argumenta proferentibus et alia quaedam*. For right to manumit at 20 years, see Paul at Dig. 40.2.15, 40.9.16.pr., 45.1.66.

[14] Honorata's marital status is unknown, but could have been relevant to the outcome, if she had previously received her share of her father Antoninus' estate in the form of a dowry as part of her marriage settlement.

[15] Dig. 37.14.17.pr. (Ulpian, *On the Lex Julia and Papia* 11). *Comperimus a peritioribus dubitatum aliquando, an nepos contra tabulas aviti liberti bonorum possessionem petere possit, si eum libertum pater patris, cum annorum viginti quinque esset, capitis accusasset, et Proculum*, sane non levem iuris auctorem, *in hac opinione fuisse, ut nepotem in huiusmodi causa non putaret*

to the 'experts' (*peritiores*) for bringing the case to their attention, and for providing a solution, backed by the authority of an opinion given by the first-century CE jurist, Proculus, 'an expert in law of no light weight'. It was not open to experts to provide answers on their own authority alone, however, and it was necessary to trace the tradition forward from Proculus to ascertain if his view still held. Here, the view of the *peritiores* received extra support from a previous decision of the emperors in response to a petition from a named individual, Caesidia Longina, and from the view expressed in a personal conversation by Volusius Maecianus, the emperors' 'friend' and former Prefect of Egypt and council member.[16]

When Paul or Ulpian cited rescripts and analysed imperial decisions, they affirmed both the personal authority of the emperor's legal advisers and the importance of the legal discipline, which formed the area of their expertise. But they also, by implication, insisted that what jurists did was different from, and independent of, the imperial power. They had an expertise necessary for emperors, but in which the emperors could not fully share – despite Severus' valiant attempts to engage with legal discourse. Their authority, therefore, did not depend only on imperial patronage, but on their ownership of a system of knowledge, the true subtleties of which could be understood only by those, who took the time and trouble to learn and understand it.

Experts under Threat: Communication and Control

Such was the jurist's representation of himself and the authority of his discipline, but was this also the perception of others? The problem of how juristic authority was to be conceptualised and sustained, independently of the status of the individual jurist, was a long-standing one, as was illustrated by Cicero's comment in 55 BCE that senatorial jurists carried weight because of their overall *auctoritas*, *not* their expertise (*ingenium*).[17] This went to the heart of the dilemma facing jurists. It was true that, to some extent, the

---

dandam bonorum possessionem. Cuius sententiam non quoque secuti sumus, cum rescriberemus ad libellum Caesidiae Longinae; sed et Volusius Maecianus, amicus noster ut et iuris civilis praeter veterem et bene fundatam peritiam anxie diligens religione rescripti nostri ductus sit ut coram nobis adfirmavit non arbitratum se aliter respondere debere (emphasis added).

[16] On Maecianus and M. Aurelius, see Cuomo 2007a; and on his role as 'procurator of libraries' at Rome, see Bowie 2013: 255–8. Compare Dig. 28.4.3, minutes of Marcus Aurelius' council on an estate forfeited to the treasury, because of a damaged will. This is reported by the jurist Marcellus, who had privileged access – and advertised the fact.

[17] Cic. De or. 1.45.198. For the argument that jurists under the Republic had an image problem, see Harries 2006: 76–9.

political and social eminence of some *iuris periti* was a positive for the evolution and perpetuation of legal expertise. A senator who had a successful political career and who was also a jurist, such as Servius Sulpicius Rufus (consul 51 BCE), Ateius Capito (consul 5 CE), C. Cassius Longinus (consul suffect 30 CE), Iavolenus Priscus (consul 86 CE) or P. Iuventius Celsus (consul II 129 CE), could expect to have a circle of influential clients, whom he assisted, and pupils, who would perpetuate his reputation through their own legal consultancies and written work. But a legal expert, who failed to differentiate his area of expertise from the values of the political culture of his non-expert contemporaries, risked the contamination of his special knowledge by more ephemeral considerations. Such was the view allegedly taken by Tacitus' consular jurist, C. Cassius Longinus, who is credited with a general contempt for the sloppy attitudes of his less rigorous contemporaries, corrupted by their role as players in the unprincipled power games of empire.[18]

Although the status of jurisprudence was defined and strengthened by the prestige of Law, it was left to the advocate Cicero to celebrate Law in the abstract as the guarantor of freedom and fairness, supporting the role of magistrates, judges and others, who 'are servants of the law, so that we can be free'.[19] Cicero, while consistent in showing respect for his own teachers, Mucius Scaevola the Augur (consul 117 BCE) and his namesake the Pontifex (consul 95 BCE), dismissed *iuris periti* in general as legal technicians, useful for specialist guidance, but not for elucidation of the 'true' law, which was based on a philosophical understanding of Law as the 'highest reason'.[20] The job of jurists, in Cicero's view, was to provide legal advice to judges and guidance on the conduct of lawsuits, along with help in the drafting of legal documents;[21] they would also have been instrumental in drafting the praetor's formulae, the definitions of the precise legal points at stake in a given case. It is perhaps a measure of Republican juristic self-confidence that they did not let themselves be drawn onto Cicero's territory, generally avoiding such inflated claims as those of Ulpian some three centuries later, at the start of his *Institutes*, that jurists were the 'priests' of their discipline and practitioners of the 'true philosophy'.[22]

A further potential challenge to the authority of jurists lay in the perceived isolation of juristic discourse from the wider cultural environment. One of Gellius' anecdotes carries a warning to *iuris periti* (or, in this case,

---

[18] Tac. *Ann.* 14.43–4. Cassius disapproved of 'new-fangled senatorial decrees' passed 'against the established usage of our ancestors', but had refrained from comment to preserve 'such authority as he had' (*quicquid hoc in nobis auctoritatis est*).
[19] Cic. *Clu.* 145–6.   [20] Cic. *Leg.* 1.14; 18b.   [21] Cic. *De or.* 1.212.
[22] Dig. 1.1.1.1; cf. Dio, *Roman History* 71.35.1–2 and 77.19.1.

iuris callentes). At a discussion of the meaning of an archaic word (*proletarius*), Gellius' legal friend, a *iuris callens*, disgraces himself first by refusing to advance an opinion on the grounds that this was a question for grammarians, and then by insisting that modern lawyers were too busy to bother with the interpretation of old, irrelevant words: 'the intellectual commitment and knowledge which it's right for me to display concerns the law and statutes and legal terminology, which we make use of today'.[23] This was not mere rudeness. Gellius, the non-lawyer, happily cites the Twelve Tables and strings of references from late Republican and early imperial jurists, often using, more transparently than some, methods of second-hand citation.[24] But if Gellius engaged with the jurists, he also expected that lawyers in their turn should debate grammar, etymology and antiquity with him. Failure to do so on the part of a 'legal expert' was not only an intellectual and a social failure, but also reflected a defect of character. While the *iuris callens* is arrogant and dismissive of a reasonable question, Gellius underlines the inadequacy of the legal pedant by his choice of an alternative guide, a passing poet, Julius Paulus, whose explanation of the term is based on law and history.

For Gellius, the *iuris callens* is not an expert in a true sense. This is not, as Gellius admits, because he lacks specialist knowledge (*scientia*). In fact, the lawyer knows all about the area of expertise, which he affects to despise, betraying himself by listing a number of disputed or obscure words, which occurred in the Twelve Tables and which puzzled ancient interpreters (as they do now).[25] Despite his learning, however, he is unable to apply it appropriately. He is an expert who makes the wrong choices, although he has the capacity to choose otherwise, failing to apply his expertise where it matters, in company with cultured Romans. As Howley puts it, for Gellius, 'the right sort of jurists ... are those who take advantage of the deep and authoritative claim to antiquarian enquiry that characterises their profession'.[26]

Perhaps the greatest threat to the independence of *iuris periti* came from the dangerous attractions of imperial service, as emperors became increasingly adept in subordinating all sources of power, including law making, to their personal control. In the first century CE, the Senate was, naturally, the first, though not the only, port of call for emperors in search of

---

[23] Gell. *NA* 16.10.8, *studium scientiamque ego praestare debeo iuris et legum vocumque earum quibus utimur.*
[24] Gell. *NA* 3.2.12–13 (Q. Mucius, at second hand); 15.27 (Labeo in Laelius Felix on Q. Mucius); 4.4.4 (Neratius Priscus); 6.15 (Labeo, Q. Mucius); 14.7.12–13 (Tubero, in Ateius Capito); 4.10.6–8; 10.16 (Capito); 13.12.1–4 (Capito versus Labeo).
[25] Gell. *NA* 16.10.8, *proletarii, adsidui, sanates, vades, subvades, viginti quinque asses* (cf. *NA* 20.1.12–13, *taliones, cum lance et liceo*).
[26] Howley 2013: 29.

legal advisers.²⁷ But from the early second century, the imperial administration no longer relied exclusively on senators and legal administrators were increasingly drawn from wealthy and educated non-senators, who owed their advancement entirely to imperial favour.²⁸ Such was the young African lawyer from Hadrumetum, Salvius Julianus, whose 'codification' of the Praetorian Edict in c. 130 on the orders of Hadrian cemented imperial control over the content of praetorian law, which henceforward could be modified only by emperors.²⁹ The three great Severan jurists, Papinian, Paulus and Ulpian, who dominate the pages of the Digest, were lawyers, not senators, and were also not Italian by origin.³⁰ All three were authors of voluminous works of legal interpretation, and instructors in their own right and, by the fifth century, they, along with Gaius and Ulpian's pupil Modestinus, formed an officially recognised canon of authorities, to which reference was routinely made by judges and educators.³¹

The threat posed by imperial patronage to the independent authority of the individual jurist became a reality from the fourth century onwards, as the separate identities of the emperors' lawyers were fully absorbed into the bureaucracy. Occasionally in late antiquity, names surface, such as those of the lawyers praised for their efforts in compiling codifications of imperial law in the fifth and sixth centuries, but, with the exception of various prominent imperial quaestors, such as Antiochus Chuzon, also praetorian prefect and prominent Christian, who was mainly responsible for the Theodosian Code from its conception in 429 to completion in 437, and Justinian's legal draftsman and adviser (quaestor), Tribonian, little is known about them.³² They were the emperor's men and the constitutions, be they edicts, letters or replies to petitions, drafted by them were issued in his name.

The process of asserting imperial control of legal experts past and present culminated in 533 CE, when the emperor Justinian, ably assisted by

---

²⁷ Senatorial jurists on the imperial advisory council (*consilium*) included M. Cocceius Nerva (Tac. *Ann.* 6.26.2; Dig. 1.2.2.48); Iavolenus Priscus (Plin. *Ep.* 6.15.3); Caelius Sabinus, cos. suff. 69 (Dig. 1.2.2.53); L. Neratius Priscus (Dig. 37.12.5).

²⁸ E.g. Titius Aristo (Dig. 37.12.5); L. Volusius Maecianus, equestrian Prefect of Egypt in 160 (Dig. 37.14.17); Q. Cervidius Scaevola (under Marcus Aurelius, no office known, on *consilium*, Dig. 36.1.23(22).pr.); Cl. Tryphoninus (Dig. 49.14.50).

²⁹ Which did not deter Julianus from composing a ninety-book commentary, the Digest, on his creation, a work of interpretation of which future emperors and lawyers would be obliged to take account.

³⁰ For an epigraphically attested Greek jurist and advocate, see Millar 1999.

³¹ Cf. the so-called 'Law of Citations' at *Cod. Theod.* 1.4.3, limiting the canon to the five named jurists and authorities cited by them, with due attention to be paid to the authenticity of the manuscripts.

³² See Honoré 1978 and 2010.

Tribonian, launched a cultural and legal coup d'état. As part of a yet more ambitious project, to codify the Roman citizen law (*ius civile*), an authoritative collection of extracts from the writings of famous jurists, known as the Digest, was issued under the emperor's name. This fifty-book magnum opus was to be used by judges in the law courts as the exclusive source of reference for legal interpretation, and in law schools of the Byzantine Empire as the sole textbook for the main syllabus. Justinian's project, lifting his extracts of the writings of the *iuris periti* from their original context and juxtaposing them with the opinions of others, asserted that emperor's ownership of the past as well as the present. Past jurists were recontextualised, their original messages distorted by the process of selection and expurgation, and by their new context. And Justinian's reach extended into the future. As all past writings not excerpted for use in the Digest were to cease to count for litigation or educational purposes, the power to emend legal interpretation was henceforward to rest not with independent jurists, but with the emperor.[33]

As we have seen, Tribonian's implementing of Justinian's instructions concerning the Digest produced not the expected 'monolith', but, through its scrupulous recording of divergent juristic opinions, a monument to the 'ambiguities' so hated by non-lawyers. One method, therefore, by which Roman imperial lawyers could survive the misguided initiatives of their masters was subversion; they made the right noises, but relied on their own expertise (and the limited attention spans of their imperial masters) to see them through. Over the longer term, however, three aspects of the activities of jurists contributed to their resilience and the continuing prestige of their discipline. One was the educational experience of lawyers; a second was the specialised (and in the view of many abstruse) nature of legal knowledge; and the third was the lawyers' construction of their own history in terms of tradition and the Roman past.

Legal Education

Cicero's definition of the job of a *iuris peritus* did not include teaching[34] or the recording of their opinions in writing, yet it was through those means that the learning of jurists was perpetuated and the authority of their

---

[33] Justinian, *Constitutio 'Tanta'* 17. See also Humfress 2005.
[34] Although Cicero and Atticus had attended the legal seminars of the Scaevolae (see Cic. *Brut.* 89, 306; *Amic.* 1.1).

discipline defined and strengthened. When the academic jurist Pomponius in the second century CE tried to make sense of the past history of jurisprudence, he concluded from the lack of early interpretative texts that lawyers were protective of their expertise and avoided making themselves generally available to those wishing to learn.[35] Pomponius' association of the teaching process with the written text is significant both for his assumption that oral teaching would be expected to translate into written form and, conversely, that the written collections of learned legal opinions current in his day were the product of the teaching process.

Informal though it was by modern standards, the system of legal education is fundamental for the understanding of how juristic culture became embedded in and accepted by the wider Roman elite. To a point, juristic methods of argument cohered with the rhetorical methodologies inculcated into the Roman youth in the secondary stages of their education.[36] Students of law would structure argument using the techniques of the rhetorical schools already familiar to them. In a treatise dedicated to his friend, the jurist C. Trebatius Testa, in 44 BCE, Cicero argued that juristic arguments were conducted on principles similar to those used by advocates. Thus, for example, he offers specimen ways of thinking about the *ius civile* in terms of argument from definition and from enumeration of its constituent parts;[37] in a nod towards his own former law teacher, he then cites Scaevola the Pontifex's definition of *gentiles*, which was offered as a series of attributes, such as membership of a *gens*, which were accumulated until, taken together, they could apply only to the thing defined.[38] Ways of thinking were also common ground; jurists and orators would both argue from analogy and similarity. Yet Cicero was forced to concede that the language of the orator and the jurist diverged; in jurisprudence, the techniques of the rhetor 'are less abundant but perhaps more subtle'.[39]

Schoolboys were conditioned to think in terms of cases, long before they chose, as young adults, to specialise in the interpretation of *ius*. Rhetorical instruction at secondary level was based on the creation of declamatory speeches for use in imaginary cases, many of which were taken from the world of history or the novel, featuring tyrants (and tyrannicides), pirates

---

[35] Dig. 1.2.2.35: *solumque consultatioribus vacare potius quam discere volentibus se praestabant.*
[36] Tellegen-Couperus and Tellegen 2013.
[37] *Topica* 9 (definition); *Top.* 28 (enumeration).
[38] *Top.* 29. For juristic use of *genus*, category and *species*, sub-category, see Gaius, *Inst.* 1.183 (four *genera* of theft); 188 (3 (Servius) or 5 (Mucius) *genera* of guardianship). For the same in rhetoric, see, e.g., *Rhet. Her.* 1.2.2. (cases); 1.4.6 (*exordia*); 1.8.12 (*narratio*).
[39] *Top.* 65. See Harries 2006: 126–32. For edition and commentary on the *Topica*, see Reinhardt 2003.

and shipwrecks, rapists and abductors, poisoners, ungrateful fathers and sons, lovers, prostitutes and wicked stepmothers, while others, a minority, drew on aspects of Roman law, such as the Lex Voconia.[40] For purposes of these exercises, a statement was made of 'what the law was', which should not be read as an accurate reflection of Roman (or any other) law as it stood at the time.[41] The student was then presented with a set of facts concerning a case and was required to argue one or other side of the case. Written statute, as represented in the teaching exercise, could be criticised as poorly worded, obscure, contrary to other laws, impracticable or inexpedient. Key to the trainee orator's success were his accurate (or convincing) identification of the legal point at issue and his analysis of the law. The boy thus learned to deploy in rhetorical argument such Roman legal terms as deposit, divorce, *lenocinium* (pimping or procuring, a popular topic), adultery and injury. However, while significant attention is devoted to the soap-opera-style antics of seriously dysfunctional families, there was rather less focus on the more austere aspects of the *ius civile*, and adherence to legal terminology is sometimes deliberately avoided.[42] This training in legal argumentation, rudimentary though it was, was essential for the future jurist as much as for the advocate and created a cultural consensus between them; both were conditioned by the system to think in terms of what the legal issues were, what kinds of facts were relevant to the legal points, and how legal texts could be read and represented.

Many of the most eminent jurists of the Late Republic and Early Empire offered informal classes within their own homes for a selected clientele. The setting itself, the town residence of a great man, inspired respect.[43] The consular jurist, C. Cassius Longinus, instructed his students in his home at Rome, remembered later as the site of the 'Cassian School' (*schola Cassiana*).[44] But the system was also flexible; classes took place in public areas or wherever the teacher happened to be (as was the case with Servius Sulpicius Rufus in hiding in 47 BCE on Samos). The subordinate status of the students was reflected in their name; they were *auditores*, 'hearers', and those who went on to be successful jurists in their turn could be identified as the *auditor* of their teachers, as Pliny's friend, Titius Aristo, for example,

---

[40] Ps.-Quintilian, *Declamations* 264.
[41] E.g. Ps.-Quintilian, *Declamations* 245, 'he who denies a deposit, let him pay fourfold'; 248, 'let a man who is convicted of involuntary homicide be exiled for five years'; 270, 'whoever is the cause of death, let him suffer capital punishment'.
[42] E.g. *ignominiosus* and *ignominia* are regularly used for *infamis* and *infamia*.
[43] Cic. *De or.* 1.199: *Est enim sine dubio domus iurisconsulti totius oraculum civitatis*.
[44] Plin. *Ep.* 7.24.

was of Cassius.⁴⁵ Yet students were not passive recipients of the rulings of the great man. Alfenus Varus (consul 39 BCE) recalled a debate at a seminar conducted by Servius on what constituted *instrumenta* (farm implements); one student cited a contrary opinion by one Q. Cornelius Maximus (which Alfenus preferred).⁴⁶ This provides also a rare insight into what could happen in a legal class. The question at issue was one of definition, a favourite preoccupation of lawyers. The contradictory opinion was advanced in the course of dialogue, perhaps even due to an intervention by the student. The student had come prepared with readings or instruction from a different authority, Cornelius Maximus. And, finally, Servius' and the student's opinions were recorded for posterity by another *auditor*.⁴⁷ Seminars, therefore, were one means by which the discipline itself could evolve and new interpretations be fed into the legal tradition.

Knowledge of how 'text' was created through the oral processes of instruction is now largely lost. Did the teacher lecture? Did he engage in 'Socratic' dialogues with his students? Were these conversations (if they happened) conducted with a selected individual or was the conversation more general? Was there any form of role play, where teacher or students 'performed' legal transactions?⁴⁸ Certainly, lawyers often wrote as if they themselves and a hypothetical 'you', the reader or student, were part of the case under discussion. To revert to Ulpian and a separate but closely related citation of the relatively obscure first-century jurist Urseius Ferox:⁴⁹

> If a slave held in common, that is belonging to me and you (*id est meus et tuus*), kills my slave, there is a place for (a lawsuit under) the Lex Aquilia (on damages), if he did it at your desire (*tua voluntate*); this was Proculus' opinion, as Urseius tells us. (Dig. 9.2.27.1)

In this, one example among many of role-play style, there is no discussion about the rights and wrongs of the killing. The issue is about damage to property and 'my' right to compensation. The question is whether 'I' can sue under the Lex Aquilia for damages. As the killer is also a slave, who is not an independent legal persona, the attitude of the master, 'you', will decide the type of legal remedy on offer. If 'you' had ordered (or condoned) 'your' slave's action, then 'you' are liable under the law and a lawsuit can be brought

---

⁴⁵ Dig. 4.8.40. For Pliny's friendship with Aristo, see Plin. *Ep.* 1.22, 5.3 and 8.14.
⁴⁶ Dig. 33.7.16.1.
⁴⁷ In the third century CE, another dutiful pupil, Herennius Modestinus, differed from his master, Ulpian, over the time limits for the lodging of an appeal concerning an unduteous will: Dig. 47.2.52.20; *Cod. Just.* 3.28.36.2.
⁴⁸ On the dynamics of the modern US law classroom, see Mertz 2007: 43–140.
⁴⁹ For more on Urseius, see the section on 'Legal Knowledge', below, p. 96.

against 'you'. If 'you' wish to question my opinion, 'you' will find it harder, because 'I' can cite Proculus' learned opinion and where I found it (Urseius). In a real case, however, (and, potentially, in a real legal seminar), 'you' may be smart enough to have found an opposing authority, and it would be up to the judge to decide which authority to prefer, and thus whether to hear the case at all.[50]

Absent from this is any discussion of the rights of the slave as a human being; the questions suggested and discussed reinforce a strictly delimited legal perspective. This enabled the teachers to inculcate in their students (as they do now) a set of cultural assumptions about what was relevant to law and what was not. Legal discourse did take account of social values, such as fairness, faith or honour, and scholars have argued that jurists did evolve a concept of human rights.[51] Yet it excluded, from its own point of view rightly, emotion. Thus, rape, defined as sex by violence (*stuprum per vim*), receives discussion in terms of the injury done, not to the woman but to the father of the family, in whose legal authority she was. Romans in general, it must be admitted, were not sensitive to the trauma of rape; in this case, however, the emotional damage done to the woman is irrelevant to the writer's legal purpose and is therefore ignored.[52]

It followed that juristic discourse deliberately distanced itself from the emotive aspects of rhetoric. By employing a style that was restrained and unadorned, a version of the *sermo humilis*, low style, jurists projected themselves as honest instructors, not artful persuaders. In that respect, they sought to imitate the character of ancient law, which offered no rhetorical arguments for purposes of self-justification; that this was the law was enough. They may also have been influenced by the economical style of the pontifical jurists, whose responses to questions of ritual law addressed only the legal points at issue, and rigorously avoided offering a judgement on the facts.[53]

---

[50] Alternatively, if the jointly owned slave killed 'my' slave without 'your' consent, a different legal solution is offered, which is affected by the fact of joint ownership: *quod si non voluntate tua fecit, cessare noxalem actionem, ne sit in potestate servi, ut tibi soli serviat*.

[51] Bauman 2000; Honoré 2002.

[52] Cf. Mertz 2007: 10: 'Your criminal law exam involves a hypothetical in which a woman is beaten, raped and killed... If you have yourself been beaten or raped you may find this question a bit difficult to answer. But your performance will depend on your ability to dispassionately analyze the details provided to you for traces of the "facts" needed to satisfy one or other legal test.'

[53] Contrast Cicero's presentation of 'facts' on the alleged unlawful consecration of his house in *De Domo Sua* (September 57 BCE) with the brief and restricted ruling of the *pontifices*, as recorded in *Att.* 4.2.3.

Simplicity of style enabled the exclusion of the irrelevant, but it did not equate with simplicity of content. Students of law required training to appreciate the immense complexity of legal allusions, within which the jurist operated. One example is the use of 'praetor' by second- and third-century jurists. The 'praetor' crops up with remarkable frequency, encouraging the uninitiated to conclude that, contrary to expectation, ancient Republican magistrates had a lot of business to attend to still, even under the empire. In fact, jurists almost invariably used 'praetor' to refer to the contents of the omnipresent Praetor's Edict, as codified under Hadrian, the foundation text for the *ius honorarium* and its series of impressively bulky commentaries; the reader was expected to understand that, despite the lack of explanation. So predictable was juristic language on many topics that it was even possible to make it the object of in-jokes. Cervidius Scaevola, last of a dynasty of jurists going back to the second century BCE, provided a variant on the standard Latin for 'ending a case' by arbitration, usually some version of *finire* or *finis*: Scaevola's discussion of a dispute over boundaries (*termini*) describes the election of an arbitrator to 'terminate' (*terminetur*) the case.[54] The word-play is unlikely to have meant much to a non-jurist.

The education of the future *iuris peritus* thus reflected and perpetuated both the common ground shared by an educated elite conditioned by the instruction of the *grammaticus* and the *rhetor* on the one hand and, on the other, the specialist training and discourse inculcated by the teaching methods of the juristic seminars. What they had in common was an appetite for *controversia*; each question had, potentially, at least two, equally plausible answers. Although the elementary school exercises focused on imaginary situations, they offered, even for the non-specialist, a rudimentary training in methods of legal argument and intellectual conditioning in the manner in which selected topics could (or could not) be legitimately addressed. Thus, the wise *iuris peritus* would aim to function on two levels: he would master the specialist discourse, aimed primarily at other jurists; but he would also – unlike Gellius' *iuris callens* – respect and respond to the wider elite culture, of which he was, both educationally and socially, a part.

Legal Knowledge

Despite the malign attentions of emperors – and, as we should also note, the absence of an independent judiciary – the specialist character of jurists'

---

[54] Dig. 4.8.44: ... *ut arbitratu eius res terminetur. Ipse sententiam dixit praesentibus partibus et terminos posuit.*

4 Iurisperiti: 'Men Skilled in Law'

expertise in the *ius civile*, in many ways a social disadvantage, also rendered it resistant to outside control. Law, in the abstract, was an entity which all, even emperors, were expected to respect and observe,[55] and jurists, in their more sanguine moments could represent themselves as the priestly guardians of this sacred knowledge.[56] Drawing, therefore, on the authority of law itself, the story that the *iuris periti* told about themselves and evolved over many centuries was scrupulously differentiated from, and independent of, their social or political status as consuls, senators or imperial administrators.[57]

Paul's *Decreta*, with which we began, is representative of juristic discussion of law, not in terms of abstract principle, but through consideration of the details of individual cases (as is also true in law schools today). There were practical reasons for this. Legal consultancy was conducted on a case-by-case basis. Clients came to their lawyers looking for 'remedies', because they had suffered harm or (in their view) lost out in a financial dispute over, say, property, boundaries or inheritance. But the lawyers did not offer advice only, or even mainly, on the rights and wrongs of the client's case. Instead, they devoted much effort to ensuring that it ended up in the right court before a competent judge and that the right legal action (*legis actio*) was brought in the right words.[58] Whatever the client's justification in everyday moral terms, his action would fail if the judge could not try the case or if the action was incorrectly worded.[59] In these respects, legal precision took precedence over simple justice.

Specialist legal criteria also dictated the arrangement and dissemination of legal knowledge in written form. The structure of the Twelve Tables (dated to c. 450 BCE), as understood in the first century BCE, dictated the order of cases discussed in the first master-treatise on the *ius civile*, the civil law, by the head of the pontifical college, Q. Mucius Scaevola, probably dating to the 80s BCE;[60] this became the standard order of topics for later commentators on the civil law, notably Masurius Sabinus, under Tiberius, who offered a concise commentary in three books, based on Scaevola; Pomponius, who commentated directly on 'Q. Mucius'; and later jurists, such as Ulpian, who wrote commentaries on the *ius civile* based not on Mucius or

---

[55] *Cod. Just.* 1.14.3. [56] Ulpian, *Institutes* 1, at Dig. 1.1.1.1.
[57] Note Crook 1995: 44: 'neither jurisprudence nor advocacy itself conferred high social standing'.
[58] Dealt with in Gaius' teaching book, the *Institutes*, in book 4, *passim*.
[59] Gaius, *Inst*. 4.11: the plaintiff complains that his vines (*vites*) have been cut down, but the Twelve Tables, on which the action is based, has trees (*arbores*), therefore the action fails. For a more inclusive view, see Dig. 43.27.3 (Ulpian) – that vines are included in the term 'trees', and 43.28.1.1 – that all fruits are included in the term 'acorn' (*Glandis nomine omnes fructus continentur*).
[60] The order is disputed. See Schultz 1946: 94–5 and Watson 1974: 143–58.

the Twelve Tables, but on the structure adopted by Sabinus. No less important was the tradition on the publicising of the methods of bringing actions available to litigants wishing to embark on a lawsuit. The aedile, C. Flavius, is said to have published the necessary information on legal actions (c. 300 BCE), thus breaking the monopoly of the pontiffs on this essential knowledge.[61] The story may be apocryphal, but it underlines the importance attached by Romans to making knowledge of the availability of legal remedies accessible to their fellow citizens, and thus ultimately to the evolution of the Roman *res publica* as a community governed by law.

From the time of Hadrian, Salvius Julianus' text of the Praetor's Edict, a relatively brief document in itself, provided a fixed core structure for a separate specialist category, the *ius honorarium*, the law of the magistrate.[62] Although the topics contained in the *ius civile* and *ius honorarium* overlapped, the separate arrangement was too securely rooted in legal tradition to be set aside; even in late antiquity, the Edict would provide the underlying structure for the early books of imperial legal codifications. These structures reinforced the distinct and separate character of jurisprudence with its own framework of reference. However, they also provided an agreed and universal road map for the interested non-specialist, seeking information on a particular aspect of civil or praetorian law.

Within these frameworks, successive generations of jurisprudents constructed chains of legal authorities, consisting of past jurists, to whom reference was routinely made. One unintended consequence of the process was to upgrade, retrospectively, the status of those socially obscure practitioners, whose authority derived from their status as a pupil of a more famous individual. We have already encountered the intriguingly named Urseius Ferox ('Bear Savage'),[63] probably a pupil of either Cassius or Proculus, who wrote a book of at least ten volumes, in which he recorded the opinions on various matters of Sabinus, Cassius and Proculus.[64] His writing thus became a conduit for the learned views of his betters, to be quarried for citations of Cassius and others by Salvius Julianus (consul 148 CE) in the reigns of Hadrian and Antoninus Pius, who composed a commentary *On Urseius*, and by Paulus and Ulpian under the Severans. Although his

---

[61] Cicero, *De or.* 1.186 and *Att.* 6.1.8; Livy 9.46.5; Pomponius at *Dig.* 1.2.2.7.

[62] But copies of the 'edicts of the ancient praetors' (*edicta veterum praetorum*) were still available in the Library of the Temple of Trajan, where they were consulted by Aulus Gellius (*NA* 11.17.1) later in the second century.

[63] 'Bear' on the analogy of the modern adventure-man, Bear Grylls. Urseius is in fact an adjectival name ('Bearish'), not otherwise to be found among the Roman elite.

[64] Ulpian at *Collatio* 12.7.9, cf. *Dig.* 9.2.27.11 (Sabinus and Proculus), citing 'Book 10'; *Dig.* 7.4.10.5 (Cassius); 9.2.27.1 (Proculus); 44.5.1.10 (Cassius); Paulus at 39.3.11.2 (Proculus).

4 Iurisperiti: 'Men Skilled in Law'

contribution to legal thought may have been more substantial than appears from the fragments which survive, Urseius gained lawyerly immortality because his recording of the opinions of his masters became embedded in the tradition as perpetuated by future generations of lawyers more prominent both socially and (probably) intellectually than he.

In juristic writing, authority was often substituted for detailed justification of a preferred option.[65] Jurists offered a brief summary of the content of a learned opinion and, as the chain grew longer, authorities were increasingly cited, often without acknowledgement, through intermediaries. The simplest form was the list: the Severan jurist Paul's view, for example, on unlawful purchase from a woman in guardianship, was that the 'ancients' (*veteres*) thought it a purchase not 'in good faith' (*bona fide*) and Sabinus and Cassius (two jurists but perhaps one work, as Sabinus was Cassius' teacher and they usually agreed) thought so too.[66] On a refinement of the case, Labeo thought one way, but Proculus and (Iuventius) Celsus held the opposite view. Paul's opinion, offered without further justification, was that the latter was 'right' or 'more true' (*verius*).[67] Less often, the chain of reference is more transparent: so Paul, again, commentating on the late first-century senatorial jurist, Neratius Priscus, stated that 'our colleague (Cervidius) Scaevola said that his opinion agreed with what Sabinus wrote, that (etc.)';[68] and Celsus cites Sabinus as quoted by Neratius in *Membranae (Parchments)*, Book 4.[69] And, as noted above, Ulpian, commentating on the Edict, cites Book 10 of the jurist Urseius for a report of an opinion of Sabinus on the liability of a master for damages under the Lex Aquilia for arson committed by slaves.[70]

The flourishing of book numbers and detailed references went one better than the writer who cited past authorities as a string of names, and hinted at another dimension to the author's authority, to which we will return – his library. But it may also mislead.[71] One of the earliest named jurists was a M. Cato, either the famous Censor, or his son, who lived in the first half

---

[65] As claimed by Cicero, *De or.* 1.56.239 (Galba refutes a *responsum* of P. Licinius Crassus in discussion, but Crassus counters with the 'authority' of P. Mucius Scaevola and Sex. Aelius). For satire of jurists and authority, see Cicero, *Ad Fam.* 7.10 and 7.17 (to Trebatius).
[66] Usually the jurists refer to 'Sabinus and Cassius', but for a rare effort to provide more scrupulous referencing, see Ulpian at Dig. 7.1.23.1 (...) *competere Sabinus respondit et Cassius libro octavo iuris civilis scripsit* (Sabinus' opinion derives from Cassius' recording of it in his treatise on the *ius civile*, 8).
[67] Dig. 18.1.27. [68] Dig. 3.5.18.1 (Paul, *On Neratius* 2).
[69] Dig. 8.6.12 (Celsus, Digest 23).   [70] *Collatio of Mosaic and Roman Law* 12.7.9.
[71] See Cameron 2004, esp. 89–123 on abuse of source citation (some invented, although I know of no invented jurist in the Digest).

of the second century BCE. Did his work survive as late as the second (as Pomponius claimed) or third centuries CE?[72] Paul offers an apparent direct citation of Cato's 'fifteenth' book in his commentary on Sabinus, implying that it did.[73] However, Ulpian comments that 'I read that Cato also wrote ... ', and that 'antiquity reports that Cato wrote ... ', implying that he had no access to Cato's text.[74] Moreover, an apparent direct citation of Cato's opinion on intercalary months (*Cato putat*) by Celsus early in the second century occurs in a passage which adduces other authorities, one of which, Q. Mucius the Pontifex, could be the original source for Cato (if indeed it was he and not, say, a commentary, such as that, later, of Pomponius).[75] The demonstration of (fake) erudition through ostensibly specific but in fact bogus or second-hand source citations was not confined to jurists; the practice was part of a wider culture of intellectual one-upmanship among the Roman elite.[76]

Tradition and Authority

Aspiring *iuris periti* were therefore conditioned and trained to read, react, write and think like lawyers. As the law student's knowledge of jurist-speak deepened, so too did his respect for the long tradition of which he hoped to be part, a tradition traced back to the dawn of Roman history. In the Rome of Antoninus Pius (138–161 CE), the academic jurist Pomponius, also a commentator on Q. Mucius, composed a brief guide for jurists on the history of law and lawyers. His Handbook (*Enchiridion*) was history with a purpose, to celebrate the antiquity of the legal tradition and the excellence of the 'very many, very great' men who professed legal knowledge and also 'enjoyed the greatest reputation among the Roman people'.[77] To reinforce the antiquity and authority of the legal tradition, he would recount 'from whom and from what quality of men these statements of legal right originated and were handed down'.[78]

---

[72] 'Cato' is cited by Aulus Gellius at *NA* 5.21.9–17, along with numerous others, but as excerpted by Sinnius Capito. Only two copies of speeches by Cato were available to readers in the 140s in the libraries of Apollo and Tiberius (Fronto, *Ep. Ad Caesarem* 4.5).

[73] *Dig.* 45.1.4.1: *libro quinto decimo*.

[74] *Dig.* 21.1.10.1; *Dig.* 1.1.11.12: *apud Catonem bene scriptum refert antiquitas*.

[75] *Dig.* 50.16.98.1 (Celsus, *Digest* 39): *Cato putat mensem intercalarem addicitium esse... adtribuit Q. Mucius* (etc.).

[76] On citation, see further Harries 2013: 182–4.

[77] *Dig.* 1.2.2.35: *qui eorum maximae dignationis apud populum Romanum fuerunt*.

[78] *Dig.* 1.2.2.35: *a quibus et qualibus haec iura orta et tradita sunt*. See Eshleman 2012: 187–91 on Pomponius and 'succession as accreditation'.

Pomponius was not alone in his ambition to enhance the status of his profession by providing it with a history, but, as we shall see, his approach was distinctive.[79] Indeed, some of the competition was far from impressive. Among Latin medical writers, Celsus' *De medicina* acknowledged the primacy of the Greeks, although only those of recent date; the Romans were nowhere.[80] True, Celsus conceded, medicine went back to Homer, but the doctors in the Iliad had confined their efforts to the treatment of wounds and subsequently the art of medicine was treated as a subset of philosophy by such philosopher-doctors as Pythagoras, Empedocles and Democritus, who was the teacher of the first 'true' doctor, Hippocrates. Celsus offered a list of names of successors with vague indications of chronology (*post quos... post... ex cuius successoribus*), crediting them collectively with enhancing the status of the *profession*, but his concession that real expertise was of recent date (and not Roman) suggests an author either unable or unwilling to engage with the more polemical aspects of succession-history.

Columella's history of agriculture, *De re rustica*, was more assertive of Roman ownership of the discipline. Like Celsus, he began with Greek antiquity, this time citing the authority not of Homer, but of Hesiod, and connected agriculturalists with philosophers. Thereafter, however, he organised his authorities by region, including a list of writers of unknown origin. This culminated in an outright takeover by the Romans, as, through a succession of agricultural writers loosely connected chronologically, Columella argued that *agricultura* was 'granted Roman citizenship' and 'educated' in Latin speech and eloquence by the Roman experts.[81] Through her assimilation into the Roman citizen community and her adoption of Roman identity through education, agriculture 'became Roman'. Even the greatest non-Roman authority was not immune: the twenty-eight-book manual of Mago of Carthage (the city long feared as Rome's rival for Mediterranean supremacy) forfeited its separate African character, when it was officially translated into Latin by senatorial decree.

By contrast, Pomponius ignored the Greeks entirely. His 'remembered' history of jurists and legal knowledge is designed to enhance the authority of jurisprudence by emphasising the antiquity, social importance and uniquely Roman character of the discipline. Pomponius' distant past was

---

[79] For a comparable exercise for the history of philosophers and philosophy, see Warren 2007, and for succession histories as selective commemorations of social groupings, Eshleman 2012: 177–212.
[80] Celsus, *De medicina* pr.1–11.
[81] Columella, *Rust.* pr.1.1.12–14; cf. Varro, *Rust.* 1 (Varro provides a list of authorities by genre); cf. Doody in this volume on both authors.

therefore not that of Homer or Hesiod, but of the Roman kings. Jurists, he maintained, were present at the dawn of Rome's history; one Papirius had collected together the laws of the kings, but had scrupulously added nothing of his own.[82] But the real work of interpretation had begun, as was 'natural' with the promulgation of the Twelve Tables (c. 450 BCE): 'the work of interpretation required the authority of the wise', although as yet nothing was written down.[83] Thereafter, there followed a series of names of famous jurists, connected by discipline and distinction, but not membership of 'schools'.

Unlike Celsus' Greeks or the eloquent but obscure Latin farmers favoured by Columella, the jurists of Pomponius' narrative combined Roman official distinction with learning. Thus, the Aelius brothers, Sextus and Publius (early second century BCE) were both consuls, while their contemporary Atilius was the first to be called 'the Wise' (Sapiens) by the people. Office and reputation were backed up by authorship; Sextus Aelius, who counted the poet Ennius as an admirer, wrote a book entitled the *Tripertita*, updating the Twelve Tables and providing a list of *legis actiones*; this was, we are told, still available in Pomponius' day. Books also, apparently, survived from the pen of the second-century BCE notables, such as the M. Cato discussed above, P. Mucius (ten books), Brutus (seven) and Manilius (three). The list of 'books' authored by the past jurists rolls on: Servius (180 books – in fact most if not all by his auditors); 'numerous' books by eleven named jurists, listed separately but mostly auditors of Servius; and Aufidius Namusa, yet another Servian auditor (140 books). Trebatius (Cicero's friend) wrote 'several books', which are 'not much read' (*minus frequentantur*) and the 'several books' compiled by Q. Tubero failed to find favour among Pomponius' contemporaries because of their archaic language (archaism alone could not confer authority).

Pomponius' enthusiasm for the books written by past jurists is too vaguely expressed to allow the reader to create a catalogue for his own library. However, his emphasis on learned jurists as producers of books was designed to enhance the collective authority of the profession. As a resident of Rome, he had access to the public collections of books and archival material assembled in structures of impressive architectural magnificence by successive imperial benefactors; we cannot therefore be sure how much

---

[82] The presence of a Papirius among later Republican jurists (Dig. 1.2.2.42) suggests the possibility of a family tradition, invented by the Papirii to enhance their family's profile as legal specialists.

[83] Dig 1.2.2.5: *ut interpretatio desideraret prudentium auctoritatem.*

of his bibliographical information was gleaned from browsing in the manner of Gellius.[84] However, his task as academic lawyer and teacher would obviously have been made easier, had he possessed a specialist library of his own.[85] Such an asset would also have entailed a group of users with privileged access, a community of lawyers with whom Pomponius, in the manner of all elite bibliophiles, would share his storehouse of knowledge.[86]

Pomponius was not alone in his advertisement of status through books. Later legal experts are known to have built up specialist collections. Preeminent among these was the immense library of Justinian's adviser and main mover of the Digest, Tribonian. Tribonian himself will have amassed his collection from the libraries of other lawyers gifted or bequeathed to him; his numerous books were therefore an indication of his standing in the profession, an assertion of cultural power. The fact that the Emperor Justinian himself was advertised as the principal user of his collection, which was exploited by the compilers as a source of material for the Digest, could only have enhanced Tribonian's already considerable authority.[87] Provincial lawyers, less favoured than Tribonian, assiduously collected their few precious books of law and conscientiously copied down imperial rescripts when publicly posted, thus creating useful little collections of authoritative texts for the benefit of themselves and their clients. The author of the fourth-century CE *Collatio of Roman and Mosaic Law* had access to basic works of the Severan jurists, including an extract from Ulpian's eighty-one-volume commentary on the Edict.[88] Various collections compiled in Italy and Africa of constitutions by Constantine and his sons may form the basis of their laws as collected a century later in the Theodosian Code.[89] And in the sixth century CE, Tribonian's contemporary, Boethius, who was based at Rome and not primarily a lawyer, drew on a relatively modest collection of legal textbooks for his commentary on Cicero's *Topica*.[90]

By the Late Republic, intellectual pedigrees are also in evidence. Pomponius' *Enchiridion* offered an explanation of why Servius Sulpicius Rufus became a jurist: he had been rebuked for his stupidity by Q. Mucius Scaevola (and had thereafter regularly contradicted Mucius' opinions). Trebatius

---

[84] See Neudecker 2013: 322, noting from the Scholiast on Juvenal 1.128 that Augustus' Apollo library specialised in texts on the *ius civile* and the liberal arts.
[85] Martínez and Senseney 2013: 407–16.   [86] On communities of readers, see Johnson 2013.
[87] Grafton and Williams 2006: 14–15. Honoré 1978: 147 counts 1,528 books as read by Tribonian's commission, many if not most from his library, but also drawn from other private collections. On the contents of Tribonian's library, as ascertained through citation, see Harries 2013: 180–5.
[88] For most recent text, edition and commentary, see Frakes 2011.
[89] Dillon 2012: 17–20.   [90] See Stump 1988.

was the auditor of Q. Cornelius Maximus, and was in turn the teacher of Antistius Labeo; Cascellius was the pupil of Q. Mucius Volusius. In constructing an intellectual family tree for his profession, Pomponius drew on the literature of 'succession' (*diadoche*) associated with philosophical 'schools',[91] putting forward the (erroneous) assertion that, in the reign of Augustus, two competing 'law schools' were created, headed by the jurists Antistius Labeo and Ateius Capito, who were believed generally to have disagreed with each other. Thereafter, Pomponius traced two dynasties of jurists from teacher to pupil, virtually down to his own day. While his narrative accounted for the competing two schools present at Rome in his own day (later to be known as the Sabinians and the Proculians), it also provided Pomponius' contemporaries with an identity and an authority directly connected to the great names of the past.[92]

Office and authorship, along with evidence of reputation among Romans at large, trundle in tandem through Pomponius' summary. His anecdotes ground his lawyers in the mainstream of Roman history as consuls and praetors; as ambassadors;[93] as orators and friends of Cicero or Caesar; and, but only incidentally, as adherents of schools of philosophy.[94] His jurists have a place in the public world of the Roman state. But Pomponius also chose to emphasise the juristic identity of an individual and his family, at the expense of other aspects, for which he might be better known. Thus, C. Cassius Longinus the jurist became the subject of two contrasting literary constructs. He appears in Tacitus as the relation of his namesake Cassius, the assassin of Caesar: the jurist Cassius' expertise in law coheres with his austere and conservative outlook and his banishment by Nero in 66 CE was the result of suspicions that he might imitate his ancestor by disposing of the present emperor.[95] For Pomponius, he was the son of the daughter of the consular jurist Tubero, who had married Servius' daughter;[96] he was therefore, through the female line, a member of the house of Servius, not of Cassius the tyrannicide. This was not simply a matter of a label. Pomponius' choice of family identity for Cassius claimed him for jurisprudence, and distanced him from the Tacitean heir to the elder Cassius' Republican legacy.

---

[91] See Warren 2007: 136–7, 141–2. [92] For evolution of the 'schools', see Liebs 1976.
[93] Q. Mucius, to the Carthaginians, Dig. 1.2.2.37.
[94] Dig. 1.2.2.40: *ille stoicus Pansae auditor.* [95] Tac. *Ann.* 16.7–9.
[96] Dig. 1.2.2.51–2: *plurimum in civitate auctoritatis habuit.* The story of the exile is included, but without explanation.

Conclusion

The sublime authority of Justinian's Digest obscures the fact that, for most of their history, men skilled in law had to compete for public recognition with other, noisier professionals, not least advocates. Jurists, while protective of their separate and distinctive expertise, were bound to engage with the wider culture of the Roman elite. Lawyers, philosophers, grammarians, historians and men of general culture could meet on the common ground of etymology or the relevance of the history of ancient law to their own day. Aulus Gellius' 'good jurist knows far more than the law'.[97]

Legal specialism carried the risk of cultural isolation in other ways. As we have seen, laymen interpreted lawyers' controversies as a love of 'ambiguity' and of obscurity as an end in itself. The jurists' canon of juristic authorities was self-created, self-sustaining and esoteric. While the style of juristic writing was simple, it was also allusive, packed with specialist terms and at times so compacted as to be incomprehensible to the outsider. Even those politicians and emperors who most valued and needed the expertise of *iurisperiti* also sought to control and take the credit for what they did. And in a world which valued those who excelled in a culture of performance and spectacle, the backstage nature of the work of the legal consultant was not productive of prestige to its practitioners.

The response of the lawyers was to make a virtue of necessity. As Pomponius argued, the history of law was embedded in that of the Roman state and many of her leaders had been lawyers too. To understand legal discourse, effort (plus leisure and money) was required; the profession therefore carried the attractions of exclusivity. Men skilled in law were necessary for the making of laws and the functioning of the entire judicial system; they were therefore indispensable. Their voluminous publications – and their collections of the writings of others – were not aimed solely at the altruistic dissemination of knowledge. Like other elite Romans, they advertised and underlined their expertise through their writings, but a Paul or an Ulpian, with the authority of an imperial adviser, could also anticipate that his published opinions would be cited by advocates, consulted by judges and affect the outcomes of litigation. The prestige of publicity was deceptive; the emperor's lawyers – whose very names were, by late antiquity, lost to the public record – could exploit and profit from the secrecy endemic

---

[97] Howley (2013): 13 on Gell. *NA* 20.2, where a jurist, Ateius Capito, makes up the deficiencies of a 'bad' grammarian.

in the imperial autocracy. And, despite the pressure exerted by emperors, lawyers were to have the last word. At the head of the Digest is the name of the imperial autocrat Justinian; its creators, the gatekeepers of juristic law, are Tribonian and his colleagues; and the text would flourish through the centuries as a statement, not, as Justinian intended, of imperial power and control, but of the law, to which even emperors were subject.

5 | Making and Defending Claims to Authority in Vitruvius' *De architectura*

DANIEL HARRIS-MCCOY

Introduction

The ways in which authority has been studied in connection with Vitruvius' ten-book architectural treatise, the *De architectura*, mirror broader trends in how technical and compilatory texts have historically been interpreted. Scholars have looked at *auctoritas* as a trait that Vitruvius attributes to particular types of actual architecture. They have also examined the relationship between Vitruvius' treatise and the political authority of Octavian/Augustus. My contribution to this discussion, which has so far tended to emphasise content and context,[1] will be to look at Vitruvius as an author of a book of knowledge and, in particular, how he defines and defends his 'editorial authority' – that is, excellence in the compilation and dissemination of knowledge.

In addressing the large subject of editorial authority, both in Vitruvius and elsewhere, some guiding questions to ask might include: how does a given text define authority and expertise in its field? Does it claim this expertise can be reproduced in textual form? If so, does it believe it has succeeded in its enterprise? What strategies are adopted in defending these claims to authority, for example, in the editor's self-presentation or his handling of contributing, and hence competing, literature on his subject?

Authority in information-centric texts, including encyclopedias, almanacs, textbooks, dictionaries or catalogues, often relates to their aspiration towards completeness and objectivity as generic ideals. This is also true for Vitruvius, who defines the quality of the *De architectura* primarily in terms of its completeness and organisation. That said, the composition of an actually comprehensive work is beyond the capacity of mere mortals

---

[1] For an analytical review of the different scholarly approaches to technical literature, see A. König 2009: 32 n. 4, who identifies a first phase of interest in items addressed in this literature, but relatively little interest in the texts themselves, and a second and still current phase in which greatest consideration is given to the relationship between technical literature and its historical and, in particular, political context. Cf. Harris-McCoy 2008b. I have also followed König's use of 'Octavian/Augustus' to indicate the uncertain date – pre- or post-27 BCE? – of the *De architectura*. This chapter is partly based on my unpublished dissertation, Harris-McCoy 2008a.

due to the multiple limitations inherent in rendering all knowledge in textual form.² The tension between the narrative ideal of completeness and its execution is, in fact, a motif that is traceable to Homer's Catalogue of Ships (Hom. *Il*. 2.488–92).³ The defence of any claim to encyclopedic status is thus as much an act of rhetoric as it is one of copious data-collection and organisation.⁴

Vitruvius, however, projects – with a few caveats – an aesthetic of completeness and control in the *De architectura*. The treatise is, according to Vitruvius' repeated claims, comprehensive and definitive, and is therefore *the* authoritative work on the subject. This is not just a rhetorical stance. For we shall see that Vitruvius develops a robust set of strategies throughout the text to defend the claims to editorial authority and pre-eminence that appear throughout the *De architectura*.

In particular, he gives the impression that he is writing a closed, internally coherent text that is complete unto itself. As its editor, he does so, first, by carefully managing how source material is presented. He rarely directs the reader outside of the *De architectura* by controlling references to external architects or treatises. Vitruvius also diminishes the reader's sense of the complexity of architectural knowledge, which could threaten the coherence and completeness of his text, by emphasising his synthesising role as an editor who has resolved the differences between disputing architects. In order to support his claims to intellectual and editorial authority, Vitruvius compares himself, as editor of the *De architectura*, to Nature and natural processes and, in particular, her inherent tendency to form holistic order out of disparate chaos. In taking Nature as a model for his own editorial practice, Vitruvius is better able to defend the highly unified, non-relativist,

---

² On omniscience as a trait that is consistently inaccessible to mortals throughout world mythology, see Harris-McCoy 2013: 158 n. 12.

³ Cf. Whitmarsh 2004a: 442–4, who observes that the tension between acknowledging the grandeur of one's subject and the limited capabilities of the narrator is a *topos* in epic, the hymnic tradition and encomium, as well as various prose genres. But, as his study of Aelius Aristides' *Sacred Tales* shows, how this *topos* is handled can reflect an author's particular definition of and goals for his work. See also Pliny the Elder, who, in spite of his 20,000 facts derived from a reading of 2,000 volumes, declares that many facts had surely eluded him, 'for we are mere mortals and occupied by our duties' (*homines enim sumus et occupati officiis*, HN pr.18).

⁴ Various aspects of encyclopedism are addressed in König and Woolf 2013, as well as Arnar 1990: 53–6; Clark 1990; Mendelson 1976; Harris-McCoy 2008a; Howe 1985; and Swigger 1975. For comprehensive knowledge as an ancient intellectual ideal, see: Cic. *De or*. 7; Columella, *Rust*. 1.pr.22; Quint. *Inst*. 12.11.23–4; Plin. *Ep*. 3.5.8. As a literary goal, see among others: Plin. HN pr.17, 33.130, 35.1–2, 37.1; Col. *Rust*. 1.pr.21; Frontin. *Aq*. 1.3; Artem. *Oneir*. 2.pr. For a concise history of the organisation of knowledge by Greek and Roman authors and their encyclopedic aspirations, see Grimal 1965–66.

non-controversial presentation of architecture that he gives us in his treatise.

Authority in the *De architectura*

The term *auctoritas* pervades the *De architectura*, appearing twenty times in the text. Its typical meaning is 'authority' and, in particular, 'excellence' – including but not limited to intellectual or political excellence – that, importantly, is acknowledged by others.[5] Moreover, because it requires acknowledgement, authority is an insecure commodity. It can be acquired, but also transferred and lost because excellent men too often go unrecognised or, even worse, because excellence is attributed to the undeserving. In a discussion of the wisdom of Socrates, Vitruvius expresses the unrealistic wish that the contents and quality of people's minds be *perspicuae et perlucidae* – that is, clear for all to see – 'so that an outstanding and steadfast authority might be attributed to the learned and the wise' (*et doctis et scientibus auctoritas egregia et stabilis adderetur*, 3.pr.1). The implication is that those who possess wisdom and are deserving of authority rarely actually get it in the form of recognition.

Building projects are able to function as demonstrations of authority, architectural or otherwise, in the *De architectura*. Buildings seemingly grant *auctoritas* because they are concrete symbols of power as well as of technical and intellectual skill.[6] For example, Vitruvius considers Octavian/Augustus' empire-wide building projects as physical affirmations of his more general authority (… *ut maiestas imperii publicorum aedificiorum egregias haberet auctoritates*, 1.pr.2). And one of Vitruvius' goals is to convey a sense of authority to the objects he designs and builds, which will be patently obvious to the viewer and in turn give authority to Vitruvius himself. At the end of Book 6, for example, Vitruvius mentions the glory that goes to the architect due to the 'authority' of his creations on account of their beauty, symmetry and harmony (*venuste proportionibus et symmetriis habuerit auctoritatem, tunc fuerit gloria area architecti*, 6.8.9). He then observes that 'all men and not architects alone are able to determine what is good' (*namque omnes homines non solum architecti, quod est bonum, possunt probare*, 6.8.10).

---

[5] The term *auctoritas* appears in the *De architectura* at: 1.pr.2; 1.1.2 (twice); 1.1.18; 1.2.5; 3.pr.1 (twice); 3.3.6; 3.3.8; 3.5.10; 5.pr.1 (twice); 5.1.10; 5.4.3; 6.8.9; 7.pr.17 (twice); 7.5.4; 7.5.7; 9.pr.17.

[6] Cf. Gros 1989, who assesses the meaning(s) of the application of *auctoritas* to architecture, including generally 'grand' terms such as *magnificentia* and *gravitas*.

As mentioned above, however, Vitruvius is a realist and is aware that, in an imperfect world, authority is seldom grounded solely in expertise, but is instead obscured and misdirected by people of 'uncertain judgment' (*incertis iudiciis*, 3.pr.1). Later on he notes that intelligent people who lack wealth or influence or skill in public speaking are rarely capable of being acknowledged as authorities in their fields because of the tendency of society to overlook them. And, in the preface to Book 5, he declares that authors who employ an elevated style of writing have a tendency to be credited with an authority that is potentially unwarranted.[7]

Vitruvius considers the nature of authoritative knowledge at some length in Book 1, discussing the education of the architect (1.1.1–11), the nature of knowledge generally (1.1.12–18) and the terms and divisions of the discipline of architecture (1.2.1–1.3.2). The general thrust is that becoming an architect is a complex proposition defined by the requirement of broad learning. The first sentence of the treatise proper highlights the many facets of architecture and the range of the proficient architect's knowledge: *architecti est scientia pluribus disciplinis et variis eruditionibus ornata, [cuius iudicio probantur omnia] quae ab ceteris artibus perficiuntur* ('The architect's knowledge is adorned by many disciplines and by various kinds of learning; everything brought to pass by these other skills is evaluated by his judgment', 1.1.1).[8]

Vitruvius goes on to describe the relationship between architecture and the several arts that comprise the *encyclios disciplina*.[9] Here, architecture acts as a master-discipline, both relying on and encompassing the rest. The architect must have some knowledge of writing, draftsmanship, geometry, arithmetic, history, philosophy, physiology, music, medicine, law and astronomy (1.1.4–10). This is in addition to the fact that the would-be architect must also possess knowledge of the theoretical as well as practical aspects of his field (*fabrica et ratiocinatio*, 1.1.1). Vitruvius describes such an architect using a military analogy, declaring that he is 'equipped, so to speak, with every kind of weapon' (*uti omnibus armis ornati*) and will act with greater 'authority' on this account (*cum auctoritate*, 1.1.1–2).[10]

---

[7] The claim by an author that, while he may not have written in an appealing fashion, the content of his text is unimpeachable is, of course, a *topos* in ancient technical literature: see Janson 1964: 98–100, 124–49 and, for example, Thucydides' well-known defence of prose historiography (1.21). That said, Vitruvius uses this *topos* to develop the larger theme of *auctoritas* as a portable object that can easily be lost and hence requires defence.

[8] All translations of Vitruvius and other authors are my own.

[9] For further discussion, see Romano 1987.

[10] On the importance of military imagery, not least as a means of self-characterisation by Vitruvius, see A. König 2009, esp. 48–50.

Vitruvius' observation that different areas of expertise are intertwined and that their practitioners must have knowledge of a range of fields is found in Cicero and, later, becomes a common trope in Latin technical works.[11] In Vitruvius, however, the trope is unusually well developed. He is not just stating that architecture is a multi-faceted discipline, but gives it an idealised status. To know architecture is, in a sense, to know (at least something) about (nearly) everything.

When Vitruvius describes architecture as 'adorned' by several disciplines and broad expertise (*pluribus disciplinis et variis eruditionibus ornata*) in the opening passage, he is alluding to Cicero's description in the *De officiis* of that 'most rare category' (*maxime rarum genus*) of men who are gifted either with exceptional natural abilities or with an outstanding education or both: ... *aut excellenti ingenii magnitudine aut praeclara eruditione atque doctrina aut utraque re ornati* ('... either gifted with outstanding intelligence or superior learning and education or some other thing', 1.33). In both passages, language of adornment (*ornata/ornati*) and learning (*eruditionibus/eruditione*; *disciplinis/doctrina*) appears, suggesting that Vitruvius is consciously translating the ideal orator of Cicero onto his own vision of the ideal architect, and is thus claiming that architecture belongs to a similarly exclusive category.

At the end of his discussion of how the various disciplines contribute to architectural knowledge, Vitruvius admits that complete knowledge of all disciplines is beyond the reach of the vast majority of architects, and that such a figure in fact 'becomes' or 'is transformed into' a *mathematicus* due to his comprehensive expertise (*praetereunt officia architectorum et efficiuntur mathematici*).[12] But Vitruvius' statement hardly challenges the overall aesthetic of comprehensiveness that pervades the *De architectura*, nor does it overturn Vitruvius' tendency to represent himself as a polymath architect who has written the most comprehensive treatise on his subject available. Vitruvius describes the polymath, i.e. the omniscient 'mathematician' referred to above, using military terminology. He is said to 'be armed with missiles from a great many disciplines' (*pluribus telis disciplinarum sunt*

---

[11] In addition to the passage from the *De officiis* below, a similar impulse is found in Cicero's *De oratore* (3.126–47). Columella likewise observes in the *De re rustica* that the broad knowledge required to farm competently – and to write a competent manual on the subject – is so vast that it could not possibly be circumscribed in a text (1.pr.22). Cf. Quint. *Inst.* 1.10.

[12] The idea that architects whose learning has become truly comprehensive somehow transcend their discipline and become mathematicians (*mathematici*) may seem odd. It is, however, harmonious with Vitruvius' belief in the mathematical underpinnings of his discipline and presumably others, as well as his worldview more generally speaking, which tends towards the Pythagorean: cf. McEwen 2003: 39–54.

armati, 1.1.17). As we saw above, this resembles the good architect who is well versed in both the theoretical and practical dimensions of his discipline (*uti omnibus armis ornati*). The architect and the omnimathic *mathematicus* are not so very different after all. Moreover, throughout the *De architectura*, Vitruvius makes a concerted effort to demonstrate that he possesses a sophisticated knowledge of these ancillary disciplines, therefore seeming to come close to his polymathic ideal if not actually achieving it.[13]

Vitruvius also explicitly claims he possesses authority literally to a superlative degree. At the end of the section that outlines the remarkable amount of knowledge required to become an architect, Vitruvius states that he holds the 'greatest authority' to write a definitive treatise on architecture, which will serve not only builders, but the broader intellectual community: *polliceor uti spero, his voluminibus non modo aedificantibus sed etiam omnibus sapientibus cum maxima auctoritate me sine dubio praestaturum* ('I promise that, as I imagine, in these volumes I will show myself to be of the greatest possible authority not only for those working in construction but even for all men of learning', 1.1.18).

Moreover, just as Vitruvius' claim to authority comes at the end of the description of the vast scope of architecture, highlighting his own broad learning, Vitruvius believes the authority of the *De architectura* similarly depends upon its comprehensive scope and promises throughout the text to furnish a totalising account of his discipline.

A claim of this sort appears at the very start, where Vitruvius outlines the purpose of the *De architectura*. In the preface to Book 1, addressed to Octavian/Augustus (*Imperator Caesar*), Vitruvius praises the emperor's imperial ambitions and, in particular, his attention to building public works projects throughout the empire (1.pr.1). The glory of Rome has been augmented by his annexation of the provinces and, Vitruvius notes, the building projects he has ordered will serve as markers of Octavian/Augustus' fame in both present and future times (1.pr.2–3). These building projects will be guided, so Vitruvius says, by the *De architectura*, which will furnish Octavian/Augustus with a complete, all-encompassing treatise on architecture, which he can consult as he proceeds with his building plans: *conscripsi praescriptiones terminatas ut eas adtendens et ante facta et futura qualia sint opera, per te posses nota habere. Namque his voluminibus aperui omnes disciplinae rationes* ('I have collected these instructions, which make use of

---

[13] The disciplines that belonged to the curriculum of the *liberales artes/enkyklios paideia* differ from author to author: see Marrou 1938: 216–17 for a useful chart organised according to author.

technical terminology, so that, by looking at them, you would know well the quality both of those projects that have already been completed and those that will come into being in the future. For in these volumes I have revealed all the principles of the discipline', 1.pr.3).[14]

Similar statements, placed throughout, reiterate these claims to comprehensiveness. They often appear in the prefaces to individual books and, in the process, take on a thematic quality.[15] Furthermore, just as the *De architectura* opens with a claim to comprehensiveness, this claim is recapitulated at its close, thus book-ending the work (10.16.12):

> *quas potui de machinis expedire rationes pacis bellique temporibus et utilissimas putavi, in hoc volumine perfeci. In prioribus vero novem de singulis generibus et partibus conparavi, uti totum corpus omnia architecturae membra in decem voluminibus haberet explicata.*
>
> In this volume, I treated the principles of machines as completely as I could, both in times of peace and in times of war, and those principles I considered to be most useful. In the previous nine volumes, I gathered together material about the disparate types and parts of architecture, so that the entire corpus might have all the parts of architecture arranged in ten volumes.

Vitruvius' claims to have written a complete and authoritative account of architecture need defence. For, as we have seen, authority according to Vitruvius' own account is all too easily lost or misdirected. For Vitruvius to have written such as treatise – that is, a body of knowledge that is complete and resists challenge – he must take into account the development of the field of architecture over time, synchronic debates within particular fields of architecture and the potential obsolescence of his particular point of view.[16]

---

[14] A. König (2009) argues that the relationship between Vitruvius and Octavian/Augustus is not nearly as subservient as has previously been claimed and that Vitruvius demonstrates the emperor's reliance on him in various ways while avoiding potential censure.

[15] Prefatory claims to comprehensiveness appear in the *De architectura* at 4.pr.1; 4.pr.5; 6.pr.7; 7.pr.14; 10.pr.4. Several book-end claims appear as well: 7.14.1; 9.8.15; and 10.16.12. Vitruvius also habitually remarks that he has treated particular subjects fully at the close of each book.

[16] These challenges do not seem to daunt Vitruvius. For example, he believes his text can transcend cultural boundaries. Its opening lines describe the *De architectura* as a gift to all the peoples of the world: *quas ob res corpus architecturae rationesque eius putavi diligentissime conscribendas, opinans in munus omnibus gentibus non ingratum futurum* ('For these reasons I determined to compose a complete treatise on architecture and its principles as thoroughly as I could, believing that it would become a not unwelcome gift for all peoples', 6.pr.7). Given the context, Vitruvius is clearly speaking to an imagined international audience, synonymous with the populations contained within the Roman Empire. The term *gentes* is used to describe the various groups of people throughout the entire Earth.

In what follows, we will look at some of the strategies Vitruvius uses to this end.[17]

Citing Sources in the *De architectura*

As a collector and compiler of information, Vitruvius adopts a number of strategies to reinforce his own authority. Some of these strategies relate to how he handles the sources that inform his text, i.e. competitor treatises on architecture. In particular, he tends to avoid directing the reader to architectural sources and debates outside of the *De architectura* text in order to preserve its coherence and completeness.

Vitruvius is quite aware of prior treatises on architecture. He includes a bibliography of works used in composing the *De architectura* and expresses his gratitude to all the authors who have informed his work (*omnibus scriptoribus infinitas ago gratias*, 7.pr.10). And, at the start of Book 4, Vitruvius assesses the 'maxims and volumes of commentaries on architecture' (*de architectura praecepta voluminaque commentariorum*, 4.pr.1). But the bibliography rarely spills into the text itself, and his brief assessment of these past writings on his subject in Book 4 is similarly self-contained.[18]

Of the thirty-nine authors Vitruvius cites in his bibliography, less than half appear anywhere else within the *De architectura*: Democritus (2.2.1, 9.2, 9.5.4, 9.6.3); Anaxagoras (8.pr.1, 9.6.3), Chersiphron (3.2.7, 10.2.11); Metagenes (10.2.12); Hermogenes (3.2.6, 3.3.8, 4.3.1); Arcesius (4.3.1); Pythius (1.1.12, 4.3.1); Diades (10.13.3); Archytas (1.1.18, 9.pr.); Archimedes

---

[17] My approach to Vitruvius shares several points of connection with McEwen's 2003 monograph *Vitruvius: Writing the Body of Architecture* 2003, which I gratefully acknowledge. In addition to a shared tendency to read the *De architectura* as something other than architectural history, her book, like this chapter, examines issues of memory (esp. 80–8), the Aristophanes episode (84–5) and Nature as a theme in the text (e.g. 40, 238). That said, McEwen tends to look at how themes in the text relate to and serve Octavian/Augustus' historical context and political program, while I emphasise Vitruvius' agenda as editor of a knowledge-book, leading to quite different readings of the various episodes. McEwen also addresses *auctoritas* directly. Here, too, she emphasises historical context, looking at the history of the word as a legal term, its relation to epigraphy and monumental architecture, inclusion of the emperor in the text, and the act of systematising knowledge. What occurs less often is a close reading of the references to *auctoritas* in the text, which usually appear in connection with intellectual and not political figures, and an analysis of how Vitruvius bulwarks the authority of his literary project per se (32–8).

[18] Cf. Cuomo 2011: 316 on the 'vague and collective' manner in which Vitruvius cites his sources in his bibliography. Novara 2005: 58–75 also compares the well-organised bibliography of Pliny to that of Vitruvius, as part of a much longer analysis that Vitruvius is attempting to demonstrate a clear line of progress from past to present.

(1.1.7, 1.1.1.7, 8.5.3, 9.pr.13–14); Ctesibius (1.1.7, 9.8.2, 10.7.4, 10.14.1); Polydus (10.13.3); and Varro (9.pr.17). This represents a mere twenty-three references to names that can be related to any kind of written treatise within the lengthy scope of Vitruvius' ten-book document. And, generally speaking, references to competing texts on architecture, whether or not they appear in the bibliography, are quite rare.

Furthermore, most of Vitruvius' citations belong either to the preliminary books or, in particular, the later specialised books on water, sundials and machines that close out the *De architectura*. Relatively few – only seven in total – are found within the core books that pertain to the subject of building or *aedificatio*; that is, Books 2 to 7. Let us consider these citations now.

Vitruvius treats the authors cited both in his bibliography and in the body of Books 2 through 7 in a way that renders them negligible or at least does not distract from his treatise. Democritus is acknowledged as a founder of atomic theory and does not count as a rival architect (2.2.1). Chersiphron is cited in passing as the designer of an Ionic temple of Diana, about which he wrote a treatise (3.2.7). Arcesius, Hermogenes and Pythius fare less well. Citing these architects as examples, he states that 'some ancient architects' wrongly condemned Doric style temples because the proportional system was inevitably 'incorrect and inconsistent' (*mendosae et disconvenientes in his symmetriae*, 4.3.1). Only Hermogenes eventually realised the truth. And a little later, Vitruvius describes their avoidance of the Doric form as unclear and irrational and says that, drawing upon the wisdom of his 'teachers' (*a praeceptoribus*), he will furnish instructions for designing a Doric-style temple that will be 'without fault' (*sine vitiis efficere possit aedium sacrarum dorico more*, 4.3.3).

Vitruvius' reference to and condemnation of this factoid from architectural history serves his authority in at least two ways. First, it demonstrates his ability to solve problems that plagued authors from the past, and thus contribute to an evolutionary view of knowledge found throughout the text, the implication being that Vitruvius, as the most recent author on architecture, is therefore also the best author on that subject.[19] Second, the fact that Vitruvius negatively refers to authors found within his bibliography casts aspersion on treatises that might be in competition with Vitruvius' own.

---

[19] The most obvious instance of Vitruvius' evolutionary view of knowledge appears in 2.1, where he outlines the progression of dwellings from primitive caves and groves to houses through the discovery of fire's usefulness and subsequent development of communal living and common language, as well as the imitation of nature (e.g. houses modelled on the nests of birds). Vitruvius' bibliography also takes a developmental view of intellectual production (cf. Novara 2005: 58 ff.).

It should also be noted that this leaves just two references to Hermogenes in Books 2 through 7, whose ideas Vitruvius wholly agrees with as they form the basis of his system of column ratios. Otherwise, if the reader hopes to find citations that suggest that useful, practical information exists outside of the *De architectura* or ones that acknowledge as potentially useful ideas that disagree with those of Vitruvius, their hopes will be in vain.

Looking at other technical and compilatory authors, we find that references to previous authors who have written on their subjects are similarly used to serve their literary programmes. We also find that, unlike Vitruvius, these authors are willing to refer their readers to external sources so long as it contributes to their goals. In the *De re rustica*, for example, Varro refers his addressee wife Fundania to a list of 'more than fifty' specialized Greek treatises on agriculture, which Varro invites his wife to refer to should she want to consider any single point in greater detail (*hi sunt, quos tu habere in consilio poteris, cum quid consulere voles*, 1.1.8). Varro uses his list of external authors on agriculture to give definition to his literary project and, in this sense, recalls Vitruvius. But, unlike Vitruvius, Varro encourages his addressee and, by extension, his reader to seek out these texts. The effect of this encouragement is that the generalising, useful nature of the *De re rustica* is confirmed. It was purportedly written to give practical instruction to one who has just bought a farm (1.1.2) and draws upon book learning, but also on personal experience and interviews with experts (1.1.11).[20] If the reader requires more specialised information than Varro has provided, he is welcome to seek it out, but should recognise he will be handling a different type of text.

Like Vitruvius, Pliny the Elder acknowledges his commitment to intellectual honesty in the preface to his *Natural History* and says that he will provide a list of works cited (*auctorum nomina praetexui. est enim benignum, ut arbitror, et plenum ingenui pudoris fateri per quos profeceris*, HN pr.21). But, unlike Vitruvius, Pliny provides both a massive bibliography of all his sources in Book 1 of the *Natural History* as well as extensive in-text citations. Pliny's goal in constantly referring to external sources is to confirm his own authority, not as technical expert like Vitruvius, but as a pre-eminent compiler of information, a fact he emphasises in citing the sheer numbers involved in the *Natural History*'s production: 2,000 volumes read and 20,000 facts extracted (pr.17).

Instead of citing architectural treatises, Vitruvius more often cites non-architectural personalities in relation to gnomic or philosophical ideas only

---

[20] Cf. Doody in this volume.

loosely connected with the technical content that forms the bulk of the *De architectura*. These personalities are wide-ranging. Philosophers, including certain of the Seven Sages (8.pr.1), loom largest of all, followed by a host of politicians, authors of non-architectural genres, as well as random characters like Egyptian priests and Etruscan liver readers.[21] In fact, a conspicuous feature of Vitruvius' ten prefaces is that they present moral lessons that are only tangentially related to architecture using stories about famous individuals from antiquity.

In the preface to Book 2, for example, Alexander the Great's semi-comical architect, Dinocrates – this one architect is admittedly included, but he is presented as somewhat of a buffoon – is used to demonstrate the need for practicality; in Book 3, Socrates and several visual artists show the fickleness of fame; in Book 5, Pythagoreanism reveals a method for admirably succinct writing.[22] The effect is, in the absence of references to the discipline of architecture, that Vitruvius' worldview becomes associated more with figures and schools from a broad intellectual and artistic history and does not get bogged down in the particulars of architectural debates, past or present.

When Vitruvius cites authors within his text, he occasionally refers to non-specific 'teachers' whose instructions have informed various aspects of his treatise. The term used is generally *praeceptor*. For example, he writes: *nos autem exponimus, uti ordo postulat, quemadmodum a praeceptoribus accepimus...* ('Next, we will present the matter, as our plan dictates, in the manner we received it from our teachers...', 4.3.3) and he goes on to provide a detailed discussion of the Doric order. And, later, he cites his teachers as the source of his information on the fundamentals of astronomy: *uti a praeceptoribus accepi, exposui* ('I have presented it just as I have received it from my teachers', 9.1.16). Variations on *quemadmodum/ut a praeceptoribus accepi/accepimus* appear several times in the *De architectura* (cf. 2.10.3, 5.pr.2, 6.pr.4, 7.pr.2, 9.1.16, 10.11.2, 10.13.8).

Vitruvius' acknowledgements of his intellectual debt to his teachers are usually made in passing. They are unspecific as to their identity and usually appear at the beginning or end of a major section. This, in addition to the general vagueness of the identity of these *praeceptores*, suggests that they are not meant to function as actual citations of practical use to the reader. Instead, they allow Vitruvius to highlight that he is part of an intellectual

---

[21] Philosophers appear at 2.2.1, 3.1.5, 6.pr.3, as well as several other passages; Euripides is cited at 9.1.13; priests and haruspices appear at 8.pr.4 and 1.7.1, respectively.
[22] Cf. Cuomo 2011: 318 on how rare it is for Vitruvius to associate himself with an actual architect and McEwen 2003: 92–129 on Vitruvius' ambiguous opinion of Dinocrates.

lineage and acknowledge his intellectual debt while not having to cite specific personalities whose names, texts and ideas might undermine the impression of the comprehensiveness and finality of the *De architectura*.

Furthermore, by developing his relationship with these teachers as a theme, Vitruvius is able to attribute to himself the moral values that are inherent in the ideal teacher-student relationship as defined by him.[23] Vitruvius discusses his own education in the preface to Book 6. Vitruvius says that his teachers – following Athenian methods – provided him with an education that was practical, directed towards a particular skill, but also complemented by a general knowledge of the branches of learning: *cum ergo et parentium cura et praeceptorum doctrinis auctas haberem copias disciplinarum, philologis et philotechinis rebus commentariorumque scripturis me delectans eas possessiones animo paravi, e quibus haec est fructuum summa...* ('Therefore, when, due to the attention of my parents and the instruction of my teachers, I obtained a plenitude of knowledge, and due to my love of literature and technical subjects, and of writing treatises, I have furnished my spirit with many stores, of which the most nourishing is this...', 6.pr.4). Vitruvius' association of his teachers with both a practical and liberal-arts education reinforces his belief, discussed above, that an architect must have mastery over the theoretical and practical dimensions of his trade and must have at least some knowledge of the wide range of disciplines with which it is interconnected, thus bulwarking his intellectual authority (1.1.1).

In another passage, Vitruvius associates his teachers with unimpeachable moral values. For example, he observes that his distaste for wealth is something he learned from his instructors: *ceteri architecti rogant et ambiunt, ut architectent; mihi autem a praeceptoribus est traditum: rogatum, non rogantem oportere suscipere curam, quod ingenuus color movetur pudore petendo rem suspiciosam* ('Other architects troll around and beg to practise architecture; however, it was taught to me by my teachers to take up a job by being asked, not by asking – something that would make a gentleman blush for shame on account of seeking something suspicious', 6.pr.5). By citing teachers as one of his sources, and then nuancing the way in which these teachers are understood by the reader, Vitruvius gives specific connotations to his citations of them. His teachers taught him not only how to be an architect, but also how to be a free-born gentleman (*ingenuus*). Thus, when he cites his

---

[23] On Vitruvius' attempts to redefine the figure of the architect and, in particular, his emphasis on his public and moral dimensions, see Cuomo 2011: 327; cf. Gros 1994: 89–90, who argues that while Vitruvius is engaged in self-promotion, he conceives of himself as an assisting figure.

teachers, Vitruvius is able to direct his readers' attention away from potential debates about a particular topic and curtail their desire to read outside the *De architectura*, as well as reinforce his own moral character at the same time.

Another group, cited moderately often, is the *maiores*, translatable as 'predecessors' or 'forebears'. The *maiores* are previous writers on architecture, Greek and Roman, of whom Vitruvius approves. In citing this group, Vitruvius is able to refer to all past writers on his subject at once, simplifying a potentially complex intellectual history. His attitude towards these predecessors is, unsurprisingly, usually one of admiration, and he recommends that architects and Romans more generally follow their lead. For example, Vitruvius cites the *maiores* as a source of knowledge pertaining to city-planning (5.9.8) and the construction of the hodometer (10.9.1). But, like the *praeceptores*, the *maiores* are also a source for correct intellectual and cultural practices and values. For example, Vitruvius states that, in writing the *De architectura*, he will mimic the *maiores* in their adoption of short, clearly divided treatises (5.pr.5). Vitruvius also recommends that the Romans return to an ancestral sheep sacrifice as practised by the *maiores* (1.4.9) and, in Ephesus, he cites a good law passed down by their ancestors regarding the payment of architects (10.pr.1; cf. 9.pr.1 for Greek *maiores*).[24]

Vitruvius' Self-Presentation as Editor

It is also worth considering how Vitruvius relates personally to his sources as editor of the *De architectura*. Rather than present himself as a detached librarian whose task it is to collate and, to some extent, organise numerous external sources and debates relating to architecture, we find that Vitruvius comes across as confidently situated at the centre of the data that makes up the *De architectura*, functioning as a synthesising figure who has mastered and unified his sources within the space of his treatise. This active role grants Vitruvius authority as a master of literature on architecture, but it also helps smooth over debates that could complicate the information found within his text, thus giving it the impression of greater coherence and self-sufficiency.

One of the salutary tales Vitruvius tells at the start of each of his books pertains to the proper handling of pre-existing knowledge. It appears in the

---

[24] Vitruvius also sometimes cites the *veteres architecti*. For example, at 5.3.8 he refers to them as a source of knowledge on harmonics and hence theatre construction. And Vitruvius declares his general desire to imitate 'ancient thinking' (*veterem... rationem*) when selecting a city site (1.4.9).

preface to Book 7 and looks ahead to Vitruvius' presentation of his own bibliography. This preface is relatively unique.[25] It is the longest and rhetorically most ornate of the ten, and also appears at a significant juncture, namely, the end of the last book on *aedificatio* or construction of buildings proper. As such, it is an appropriate place to introduce a theme of special importance.

In this story, six 'well-read judges' (*iudices litterati*) are said to have been assembled to oversee a poetry contest. The bibliophile Aristophanes of Byzantium was selected as the seventh. And, while the six judges chose as the winner the poem that had most pleased the audience, Aristophanes chose the least pleasing poem, having detected that all but one of the poets in the contest were guilty of plagiarism (*furta... scripta*, 7.pr.7). Aristophanes was able to detect this crime due to his deep reading, amounting to the seeming memorisation of all the contents of the Library at Pergamum: *tunc ei dixerunt esse quendam Aristophanen, qui summo studio summaque diligentia cotidie omnes libros ex ordine perlegeret* ('Then [the librarians] noted that there was a certain Aristophanes who, with the greatest zeal and stamina, daily read through all of the books in order', 7.pr.5).

This story, about a man whose extensive reading allows him to detect plagiarism, functions as a moral tale and signals Vitruvius' commitment to acknowledging his sources, something he does later on in the same preface by including a bibliography (7.pr.10).[26] But Aristophanes also represents Vitruvius' vision of the ideal handler of knowledge in a broader sense. His ability to evaluate contemporary poetry reflects Vitruvius' stated intention for the *De architectura*, namely, that his addressee Octavian/Augustus will become better able to evaluate good and bad architecture by reading Vitruvius' treatise: ... *ut eas adtendens et ante facta et futura qualia sint opera, per te posses nota habere* ('so that, by looking at them, you would know well the quality both of those projects that have already been completed and those that will come into being in the future', 1.pr.3). Elements of Aristophanes' expertise, and of the story more generally, also resemble Vitruvius' characterisation of *auctoritas*. While the majority of judges granted the highest prize to the poem that most pleased the audience – recall Vitruvius' complaint that authority is usually misattributed – Aristophanes was able

---

[25] Cf. Cam *et al.* 1995: xv–xxv.
[26] Authors of information-centric texts often accuse their rivals – who typically go unnamed – of having committed plagiarism while avowing that they have not. Pliny the Elder, whose self-defence is supported by the inclusion of an extensive combined table of contents and list of works cited that comprises all of Book 1, is a well-known example (*HN* pr.21–2). For a concise study of the treatment of external sources in the prefaces to Greek and Roman scientific treatises, see Alexander 1993: 79–80. On plagiarism as a prefatory theme, in particular, see McGill 2012: 31–73.

to assess the poems on the basis of their inherent, internal merit due to his intelligence and wide reading (cf. 3.pr.).

The acknowledgement of an item's excellence – that is, the conferral of authority upon it – thus depends on the presence of an equally outstanding evaluator. Furthermore, Aristophanes' ability to serve as an authoritative judge rests on his knowledge of and control over an enormous corpus of literature – literally an entire library's worth – contained in his astonishing memory: *admirante populo et rege dubitante, fretus memoriae certis armariis infinita volumina eduxit et ea cum recitatis conferendo coegit ipsos furatos de se confiteri* ('When the populace marveled [at Aristophanes' accusation], and the king remained doubtful, trusting in his memory, Aristophanes pulled endless volumes out of certain cabinets, and comparing their works with the ones he had recited he compelled the plagiarists to confess their crime', 7.pr.7). While Vitruvius accepts that it is commendable that the Attalids and Ptolemies built great libraries at Pergamum and Alexandria, such libraries' contents must be internalised by figures like Aristophanes, who are capable of directing the knowledge contained in these libraries to their own purposes.

The implication is that Vitruvius, even though he has received information from a number of different sources as indicated by his bibliography, has interacted with these works in an organic and personal way. He resembles Aristophanes insofar as he, too, possesses a library's worth of knowledge, but also because he has internalised it. This information is not only at hand, but in his head.

Just as memory is presented in the Aristophanes episode as one aspect of the proper collection and treatment of pre-existing information, memory is cited as a source of the knowledge underlying the composition of the *De architectura*.[27] This is most often expressed using the verb *succurro* (meaning 'to come to mind' or 'occur to') and such statements occur at several points throughout the work. A typical example appears at the end of Book 5 on public spaces: *quae necessaria ad utilitatem in civitatibus publicorum*

---

[27] Vitruvius synthesises past treatises on architecture into something new through their retention in his memory and, for the strategic purposes outlined in this chapter, then fails to mention them in his text. His work is the only one that counts. By contrast, Pliny the Elder mourns the gradual loss of the 'accomplishments of the ancients' (*recte facta veterum*) and proposes his *Natural History* as a means of reviving them (25.1–2). Cf. Conte 1994, who observes Pliny's recognition of his distance from original research: '... [Pliny] has no intention of passing himself off as a hero of science. Heroes of this sort seem to belong to other times, by now past, and, if anything, Pliny wishes, like a good master of ceremonies, to introduce the larger public to the spectacle of their exploits' (70). This stands in stark contrast to Vitruvius, who presents himself as a (formerly) working architect and active producer of knowledge.

locorum succurrere mihi potuerunt, quemadmodum constituantur et perficiantur, in hoc volumine scripsi ('I have written down in this volume that which I could recall as being crucial to the functioning of public spaces in cities – in what manner they are constructed and brought to completion', 5.12.7).[28]

But the knowledge Vitruvius possesses cannot remain discrete and associated with particular individual authors and texts, in part because they would distract from the self-contained quality of the *De architectura*. Vitruvius thus employs a complementary metaphorical image, which appears just after the Aristophanes story, to emphasise that knowledge, once it has been consumed, must then be digested and synthesised. To demonstrate this, Vitruvius uses the culturally robust image of confluent streams (7.pr.10):

> *Ego vero, Caesar, neque alienis indicibus mutatis interposito nomine meo id profero corpus neque ullius cogitata vituperans institui ex eo me adprobare, sed omnibus scriptioribus infinitas ago gratias, quod egregiis ingeniorum sollertiis ex aevo conlatis abundantes alius alio genere copias praeparaverunt, unde nos uti fontibus haurientes aquam et ad propria proposita traducentes facundiores et expeditiores habemus ad scribendum facultates talibusque confidentes auctoribus audemus, institutiones novas comparare.*

> But in truth I, Caesar, have neither put forth a text with my name on it while removing the signs that it belongs to another, nor by maligning the thoughts of another have I attempted to gain esteem for myself. But I give infinite thanks to all authors, because, through the outstanding talent of their minds, they have prepared for us overflowing riches gathered from history, each author a different kind, from which we, as if drinking water from springs, and turning them to our particular projects, will possess a richer and more ready ability to write, and trusting in authors of this sort, will dare to gather together new teachings.

The image of Vitruvius' text as produced through a process of ingestion – drinking water from a variety of intellectual springs – enabling the creation of new material (*institutiones novas*) through a process of adaptation or translation (*traducentes*), recalls a broader literary discussion in the Greco-Roman world that Vitruvius seems to be participating in.[29]

---

[28] Lewis and Short, *Latin Dictionary* s.v. *succurro*, II.B. For other instances of citations using *succurrere* in the *De architectura*, see 7.5.8, 7.9.6, 7.14.3, 8.5.3. McEwen 2003 discusses memory as a wide-ranging theme in the *De architectura*, including its awareness of a collective historical memory and its own memorable or, rather, memorisable nature due to it being composed according to mathematical principles: see esp. 80–8.

[29] Vitruvius uses the drinking image a second, earlier time in relation to his architectural predecessor Hermogenes: *Quare videtur acuta magnaque sollertia effectus operum Hermogenis*

The use of a water source, in particular, to describe intellectual or literary borrowings is common in Latin literature. Plautus speaks, for example, of the font of the Muses (*fontem Pireneam, Aul.* 559). And it appears in a variety of disciplinary contexts, for instance, in connection with philosophy and the arts.[30] A more specialised account of drinking from a multiplicity of sources can be found in Cicero's *De oratore*, which addresses the nature of oratory as a unitary versus fragmented concept. Cicero says that the dull-witted are enticed by minute aspects of the subject, which he likens to 'rivulets' (*rivulos*), but are not aware of the single 'spring' (*fons*) from which oratory's sub-parts flow.

The crucial aspect of Vitruvius' image, which relates to his attitude towards the collection of knowledge, is his reliance on multiple sources and their subsequent digestion to produce new knowledge. Images of literary consumption and digestion likewise abound in Greek and Latin texts. The bee sipping nectar is used by Isocrates to describe the 'cultured life', where one gathers from the best of all things and leaves nothing untried (*Ad Demonicum* 52). Quintilian uses eating imagery in reference to reading the poets, representing a process of synthesis (*Inst.* 10.1.19):

> *Lectio libera est nec ut actionis impetus transcurrit, sed repetere saepius licet, sive dubites sive memoriae penitus adfigere velis. Repetamus autem et tractemus et, ut cibos mansos ac prope liquefactos demittimus quo facilius digerantur, ita lectio non cruda sed multa iteratione mollita et velut [ut] confecta memoriae imitationique tradatur.*

> Reading is an independent act nor does the speed of its performance outpace us, and you can go back over a passage repeatedly if you have questions or if you wish to lodge the material deeply in your memory. Let us then review our texts and ponder over them, just as we only swallow our food once it has been chewed almost liquefied, in order that it might be more easily digested. In the same way, let our reading be consumed for the sake of memory and imitation, not raw, but, so to speak, softened by much repetition and thus made suitable for us.

Finally, Seneca's *Letter* 84 ponders the nature of honey production as a metaphor for literary production where the crucial question is whether

---

fecisse reliquisseque fontes, unde posteri possent haurire disciplinarum rationes ('Wherefore it is evident that Hermogenes, due to his keen and robust skill, has had a great effect through his works and has left behind him fonts of knowledge, from which those who have come after him can imbibe the principles of the discipline', 3.3.9). This paean looks ahead, of course, to Vitruvius' later statement that, in composing his treatise, he has drunk from the springs of past writers, which would include, but is not limited to, Hermogenes.

[30] E.g. Cic. *De or.* 1.12; 1.42.

something new is produced through the consumption process (84.3–8). Both in Seneca and in Vitruvius, the act of synthesising previous knowledge creates a single body of authoritative knowledge that has, moreover, been internalised and is therefore uniquely associated with its possessor. This negates potential differences in opinion between Vitruvius and his sources and helps explain Vitruvius' simultaneous but seemingly paradoxical impetus to acknowledge his debt to the historical literature on architecture, while, within the body of his treatise, avoiding the citation of particular authors.

Taking this a step further, Vitruvius associates the act of collating and synthesising previously disparate entities into a unified whole with the supremely authoritative figure of Nature and, in particular, the tendency of Nature to cohere disparate atoms in a rational and providential manner.[31] This can be seen in the preface to Book 4 (4.pr.1):

> *cum animadvertissem, imperator, plures de architectura praecepta voluminaque commentariorum non ordinita sed incepta, uti particulas errabundas reliquisse, dignam et utilissimam rem putavi antea disciplinae corpus ad perfectam ordinationem perducere et praescriptas in singulis voluminibus singulorum generum qualitates explicare. Itaque, Caesar...*

> When I had realized, Emperor, that many had left behind maxims and volumes of commentaries on architecture that were in no way organized but merely started, as if randomly wandering atoms, I thought it would be a worthwhile and exceedingly useful thing, first, to go over the field in its entirety and then, in the individual volumes, to explain the qualities of its individual parts. And so, Caesar...

This statement relates to Vitruvius' views on the history of treatises on architecture, a connection Vitruvius encourages us to make on the basis of an allusion that appears later in the treatise. In the preface to Book 7, Vitruvius declares that he has organised the 'several volumes published by the Greeks' (*ab Graecis volumina plura edita*) into a 'single body' of knowledge (*in unum coegi corpus*, 7.pr.10). This passage looks back to the one cited above, in which Vitruvius likens the previous state of knowledge on architecture and, in particular, previous architectural treatises (*praecepta voluminaque commentariorum*) to the primordial elements, with the terms *plures/plura* and *volumina* appearing in both. Vitruvius seems to want the reader to see his activities as collator-editor as similar to that of Nature as arranger of wandering atoms.

---

[31] For further discussion of that theme, see esp. Courrént 2011 and McEwen 2003: esp. 47–71 and 142–8.

An additional self-comparison to Nature is made in the preface to Book 5, where Vitruvius describes the authors upon which he draws using atomic language when he refers to the 'far wandering writings of the authorities' (*praeceptorum late evagantes scripturae*, 5.pr.2). Vitruvius, in contrast, will write his text in a more collected and composed fashion.

The term *particula* has a number of potential valences. It can refer to a grammatical clause or particle.[32] And, in some sense, Vitruvius must be relying on the multiple significations of this word when he applies it in a literary context. The most obvious reference, however, especially given its appearance within an explicit simile, is to the elemental particles of the atomist philosophical tradition. It is not surprising that Vitruvius would represent the compilatory aspect of his editorial project in terms of a union of atoms. Democritus and other atomists including Lucretius are explicitly named in the *De architectura* and provide the foundation for Vitruvius' anthropology in Book 2.[33] Vitruvius also borrows heavily from Lucretius' *De rerum natura* in the composition of the *De architectura*. The paean to Octavian/Augustus in the preface to Book 1 contains a large number of parallels with the so-called *laudes Epicuri* of *De rerum natura* 1, 3 and 5, and its philosophical exempla and vocabulary reappear in Vitruvius as well.[34]

Tying atoms, Nature and literature together, it is notable that, as part of his discourse on the atomic foundations of the universe, Vitruvius refers to the origins of human language as a gradual process that likewise parallels atomic coalescence. According to Vitruvius, once human communities were established, people first communicated by signs, then words and full speeches (2.1.1).[35] In this sense, Vitruvius is mimicking the gathering and organising activities that occur in and by Nature at the macro- and micro-levels. As editor, he, too, is a collector and organiser, not of actual atoms, but of the 'wandering particles' of previous treatises on architecture, which he has made into a unified treatise that, as we have seen, largely lacks any sense of fragmentariness due to relativism or unresolved controversies in the field.

---

[32] Lewis and Short, *Latin Dictionary* s.v. *particula* II. A. and B.
[33] Democritus appears at 2.2.1 (twice); 7.pr.2; 7.pr.11; 9.pr.2; 9.pr.14; 9.5.4; 9.6.3; Lucretius appears at 9.pr.17.
[34] Lucretius uses the term *particula* as one of his many terms for the atoms to refer to the basic building blocks of Nature and more specifically in the description of the atomic basis of smells, the perception of colour and the experience of cold. E.g. Lucr. 2.833 (for *particulae* as building-blocks of Nature); 4.692 (as basis of smell); 2.826 (colour); 4.259 (cold).
[35] See Cole 1990 on Democritus as a source for Vitruvius' anthropology and the theme of human evolution in atomist philosophy generally.

Ancient authors differed in the extent to which they used Nature as a guide for the composition of their text. Varro states that the organisation of his text will mirror Nature: 'Therefore I will first demonstrate that which it is right to leave out from our discussion; then I will speak about [topics worthy of mention] based on their natural divisions' (*Itaque prius ostendam, quae secerni oporteat ab ea, tum de his rebus dicam sequens naturales divisiones*, *Rust.* 1.1.11). The idea of treating a subject according to natural divisions speaks to a sense of content that is static – as the subject possesses a Platonic reality, so the book will reflect this reality. This notion of a subject possessing its own, 'natural' order is also present in Columella (*Rust.* 1.pr.33):

> *De cuius universitate nihil attinet plura nunc disserere, quandoquidem cunctae partes eius destinatis aliquot voluminibus explicandae sunt, quas ordine suo tunc demum persequar, cum praefatus fuero, quae reor ad universam disciplinam maxime pertinere.*

> About the entirety [of the subject] there is no opportunity to treat it in detail now, since all of its parts will be explained in the individual volumes dedicated to each subject, which I will go through, each in its own order, after I have treated as a preface the things that I believe are most pertinent to the subject as a whole.

This passage reiterates Columella's goal of providing a comprehensive account of his subject and claims that the book will be a mirror of the order inherent in the subject, which will appear in its own order (*suo ordine*).[36] And yet Pliny the Elder asserts that humans are fundamentally incapable of achieving total comprehension of the universe. This is particularly clear in his definition of the *mundus*, which, like Nature herself, is synonymous with totality (*HN* 2.1–2):

> *Mundum et hoc quodcumque nomine alio caelum appellare libuit, cuius circumflexu degunt cuncta, numen esse credi par est, aeternum, inmensum, neque genitum neque interiturum umquam. huius extera indagare nec interest hominum nec capit humanae coniectura mentis. sacer est, aeternus, immensus, totus in toto, immo vero ipse totum, infinitus ac finito similis, omnium rerum certus et similis incerto, extra intra cuncta conplexus in se, idemque rerum naturae opus et rerum ipsa natura.*

> The world and whatever it is that we have, by another name, opted to call the heavens, within whose arch all things exist, are rightly thought to be a god, eternal, immeasurable, never born and never dying. That which

---

[36] Cf. Vitr. *De arch.* 7.pr.18.

lies outside it is not appropriate for men to investigate, nor can the speculations of the human mind grasp it. It is sacred, eternal, immeasurable, whole unto itself because it is the whole, infinite yet also resembling the finite, certain in every respect and yet resembling the uncertain, embracing all things, within and without, inside itself, both the work of Nature and Nature herself.

This passage is relevant to Pliny's composition given his stated task of investigating all aspects of Nature and the pre-eminent position of this quotation at the start of Book 2 (the beginning of Pliny's catalogue proper). The *mundus* is here equated with Nature (*natura*). *Mundus/natura* are also synonymous with the boundaries of the universe, which human intelligence is incapable of circumscribing. Unlike Vitruvius, who is acting in Nature's image, Pliny presents himself as a human butting up against the infinitude of Nature, claiming simply to honour Nature through the compilation of vast quantities of disparate knowledge. Vitruvius is Nature's imitator, taking it as an editorial model as he composes his perfect text.

Conclusions

In an analysis of the history of encyclopedic texts, Roland Barthes has observed that, while early knowledge-books claimed to contain all that could be known, over time such claims decreased in frequency as humans became more aware of the vast scope of knowledge that was out there, their epistemological limitations and the limitations of the technology of the book as a container of knowledge.[37]

This is likely generally true in terms of overall trajectory. It is no longer possible for a would-be Vincent de Beauvais to write a textual 'mirror' of the universe, that is, his monumental *Speculum maius*. Nevertheless, knowledge-books with more and less self-confidence are found in all historical periods. Artemidorus' compendium of dreams and their outcomes, the *Oneirocritica*, written in the second century CE, is openly critical of the encyclopedic project; the contemporary *World Book Encyclopedia* is remarkably self-assured given the ever-expanding scope of human knowledge and the numerous possible methods we possess for its organisation and transmission.

Using Vitruvius' own claims to have written an authoritative knowledge-book has led us on a tour of several aspects of the *De architectura* – his use

---

[37] Barthes 1987: 93–5.

of in-text citation and bibliography, metaphors used in his self-presentation as editor – that relate directly to Vitruvius' project as a compiler of information, but nevertheless are understudied or else are usually looked at in relation to the historical context of the *De architectura*.

Indeed, the reader will have noticed that relatively little consideration has been given to historical figures and events in this chapter. This was purposeful. As is now well known, classicists until recently tended to study technical and compilatory texts not as literature, but as sources of information about some other subject of primary interest: for example, Vitruvius as a source on machines used in war. The next phase in the scholarship on these texts has been an emphasis on historical context and specifically the dialogic relationship between this seemingly innocuous genre and their political milieu: for example, Vitruvius as a partner or competitor with Octavian/Augustus and his imperial regime. This has been a fruitful approach for several reasons, not least because it encourages closer and more holistic readings of the texts themselves and makes them meaningful to a broader audience.

But, in this chapter, I hope to have contributed in a small way to our understanding of how technical and compilatory literature is read by encouraging even closer engagement with the texts themselves as bodies of knowledge, as well as their literary milieux. By focusing on the guiding questions outlined in the introduction, questions about how authors like Vitruvius define their works as compendia of information, it is my hope that we have found and will continue to find our way towards even more focused and engaged readings of the *De architectura* and texts like it, as well as contribute to writing the long history of information science.

6 | Fragile Expertise and the Authority of the Past: The 'Roman Art of War'

MARCO FORMISANO

Quis enim caneret bella melius quam qui sic gerit?
(Quintilian, *Institutio oratoria* 10.1.91)

Introduction

The ancient art of war is one of the most discussed subjects both within the history of the reception of classical antiquity and in current popular perceptions of ancient civilizations. The association between antiquity and war has been and still is very widespread in historical novels, movies and popular culture on the one hand, and in scholarship on the other.[1] This interest in ancient warfare in its historical, material and strategic aspects has also been nourished by the growing discussion on warfare in other contexts, in particular the ancient Chinese art of war. But despite all of the attention directed to this field of knowledge, relatively little attention has been directed to the ancient 'art of war' as a literary genre. Indeed, this term is usually used in a vague and general way to refer not only to knowledge of various aspects related to war, such as fortification, strategy, tactics, engineering etc., but also to texts. It is in this last sense – dating back to the Renaissance, when an astonishing number of works on warfare were written in Italy and elsewhere in Europe – that I use the phrase 'art of war' in this chapter.[2]

While war constitutes the main topic of so many ancient texts, both in poetry and prose, it is safe to say that in two kinds of text in particular it is prominently represented as a field of knowledge which can be learned and taught: historiography and technical treatises. Since the Hellenistic military revolution, as it has been described by Serafina Cuomo, war was increasingly associated with technical knowledge and expertise, and this development

---

[1] It would be impossible to quote here all relevant bibliography; see volume 2 of the *Cambridge History of Greek and Roman Warfare* (Sabin *et al.* 2007) for a systematic treatment of warfare in various periods of ancient history.
[2] See Formisano 2014 for further discussion. Even the just quoted *Cambridge History of Greek and Roman Warfare* does not contain any treatment of ancient military treatises in their own terms, but only as (questionable) sources of historical reconstruction.

consequently modified mental attitudes towards war.[3] From this perspective, the Romans in particular have been seen as the perfect embodiment of this new culture of war, since they erected a system of warfare based on both military discipline and mastery of technological knowledge, as was noted centuries ago by that admirer of Roman politics, Polybius.[4]

Given the importance of war within ancient culture, an obvious question is how authority is constructed and represented within the texts. One might expect that texts dealing with war as a teachable and learnable field of knowledge, one that was so vital for Roman identity and expansionistic politics, would have a clear connection to this mental attitude and that they would constitute a homogeneous corpus containing not only a body of rules applicable in all possible situations, but also sophisticated reflections on what war is: in short, a textual corpus of the 'Roman tacticians and strategists'. But nothing like that exists. Instead, we are confronted with a relatively small number of works which are not consistent with each other. As Maurice Lenoir observes, the number of military treatises written in Latin between the second and fourth centuries AD, as far as we can reconstruct them, is surprisingly low: Cato, Celsus, Frontinus, Vegetius.[5] Of these, only Vegetius' *Epitoma rei militaris* survives, and all our information about the Roman art of war as a literary genre is to be found in this text.[6] If Vegetius' *Epitoma* had not survived, we would know almost nothing about the others. More interestingly, these authors do not seem to belong to the same tradition and their perspectives on the art of war vary according to the status of the authors and the nature of their texts. Cato writes in order to instruct his son Marcus, while Celsus' treatise was part of a larger encyclopaedic project, also including agriculture, medicine and rhetoric, so that it is just as improbable that he was an experienced soldier as that he was a practising physician.[7] Frontinus was an imperial functionary with several fields of competence and Vegetius was a high-ranking official writing from the point of view of a Roman administrator.

Paradoxically enough, the situation looks better for the Greek military authors who produced their texts under the Roman Empire: Onasander, Arrian, Aelian and Polyaenus. All of these writers dedicated their work to

---

[3] Cuomo 2007b: 73.   [4] See Cuomo 2007b: 73–5.

[5] Lenoir 1996: 78–9. There is also Paternus, but his treatise dealt with military law rather than with war conduct and strategy. See also Giuffré 1974.

[6] Veg. *Mil.* 1.8 and 2.3. Frontinus' *Strategemata* has survived, but this text has the nature of a catalogue of examples rather than being a treatise or art of war; Frontinus' earlier *De re militari* has not survived: see A. König in this volume for further discussion.

[7] See Quintilian 12.11.24 and Serbat 2003: xii.

Roman rulers in order to help them to preserve and increase their military hegemony. Implicitly but clearly, they present the military skills of the Greeks as an ideal model for the Romans to follow in their military enterprises.

In this chapter I will focus on the Roman art of war from a different angle. In the first part I briefly discuss what in my opinion is the conceptual core of the ancient discourse of war, the tension between theory and practice. Here I refer to three quite diverse texts – by Sallust, Cicero and Ovid – in which this tension is thematised. The second part is dedicated to two important military texts, the *Strategikos* by the early imperial Greek writer Onasander and the *Epitoma rei militari* by the late Roman Vegetius. These texts, though having different designs and serving different purposes, are exemplary for the entire genre. It is my argument that precisely the tension between theory and practice, which is inherent in any discourse of war, ancient or modern, destabilises the very authority of those who write. The kind of expertise that emerges from these two different examples ends up being far removed from the technical subject matter and instead shows itself to be part of literary discourse, which characteristically places reflection on textuality at its centre.

Deeds or Words? Cicero, Sallust and Ovid

An anecdote, precisely by virtue of its exemplary quality, is often the best way to illustrate theoretical questions. In the second book of the dialogue *De oratore*, Cicero depicts a situation and describes several characters who play no small role in the construction of the art of war and who are quite frequently cited and discussed as models in Renaissance works. Before I discuss one passage in particular, it may be useful to note that Cicero's text in many ways revolves precisely around the relationship between theory and practice in oratory. The dialogue is a very complex theoretical work, and its complexity is increased by the dialogical form itself. Every person in the dialogue plays a role which is contradicted by someone else, so that any assertion or claim is relativized by the assertions and claims of others.[8]

In Book Two, Antonius, the apologist of oratorical practice, argues for practical experience and against a purely theoretical approach in the

---

[8] See the commentary by Leeman *et al.* 1981–96, Wisse *et al.* 2008 and Fantham 2004. Surprisingly enough, the dialogic dimension of *De oratore* does not seem to have been given any special attention by later Roman readers, who had a tendency to attribute to Cicero whatever the various personae of the fiction say. Aude Doody in this volume points out an analogous situation for the reception of Varro's *De re rustica*.

education and training of the orator (from 2.64 onwards); in this he is opposed to the main character Crassus, who emphasises the role of theoretical preparation (philosophy) for the ideal orator. Catulus, a minor persona of the dialogical fiction, asks Antonius to illustrate methods (*rationes*) and rules (*praecepta*) with which the orator can successfully achieve his aims. Catulus, however, asks this not in order to practise these rules himself, but purely for the sake of knowledge (*tantum cognoscendi studium*). At this point Catulus relates the famous meeting of Phormio the thinker and Hannibal the warrior:

> *ut Peripateticus ille dicitur Phormio, cum Hannibal Carthagine expulsus Ephesum ad Antiochum venisset exul proque eo, quod eius nomen erat magna apud omnis gloria, invitatus esset ab hospitibus suis, ut eum quem dixi, si vellet, audiret; cumque is se non nolle dixisset, locutus esse dicitur homo copiosus aliquot horas de imperatoris officio et de omni re militari. tum cum ceteri, qui illum audierant, vehementer essent delectati, quaerebant ab Hannibale, quidnam ipse de illo philosopho iudicaret. hic Poenus non optime Graece, sed tamen libere respondisse fertur multos se deliros senes saepe vidisse, sed qui magis quam Phormio deliraret vidisse neminem.*

> And I don't need some Greek teacher who will reel off for me the common precepts, without ever having set eyes on the forum, and without ever having seen a single trial – as in the story of Hannibal and Phormio the Peripatetic. Hannibal banished from Carthage, had come in exile to King Antiochus in Ephesus and, in keeping with the great glory that his name enjoyed among all, had been invited by his hosts to hear a lecture by Phormio, if he wished. When Hannibal indicated that he was not averse to the suggestion, the eloquent fellow is said to have talked for several hours on the duties of a general and on every aspect of military affairs (*de imperatoris officio et de omni re militari*). The rest of the audience, having thoroughly enjoyed themselves, then asked Hannibal what he thought of this philosopher. The Carthaginian is reported to have replied, not in the best of Greek (*non optime Graece*), but nevertheless frankly, that he had often seen many raving old men, but none who raved more madly than Phormio. (Cic, *De or.* 2.75)[9]

Commenting on the episode, Catulus confirms Hannibal's words:

> *quid enim aut adrogantius aut loquacius fieri potuit quam Hannibali, qui tot annis de imperio cum populo Romano omnium gentium victore certasset, Graecum hominem, qui numquam hostem, numquam castra vidisset, numquam denique minimam partem ullius publici muneris attigisset, praecepta de re militari dare?*

---

[9] The English translation of *De oratore* is by May and Wisse 2001.

> For could there be a worse case of loquacious arrogance than for a Greek, who had never laid eyes on an enemy, who had never seen a camp, and who had never even played the slightest role in any public function, to give precepts on military affairs to Hannibal, who for so many years had vied for supreme power with the Roman people, the conquerors of all nations? (2.76)

The philosopher Phormio, otherwise unknown, is depicted as the Greek par excellence: being free from any civic duty he has time enough for *otium*, discussion of which represents one of most relevant topics of *De oratore*. Moreover, he perfectly embodies what Crassus has previously described as the most typical Greek vice: being *ineptus*, i.e. *non aptus* (Cic, *De or*. 2.17–18). In this passage Crassus explains that the very fact that Greek language does not have a word for that concept should be considered a proof of the 'ineptitude' of the Greeks.

What obviously bothers Hannibal is precisely the language spoken by Phormio: Greek is the language of philosophy, not of warriors. How can a person who never went to war or fought on the field have the authority to teach about war, in the presence of a great general at that? Hannibal's train of thought is typical of every art of war, ancient and modern, where the question ultimately asked always is this: how can a book possibly teach one how to conduct war?

Although in his dialogue Cicero often puts oratory in relationship to the art of war, and although I would argue that this anecdote illustrates a very important aspect of the construction of knowledge in the Roman context, this passage has not received much attention; even the commentators seem to sidestep it, simply reminding their readers that Phormio is otherwise unknown.[10] And yet the anecdote was very well known to Renaissance authors of the art of war, to the extent that one can find it quoted by almost all of them. What can be inferred from this passage is extremely interesting for the construction of authority within the art of war. The soldier Hannibal, a great strategist and invincible commander, is confronted with another way of conceiving of his profession, i.e. with a theoretical approach to war, a philosophical abstraction which produces rules and general criteria. In fact, even though it is not said, we can easily guess what irritated Hannibal so much about Phormio's performance: his tendency to construct general rules and to give theoretical standards. The art of war of the warrior does not correspond in this case with that of the philosopher. Hannibal, with his broken Greek, is contrasted with his philosophical *alter ego*.

---

[10] See Leeman *et al.* 1981–96: Vol. 2, *ad loc.*

Besides this episode there is another character who is frequently cited in Renaissance treatises on war, namely Sallust's Marius from the *Bellum Iugurthinum*. In his famous speech to the assembly (*contio*) in chapter 85, Marius presents himself as a proud soldier and as a *homo novus* who is contemptuous of Roman nobility, in particular for its inability in action. He says:

> *conparate nunc, Quirites, cum illorum superbia me hominem novom. quae illi audire aut legere solent, eorum partem vidi, alia egomet gessi; quae illi litteris, ea ego militando didici. nunc vos existumate facta an dicta pluris sint. contemnunt novitatem meam, ego illorum ignaviam; mihi fortuna, illis probra obiectantur.*

> Compare me now, fellow citizens, a 'new man', with those haughty nobles. What they know from hearsay and reading, I have either seen with my own eyes or done with my own hands. What they have learned from books I have learned by service in the field; think now for yourselves whether words or deeds are worth more. They scorn my humble birth, I their worthlessness; I am taunted with my lot in life, they with their infamies. (Sall. *Iug.* 85.13–14)

From the warrior's perspective, action is superior to words, *facere to dicere*. This distinction gives Marius an advantage over the nobles and puts him on an equal footing with their ancestors. Through action he attains the higher social level of his opponents, because he addresses the very origin of valour: 'But if they rightly look down on me, let them also look down on their own forefathers, whose nobility began, as did my own, in manly deeds' (*quod si iure me despiciunt, faciant item maioribus suis quibus, uti mihi, ex virtute nobilitas cepit*) (Sall. *Iug.* 85.17).

Marius is also emphasising that his education is not based on literary study, but on the practice of war. He does not set much store in his noble opponents' rhetorically sophisticated speech, nor he is able to speak and read Greek:

> *equidem non ignoro, si iam mihi respondere velint, abunde illis facundam et conpositam orationem fore… non sunt conposita verba mea: parvi id facio. ipsa se virtus satis ostendit; illis artificio opus est, ut turpia facta oratione tegant. neque litteras Graecas didici.*

> I am of course well aware that if they should deign to reply to me, their language would be abundantly eloquent and elaborate… My words are not well chosen; I care little for that. Merit shows well enough in itself. It is they who have need of art, to gloss over their shameful acts with words. Nor have I studied Greek letters. (85.26 and 31–2)

Action and military practice have made him different from the nobles. His body itself in effect proves it; it is evidence of his valour:

> *non possum fidei causa imagines neque triumphos aut consulatus maiorum meorum ostentare; at si res postulet hastas vexillum phaleras, alia militaria dona, praeterea cicatrices advorso corpore. hae sunt meae imagines, haec nobilitas, non hereditate relicta, ut illa illis, sed quae ego meis plurimis laboribus et periculis quaesivi.*

> I cannot, to justify your confidence, display family portraits or the triumphs and consulships of my forefathers; but if occasion requires, I can show spears, a banner, trappings and other military prizes, as well as scars on my breast. These are my portraits, these my patent of nobility, not left me in heritance as theirs were, but won by my own innumerable efforts and perils. (85, 29–30)

The glorious past of the noble stock is here something for displaying in portraits, parades, rhetorical skill and the Greek language. In fact, a talent with words seems here to be used for masking lies. To these signs of social distinction Marius contrasts his own countersigns, namely weapons and standards, but also the scars on his body: these features are based on hard facts and are concrete evidence of his valour. What Sallust via Marius introduces is a profound epistemic difference between the two parties: action is not a legacy and consequently it cannot be transmitted like blood. The qualities which derive from risks and strenuous action belong exclusively to the person who ventures them and cannot ultimately serve to glorify other persons, not even family members.

But besides the historical and social context of Marius' speech there is something more interesting for our discussion of the construction of authority in the art of war. Marius enacted heroic deeds which others can only read in books or other reports on their ancestors. For these *literati*, actions are something to write and read about. Writing and reading thus grant them the same authority as that of the actual agents, and perhaps an even greater one. As we will shortly see, this thought constitutes the conceptual basis of Onasander and Vegetius as for any other writer *de re militari*.

Before coming to the discussion of these authors, I will briefly refer to another famous model of the 'struggle between weapons and literature', to use an anachronistic expression which was, again, very much in vogue in the Renaissance. While Cicero refers to an imaginary anecdote and Sallust to the very historical figure of Marius, in the following passage we have to do with a well-known mythical event. In the thirteenth book of the *Metamorphoses*,

Ovid stages with his typical irony the dispute between Ajax and Ulysses over Achilles' weapons, the famous *iudicium armorum*.[11] Both are required to deliver a speech before Agamemnon and the assembly of the Greek leaders.

Ajax represents himself as a soldier, one who fights the enemy with his hands and does not confine himself to words:

> *tutius est igitur fictis contendere verbis,*
> *quam pugnare manu! sed nec mihi dicere promptum,*
> *nec facere est isti, quantumque ego Marte feroci*
> *inque acie valeo, tantum valet iste loquendo.*

> It is safer, then, to fight with lying words than with hands. But I am not prompt to speak, as he is not to act; and I am as much his master in the fierce conflict of the battle line as he is mine in talk. (Ov. *Met.* 13.9–12)

Although it is a quite different context, we can recognise in this quarrel the same distinction between words and action that Sallust's Marius makes. Ajax underscores the paradox of Ulysses' pretension: *optima (scil. arma) num sumat, quia sumere noluit ulla?* ('And shall this man have the world's best arms, who wanted none?') (13.40).[12] Ajax also seems to resist the very method chosen for the competition for Achilles' arms – words, which are not his specialty, as he well knows: *denique quid verbis opus est? spectemur agendo!* ('In short, what need is there for words? Let us be tried in war!') (13.120).

At this point Ajax ends his speech (118 lines) and Ulysses presents his own motives for acquiring Achilles' weapons in a speech which is twice as long (250 lines). His arguments are very strong and seem certain of achieving success for one reason in particular: he emphasises the fact that words can be just as effective as actions. It should not come as a surprise that Ulysses is repeating here arguments we already know from Sallust's Marius.

Ovid's Ulysses does not hesitate to call his rival *hebes* ('stupid', 13.135), and he defends the worth of his *facundia* as a personal quality:

> *huic modo ne prosit, quod, ut est, hebes esse videtur;*
> *neve mihi noceat quod vobis semper, Achivi,*
> *profuit ingenium, meaque haec facundia, siqua est,*
> *quae nunc pro domino, pro vobis saepe locuta est,*
> *invidia careat, bona nec sua quisque recuset.*

---

[11] On the place of this episode within the architecture of the *Metamorphoses*, see Hardie 2015: 213–18.

[12] *Sumere* has here two different senses: to 'take his (Achilles') arms' and 'to take up arms': cf. Hardie 2015: 224.

> Only let it not be to this fellow's profit that he seems to be, as indeed he is, slow of wit; and let it not be, o Greeks, to my hurt that I have always used my wit for your advantage. And let this eloquence of mine, if I have any, which now speaks for its owner, but often for you as well, incur no enmity, and let each man make the most of his own powers. (13.135–9)

Ulysses does not care about noble lineage, because he is using his own personal qualities to achieve glory (140 ff.). His argumentative strategy is quite clear: he elevates to a higher plane his successful verbal subterfuges. And in doing this he appeals to the kind of rhetoric that we might rather have expected from Ajax:

> *ergo operum quoniam nudum certamen habetur,*
> *plura quidem feci, quam quae comprendere dictis*
> *in promptu mihi sit. rerum tamen ordine ducar.*
>
> So then, since it is a pure contest of deeds, I have done more deeds than I can well enumerate. Still I will tell them in their order. (13.159–61)

Ulysses is reminding the leaders that he wants to be judged for his 'naked' (*nudum*) actions, which are not easy to recount in words. But with his typical skilled irony, Ovid has his character nonetheless proceed to make what he describes as a well-ordered speech (*rerum tamen ordine ducar*).

In the following lines Ulysses ascribes to himself the glorious exploits of Achilles (13.165 ff.), particularly in 13.171: *ergo opera illius mea sunt!* ('So then, all he did is mine!'). Ulysses thus assigns to himself the most important role in the war of Troy, emphasising his ability and skill in setting traps, encouraging his companions and comforting them. With this premise he directs a quite paradoxical accusation against Ajax: *quid facis interea, qui nil nisi proelia nosti?* ('What were you doing in the meantime, you whose only knowledge is of battles?') (210). The logic of the eloquent hero becomes even derisory and sardonic. In line 290 he defines his adversary as *rudis et sine pectore miles*. Franz Bömer notes that here *sine pectore* means approximately 'lacking good sense',[13] i.e. an intellectual lack, and Philip Hardie in his commentary points out that *rudis* recalls the stereotypical contrast between the sophistication of Greek culture and the simplicity of the Romans.[14] Ajax in fact cannot even decipher and understand the figures carved in the weapons he aspires to have: *postulat ut capiat quae non intellegit arma* ('He asks for armour which he does not understand') (13.295).

Before he ends his harangue, Ulysses levels his final verbal blow. First he notes that Diomedes too has considerable merits in warfare, but:

---

[13] Bömer 1969 *ad loc.*
[14] *Rudis* is also an intratextual reference to Met. 1.7, i.e. to primitiveness: Hardie 2015: 256.

> *qui nisi pugnacem sciret sapiente minorem*
> *esse nec indomitae deberi praemia dextrae*
> *ipse quoque haec petered.*

And if he did not know that a fighter is of less value than a thinker, and that the prize was not due merely to a right hand, however dauntless, he himself also would be seeking it. (13.354–6)

Then he attacks Ajax again, besting him precisely where he feels most confident and in terms of those qualities with which he constructs his own identity as a warrior:

> *tibi dextera bello*
> *utilis: ingenium est, quod eget moderamine nostro;*
> *tu vires sine mente geris, mihi cura futuri;*
> *tu pugnare potes, pugnandi tempora mecum*
> *eligit Atrides; tu tantum corpore prodes,*
> *nos animo, quantoque, ratem qui temperat, anteit*
> *remigis officium, quanto dux milite maior,*
> *tantum ego te supero; nec non in corpore nostro*
> *pectora sunt potiora manu, vigor omnis in illis.*

Your good right arm is useful in the battle; but when it comes to thinking you need my guidance. You have force without intelligence; while mine is the care of tomorrow. You are a good fighter; but it is I who help Atrides select the time of fighting. Your value is in your body only; mine is in mind. And, as much as he who directs the ship surpasses him who only rows it, as much as the general excels the common soldier, so much greater am I than you. For in these bodies of ours the heart is of more value than the hand; all our real living is in that. (13.361–9)

Ulysses invokes the superiority of *thinking* about action over action itself. He is superior to Ajax as a captain is superior to a simple rower, or a general to a private. The ability to cogitate on action is directly connected with political supremacy. In Ovid as in Sallust, then, the theory of action has a higher hierarchical standing. In the end the Greek leaders reward Ulysses for his eloquence: *fortisque viri tulit arma disertus* ('The eloquent man carried off the brave man's arms') (13.383), whereby Ovid contrasts *fortis* with *disertus*.

In the past the *iudicium armorum* represented by Ovid has mainly been read as as a *mise-en-scène* of the victory of a new kind of rhetoric, sophisticated but morally dubious, that in the first century AD becomes superior to the traditional concept according to which oratory comes second to action.

For instance, Richard Tarrant cites this episode for the 'failure of rhetoric' typical of Ovidian poetry: here Ulysses, although he wins the contest, displays a negative and dishonest concept of rhetorical skill.[15] More recently, critics have concentrated on the literary aspects of the contest between the two Homeric heroes, taking into consideration particularly the way in which Ovid re-works various models, stemming from various genres, above all epic and dramatic poetry.[16] Both aspects merit attention, but both have the effect of minimising the important role of the tension between words and acts which characterises the discourse of war and of making of it an occasion for more or less strictly literary purposes. And yet this passage needs to be read in connection with the complex construction of a specific form of authority of writers *de re militari* based on a paradoxical expertise: in contradiction to widespread assumptions, he who skillfully writes on war need not necessarily have experience as a soldier.[17]

Onasander and Vegetius

Onasander and Vegetius wrote two different texts in two different periods for different purposes and with different argumentative strategies, and yet their works show an analogous perspective on their subject matter. In what follows I show how these authors construct their authority on the basis not of military action, but rather of their expertise precisely as writers; in this respect these texts are similar to all other ancient military treatises, and in virtue of their fortunate reception through the centuries ended up contributing to a model of authority which became standard in all other texts in the Western art of war tradition until Carl von Clausewitz, the Prussian strategist who in the early nineteenth century introduced a new way of writing on war.[18] In this, the art-of-war tradition before Clausewitz stood in tension with larger cultural expectations according to which he who wanted to write properly about war, as in many other fields including medicine and philosophy, was supposed to have been active in the field. In addition to the passages from Cicero, Sallust and Ovid quoted above, we have the opinion of a Greek historian highly revered for his seriousness. In the 12th book of his *Histories* Polybius authoritatively affirms that 'it is neither possible for a man with no experience (*empeiria*) of warlike operations to write well

---

[15] Tarrant 1995: 72.   [16] Cf. Hardie 2015: 213–68.
[17] For discussion of practice and theory of warfare within ancient historiography, see Bettalli 2010.
[18] See Formisano 2011; Heuser 2007, 2010.

about what happens in war, nor for the unversed in the practice and circumstances of politics to write well on that subject'.[19] Here, Polybius is following mainstream Greek historiographical discourse which explicitly emphasised *empeiria* and *autopsia* as *the* criteria that must be applied in order to write good history. By contrast, the Romans – as has been argued by John Marincola – tended to put at the centre of their historical method social and cultural *auctoritas*, which served as a guarantor of the truthfulness of texts.[20]

In the contemporary field of military history, by contrast, the modern correspondent of the ancient and early modern art of war, no one is expected to possess any practical experience in the field in order to achieve excellent results in military tactics and strategy.[21] In contrast to ancient and early modern expectations, modern military history as a discipline no longer tends to have its foundations in the lived, practical experience of those who write about war. Cynically, one could argue that this was in fact never the case, but within the ancient tradition authors tend to conceal their lack of direct experience in the field even as they followed the established rhetorical pattern according to which one can profitably write only about those things of which one has direct experience. In modern times, as Hans Delbrück in the preface of his monumental *Geschichte der Kriegskunst*[22] and John Keegan in the famous *The Face of Battle*[23] make clear, not only is the relationship between action and writing no longer relevant in this way, but precisely a displayed lack of experience has become a mark of the genre.

And yet, paradoxically enough, modern scholarship on ancient and early modern military treatises still considers the question of how *useful* or *practical* ancient military texts, and all other 'technical' literature from antiquity for that matter, might have been. Although some efforts have recently been taken to focus on the literary quality rather than the technical aspect of such texts, this still constitutes the principle generally governing the language and methodology of scholarship in this field.[24]

---

[19] Polybius 12.25g.

[20] Marincola 1997: 137–9. Later still, in the context of the early modern scientific revolution, scientists argued for the trustworthiness of their experiments on the basis of their authority as gentlemen: see Shapin 1994.

[21] See, e.g., Keegan 1983: 11–12, with comments at Formisano 2011: 5–6.

[22] English translation: Delbrück 1975–85. [23] Keegan 1983.

[24] For the art of war, see the important article by Campbell 1987. More generally, recent scholarship on ancient technical texts, insisting on the literary quality of those works, has directed attention to the linguistic features and argumentative strategies used by the authors in order to make their book *useful*, i.e. applicable. In particular, the German term *Wissensvermittlung* has acquired a central role within the field: see among others Asper 2007; Fögen 2009; Horster and Reitz 2003. On the instability of the principle of practical applicability in ancient texts, see Formisano and van der Eijk 2017.

The Contemplative General: Onasander

Onasander appears in the Byzantine lexicon the *Suda* as a 'Platonic philosopher', since he also wrote a commentary on Plato's *Republic*, but he is best known for his military treatise, probably composed during the 50s AD, and dedicated to the Roman general Quintus Veranius, consul in 49.[25] The Byzantine lexicon calls this text *Taktika peri strategematon*, but the manuscripts themselves always bear the title *Strategikos*, and that is the generally accepted title today. While it appears not to have had much resonance among contemporaries and only a slight reception can be inferred from late antique authors, Onasander's text enjoyed a formidable reception during early modern times.[26]

A major topic in the scholarship on Onasander (and, as we have seen, on other writers of this type) has traditionally been an attempt at establishing the contribution his text aims to make to the practical improvement of the army. Moving in another direction, ancient military historian Yann Le Bohec provocatively claims that Onasander's work has no political or military aims at all, but is a pamphlet devoted to self-promotion.[27] That view is certainly legitimate, but describes the social context rather than the text itself. If we look at the text of the *Strategikos* we find that it is concerned with various topics: Roman *virtus*, the role of fortune or τύχη, the concepts of φόβος, ζῆλος and φθόνος, the idea of *bellum iustum* and, perhaps most importantly, psychological qualities of the (ideal) general.[28] As Ambaglio has emphasised, many of these topics derive from Xenophon, Aineias the Tactician, Plato and Polybius: Onasander seems to be re-proposing Greek moral and technical norms for the benefit of Romans, among other things in order to establish a continuity between Greece and Rome.[29] This point, too, is certainly valid, but it is worth noting that the *Strategikos* does not in fact refer to any predecessors by name or contain any reference to contemporary military technology or techniques: its norms and rules have a moral and ahistorical significance.[30] Onasander seems to present them as valid *in the abstract*; and precisely this intellectual quality to his art of war makes his

---

[25] Here I partially reproduce Formisano 2011.
[26] For a list of older contributions, see Ambaglio 1981: 354, n. 10; for Onasander's reception, see 353 and 366, where Ambaglio argues that Frontinus' *Strategemata* shows the influence of the *Strategikos*. Subsequent studies include Campbell 1987, Galimberti 2002, Le Bohec 1998, Smith 1998, and the edition of the text with Italian translation, introduction and rich commentary by Petrocelli 2008.
[27] Le Bohec 1998: 178.
[28] For a list of topics, see: Ambaglio 1981: 355; Campbell 1987: 13–14.
[29] Ambaglio 1981: 362.   [30] See Petrocelli 2008: 15.

text especially relevant to the tradition of military writing culminating with Clausewitz.

I would argue, moreover, that the *leitmotiv* of this text, clearly thematised in its preface, is the tension between theory and practice and that the text provides a perhaps surprising answer to the questions raised by that tension. The *Strategikos* is also characterised by an absence of historical *exempla*, an absence which is rather astonishing for this genre, and by an absolutely generic treatment of military matters without any reference to the historical moment. In accordance with early imperial trends, when a commander was no longer a political figure, Onasander gives a depoliticised view of the military.[31]

The absence of explicit references to the exemplary past, so characteristic of other Greek and Latin texts on warfare, combined with the implicit and massive use of Greek sources as described by Ambaglio, can be seen as an *intellectualising* way of making the past present. This process of 'presentisation', i.e. of rendering the past present and projecting it into an ahistorical dimension, emerges clearly over the course of the text and, as we will see, is well expressed through the motif of *sight*, of seeing what is present.

With these considerations in mind, and asking questions that are less historical than literary in nature, let us turn to Onasander's preface, which, as has generally been recognised and is very often the case with ancient technical and scientific technical writing, has some odd rhetorical strategies that repay a closer look.[32] The first points worth emphasising are that it is written in Greek, even as it professes to be a source of utility for generals and delight for retired commanders in this time of the *pax Augusta* (pr.4), and that although Onasander declares that he wishes to provide 'a summary sketch of what the Romans have already accomplished by their mighty deeds' (pr.3), he does not actually make any reference to Roman history in the body of his text. All of this, according to Christopher Smith, 'indicates the precise area of Greek superiority'.[33] More generally, with Smith it is interesting to notice how the entire preface is dense with 'culturally loaded comment'.[34]

---

[31] Emphasising the socio-cultural significance of the text, Ambaglio 1981: 374 affirms: 'Non è sul piano dell'utilità pratica che si devono ricercare le motivazioni reali che indussero Onasandro a comporre lo *Strategikos*. L'atteggiamento dello scrittore trova infatti un'esauriente spiegazione se si intende l'opera come frutto di un'operazione eminentemente culturale-politica, nel contesto della quale i risvolti pratici – ovvero la didattica del comando – assumono, a dispetto delle apparenze, un rilievo non di primo piano.'

[32] For the prefaces of Latin technical and scientific texts, see Santini *et al.* 1990–98.

[33] Lammert 1931: 34 and Smith 1998: 156 see in some passages of the *Strategikos* possible allusions to events of Roman military history. Galimberti 2002 builds upon this rather thin argument in order to show that Onasander was actually knowledgeable about Roman military history.

[34] Smith 1998: 155.

6 The 'Roman Art of War' 143

Onasander begins by explaining why he dedicates this text on military matters to Romans, namely because they are the pre-eminent military power in the world:

> It is fitting, I believe, to dedicate monographs on horsemanship, or hunting, or fishing, or farming, to men who are devoted to such pursuits, but a treatise on military science (στρατηγικῆς δὲ περὶ θεωρίας),[35] Quintus Veranius, should be dedicated to the Romans who have attained senatorial dignity, and who through the wisdom of Augustus Caesar have been raised to the power of consul and general, both by reason of their military training, in which they have had no brief experience (ἐμπειρίαν), and because of the distinction of their ancestors. I have dedicated this treatise primarily to them, not as to men unskilled in generalship (οὐχ ὡς ἀπείροις στρατηγίας), but with especial confidence in this fact, that the ignorant soul is unaware even of that in which another is successful, but knowledge bears additional witness to that which is well done. For this reason, if what I have composed (τὰ παρ'ἐμοῦ συντεταγμένα) should seem to have been already devised by many others, even then I should be pleased, because I have not only compiled (συνεταξάμην) precepts of generalship, but have also endeavoured to get at the art of the general and the intelligence (φρονήσεως) that inheres in the precepts. I would be fortunate if I should be considered capable, before such men, of making a summary sketch (λόγῳ) of what the Romans have already accomplished by their mighty deeds. It remains for me to say with good courage of my work, that it will be a school (στρατηγῶν τε ἀγαθῶν ἄσκησις) for good generals, and an object of delight for retired commanders in this *pax Augusta* (παλαιῶν τε ἡγεμόνων κατὰ τὴν σεβαστὴν εἰρήνην ἀνάθημα). (pr.1–4)[36]

These opening words suggest that Onasander's main concern is to provide *knowledge* rather than direct *experience*. Experience is taken for granted – these Romans are 'not inexperienced in generalship' (οὐχ ὡς ἀπείροις στρατηγίας), but it is understanding and knowledge that counts: 'the ignorant soul (τὸ ἀμαθὲς τῆς ψυχῆς) is unaware even of that in which another is successful, but knowledge (τὸ ἐν ἐπιστήμῃ) bears additional witness to that which is well done' (pr.2). Interestingly enough, here and later (pr.4, 7 and 10) Onasander uses the verb συντάττω in order to establish a parallel between the main activity of a general, which is to arrange the army, and his own activity as a writer. This very word not only establishes an equivalence between 'drawing up an army' and 'writing down', present also

---

[35] *Theoria*, a key term within Greek philosophical terminology, recurs in Aelian's preface to his work, called *Tactica theoria* (pr.1–2): see Petrocelli 2008: 129.
[36] I use the English translation by Oldfather 1928 with occasional modifications.

in Aelian (pr.1),[37] but also reproduces the visual quality of Onasander's argumentation, which plays (as we will see) a fundamental role in the *Strategikos*.

After reflecting on Rome's success in dominating the known world, and rejecting the view that fortune (τύχη) alone is responsible (pr.4–6), Onasander turns to the tension between theory and experience:

> Now since all men naturally give credit for truthfulness to those who appear to write with professional experience (δι'ἐμπειρίας συντετάχθαι), even though their style be feeble, while for inexperienced writers (τοῖς δὲ ἀπείροις), even though their teachings are practicable, they feel distrust on account of their lack of reputation, I consider it necessary to say in advance, about the military principles collected in this book, that they have all been derived from experience of actual deeds (διὰ πείρας ἔργων), and, in fact, of exploits performed by those men from whom has been derived the whole primacy of the Romans, in race and valor, down to the present time. For this treatise presents no impromptu invention of an unwarlike and youthful mind, but all the principles are taken from authentic exploits and battles (διὰ πράξεων καὶ ἀληθινῶν ἀγώνων), especially of the Romans. For the expedients they used in order to avoid suffering harm, and the means by which they contrived to inflict it, all this I have collected. Nor have I failed to perceive that a writer, seeking greater praise from credulous readers, would prefer to have it appear that the source of all military stratagems he described was himself and his own shrewdness (ἑαυτοῦ καὶ τῆς ἰδίας ἀγχινοίας) rather than the sagacity of others (τῆς ἀλλοτρίας ἐπινοίας). But I do not think that the latter diminishes one's glory. For if a man after experience in the field had drawn up such a work (εἴ τις ἐν πολέμοις αὐτὸς στρατευσάμενος συνετάξατο τοιόνδε λόγον), he would not be considered to have given a less valuable testimony (ἥττονος ἠξιοῦτο μαρτυρίας) if he introduced and commemorated in his work, not only the personal discoveries of his native wit (φυσικῆς ἀγχινοίας ἰδίαν εὕρεσιν εἰσηνέγκατο στρατηγημάτων), but also the brilliant deeds of other men (τὰ δι' ἄλλων εὖ πραχθέντα); in the same way I do not consider that I myself shall win less praise, because I admit that not everything I write springs from my own intelligence (τῆς ἐμῆς συνέσεως). On the contrary, I have chosen the opposite course, that I may have praise without reproach and trust without slander. (pr.7–10)

These last paragraphs are particularly deserving of emphasis as they indicate the kind of enterprise this text aims to engage in. Again, we find the tension between experienced and unexperienced – but this time the subjects are *writers*. Adding to the complexity is that Onasander compares himself

---

[37] See Petrocelli 2008: 133.

precisely to a general who, after having collected direct experiences in the field, writes down a book, not only recurring to his personal experience, but also recalling others' deeds. Similarly, Onasander claims he is no less deserving of glory if he admits that not everything he has written down is a product of his own intelligence or understanding (σύνεσις).

In this text the reader encounters something surprising, all the more so if we consider what seem to have been the general expectations of ancient readers: the author explicitly and self-consciously does *not* base his work on his own direct experience (pr.7: the principles in this book 'have all been derived from experience of actual deeds' – performed, however, not by Onasander, but by Roman commanders of the past); yet nonetheless and precisely for this reason he wants to be considered at a higher level. Two details in particular deserve greater attention. First, the very action of writing, as we have seen, is described by the verb συντάττω, which is also a technical term referring to the arrangement of troops. This text appropriates military language, consciously deploying it for its own purposes and in effect substituting writing for military action. Second, we find another surprising term when Onasander admits that not everything in this text derives from his own σύνεσις ('intelligence' or 'understanding', etymologically pointing to the ability to put things together). We might have expected something else, like a reference to the concept of experience (ἐμπειρία) abundantly present in previous paragraphs. The final section of Onasander's preface is constructed, in short, around terms referring to intellectual qualities rather than practical activities.

Then there is the term μαρτυρία (10), referring to the 'testimony' for which Onasander's hypothetical general/author would be valued (cf. pr.6, ἀμάρτυρον). I would argue that the term points to the notion of testimony through *sight*, hinting at the image of the commander's view over the battlefield, to external observation rather than action and direct involvement. Throughout this text, in fact, the reader's attention is drawn to vision or, to use the expression used by Shadi Bartsch in her 2006 book *The Mirror of the Self*, to the 'scopic regime' so recurrent in philosophical and literary culture of the early empire, when 'the stimulus of the act of seeing in an urge to emulate or to objectify' was paramount.[38] In a number of passages Onasander implicitly connects what purports to be practical advice with the theme of sight.[39]

---

[38] Bartsch 2006: 16.

[39] See, e.g., 6.7–8: 'He must send ahead cavalry as scouts to *search* the roads, especially when advancing through a wooded country, or a wilderness broken up by ridges. For ambuscades are frequently set by the enemy, and sometimes failure to *detect* them brings complete disaster to the opposing side, while their discovery, by a slight precaution, attests to the general of the

With regard to Onasander, Christopher Smith observes that he 'offers us a treatise on military science that is essentially philosophical... Onasander gives us a moral and ethical discourse which touches on the nature of the good general, and of the essence of military endeavour, which, it turns out (and here is the twist), *lies less in action than in dissembling*. Onasander's ideal campaign would be completely bloodless on both sides; two perfectly matched armies would employ so many feints and counterfeints that they would never come to blows!'[40] The attention given to 'dissembling', which in Smith's opinion turns out to be paradoxical since it would eventually prevent any action, ends up being tightly connected with the visual paradigm governing the culture and literature of the first century of the Empire.

I would argue that the contemplative and *theoretical* nature declared by Onasander at the opening of his work represents much more than a rhetorical gesture. Vision (*opsis*) becomes a substantive approach to strategy; it becomes a technical matter. The intellectual quality of this work composed by a 'Platonic philosopher' is clear. Far from being secondary, the theme of the visual deeply informs the rhetoric of the text; it changes, perhaps even destabilises, the nature of war itself, usually perceived through the lens of action rather than theory. As we have seen, the motif of sight represents an argumentative strategy aimed at supporting the 'presentisation' of the Greek past and military tradition that is characteristic of the *Strategikos*. Alice König in her chapter in this book convincingly shows that two forms of expertise emerge within Frontinus' *Strategemata* – scholarly learning and practical know-how – which, instead of harmoniously cooperating, end up competing with each other. Readers of Onasander's text are confronted with another strategy aimed at resolving the tension between these two forms of expertise: the ideal general sketched by Onasander harmonises them under the spell of theory, so that action becomes contemplation.

---

enemy great prudence on the part of his adversary. For in a level and treeless country a general *survey* is sufficient for a preliminary investigation; for *a cloud of dust announces* the approach of the enemy by day, and *burning fires light* up a near-by encampment at night'; 10.3: '... but those who are well trained in formations quickly – indeed automatically, so to speak – rush to their stations, presenting a *harmonious*, I may say, *and beautiful sight*' (emphasis added); 10.14: 'But if, while keeping his army in the same spot, he should come to a conference with the opposing general, either to make or to receive some proposal, he should choose as an escort the strongest and *finest-looking of the younger soldiers, stalwart, handsome and tall men, equipped with magnificent armour*, and with these about him he should meet the enemy. For often *from the view* of a part the whole is judged to be like it, and a general does not determine his course of action by what he has heard, *but is terrified by what he has seen*.'

[40] Smith 1998: 165 (emphasis added).

The Expertise of the Book: Vegetius

One Roman author deserves more attention than any other ancient military writer for a number of reasons, but in particular because he treated and interpreted the problem of controlling events through theoretical models in war in a more complete fashion than any other author. This author is Vegetius.[41] His *Epitoma rei militaris* was the canonical text on war in the Middle Ages and in the Renaissance; without considering Vegetius we would be unable to analyse the concept of war itself in Western culture. The *Epitoma rei militaris* has recently been rediscovered by classicists in conjunction with the general growing interest in late antiquity: in less than twenty years, for example, two critical editions have been published.[42] In the following I aim to briefly present some typical arguments from Vegetius' work and to trace its internal logic.

Within Western culture, the art of war has taken the form of literary knowledge and of a genre more often, I would argue, than that of a technique. This is not to say that genuinely technical discussions have been banished from these texts. Rather, I argue that even the most technical aspects, per se perfectly applicable in the extra-textual reality, can *also* be read, within the text that contains them, as elements of a much more complex literary discourse. But what is the basis for such an intimate connection between war and literature? Why is the relationship between these two very different human activities so strong that one might say their identities have been reciprocally constructed? To all of these questions Vegetius provides illuminating answers.

The objective of the *Epitoma* is to reintroduce Vegetius' contemporaries to the way of warfare of Ancient Rome. The goal of recollecting the past is meant to be realised basically through the written word. Vegetius' book is thus an extremely apt instrument for this goal. Earlier scholars long dwelled on his epitomatory method, trying to reconstruct a text whose peculiarity they saw mainly in close correspondence with and dependency on its sources. But this kind of approach turned out to be much less effective than

---

[41] Little is known about the life of Publius Flavius Vegetius Renatus. Although he addresses his *Epitoma* to the emperor, he does not indicate which one. Scholars are generally inclined to identify the unnamed emperor as Theodosius (late fourth century AD), but this is a conjecture. Charles 2007 thoroughly rediscusses the problem of dating and argues for a later period, namely the mid fifth century under Valentinian III.

[42] The first in 1995 by Alf Önnerfors in the Teubner series, the second in 2004 by Michael D. Reeve in the Oxford Classical Texts. On the massive reception both in the Middle Ages and Renaissance, see the monographs by Allmand 2011 and Richardot 1997, with my review of the former: Formisano 2012.

one might have thought, not least because it is characteristic of Vegetius' compositional method to make various materials coming from the tradition simultaneous and synchronic. Historical accuracy is not the criterion guiding Vegetius; rather, he provides a selective vision of the Roman army, and his aim is clearly to idealise the *antiqua legio* instead of describing it with accuracy.[43] In fact, this seems to be a way of relating to the past that is typical of the later art of war.[44]

The argument of the *Epitoma* follows two main themes that are closely linked. The first theme is 'ethical': Vegetius realises that the art of war has fallen into decay. It has been neglected because of a long peace which has induced a deficiency in recruiting and keeping trained armies:

> *sed longae securitas pacis homines partim ad delectationem otii, partim ad civilia transduxit officia. ita cura ex exercitii militaris primo neglegentius agi, postea dissimulari, ad postremum olim in oblivionem perducta cognoscitur.*
>
> However, a sense of security born of long peace has diverted mankind partly to the enjoyment of private leisure, partly to civilian careers. Thus attention to military training obviously was at first discharged rather neglectfully, then omitted, until finally consigned long since to oblivion. (Veg. *Mil.* 1.28.6–7)[45]

Such a tendency, however, is not new at all: in the glorious past of Rome it also happened that attention to military art vanished. What solution did ancient Romans conceive of for reintroducing higher standards? They first enquired into books and then they let the generals deploy that knowledge on the field:

> *haec ex usu librisque antea servabantur, sed omissa diu nemo quaesivit, quia vigentibus pacis officiis procul aberat necessitas belli. sed ne impossibile videatur reparari disciplinam cuius usus intercidit, doceamur exemplis. apud veteres ars militaris in oblivionem saepius venit, sed prius a libris repetita est, postea ducum auctoritate firmata.*
>
> These skills were formerly maintained in use, as well as in books, but once they were abandoned it was a long time before anyone needed them, because with the flourishing of peacetime pursuits the imperatives of war

---

[43] See Milner 1996: xvi.
[44] A. König in her chapter points to an analogous textual strategy in Frontinus' *Strategemata*, where exempla stemming from various historical contexts and nations are put together in such a way as to implicitly suggest an ideal synchronicity.
[45] English translation by Milner 1996.

were far removed. But lest it be thought impossible for an art to be revived whose use has been lost, let us be instructed by precedents. Among the ancients, military science often fell into oblivion, but at first it was recovered from books, and later consolidated by the authority of generals. (Veg. *Mil.* 3.10.17–18)

Already at the beginning of his work Vegetius states his intention to recall and imitate the way of the ancestors, who taught that one should write down useful and honourable knowledge. An example of this was Cato: he was not just a successful soldier, but he also wrote down military precepts and provided a starting point for the tradition to come.[46] To the general decay of military structures Vegetius opposes *diligentia*, which should characterise the leader of the army. And for Vegetius war is an ineluctable factor of civilization: *o viros summa admiratione laudandos, qui eam praecipue artem ediscere voluerunt sine qua aliae artes esse non possunt!* ('O men worthy of the highest admiration and praise who wished to learn that art in particular without which the other arts cannot be!') (Veg. *Mil.* 3, pr.3).

His way of approaching the problem is clear: he wants to restore the diligent concern for armies shown by the ancients by way of books. Thus, his expertise in our eyes might be compared to that of an antiquarian rather than a soldier. His major aim is to look for universally valid *exempla* the recollection of which is meant to offer strategic criteria valid for present war circumstances. They become normative models for concrete actions and technical instruments. In this perspective, which is peculiar to the art of war, history ceases to be the object of contemplation; it becomes instead an effective standard, to which one should hold if one wishes to be successful. Reiteration of the past in this case does not represent a simple mean in order to adorn and make interesting for a general reader technical subject matter, but it becomes the only possible criterion in order to discuss action. Even *ars* and *exercitium* are considered by Vegetius to be a pure projection of feats of the past.

The most fascinating point is the centrality which writing itself acquires. The internal logic of his work is circular; the author starts from writing, tries to support present action by means of *exempla*, but comes back inexorably to writing: *nam unius aetatis sunt quae fortiter fiunt, quae vero pro utilitate reipublicae scribuntur aeterna sunt* ('For brave deeds belong to a single age; what is written for the benefit of the State is eternal') (Veg. *Mil.* 2, 3, 7). Thus, the technical subject matter becomes tightly connected

---

[46] Cf. Veg. *Mil.* 2.3.6–8.

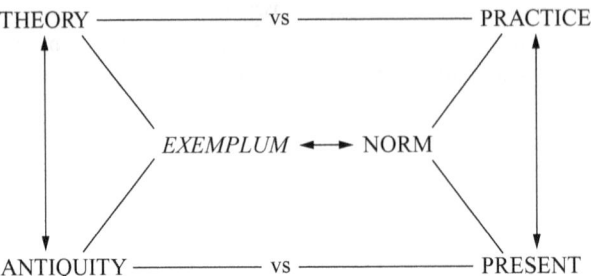

Figure 6.1 Diagram of the Relationship between Key Concepts in Vegetius' 'art of war'

with literary structures. Vegetius establishes an interdependence between *literary* past and present *action*, and he creates a language suited to thinking about war, based on a special contiguity between writing and action. This becomes the fundamental principle of the Western genre of the art of war for centuries to come. Already twelve centuries before, in connection with a reform of the army, Dutch reformers introduced the first military school in Europe. Vegetius bases an archetype of military reform on systematic reference to antiquity (see Figure 6.1).

The diagram in Figure 6.1 is meant to map out the complex conceptual constellation which characterises the 'art of war'. The pair 'exemplum' and 'norm', located at the centre, stands at the core of any discourse on war; the two turn out to be interconnected, in such a way that one generates the other. Around this core are located the pairs 'Theory vs Practice' and 'Antiquity vs Present', pointing to a cognitive tension which is central to any art of war: how to put theory into practice? Other connections exist too. On the left side of the diagram, Theory is linked with antiquity precisely through exemplarity, while on the right side Practice and the Present are intertwined through normativity. It is Vegetius, I would argue, who systematises the discourse of war by making *exemplum* and *norm* the two basic concepts. When evoked through writing, *exempla* become norms, the only possible models for present action. They set standards and are represented as ahistorical categories with a universal validity.

Yet the *Epitoma* contains many errors and historical imprecisions. For this reason Sidney Anglo, author among other works of a monograph on the first century in Machiavelli's reception, defines Vegetius as an author who entirely lacks originality and his astonishing reception in the Renaissance as the 'triumph of mediocrity'.[47] This harsh judgement may or may not

---

[47] Anglo 2002.

suit a reading of the *Epitoma* for its 'technical' accuracy or concrete 'practicability', but it certainly does not suit its theoretical core, with its creation of a contiguity of the written word and action which, I have argued, is the basis for the undeniable fact that this text became *the* model for writing on war in the West. In a more recent article, military historian Sylvain Janniard rehabilitates Vegetius' concreteness: his *Epitoma* is by no means only a bookish work but it contains eminently practicable and applicable instructions.[48] But it is important to emphasise that the appreciation of the literary quality of Vegetius' text, as of any other ancient 'technical' text, cannot be confined to the reconstruction of the intentions of the author, his technical mastery or any concrete effects the text might have. The two spheres of the text, the technical and the literary, might in fact cooperate towards a unique goal – in this case a re-birth or renaissance of early Roman military strategy – but of course with different strategies and using different languages.[49]

In short, what has been taken to be a problem concerning some 'manuals' turns out to be the theoretical core of a genre. Commenting on 'contemporary Roman handbooks' from the Late Republic and principate, ancient military historian Catherine Gilliver seems to complain that the more technical treatises dealing with mechanical constructions (such as catapults and other war engines) 'lack advice on the practical application of the weapons in the field', especially because their authors 'often reproduced material from earlier works despite it being obsolete'. Furthermore, discussing 'more general manuals on warfare', written by 'philosophers with no military experience', she observes that these texts were 'based on earlier works'.[50] Precisely this statement, i.e. that some military texts depend on earlier models, is not, I would argue, a contingent aspect of early imperial works, but in fact constitutes the single most important mark of the genre 'art of war' from antiquity through the Middle Ages to the Renaissance. Machiavelli's influential contribution to this genre (*Dialogo sull'arte della guerra*, Florence 1521) is in fact heavily based on Vegetius, who is such an authority for the Florentine that he does not even need to cite him.[51]

As a final coda I would like to briefly reconnect the ancient art of war as I have been describing it here with the debate about the usefulness of military history in contemporary Western culture. In an important discussion within a volume significantly entitled *Understanding War*, Peter Paret, a leading military historian, clearly illustrates the impasse of his discipline. On the

---

[48] Janniard 2008: 20.
[49] On this point, see my contribution in Formisano and van der Eijk 2017.
[50] Gilliver 2007: 124. [51] See Anglo 2005; Formisano 2002; Schwager 2012.

one hand, military history shares methodologies and goals with historical research in general, i.e. it aims at reconstructing aspects of the past related to wars, conflicts and the connections between these and other spheres such as politics and culture. On the other hand, military history inherently tends to create a relationship with present-day conflicts by implicitly or explicitly suggesting norms of action that could be directly applicable by generals and soldiers. Yet these two types of military history are not in harmony with each other, indeed there is a clear tension between them. Paret cites examples of both. Hans Delbrück, for example, author of the monumental *Geschichte der Kriegskunst* (1900–1920), declared in the introduction to his multi-volume work that this 'was not written for the sake of the art of war, but for the sake of world history', emphasising that his work is written 'for friends of history by a historian'.[52] Walter Millis, on the other hand, author of a famous pamphlet called *Military History* (1961), insists that military history should have the function 'to train professional military men in the exercise of their profession',[53] and he invites fellow military historians 'to turn away from a study of past wars to the study of war itself'.[54] Paret also indicates a feature of this type of history which in my opinion needs to be taken more seriously than as a mere matter of style precisely because it directly reconnects us to the discourse shaped by the Roman art of war. Military history, Paret writes, frequently turns out to be 'popular history' written in a fashion which appeals less to scholars than to a general audience.[55] The tradition of the Roman art of war, with its distinctive tensions on the point of technical expertise and authority, seems not to have faded out entirely. Fragile expertise and the authority of the past continue to be at the centre of writing on war.

[52] Quotations from Paret 1992: 219. [53] Millis 1961: 16. [54] Millis 1961: 18.
[55] Paret 1992: 214.

7 | Conflicting Models of Authority and Expertise in Frontinus' *Strategemata*

ALICE KÖNIG*

To All Military Souls of the English Nation

Tis for your Perusal that this Treatise is publish'd ... if the English Courage alone, without the Assistance of Art, hath been so Victorious, what Wonders would it not be able to perform, if it were seconded by Policy and Craft? I conceive therefore it may not be useless to you, my Brave Countrymen, to have an Abstract, or a Collection in your own Language of the Stratagems which have been practiced in War by the most experienced Commanders... For that purpose I have Translated FRONTINUS, who, being a ROMAN Warriour, and of the Order of the Consuls, Collected the most remarkable Stratagems of the PERSIANS, GREEKS, ROMANS and CARTHAGINIANS.

> Marius d'Assigny (1686) *The Stratagems of War, or, A collection of the most celebrated practices and wise sayings of the great generals in former ages written by Sextus Julius Frontinus, one of the Roman consuls; now English'd*

For more than sixteen centuries, from the late first century AD down to the end of the Renaissance and beyond, Frontinus was regularly referred to as an expert and authority on military matters. The twelfth-century scholar John of Salisbury, for instance, drew on Frontinus' writing as much as he did on the works of Virgil, Plato and a host of other classical authors in his formulation of a new political philosophy in the *Policraticus*.[1] Excerpts of Frontinus' *Strategemata* turn up in mediaeval crusading manuals, such as Marino Sanudo Torsello's *Book of the Secrets of the Faithful of the Cross*. In 1417, one Jean Gerson, tutor to the then Dauphin of France, lists the *Strategemata* alongside the Bible, other Christian texts, and works by Aristotle, Sallust, Livy, Valerius Maximus, Seneca, Vegetius and Augustine (inter alia), as a kind of literary 'Ark of the Covenant' that the young prince should absorb and carry metaphorically about with him 'through the desert

---

* I am grateful to The Leverhulme Trust for the Research Fellowship during which this chapter was written; also to Jason König, for his incisive feedback and advice.
[1] Martin 1997; cf. Nederman 1990: xx.

of this world'.² Christine de Pizan (a scholar well known to Gerson, who also proffered advice to those in power in fourteenth- and fifteenth-century France) quotes extensively from the *Strategemata* throughout the second part of her *Book of Deeds of Arms and Chivalry*.³ Although Machiavelli never mentions Frontinus by name, he too seems to have borrowed from the *Strategemata* in more than one of his works.⁴ And translations of the text abounded across Europe, including one addressed to Henry VIII, which promised to support that 'moste high, excellente, and myghtye Prynce' by inspiring and instructing his military captains (who 'have oft declared that they lytell nede any instructions, any bokes'), just as the *Strategemata* claimed to inspire and instruct its original readers.⁵

This chapter is about the models of authority and expertise that the *Strategemata* itself projects and prompts reflection on. But it is instructive to begin by looking at what it was about the text (and its author) that prompted later readers to deem it authoritative; not only because some of them continue to shape our approaches to the *Strategemata* today, but also because the differences between their responses to it point up some fascinating contradictions and tensions within the treatise – tensions which are revealing of the challenges and opportunities that many scientific/technical/didactic authors have wrestled with in constructing and parading authority and expertise, especially in the military sphere.

It was arguably Vegetius, writing in the late fourth or early fifth century AD, who cemented Frontinus' status as an authority on military matters by citing him as one of his most important sources (Vegetius, *Mil.* 1.8 and 2.3).⁶ Vegetius builds his own authority and expertise on scholarly foundations, and presents himself as writing within an established and important literary tradition: one (he claims) that had begun in Ancient Greece, but had been honed and was now dominated by Roman writers, who had helped to transform military practice into a scientific discipline.⁷ Cato the Elder is

---

[2] Gerson, *Au précepteur du Dauphin, Constance, vers Juin 1417* (Gerson 1960: 203–15; cf. Thomas 1930: 30–55, who dates the letter rather to 1408–10; also Mazour-Matusevich and Bejczy 2007).
[3] Forhan 2002: 150–7; le Saux 2004; Willard 1995.
[4] In particular, in his *Art of War* (Lynch 2003: xiv–xv); see also Wood 1967, on the possibility that Frontinus' *Strategemata* influenced Machiavelli's didactic method in his *Discourses on Livy*.
[5] Morysine 1539.
[6] On the relationship between Vegetius and Frontinus' survival, see esp. Allmand 2009; also Allmand 2011: 48–61; Lenoir 1996: 81.
[7] See esp. the prologues to Books 1 and 3, and 1.8, where the act of writing is associated with the development of military knowledge; also Formisano in this volume on the authority of literature/writing (as opposed to practical experience) in Vegetius' *Epitome* and other military manuals; and Formisano 2009: 333–5 on Vegetius' role in making the 'art of war' a fundamentally literary phenomenon.

identified not just as an early example, but as a forthright champion of this tradition:

> Cato, because he was invincible in battle and had often led the army as consul, believed that he could benefit the republic further by setting down in writing his military learning. For things that are done bravely last one generation; but things that are written down for the genuine benefit of the republic last forever. (Veg. *Mil.* 2.3)

Frontinus is then singled out as following in Cato's footsteps: 'Several others did the same, but in particular Frontinus, whose industry in this regard was approved by Trajan.' Frontinus, like Cato, had not just written about soldiering, but had seen plenty of military service himself; he had also served as consul no fewer than three times, and was doubtless 'approved' by Trajan far more for the role he played in securing Nerva's adoption of Trajan as his heir than for anything that he wrote.[8] Nonetheless, Vegetius' overriding interest in the written word leads him to suggest that what makes Frontinus a significant figure in military history (someone worth citing and connecting oneself with) is his literary activity, not his practical experience. For Vegetius, Frontinus is authoritative (and was so for his contemporaries) above all because he was a Roman author writing in (and able to link Vegetius to) a long line of earlier authors who together had refined and disciplined military knowledge.

Aelianus Tacticus, writing much closer in time to Frontinus but from a Greek perspective, offers a different analysis. The prologue to his *Tactical Theory* stages a stand-off between Greek military 'science' (*mathema* and *theoria*) and contemporary Roman military practice (*dunamis* and *empeiria*). Inspired by the former, but (supposedly) daunted by the momentum of the latter (which was achieving unparalleled successes under the aegis of Aelian's dedicatee, Trajan), Aelian claims that he initially hesitated to write about 'a science forgotten and moreover long out of use since the introduction of the [Roman] system' (*Tactical Theory* pr.2);[9] until, that is, an encounter with Frontinus, whose own interest in Greek military theory encouraged Aelian to proceed:

> I was able to spend some days at Formiae with the distinguished consular Frontinus, a man of great reputation by virtue of his experience (*empeiria*) in war. Discovering in conversation with him that he had no lesser regard for Greek tactical science, I began not to despise their tactical writing,

---

[8] On Frontinus' career: Eck 1982: 47–52; Rodgers 2004: 1–5.
[9] This and the following translation are from Devine 1989.

thinking that Frontinus would not pay so much attention to it if he indeed considered Roman tactical usage superior. (*Tactical Theory* pr.3)

Later on (*Tactical Theory* 1.2), Aelian identifies Frontinus explicitly as an author (as he asserts the literary and scholarly foundations of his own expertise). Indeed, Frontinus stands out as the only Roman in a list of notable military writers whose works Aelian has read; and it is possible that the very format of his *Tactical Theory* was influenced by the *Strategemata*.[10] It is significant, however, that in the story of their meeting it is Frontinus' consular status and his practical experience that Aelian chooses to highlight. His literary activities are implicit in the background, but the hands-on connotation of the word *empeiria* identifies Frontinus' campaigns in Britain and elsewhere as the foundation of his authority and expertise and the main reason why his views on military matters (and Aelian's literary project) might carry some weight. Frontinus' endorsement of Aelian's *Tactical Theory* is not authoritative because he is a leading light in a long scholarly tradition. Indeed, as flattering as his inclusion in that list of Aelian's sources might look, it serves primarily to point up the overwhelming dominance of Greek learning on the subject and to undermine Rome's contribution.[11] Rather, Frontinus' supposed support of Aelian is meaningful because of his political prominence and connections; and also (especially) because Frontinus can serve as an embodiment of the contemporary Roman military practice/prowess that Aelian's Greek 'science' is trying to compete with.

In the preface to his edition of the *Strategemata* quoted above, Marius d'Assigny identifies Frontinus both as an author (whose text has much to teach the valiant English about the 'art' of war) and as a 'warrior' and consul. He is ROMAN, and the authority of the Classical past is one of the things that makes him worth reading; d'Assigny, like Vegetius, turns to Frontinus in part because he hails from the height of the Roman Empire. But it is also the access that he gives to other historical figures, PERSIANS, GREEKS, ROMANS and CARTHAGINIANS, that attracts d'Assigny. For him, the authority of the *Strategemata* lies as much, if not more, in the expertise of the 'experienced commanders' whose 'remarkable stratagems' Frontinus has collated. The author's scholarly, political and military credentials count for something; but readers will learn even more from the characters who inhabit the text and whose deeds Frontinus (and now d'Assigny) has

---

[10] For instance, commentators often see Aelian's use of sub-headings as his own innovation (Devine 1989: 32; Stadter 1978: 118), but he may have been copying Frontinus in this.

[11] In fact, Frontinus is characterised there (somewhat unjustly) as a commentator on other writers, not a theorist in his own right.

helped to immortalise. Indeed, they will learn from others too: for d'Assigny appends to his translation of the *Strategemata* 'A Collection of the Brave Exploits and Subtil Stratagems of several famous Generals since the Roman Empire' and, to follow that, 'A Discourse of Engines used in War'. Like de Pizan, among others, he not only 'Englishes' FRONTINUS, in other words; he updates him, leaning on his various layers of authority to generate some of his own, but also alerting us to its limits – to the fact that Frontinus (like d'Assigny himself) does not have a monopoly on military know-how, but is one step in an ongoing process of pooling and re-circulating many people's (different forms of) expertise.

The story of the *Strategemata*'s reception is much longer and wider-ranging than that; but these three episodes offer a taste of the variety of responses to and uses made of the text by later readers. Between them (and this is why they were chosen) they also expose the multiplicity of axes along which authority and expertise are constructed – or at least explored – in the *Strategemata* itself. For as this chapter will show, textual and scholarly authority share the stage with hands-on experience and the native wisdom of men from days gone by. That combination is not uncommon; indeed, it is evident in several of the other texts discussed in this volume. However, this chapter will argue that in Frontinus' *Strategemata* scholarly learning and practical know-how end up in tension more than in partnership with each other in particularly thought-provoking ways.[12]

Textual Authority and 'Scientific' Expertise

The first kind of authority that Frontinus lays claims to at the start of the *Strategemata* is of the scholarly, textual, 'scientific' variety. By way of introduction, he reminds us of the existence of his (now lost) *De re militari*, which (he claims) is what inspired the present text:

> Since I, alone amongst those studying it, have attempted to draw up (*instruendam*) a science (*scientia*) of military matters, and since I seem to have achieved my objective, as far as my efforts could manage, I feel that the project I have begun still requires me to collect together in a serviceable handbook (*expeditis amplectar commentariis*) the clever deeds of generals (*sollertia ducum facta*) which the Greeks have gathered together under the one name 'strategemata'. (Frontin, *Str.* 1 pr.1)

---

[12] Cf. Formisano in this volume on the tension between practical experience and textual authority as a central feature of ancient and modern writing on strategy.

His industry, which borders on perfectionism, is impressive, but on its own does not render him particularly authoritative.[13] More significant is the systematisation of material implicit in his use of *instruo* and *scientia*: for these set Frontinus up not merely as a conveyor, but as a refiner of knowledge, a theorist even. In addition, his suggestion that he is one of the first to discipline military know-how thus (despite being a stock and highly debatable claim) stamps his mastery over it yet more forcefully.[14] By dint of his earlier writing, and the proclaimed originality of the 'scientific' approach that underpins it (which itself taps into a wider trend of systematisation of knowledge that had long been associated with authority and expertise), Frontinus figures at the start of the *Strategemata* as a – if not *the* – contemporary expert on military matters.

In alerting us to his own achievements as an author, he also positions himself vis-à-vis other literary and scholarly authorities. His use of Greek terminology to explain what he means by 'the clever deeds of generals' (*sollertia ducum facta*), for instance, invokes a strand of the Greek military writing tradition and identifies the *Strategemata* as a descendent of it.[15] But it perhaps also invokes a Roman tradition too, that of supplanting Greek models with new Roman equivalents. Frontinus' sentence structure here surrounds the Greek military writing tradition (*quae a Graecis una* στρατηγημάτων *appellatione comprehensa sunt*) with his (very Roman) new version of it (*sollertia ducum facta ... expeditis amplectar commentariis*[16]), hinting – not least through the suggestive military metaphors lurking in *expeditis* and *amplectar* – at the possibility that he is not merely adopting a Greek model, but besieging and taking it over.

The historiographic tradition is another coordinate that Frontinus uses to characterise his *Strategemata* and assert its authority. He points out overlaps (from which his own text derives some associated validity and status): instructive *exempla* can be found in many historical texts. Indeed, much of the material in his treatise has already been recorded elsewhere: 'I neither ignore nor deny the fact that in the course of their works historians have also included this feature, and that all significant examples have already been set down by writers in one way or another' (*et ab auctoribus exemplorum quidquid insigne aliquot modo fuit traditum*) (*Str.* 1, pr.2). The juxtaposition

---

[13] Although, as Wietzke in this volume argues, industry itself – particularly in a work that could be said to benefit the community (cf. Frontin, *Str.* 1 pr.3) – confers a degree of social authority on the author.

[14] Lenoir 1996: 82; Santini 1992: 984–5; Wheeler 1988: 19–20.

[15] Cf. Valerius Maximus 7.4, whose suggestion that there is no equivalent Latin term for *strategemata* also implies that 'stratagems' (or at least writing about them) are originally a Greek tradition; also Wheeler 1988: 25–49 on the Greek history of the genre.

[16] The *commentarius* is a particularly Latin genre, of course.

here of *ab auctoribus* and *exemplorum* indicates that as well as historians Frontinus is keen to connect his work with a specific off-shoot of historiography, the *exempla* tradition. In fact, an observant reader will notice that what he goes on to say – about what distinguishes his *Strategemata* from these other texts – invites particular comparison with Rome's most famous exponent of that genre, Valerius Maximus. For, when Frontinus claims that his *Strategemata* will spare readers the tedious task of sifting through the vast body of historical writing to look for scattered examples themselves, he is echoing Valerius Maximus' own prefatory remarks.[17] Frontinus goes on to assert the *Strategemata*'s superiority over even Valerius' kind of writing, however; for he argues that authors of *exempla* collections, no less than historians, still 'confound' the reader with the volume of their material, despite the fact that they are in the business of excerpting from histories ('those, too, who have made selections of notable examples have overwhelmed the reader with, as it were, a great heap of information'). He thus highlights links between his treatise and other well-established genres, and derives some associated authority from them (and the overlaps that he points to between his writing, historiography, and so on, serve as a useful reminder that distinctions between 'technical' and other kinds of text/genre were not nearly so clear-cut in antiquity as many studies suggest). However, he also claims to stand out from them, to offer his readers something different.

Systematisation comes to the fore again as Frontinus explains what it is that sets his *Strategemata* apart. Rather than far-flung anecdotes or an overwhelming mass of material (the imagery here makes quite an impression), he promises a collection of examples that has been organised, fittingly, with military precision:

> My effort centres around the challenge of setting out precisely whichever example is required, in any given circumstance, as if in response to questions. For, having surveyed the categories, I have prepared a set of suitable examples as one might prepare a plan of campaign. Moreover, so that they might be divided up and organised according to the variety of their subject-matter, I have separated them into three different books: in the first are examples that relate to pre-battle activities; in the second, those that pertain to the battle itself and to the resolution of conflict; the third will contain 'stratagems' for the formation and breaking of sieges. (*Str.* 1, pr.2)

The very military-ness of this layout and Frontinus' authorial approach lends both text and writer an air of martial expertise; indeed, the military

---

[17] Comments that Frontinus makes later in the preface (*Str.* 1.pr.3) on the theme of not being exhaustive are also reminiscent of Valerius Maximus' preface.

metaphors that he uses to describe the *Strategemata*'s organisation perhaps hint at his personal talent for or experience of command. But (as in the case of the *De re militari*) it is also the text's methodical discipline, the systematic nature of its composition, that generates authority – a feature which subsequent prefaces flag up, and which is reinforced too in the lists of business-like section headings that begin each book and take us step by step through every stage of battle by directing us to the relevant set of anecdotes for each one.[18] In comparison with the texts and literary traditions on which the *Strategemata* has drawn, Frontinus' authorship seems (or is meant to seem) not only considerate to his readers, but impressively rigorous.

This returns us to the image of Frontinus as a cutting-edge author with which the preface began. And if we wanted further proof that he is keen to establish himself as a big name in the world of military writing, we have only to turn to the start of the text proper, where Cato the Elder pops up as Frontinus' first *exemplum* (*Str.* 1.1.1). We meet him in action, in his capacity as one of Rome's most successful commanders; but he is also writing – albeit letters, designed to outwit any Spanish rebels, and not the texts for which Vegetius would later revere him. Nonetheless, his presence heading up the very first section of the *Strategemata* (leading us into battle as it were, if we want to pursue the text's penchant for military metaphors) is significant. It signals to his readers that Frontinus knows what he is doing (who else should a Roman military writer worth his salt start with?), and that his text takes its inspiration first and foremost from the man credited with establishing Rome's military writing tradition. However (as in the case of Frontinus' self-positioning vis-à-vis the Greek *strategemata* tradition), it may also do more than that; for in returning Cato to the field of battle (rather than explicitly foregrounding his literary achievements, as Vegetius did), Frontinus may be subtly (even subconsciously) suggesting that he is a successor of Cato's whose own writing on tactics could eclipse that of his eminent and learned predecessor.

Non-Textual Authority and Unscientific 'Expertise'

Cato's characterisation in the opening *exemplum* of the *Strategemata* as a (literate) general, a doer not just an author, is interesting also because of the

---

[18] On both the pragmatics and rhetoric of 'tables of content', see esp. Riggsby 2007 (who overlooks Frontinus' *Strategemata* in his discussion).

way in which it shifts the text's emphasis from scientific rigour and book-learning (the themes that have dominated the preface – and that are foundational to so many other military treatises' claims to authority) to less scholarly phenomena, such as native intelligence and practical reasoning. In fact, unscientific 'expertise' dominates the bulk of the text. For, having set himself up as an authority in the preface, Frontinus departs the arena and leaves it to the generals who populate each section to provide the instruction. And, far from drawing on any textual tradition, they rely on their wits. For all its literary-scientific foundations, the *Strategemata* promises its readers illustrations of what commanders have done by 'ingenious resourcefulness' (that is the force of *sollertia ducum facta*); 'for in this way future commanders will be surrounded by examples of both *consilium* ("deliberation" or "judgement") and *providentia* ("forethought"), and these will nurture their own ability to think up and execute similar deeds…' (*Str.* 1, pr.1).

Take the first section of Book 1, which contains illustrations of the ways in which commanders have successfully concealed their plans. Here, as throughout the treatise, each *exemplum* begins with the name of the commander whose stratagem is being recorded, reinforcing the sense that it is they who are real authorities here, in both a military and a didactic sense. Their dominance of the narrative (as well as kick-starting every *exemplum*, they are the subject of most of the main verbs) attests to their ascendancy on the field of battle; but it also establishes them not just as the tactical lessons to be learned (the models to emulate), but as the readers' teachers. They are the figures whose thoughts and voices we get to hear, judging, deciding, coordinating and commanding; Frontinus, by contrast, almost never interjects to offer any commentary of his own (a feature we will come back to). His style of narrative does influence the way in which we react to them, however. For in example after example, we move rapidly from a commander thinking or wanting something to him acting and achieving it – with no reference, usually, to the episode's wider context, or to any historical precedent or future repercussions for that matter. And this brevity and simplicity, the reductive economy with which Frontinus recounts each anecdote, repeatedly presents tactics as a matter of on-the-spot intuition, wisdom and decision.

At *Strategemata* 1.1.1, for example, we learn that Cato no sooner 'reckoned' (*existimabat*) that the Spanish cities that he had vanquished might rebel against him than he took steps to prevent them from doing so.[19] He wrote to each, ordering them to destroy their fortifications, and threatening

---

[19] *Existimo* is a verb which conveys some of the imprecision of mental reasoning, flagging up Cato's agency in judging, deciding.

war unless they obeyed straightaway; and he ordered the letters to be delivered to all the cities on the same day. No details of the wider campaign are provided; we get only this compressed description of Cato's concern and the stratagem that he came up with in response to it.[20] Combative verbs abound as he switches from thinking to doing; and the brevity of clauses and rapid alternation between Cato's actions (*scripsit, minatus, iussit*) and the activities that he demands of the Spanish cities and his envoys (*diruerent, obtemperassent, reddi*) conveys both the speed with which he will respond if his instructions are not quickly obeyed and his decisiveness in penning and dispatching them. The result (apparently) is instant. In reality, the coordination of their delivery would have delayed the letters' arrival until those destined for the farthest cities had had time to reach their goal (as Appian pointed out, *Hisp*. 41). In this account, the letters are no sooner sealed and sent than they are received – and acted upon: 'Each of the cities thought that the order had been for them alone; if they had known that the same message had been sent to all of them, a joint refusal would have been possible.' Thanks to his foresight, and with a few strokes of his pen (and that of Frontinus), Cato has tricked every city in the region into swift capitulation.

And so it goes on. Again and again, over the course of four books, fifty sub-sections, and nearly 500 more or less formulaic *exempla*, commander after commander notices/realises/discovers/believes/fears (*animadverto, sentio, intellego, compero, didico, vido, scio, credo, timeo, vereor*); he quickly thinks/decides/plans/desires (*arbitror, statuo, constituo, peto*); and then he acts – and invariably succeeds with immediate effect. Occasionally the protagonists are whole nations ('The Romans', 'The Athenians', 'The Thracians'[21]) or groups of commanders ('certain Spartan generals', 'the survivors of the Varian disaster'[22]); and in Book 4 commanders sometimes collaborate with the Senate or consuls.[23] For the most part, however, the *exempla* concern individuals who can take sole credit for their triumphs (at least as Frontinus narrates them). And it is their reasoning, judgement, common sense, use of logic, wisdom, intelligence, cleverness, resourcefulness, inventiveness and cunning that wins the day. As noted above, Frontinus' authorial absence from the main body of the text means that he rarely comments

---

[20] Cf. Livy 34.17 and Appian, *Hisp.* 40–1, where we discover, e.g., the location of the cities in question, and learn more about Cato's motivations.
[21] This is particularly true in a couple of sections of Book 3: e.g. *Str.* 3.15.1–3 and 5, 3.13.1–2, and 3.18.1–3; also, e.g., 1.3.4; 3.17.1.
[22] *Str.* 1.4.12 and 3.15.4; also, e.g., *Strat.* 3.13.3–5, although these *exempla* may be later interpolations.
[23] E.g. *Str.* 4.1.18, 20, 24, 25, 28, 38.

explicitly on a stratagem or a commander; however, the vocabulary that he uses to characterise them in passing – *consilium, sententia, prudentia, ratio, calliditas, sollertia* – tells a consistent tale.

There are exceptions. In a couple of anecdotes (1.10.1 and 4.7.6) we are told that a commander was prompted towards a particular stratagem by experience (*experimentum*). Another (2.3.7) employs veteran troops that had been 'long trained' (*diu edocto*) and were 'practised' (*peritus*) – one of the few references in the text to training/instruction (of troops, of course, not commanders). A few *exempla* later (2.3.15), Mark Antony has recourse to a technical manoeuvre (the *testudo*). These references to experience, training and specialist methods are unusual, however. In the preface to Book 3, Frontinus explicitly rules out discussion of technological operations (and with it, the need for any associated specialist learning) on the grounds that military engineering has nothing to contribute (any longer) to the formulation of stratagems. His commanders rarely base their schemes around set-piece manoeuvres and typically depart from, rather than follow, conventional practice. And though in one *exemplum* (2.6.10) the 'hero' (Pyrrhus) is the author not of a cunning deed but of a collection of 'precepts on generalship' (*praecepta imperatoria*) from which one specific stratagem is drawn, textbooks play no formative part in any of the stories that the *Strategemata* sets out.[24] At no point do we ever see a commander reading a military manual – or any kind of commentary, or history, philosophy, or even epic, for that matter (just the odd letter, which invariably outwits them).[25] Nor do we see one devising a plan by copying a precedent.[26] Established procedures and principles pop up from time to time (often to be bypassed or adapted); but their input is drowned out by the volume of stories that showcase off-the-cuff, out-of-the-box, non-specialist, 'unscientific' intelligence. As the bulk of the text post-preface presents them, military stratagems – and, by extension, generalship itself – have little to do with learning or indeed teaching; they rely on individual nous.

Between *Sollertia* and *Scientia*

The scholarly, almost 'scientific' authority that Frontinus establishes around himself at the start of the *Strategemata* (and that is typical of many a

---

[24] Pyrrhus' *Art of War* is mentioned by Cicero (*Ad. Fam.* 9.25.1) and Polyaenus (6.6.3).
[25] *Str.* 1.1.1; 1.4.13; cf. 3.1.7–8. Cf. Veg. *Mil.* 3.10.17–18 (discussed by Formisano in this volume).
[26] By contrast, exemplary figures in Valerius Maximus are sometimes described as imitating or learning from other exemplary material (Langlands 2008: 163, n. 14).

military author) thus gives way to a very different kind of 'expertise' over the course of the text (one often celebrated in more historical texts) – if the innate 'cleverness' of lots of different generals can indeed be called 'expertise'. In fact, that is one of the questions which the *Strategemata* raises. What does 'expertise' in the strategic context consist of – a high degree of prescribed, specialist knowledge, or a less disciplined, less fathomable and less acquirable kind of skill? Does the text endorse the shared authority of the scholarly tradition to which it claims to belong? Does it give more weight to a more solitary, intuitive kind of 'know-how' that does not arise out of that tradition? Or does it champion both – or neither? To put it another way, how do the text's different authority figures – its erudite, systematising author and the hundreds of adroit but not conspicuously learned generals to whom he entrusts the task of instructing his readers – relate to each other? Do the different models of expertise and authority that they embody work in partnership, in parallel or in tension with one another?[27]

Cleverness and cunning were identified as a crucial feature of successful generalship from Homer onwards, of course. Indeed, the 'wiles' of Odysseus have long been shorthand for the whole of the more cerebral side of war, the antithesis of plain might or simple valour. And many texts testify to a widespread assumption in Roman society in particular that generalship was more a matter of practice, character and innate ability (as well as social status) than scholarship or science. (It is that kind of assumption that Rycharde Morysine is also arguing against in the preface to his sixteenth-century translation of the *Strategemata* quoted above.) In the *Pro Fonteio*, for instance, Cicero hails the generals of former days (who represent an ideal whom today's lesser men would do well to emulate) for their *virtus* ('valour'), *industria* ('energy') and *felicitas* ('good fortune') in military matters, and states outright that these men, highly skilled in waging war, were not trained in any military science that came from books (*non litteris homines ad rei militari scientiam*...), but by their own deeds and successes (*sed rebus gestis ac victoriis eruditos*).[28] Similarly, when celebrating Pompey's extraordinary military prowess in his speech *On Pompey's Command*, Cicero emphasises not only its practical (as opposed to theoretical) foundation (going into the army straight from school, Pompey could boast more encounters with the enemy than any other man; indeed, he had conducted more campaigns than other men have read of), but also its basis in that

---

[27] Cf. Formisano in this volume, who identifies the tension between theory and practice as a recurring feature of the discourse of war, and one that has a habit of destabilising the authority of texts.

[28] *Pro Fonteio* 42–3. On this and some of the following passages, see Campbell 1987: 21–3.

more elusive phenomenon, ability – and its natural consequence, success: 'as a young man he became learned (*erudita*) in the science of war (*ad scientiam rei militaris*) not through other men's prescriptions but through his own commands, not through the set-backs of battle but through victories, not through mere service but through triumphs'.[29]

That is not the whole picture, of course; it was widely recognised that most generals acquired at least some of their know-how from sources external to their own experience. In *Epistles* 8.14.4–5, Pliny the Younger (a contemporary of Frontinus) looks nostalgically back to the time when it was the established custom for aspiring young commanders to learn from their elders (a practice now problematised, he claims, by the lack of *virtus* in all generations under Domitian): 'It used to be the custom in days gone by that we would learn from infancy upwards from our elders, not only by listening but also by watching, and so acquire a sense of the things that we ourselves must do and pass them on in turn to our juniors. Thus young men were initiated into military service right away, so that they might get used to commanding by obeying, and to leading by following.'[30] In Tacitus' *Agricola* (5), we see Agricola himself acquiring skill (*ars*), practice (*usus*) and ambition (*stimulus*) not only through his own early hands-on experiences, but also by 'learning from the skilful, and following the best'. And Cicero praises another general, Lucullus, not only for his 'talents', but also for his 'industry' and 'enthusiasm', which led him – when posted to Asia to campaign against Mithridates – to spend the whole of his journey there 'questioning experts on the one hand and reading about past deeds on the other'. 'Thus he arrived in Asia as a finished commander, despite having been unversed in military matters when he set out.'[31] The potential of book-learning was acknowledged, in other words, alongside other instructive forces. Indeed, if Vegetius is to be believed, that was the impetus behind Cato's (practically inspired) *De re militari*.

The reading that Lucullus does (like Cato's writing) is essentially an extension of the oral tradition that Pliny romanticises – the kind of book-learning associated with historical texts and the *exempla* tradition (and with

---

[29] *De imp. Cn. Pomp.* 28. See also, e.g., *Balb.* 47, where Cicero implicitly contrasts Gaius Marius' military know-how with a more theoretical kind of study; and Sall. *Jug.* 85 (discussed by Formisano in this volume).

[30] Pliny is romanticising the transmission of what might be termed 'tacit knowledge', 'things that cannot be articulated in a written form, and whose transmission requires socialisation with the expert, or with the expert community' (Cuomo 2011: 327); this is different from the kind of intuitive, inborn 'cleverness' that the *Strategemata*'s generals tend to display.

[31] Cicero, *Lucullus* 1.1–2.

the education of the young by their seniors), not with theoretical or scientific works.[32] Other authors promote the relevance of more specialist, systematising, even 'technical' texts, however; in fact it is clear that, although innate ability and practice were highly valued, they were often seen as something that could be combined with more formal learning. In a devastating critique of some Achaean generals, for instance, Polybius famously argues that there are three routes available to those who want to acquire an understanding of the art of generalship: the first is the study of memoirs and the campaigns narrated in them; the second is the study of the systematic doctrines of experienced men; and the third is personal experience and practice. (The Achaean generals, he claims, were ignorant of all three.[33]) Xenophon's earlier (mid-fourth-century) *Discourse on the Command of Cavalry* certainly flirts with the idea that a military commander might learn his trade at least in part by following prescriptions set down in a treatise by an expert; although, at the same time as propounding some universal principles and even a degree of technical expertise, the text acknowledges the limitations of books and the relevance of both experience and ingenuity.[34] At about the same time, in his only surviving treatise, *How to Survive under Siege*, Aeneas Tacticus weaves *exempla* together with a more 'scientific' approach by inviting readers to contemplate past practice (via lots of illustrative anecdotes) at the same time as establishing a canon of definitive methodologies and directing them to other treatises that he has written (e.g. 7.4; 8.2–5; 21.1–2).

From early on in the Greek military writing tradition, in other words, 'scientific' learning, experience and nous were brought into (a shifting) dialogue with each other; and that trend was not restricted to Ancient Greece. In the tenth book of Vitruvius' treatise *On Architecture*, for example, a series of architects and engineers outwit various military commanders and win decisive victories for their own generals and countrymen by employing a mixture of *scientia* (precisely the kind of specialist, technical knowledge that Book 10 claims to transmit) and *sollertia* and *consilia* (native cunning and shrewd judgement).[35] As Serafina Cuomo has pointed out, when Vitruvius was writing, the question 'what makes a good military leader?'

---

[32] On the likelihood that Cato's military writing was part of his set of instructions destined for the education of his son, see Lenoir 1996: 84; also Astin 1978: 184–5.
[33] Polybius 11.8.1–2.
[34] See esp. the treatise's closing section, 9.1–2; also 8.1–3, on technical expertise; 5.4, on experience; 5.1–3, 5.9–11 and 7.1, on ingenuity and ruses; cf. *Memorabilia* 3.3.
[35] Vitr. *De arch.* 10.16, where one of the generals outwitted by this combination of *scientia* and *sollertia* is none other than Julius Caesar (A. König 2009: 49–50).

(birth, virtue, experience and/or specialized knowledge?) had become particularly urgent, following a rise in the prevalence and importance of technical expertise in recent conflicts, and we can see authors like Caesar grappling with it too.[36] Eighty years later, the ideal general that emerges from Onasander's *Strategikos* also represents a finely tuned balance of native qualities and acquired expertise. This text opens with a particularly forthright exposition of the principle that successful generalship depends at least in part on character (*Strategikos* 1–3). It returns time and again to the importance of cleverness, intelligence (ἀγχινοία, γνώμη) and experience (ἐμπειρία).[37] But it also toys with the possibility that collective strategic wisdom can be usefully systematised and handed on as a 'science'; indeed, it attempts to distil from past practice (στρατηγήματα – illustrative 'strategems') an overarching 'theory of generalship' (στρατηγικῆς δὲ περὶ θεωρία) that aims 'to get at the art of the general and the wisdom that inheres in the precepts' (*Strategikos* pr.3).[38] Time and again, in other words, ancient texts (and many subsequent ones too, for that matter) present generalship as an endeavour that operates somewhere between *sollertia* and *scientia*. As the appeal of systematisation and rules competes with (or asserts itself into) the complex reality of warfare, native wit, collective experience and more scholarly approaches are seen to complement – and even be indispensible to – each other.[39]

Unpicking Theories in *Strategemata* 1–3

In theory, Frontinus' *Strategemata* fits into and perpetuates that trend. Frontinus had almost certainly read and may well have been influenced by the likes of Xenophon, Aeneas Tacticus, Polybius, Vitruvius and Onasander, inter alia. Of course, we can only guess at the format and contents of his earlier *De re militari*; but his collection of 'clever deeds of generals', designed as it was (or so we are told) to supplement that earlier *scientia*, ought by its own reckoning to sustain the collaboration that other authors had long been mooting between unlearned strategic know-how/experience and more

---

[36] Cuomo 2011: 323–6; on the increasing importance of technical skill, see also Cuomo 2007b: 73.
[37] E.g. *Strategikos* pr.7; 21.3–4; 24; 32.9; 33; 42.10.
[38] On Onasander's text, see esp. Formisano in this volume.
[39] Even Aelian's highly technical *Tactical Theory* recommends combining precepts with practice (21.2–3); see also 3.4, where he distinguishes between the 'science' set down by Aeneas Tacticus and the more hands-on kind of training (*paideia*) that Polybius appears to advocate. Formisano 2009: 228–30 offers a particularly succinct survey of ancient military writing, and its oscillation between technical theories and evocative *exempla*.

specialist, disciplined, 'scientific' learning. Perhaps the two texts did complement each other, in all sorts of ways that we will never know about. Even if they did, however, there is no getting away from the fact that some aspects of the *Strategemata* work against, in real tension with, the momentum of more systematising, 'scientific' endeavours. In fact, in distilling *sollertia ducum facta* from both historical and more 'technical' sources and in rearranging them together (stripped of their contexts) into a new textual space, the *Strategemata* potentially unpicks the efforts of a huge range of texts – both scientific and unscientific – to theorise and idealise about generalship. It raises questions about the authority of wider literary and intellectual traditions (like historiography and epic) – in other words, not just military writing. And in the process, it challenges assumptions about the provenance, nature and status of individuals' strategic expertise and authority.[40]

For, despite its superficial organisation, a destabilising sense of chaos emerges as one reads the *Strategemata* through. As I noted above, the text's classification of *exempla* according to the various stages of conflict that a general might face makes an authoritative, rationalising impression. However, this organisation of material does not simply place like stories alongside each other in ways that illuminate particular strategic themes; it also juxtaposes anecdotes in various disorientating ways.[41] For instance, we repeatedly see stratagems that proved successful in one encounter being overturned (or adapted and turned back on the enemy) a few *exempla* later. Similarly, victorious generals are frequently defeated by others in turn – sometimes by the very foe we had just seen them vanquish. This emerges particularly clearly in Frontinus' presentation of Punic *exempla*: Scipio, Hannibal and a host of other Roman and Carthaginian commanders are frequently seen foiling or adapting each others' stratagems in quick succession, in a disconcerting back-and-forth between victory and defeat. No matter how much know-how or *sollertia* they have at their fingertips, the 'heroes' of the text are always on the verge of being outmanoeuvred themselves, with luck often playing a part. It is not simply that reliable patterns and methodologies fail to emerge (and are even undermined) as one reads each section through; the anecdotes are interspersed with each other in

---

[40] The arguments that follow are explored at greater length (and with more detailed discussion of illustrative passages of text) in my forthcoming book on Frontinus. They have something in common with Kronenberg's approach in this volume to Varro's *Antiquitates rerum divinarum*, in their openness to the possibility that Frontinus (like Varro) may be exposing to scrutiny – if not satirising, precisely – some of the literary and scholarly traditions that his *Strategemata* supposedly derives from and contributes to.

[41] On the tendency of readers to search for order, coherence, themes and subtexts even in miscellanistic writing, see J. König 2007.

ways which underline the profound unpredictability and uncontrollability of warfare. In the to-and-fro of battle (that emerges so powerfully from the to-ing and fro-ing in each section between Roman, Carthaginian, Spartan, Athenian, Sullan, Sertorian, Pompeian and Caesarian victories) even experience and on-the-spot ingenuity sometimes count for nothing – or emerge, at least, as having only ephemeral effects. Expertise of all kinds proves far from infallible, while generalship (it becomes clear) involves a good deal of chance.[42]

That message has obvious ramifications for a host of military treatises, particularly those with strong rationalising tendencies (many of which have a tendency to downplay the significance of chance and the unpredictability of warfare); but the internal dynamics of the *Strategemata* pose a challenge to other literary traditions too. The text's constant back-and-forth between different time periods dismantles familiar historical narratives, for instance. *Exempla* from Rome's various conflicts with Carthage, for example, are scattered all over the text, with episodes from the first, second and third Punic wars even merging into each other in ways that frustrate attempts to identify progress, decline or any kind of periodisation. Material that readers would normally encounter in historical works (as Frontinus himself points out in his introduction) is rearranged a-chronologically, according to military time – what happens when in a battle – in a way that foils many of the conventional and ideological interpretative moves readers are accustomed to make when trying to assess it. That is destabilising in a general way, but particularly so because such historical texts served both as an alternative and a complement to more technical military treatises. Thus, both traditional sources of strategic instruction and inspiration – the historiographic and the scientific – are being challenged here.

Even more disconcerting is the text's failure to observe or preserve geographical, ethnic or cultural divisions. Many of the surviving military treatises that set out to explore or establish enduring military principles, especially those written under the Roman Empire, combine that universalising project with one that asserts the distinctiveness of different nations. Onasander's *Strategikos*, for instance, invites readers to reflect on Roman (in comparison with non-Roman) military models;[43] and Frontinus' near-contemporaries Aelianus Tacticus and Arrian (in his *Ars Tactica*, for

---

[42] Of course, Frontinus is not alone in acknowledging the role played by chance in – and the unpredictability of – warfare; the *Strategemata* is unusual, however, in foregrounding both so prominently.

[43] *Strategikos* pr.3–4, where he claims that 'we shall consider above all the valour of the Romans'. Cf. Ambaglio 1981: 362–5, who argues that Onasander promotes specifically Greek principles

example) both differentiate between Greek and Roman military methods (in different ways, and with different agenda).⁴⁴ Of course, ethnic and cultural differences are a recurring (indeed, often a structuring) topos in ancient historiography too; in fact, questions about national identity (and the desire to define it) informed the composition and consumption of many ancient texts in a huge variety of genres.⁴⁵ Valerius Maximus' decision to distinguish between his Roman and non-Roman *exempla*, in other words, would have seemed more conventional to ancient readers than Frontinus' decision not to. For as well as jumping here, there and everywhere chronologically, Frontinus' *exempla* are organised in a way that criss-crosses all over the Greek and Roman worlds. From time to time, Frontinus identifies with Roman forces, referring to them as *nos*, or *nostri*; but the text does not promote national ideals or support Romano-centric historiography in the way that many others do.⁴⁶ Plenty of Roman stratagems arouse admiration, of course, and we see the borders of the empire being extended and defended. However, the to-and-fro of the text means that linear narratives of conquest and expansion give way to a more complex kaleidoscope of images that emphasises the frequent back-and-forth and convoluted inter-relations between Romans, Italians and other allies or subjects. Additionally, interspersed with *exempla* from inter-state conflicts is a significant number of anecdotes from Rome's various civil wars – another way in which Frontinus' text differs from that of Valerius Maximus, who explicitly eschews reference to civil strife (3.3.2).

Many of the macro-narratives that we are familiar with about Athens, Sparta, Persia, Thebes, Macedon and so on, are similarly broken up by the text's constant oscillation (within individual sections and across the collection as a whole) between different theatres of war; and also by the way in which that oscillation underlines the multinational dimension of many conflicts (and nations' histories). We repeatedly see Spartans, Gauls, Macedonians, Iberians and so on fighting on different fronts, with different allies and enemies, in *exempla* that are juxtaposed with each other. And the

---

and *exempla* to his Roman readers in order to establish the ongoing significance of Greece in Rome; and Formisano 2011: 45, who notes the lack of references to Roman history.

⁴⁴ Bosworth 1993; Stadter 1978: 41–5.

⁴⁵ As Harries notes in this volume, an emphasis on the Greekness or Romanness of a particular body of knowledge was often deployed in specialist texts to enhance the authority of the expertise they promised to share.

⁴⁶ Cf., e.g., Valerius Maximus (2.7.pr.), who identifies military discipline as largely the prerogative of his own race, 'the chief glory and mainstay of the Roman empire' no less. Overall, Roman *exempla* slightly outnumber non-Roman ones (see Campbell 1987: 15, n. 11 for the figures), but (as Gallia 2012: 197, n. 56 points out) '... the clever stratagems of Roman generals make up only 56 percent of the total'.

parallels and paradoxes that emerge from these jumps and juxtapositions, as different histories are brought into proximity with each other, expose history itself as a bewildering tangle of unpredictable oscillations between different peoples. In excerpting episodes from a huge variety of texts, in other words, and in rearranging them alongside other thematically related anecdotes from different times and places, the *Strategemata* does not simply criss-cross but somehow erodes (or exposes as inherently unstable) political and cultural boundaries. It teaches us that stratagems and successful generalship are not the preserve of one particular race more than any other; and in so doing, it also unpicks many other stories that are conventionally told about the Mediterranean past, obscuring rather than reinforcing a host of ideas that have built up about national distinctions and identities. Thus, it is not simply histories that are deconstructed here; fundamental aspects of foundation myths and national ideologies – the stuff of epic even – have their authority challenged; some of the very narratives, indeed, that often contribute to the construction of an individual general's or nation's strategic authority.

This may not be intentional; indeed, it is (in part, at least) an inevitable consequence of the excerpting process, what often happens when material is lifted from lots of different sources and arranged together according to new criteria (although Valerius Maximus offers a telling counter-example of a compilatory text that operates rather differently, as does Polyaenus' later *Strategemata*, which groups stratagems chronologically and geographically). Even so, the effect is disorienting – as is Frontinus' authorial absence. There are a few rare occasions when Frontinus slips in the odd word of personal analysis. He might note that a stratagem was carried out *ita perite* ('with such skill'), for instance (3.13.6), or he might identify a general's 'steadfastness' or the 'imprudence' of certain troops as significant (e.g. 2.7.11 and 12, where we see the ghost of some authorial interpretation). For the most part, however, he almost never intrudes to offer any guiding commentary – despite his own significant strategic experience, reference to which would have lent his voice extra authority. While Valerius Maximus' regular expressions of approbation, condemnation, exclamation and exhortation steer his readers towards particular interpretations, Frontinus suppresses the author's analytical, expert potential, leaving his material to speak for itself and his readers to identify what connections or themes they will.[47]

---

[47] As Langlands 2008: 162, n. 10 notes, authorial comment does not necessarily limit the potential multi-valence of *exempla*; lack of authorial comment, on the other hand, further enhances their 'undecidability' (Lyons 1989: 8–15).

Of course, much of Valerius Maximus' assertive interpretation is focused around his interest in exploring right and wrong, virtue and vice. Ethics are foregrounded in Onasander's *Strategikos*, too, as something that should guide a general in his military decisions and activities (and lend him a different kind of authority alongside his strategic expertise). In fact, many ancient discussions of warfare and generalship touch on ethical issues, be they historical, philosophical or more 'technical' accounts.[48] Though many before him had acknowledged that stratagems especially were often morally debatable,[49] this is not an angle we are encouraged to pursue in Frontinus' *Strategemata*. Unlike Valerius, Frontinus tends to avoid morally emotive vocabulary like *dolus* (trick), *fallacia* (deceit) and *vafritia* (craftiness), preferring the more intellectual language of *consilium* or *ratio* even above words like *sollertia* and *calliditas* when describing the thought processes of his generals. His lack of authorial commentary also means that accounts of the clever use of terrain or the deft way in which a lack of resources was overcome are narrated with the same compressed neutrality as, for instance, an episode in which Roman soldiers are induced to gorge on raw meat so that they may be attacked later that night while struggling to digest it (2.5.13).[50] Very occasionally, a character from within an anecdote questions the ethics of a commander's behaviour (at 3.12.2, 3.11.1 and 2.5.41, for instance); however, rather than pursuing or endorsing their protest, the text focuses on (and holds up for approbation) the end result of the stratagem, not its morality.[51] Ethics are thus subordinated to tactics, as virtuous and unscrupulous commanders triumph alike. Of course, this is the reality of many a history; but in the case of the *Strategemata*, which presents these *exempla* as authoritative models for readers to imitate, it may be a little more disconcerting – not least because they would be used to, or aware of, other military manuals that are more ethically judgemental or prescriptive. It is not simply that ethics are overlooked or subordinated, in other words; readers of the *Strategemata* are invited to contemplate acting unethically

---

[48] E.g., Thuc. 1.71–9 and 3.36–49; Plato, *Alc.* 1, 107c–9d; Plato *Rep.* 5.470c–1c; Plato, *Laws* 1.628c–e; Arist. *Eth. Nic.* 3.5–9 (1115a3–17b23); Arist. *Pol.* 7.2 (1324b2–5a10); Polybius 18.14–15 and 35–7; Cicero, *Off.* 1.34–41 and 61–8; Virgil, *Aen.* 12.919–52. See also Most 2011 on justice and war in Hesiod.

[49] Esp. Val. Max. 7.4; cf. Wheeler 1988: 2–18; 21; 109.

[50] Compare also Valerius Maximus' criticism of Carthaginian deceit (at, e.g., 7.4.ext.2) with Frontinus' authorial detachment at *Str.* 2.2.7 (where Hannibal's positioning of his troops so that the Romans are troubled by wind and dust, far from being deplored, is set alongside *exempla* in which Romans and Thebans deploy the same trick – 2.2.8 and 12 – undermining the distinction that Valerius tries to draw between Roman and non-Roman behaviour).

[51] See also, e.g., 2.11.5–7; and Wood 1967: 247 on this feature.

from time to time themselves (that is, to dispense with some of the moral parameters that, in dialogue with other social, cultural and political discourses, often boosted a general's wider authority, at least in retrospect), as the text prompts them to visualise themselves not just as another Alexander, or another Scipio, or Cato, but as another Tarquinius Superbus, or another Iphicrates, or Coriolanus.[52] More than that, they are (once again) prompted to reflect on the habits and validity of historiographic traditions that have long been valued for their didactic potential.

The variety of people whom the *Strategemata* suggests readers might emulate is potentially disconcerting in other ways too. No distinctions are drawn between kings and slaves, consuls and rebels, emperors and bandits, magistrates and mercenaries, and in a couple of *exempla* women show themselves as capable as men when it comes to devising stratagems. Perhaps more significantly, non-experts and one-day wonders triumph alongside experienced or virtuoso generals, and occasionally even overthrow them. This macro-trend brings us back to the impression that emerges out of individual *exempla*: that learning and training are less likely to determine a commander's chance of success than the wits he was born with and his on-the-spot 'cleverness'. But as the present section has shown, that impression is only one of the ways in which the *Strategemata* undermines the authority of some of the more systematising, theorising, idealising texts and literary traditions that touch on warfare or generalship and shape ideas about what it is that can make a general seem particularly authoritative (in his own day, and beyond it). For en masse, and through their organisation and presentation as well as their selection, Frontinus' *exempla* unpick historical, epic and philosophical narratives about political periods, national identities and behavioural boundaries that so often feed into (and are in turn sometimes influenced by) the narrower literary-theoretical tradition that the *Strategemata* claims to be part of: the 'drawing up' of a transferable 'science' of generalship. All strands of military writing (not just the technical tradition) are exposed to scrutiny here.

From Reading to Doing

As Marco Formisano has pointed out, we do not need to wait until the nineteenth century to find examples of military treatises that reflect as much on

---

[52] le Saux 2004: 99 notes earlier readers' anxiety about this feature of the *Strategemata*; cf. Livy (1.pr.10) and Valerius Maximus, who present *exempla* for both imitation and avoidance.

the 'essence' of war as they do on how to wage it; indeed, a number of ancient texts could be said to fall into that category.[53] The *Strategemata*, I suggest, is one of them: whether intentionally or not, its collation and presentation of strategic *exempla* across all four books teaches us as much about the chaotic nature of battle as what stratagems or strategy a general might adopt to try to control it.[54] In so doing, however, it also prompts reflection on some of the wider literary movements that it draws on and engages with – especially the systematising, 'scientific' side of the military writing tradition. Through the military precision of their internal organisation, Books 1 to 3 appear to distil history into a comprehensive set of instructive illustrations for every strategic eventuality. And yet the volume, variety and disorienting to-and-fro of the *exempla* within each sub-section combines with an emphasis on on-the-spot intuition, intelligence and native wisdom (rather than learning, training or experience) to frustrate attempts that we might make to relate them to – let alone derive from them – any coherent, definitive, idealising or self-promoting theory of tactics or generalship. Then Book 4 is tacked on, and sections 4.1 to 4.6 – which organise their *exempla* into moral categories such as 'On Discipline', 'On Justice' and 'On Steadfastness' – try out a different methodological approach that brings us closer to a more prescriptive, theorising (and indeed ethically oriented) kind of treatise, at the same time as moving us further away from the idea that such a treatise could ever be conclusive or completely convincing. (Book 4, after all, is presented as a supplementary afterthought to an already supplementary text, part of an ongoing expansion of Frontinus' original literary project).[55] Finally, section 4.7 (*de variis consiliis*) steers us away from virtues and back to 'sundry' examples of *consilium* (Frontinus' favourite synonym for *sollertia*, strategic 'cleverness' or 'cunning'). Intuition, intelligence and native wisdom (those potent but elusive qualities) take centre stage again, but in a miscellany of *exempla* that return us to many of the themes treated in Books 1 to 3 and emphasises more than ever the bewilderingly infinite array of

---

[53] Formisano 2011: 40–1.
[54] The authorship of *Str.* 4 was questioned in the nineteenth century (Wachsmuth 1860; Wölfflin 1875); but Bendz 1938 argued in favour of identifying Frontinus as its author, and has since been followed by the majority of commentators (e.g. Campbell 1987: 15; Campbell and Purcell 1996: 785; Gallia 2012: 204; Goodyear 1982; Malloch 2015; Turner 2007: 432; Wheeler 1988: 20).
[55] *Str.* 4 pr. distinguishes between Book 4 and the rest of the treatise, and stresses the potential incompleteness of the author's efforts: 'I shall now set out in this book examples that did not seem to come under quite the same category as those in the earlier books... namely examples rather of generalship than stratagems – as a way of completing the project of the three earlier books (if indeed I have completed them).'

situations that any general may face, and the near impossibility of employing even well-established strategic approaches in predictable ways.[56]

The very last *exemplum* in the book encapsulates that lesson in unpredictability beautifully. Quintus Metellus, the 'hero' of the tale, is in Spain, about to break camp; and in order to keep his troops in line, he deceives them, telling them that he has discovered that ambushes have been laid by the enemy. We are told that he did this *ex disciplina* – 'to maintain discipline'; but, by chance (*forte*), they then did meet with an ambush, and because the soldiers were prepared for one they were unafraid. In being unexpectedly providential, Metellus' approach advocates the strategic advantages of *disciplina* (something often promoted in technical, historical, ethical and epic descriptions of generalship); but it also reminds readers of the accidental, unforeseeable, uncontrollable forces of war, and of the chance that is involved in strategic success (or failure). In that sense, 4.7.42 offers a compelling conclusion to the *Strategemata*, insofar as it alerts readers one final time to a message that the entire collection has been building towards (right from Frontinus' very first preface): that, no matter how many supplementary bits are added, no text, not even one as expansive as the *Strategemata*, can hope to offer complete closure on or be definitively authoritative about tactics or generalship, because in war the unexpected will always happen.[57] As well as completing and complementing his earlier *scientia rei militaris*, in other words (for instance, by reinforcing *disciplina* as a principle), this vast and destabilising collection of *sollertia ducum facta* threatens to undermine it – and the authority of many other prescriptive, theorising, idealising or romanticising discussions of generalship and warfare – by making it clear how many endless permutations of strategic cleverness there are (and must be) because war itself presents an infinite variety of challenges.[58]

That is not to say that the *Strategemata* does not teach its readers how to think up stratagems or operate as a general. (In arguing this, I should stress, I am not suggesting that the *Strategemata*'s readers were all or exclusively interested in learning the hands-on practicalities of generalship. Many would have been reading with broader intellectual agenda – to enquire into the nature of military strategy, perhaps, or the history of generalship.

---

[56] In this breakdown of categories and textual organisation, Frontinus may even be satirising attempts by other writers to systematise military know-how successfully.

[57] Smith 1998: 163, n. 52 is thus wrong to suggest that Book 4 'tails off limply'.

[58] This takes one step further a disclaimer often made in 'technical', 'scientific' or encyclopaedic texts, that they cannot hope to be fully comprehensive. Cf. Doody's suggestion in this volume that Varro's *De re rustica* is another text which questions the authority/efficacy of writing when it comes to learning about agriculture.

Because the text casts all of its readers in the role of aspiring generals, however, we are all invited to consider what practical lessons we potentially learn from it, whatever our other agenda.) In fact, it is profoundly instructive even as it exposes its own limitations and those of the wider military writing tradition – precisely because it confronts readers with the fact that they can never hope to acquire an exhaustive set of easy-to-implement prescriptions and forces them into a different model of learning from the one that so many other texts default to (whereby an expert passes on general precepts to the uninitiated). At first glance, it might look as if Frontinus establishes a very simple relationship between his text (and the know-how that it embodies) and his readers (and the know-how that they supposedly aspire to): one that sets the *Strategemata* up as a direct conduit of strategic expertise between former and future commanders and implies a single step between reading about successful stratagems and performing some oneself.[59] The language that he uses, however, to characterise that relationship in the preface to Book 1 reveals that it is rather more complex than that, and that reading does not necessarily translate immediately or straightforwardly into an ability to do. This is what he says, after he has explained his reasons for writing the *Strategemata*:

> Future generals will thus be surrounded [or 'girded up' – *succincti*] by examples of wisdom and foresight, which will nurture (*nutriatur*) their own ability to think up (*excogitandi*) and generate (*generandi*) 'similar deeds' (*similia*). It will also follow that a general need not be nervous about the outcome of his own inventiveness (*inventionis suae*), because he will be able to compare it with experiments that have already been tried out. (*Str.* 1.pr.1)

This is explicitly nourishing, supportive language – indeed, the kind of language that might characterise a commander's relationship to his troops, for *succingo* can be used for 'equipping' an army. On one level, then, Frontinus as author (or commander in chief) is supplying his readers (or generals) with the 'equipment' they need to succeed as military leaders. However, the emphasis on 'thinking up', 'generating' and 'inventiveness' also puts the onus on them not merely to copy, but to come up with similarly successful stratagems of their own devising – and that inserts another, far more challenging step into the process. It is not simply a case of taking up the proffered 'equipment' that the text offers, in other words, and putting it into practice. The *Strategemata* demands something more from aspiring

---

[59] This is how many read it; e.g. Gallia 2012: 193.

generals/readers, something whose difficulty (as we have seen) the text itself underlines. For, although the vast litany of 'clever' commanders that the *Strategemata* parades before our eyes reminds us that a huge variety of people in the past have been successful strategists – and dangles the prospect of further 'like deeds' (*similia*) tantalisingly in front of us – its emphasis on the role played by both chance and native, on-the-spot, unlearned 'cleverness' makes the likelihood of achieving any comparable strategic triumphs of our own look dauntingly remote, even with a handbook to 'help' us.

We need not despair, however. The verb *succingo* can mean to 'hem in', or even 'besiege', of course, as well as to 'surround' or 'equip', and that is fitting because in a sense that is what the *Strategemata* does to its reader. It transports us to lots of tricky battlefield scenarios, and instead of offering us hard-and-fast rules it encircles (one might almost say ambushes) us again and again with an array of strategic *exempla* that undercut many of our assumptions about the foundations of generalship. And yet that process is itself didactic in quite a sophisticated way: for instead of offering us prescriptions in the manner of many 'technical' treatises, Frontinus gives his readers an experience of the on-the-spot processes of decision-making in unpredictable circumstances that a successful general needs to be expert in. In plunging us into these situations (which take us steadily through the different stages of battle) and surrounding us with so many different (and differently valid, sometimes controversial) strategic options, the text gives us the opportunity to hone our skills of generalship (or observe the nature of it) – to confront the difficulty of decision-making and to practise exercising our judgement, drawing on our nous and being inventive in conditions which are constantly shifting – rather than just offering a set of *exempla* that can be straightforwardly imitated.[60] That is similar to the way in which Rebecca Langlands has suggested Valerius Maximus' collection teaches us ethics;[61] but it also takes us back to a long-established and much romanticised non-textual tradition of learning about generalship that I mentioned above, the kind that Cicero talks about in his text *On Pompey's Command*, where commanders learn about generalship *by commanding* (victoriously/vicariously).

---

[60] Cf. Wood 1967: 246–8.
[61] On this, see esp. Langlands 2008, who draws attention to the interaction between strings of *exempla* in Valerius Maximus (which expose ethical complexity, not clarity) and argues (160) that Valerius' arrangement of ethical *exempla* 'is designed to tell Roman readers not simply what to think but *how* to think ethically, enabling Roman readers both to explore the scope of these moral categories and to develop their skills of moral reasoning'. See also Langlands 2011: 23; J. König 2007 on Plutarch's *Sympotic Questions*; and Rimell 2007 on Petronius' *Satyricon*.

The *Strategemata*, then, is not so much a nourishing as a challenging text – but that is what makes it instructive, in all sorts of respects. As we have seen, it brings different manifestations of 'expertise' (systematising, 'scientific' authors, and intuitive, unlearned but triumphant former generals) not only into collaboration, but into tension with each other. In so doing, it raises questions not just about what authorial and strategic expertise respectively consist of, but also about the authority of a whole range of texts which depict generalship, both within and beyond the military writing tradition (historical, philosophical and even epic, as well as 'technical' works). In fact, its overlaps and engagement with non-technical as well as more 'scientific' texts draw attention to the shifting identity of (especially Latin) military writing and to its lack of consistency as a genre.[62] More than that, in exposing the limitations of all such texts (itself included) to communicate reliably or conclusively about the nature of generalship, it prompts reflection (not necessarily consciously) on the relationship between writing, reading and doing – on the interplay between texts and the world that they aim/claim to reflect and influence. The *Strategemata* draws attention to a gap, on the one hand, between words and deeds – between the Art of Generalship as an intellectual, textual discipline and the act of generalship as a practical phenomenon. At the same time, however, the didactic method that it adopts (that is reminiscent of some non-textual approaches to teaching and/or learning) goes some way towards bridging that gap, by providing a textual space in which its readers may absorb and even practise the very skills that could enable their or others' future deeds to go down in history (books) as successful strategic *exempla*. A treatise that overturns assumptions about authorial expertise and textual authority, in other words, ultimately proves itself expert and authoritative in instructing its readers – on a number of different levels: practical, theoretical and ideological – in the elusive 'art' of generalship.

Authors, Generals and Emperors

The relationship between writing and doing brings us back one final time to the authorial persona adopted by Frontinus in the *Strategemata*, which is strikingly different from the one he employs in a later, Nervan treatise, the *De aquis*. In that later text, Frontinus is a dominant and indeed exemplary

---

[62] Formisano 2011: 42 and Lenoir 1996: 85 both note a lack of continuity between the surviving Latin military treatises.

presence: from its preface, where he parades his proximity to the Emperor Nerva; through catalogue after catalogue of facts and figures (in which autopsy and Frontinus' personal interventions are repeatedly flagged); to the closing section, where the weight of the expertise and authority that he has accrued over the course of the text (both through its preparation and its publication) gives him the power to adjudicate on the emperor's behalf, as his representative not just his right-hand man.[63] Doing and writing come together to reinforce each other in lots of ways in that treatise. In the *Strategemata*, by contrast, Frontinus absents himself from the bulk of the text, as we have seen, handing the limelight over to his vast collection of exemplary commanders and rarely intruding with any commentary of his own. Perhaps the most striking aspect of his authorial self-effacement is the fact that this experienced commander (whose text revolves around the didactic potential of experienced commanders) makes almost no reference to his own experiences as a commander.[64] For Frontinus was an actor not just an author in the military sphere (indeed, in addition to his involvement in at least one German campaign, Frontinus had served as Governor of Britain just before Agricola and succeeded in subduing parts of Wales during his tenureship).[65] In Book 4 we find one *exemplum* which discusses a campaign in which Frontinus was involved (4.3.14 – and that anecdote deserves a whole chapter to itself). Aside from that, there is no clear reference anywhere else in the text to any stratagems for which Frontinus was responsible. This failure to refer to his own strategic experiences is surprising particularly because there was a widely held assumption in many strands of ancient technical and scientific writing, not least in the military writing tradition itself, that an author's personal experience could lend his writing an extra layer of authority. That, after all, is one of the things that makes Cato an authoritative military writer.[66]

Of course, Frontinus may have referred to his own military experiences more in his lost *De re militari*. But there may also be a number of social and political explanations for his failure to bring his practical expertise and 'scientific'/scholarly authority into greater dialogue with each other in the

---

[63] A. König 2007.

[64] Cf. Varro in the *De re rustica*, who (according to Doody in this volume) steps back from claims he makes in the prefaces about the extent and authority of his practical experience/know-how. Cf. also the preface to Onasander's *Strategikos* (and Formisano in this volume), where the value of authorial experience is debated.

[65] Tac. *Agr.* 17.

[66] See also, e.g., Polybius 12.25g; and Marius' speech in Sallust, *Iug.* 85.12. On the use of the first person (and indeed the second-person) in technical/scientific texts, see esp. Hine 2009; also Nutton 2009.

Strategemata.⁶⁷ His textual self-presentation may have been closely linked to his public status, for instance. It is perhaps no surprise that he promoted himself and his activities so boldly in the *De aquis*, for by then he had become a triple consul and one of Rome's most influential statesmen; moreover, he was writing from the heart of – almost as a spokesman for – the imperial government. When he wrote the *Strategemata* (under Domitian) he had a lower public profile and was writing in a private capacity; and the persona that he adopts in that text may have been crafted in part to be commensurate with that: modest (as far as hands-on experience went), to suit a relatively modest CV. It was also the case, of course, that authors constructed and paraded authority and expertise somewhat differently under different emperors. Arguably, Frontinus' reticence in putting himself forward as an *exemplum* (not just a scholar) in the Domitianic text had something to do with the political context in which he was writing – just as his assertiveness in the *De aquis* exploited the Nervan-Trajanic regime's supposedly more liberal atmosphere.⁶⁸ The currency of doing/experience versus writing/learning – and the kind of authority that an author might derive from each – fluctuated in response to many factors.

By way of conclusion, it is worth considering the other side of that coin too: the ways in which textual images and ideas of authority and expertise shaped wider discourses. For Frontinus' presentation of generalship and interrogation of different kinds of authority and expertise within the *Strategemata* had implications, potentially, for Domitian and the principate itself, raising all sorts of questions about both. For instance, how did the *Strategemata*'s characterisation of generalship as an activity that almost anyone could turn their hand to but also something that was tantalisingly difficult to guarantee any success in (because native wit and chance were of more significance than learning or virtue) fit – or conflict? – with Domitian's attempts to inflate his own military expertise and so further cement his imperial authority? Might the unpredictable to-and-fro of victory between a host of Roman and non-Roman *imperatores* have reminded readers how unexpected, short-lived and unstable command could be, and how brutal the process was when it was challenged? (For Domitian and his contemporaries – whose memories of the conflicts and chaos of AD 68 to 69 would still have been fresh, and whose anxieties about the future were equally pressing – that would naturally have been a destabilising message, one reinforced

---

[67] This is a topic I pursue in my forthcoming book.
[68] Cf. Malloch 2015 and Turner 2007 on the politics of Frontinus' self-presentation in the *Strategemata*.

perhaps by the volume of civil war *exempla* in the text.) And what might the text's rejection of authoritative (indeed authorising) epic teleologies and moral fundamentals have done to the image and authority of the Roman Empire?

Such political questions are not the only ones we might ask; but it is fitting to end by gesturing towards the wider context in which the *Strategemata* was written. For constructions and expressions of expertise and authority within individual disciplines always intersect with constructions and expressions of expertise and authority in other spheres of activity (literary, cultural, social and political), in thought-provoking and sometimes challenging ways.

8 | The Authority of Writing in Varro's *De re rustica*

AUDE DOODY

non omnes qui habent citharam sunt citharoedi
Owning a guitar doesn't make you a guitarist
(Varro, *De re rustica* 2.1.3)

In Varro's *De re rustica*, owning a farm does not make you a farmer, any more than owning a cithara means knowing how to use it. It is a playful analogy, comparing the business of Roman farming with the pleasures of Greek music, and one that points to a central theme in Varro's dialogues on agriculture: the relationship between the elite Roman owners of country property and the knowledge they need in order to enjoy and profit from the land. Each of the three books of the *De re rustica* is addressed to someone who owns a farm, but apparently lacks knowledge on how to run it successfully. The first book, addressed to his wife Fundania, deals with arable farming, and is set at a dinner party that is called off abruptly when news comes that the absent host has been murdered in the forum. The second, on large livestock farming, is addressed to Turranius Niger, and set in Epirus when Varro was acting as a general in the war with the pirates. The final book is addressed to Pinnius, and deals with smaller livestock, such as birds, bees and game, and is set in the Villa Publica during the aedile elections. Varro writes of himself as a character in each of these dialogues, a participant in conversations with other elite Romans, both real and fictional, who exchange knowledge about farming in time borrowed from other responsibilities.

The first word of Varro's *De re rustica* is *otium*, a very Roman concept which connotes the range of cultured pursuits a member of the elite might pursue when not engaged in business. It is perhaps an odd word to choose to begin a work on agriculture, but it signals to us something crucial about the nature of Varro's own text and his attitude to his subject. Farming is about pleasure as well as profit, as Tremelius Scrofa is made to say:

> *agricolae ad duas metas dirigere debent, ad utilitatem et voluptatem. Utilitas quaerit fructum, voluptas delectationem; priores partes agit quod utile est, quam quod delectat.*

> Farmers should aim for two goals: usefulness and pleasure. Usefulness is directed towards profit, pleasure towards enjoyment. The useful aspects are more important than the enjoyable ones. (Varro, *Rust.* 1.4.1)

Varro's dialogue also seems to have two goals, to amuse and to instruct the reader, and recent work on the text has emphasised that the *De re rustica* is a self-consciously literary text, exploring its ludic, allegorical and literary features.[1] Choosing to employ a genre of writing closely associated with Greek philosophy to discuss Roman farming is already a sort of joke. The escapism and irony of the first book in particular, where the characters discuss farming while their host is being murdered in the forum, shares a perspective with other Roman writing which presents an idealised view of the countryside from the perspective of an urban elite.[2] The dialogue form frames agriculture as a subject for conversation between elites, at the same time as it offers the reader a highly stylised representation of speech, within its own literary conventions.

In Varro's carefully written dialogues, writing about farming emerges as a theme, and it is possible to find a certain wry humour about the authority and the usefulness of writing about farming. Varro's choice of genre allows for a more subtle commentary on the authority of written texts, inherited knowledge and contemporary practice than we find in other scholarly writing on agriculture at Rome. In the preface to Book 1, Varro tells us that his own knowledge about agriculture comes from three 'roots': what he has learned from working his own land, what he has read, and what he has heard from experts (*a peritis*) (*Rust.* 1.1.11). In the dialogues themselves, characters claim authority to speak on agriculture for a wider range of reasons, including punning affinities between their names and the topic at hand, or affinities with particular regions that are famous for particular forms of agricultural production. The two characters who produced books on agriculture, Tremelius Scrofa and Varro himself, are portrayed in ways that seem to subtly undermine their standing as men of practical knowledge. And when the work of Roman authors is discussed, their failings as writers seem to be mirrored by the tendentious and elliptical readings of their work that the dialogue dramatises. Questions about who writes about agriculture, where their knowledge comes from, how their work is read and by whom, thread through the discussions Varro constructs between himself and his peers.

The questions of who knows best about farming and where that knowledge comes from resonate within a wider Roman discourse on agriculture

---

[1] See Diederich 2007: 172–208; Green 1997; Kronenberg 2009: 76–131.
[2] Perhaps most obviously, Horace, *Epode* 2. On irony in Varro, see Kronenberg 2009: 76–131.

and the land. Roman agricultural writing, from Cato onwards, is mainly concerned with the management of large estates that relied on a slave workforce, supplemented when necessary by hired labour.[3] To some extent, producing books about how to farm was to suggest that the days of the self-sufficient aristocratic farmer and the hardy Italian peasant were consigned to history, or at least to cultural memory. Varro begins Book 2 by remarking that, these days, heads of families are more likely to be seen in the theatre and circus than out in the fields or the vineyards (Varro, *Rust.* 2.1.3). Pliny repeats for us the legend of fifth-century Cincinnatus, the senator called from the plough to lead the Roman army, who is used also by Cicero as an emblem for a past where senators farmed and fought for Rome with the same austere commitment – but only to contrast this nostalgic vision of the past with a present in which slaves work the land, an affront to Mother Earth.[4] Modern scholarship has developed a more complex picture of land-use and the relationships between slave and non-slave labour in ancient Italy than is obvious from our texts, where the emphasis is on elite owners and slave workers.[5] Their relative silence about the continuing work of the peasant farmer and free labourer is perhaps not an innocent one: Brendon Reay has argued that Cato's *De agri cultura* attempts to shore up the traditional role of the statesman-farmer by portraying the slave as an instrument of the master's will, denuded of humanity and agency.[6] Although Varro appeals to personal experience in his text, there is a potential crisis of authority for the Roman landowner, if too much weight is placed on the experience of working on the land. Writing, and the transfer of knowledge from writer to reader, has a powerful role in controlling and mediating the authority of experience in Varro's text. At some points in the *De re rustica*, the ultimate recipients of the weight of this written tradition are slaves, whose practice is directed by the presence of texts which instruct them in the owners' absence.

The premise of Varro's work on agriculture, made explicit in his three prefaces, is that it is possible to learn how to farm by reading about it in a book. Within the dialogues themselves, the politics and the practicalities of this are brought into question. My aim here is to explore the ways in which the authority of writing about agriculture plays out in Varro's dialogues by

---

[3] White 1970: 332–83.
[4] Plin. *HN* 18.20; Cic. *Sen.* 56; Cincinnatus is coupled by both authors with Manius Curius Denatus, whose example is supposed to have impressed Cato: Plut. *Cat. Mai.* 2.1–2. On the relationship between nature and the divine in Pliny, see Beagon 1992: 26–54.
[5] On non-slave labour and small-scale farms in Italy, see: Erdkamp 1999; Foxhall 1990; Goodchild and Witcher 2010; Kron 2008; Rasmussen 2001; Rathbone 2008; Rosafio 1994.
[6] Reay 2005.

examining three aspects: the implications of the ways in which slaves are represented as readers; Varro's treatment of earlier Greek and Roman writers on farming; and the characterisation of Varro and Tremelius Scrofa, the writers in the text. As we will see, writing in the *De re rustica* seems oddly vulnerable to fragmentation, dismantlement and misunderstanding, at the same time as it asserts a claim to secure knowledge, and authority over its readers.

Slaves as Readers

In the first book of his *Res rustica*, Columella tells us that he sometimes asks slaves for their advice on the work that they are doing:

> *Iam illud saepe facio, ut quasi cum peritioribus de aliquibus operibus novis deliberem, et per hoc cognoscam cuiusque ingenium, quale quamque sit prudens. Tum etiam libentius eos id opus aggredi video, de quo secum deliberatum et consilium ipsorum susceptum putant.*

> What I often do now is discuss whatever the new work is with them as though they were the more expert, and by doing this I can find out each one's abilities, and what their skills might be. And also I find that they approach the work more enthusiastically, when they believe they have been consulted and their advice taken on it. (Columella, *Rust.* 1.8.15)

Columella raises the spectre of slaves who might be more expert than their master, and whose practical expertise might be valuable to the running of the farm, only to quash the thought. Here, the master asks the slaves for their advice as a means of judging the slaves' expertise, as a clever ploy to raise productivity, not because he knows less than them. Elsewhere, Columella stresses the importance of knowing as much as your estate manager (*vilicus*), so that you can properly instruct him and judge his work, in terms that make clear that contemporary Romans are not always in a position to do so (Columella, *Rust.* 11.1.4–7). He cites Xenophon and quotes Cato, who states the problem with trenchantly elegant economy: *Male agitur cum domino, quem vilicus docet* ('It doesn't go well for the master who is taught by his *vilicus*') (*Rust.* 11.1.4–5). Part of the point of these works on agriculture is to pass on knowledge, from elite Roman to elite Roman, so that the landowner has both the authority and the expertise to instruct his manager, who in turn instructs the workers on how to farm the land.[7]

---

[7] On the hierarchical relationship between writer and reader in Columella's work, see Henderson 2002: 118–19.

Agricultural literature allows knowledge to flow downwards, from the master to the *vilicus*, and finally to the workers, who in these texts are almost always slaves.[8] When slaves become readers, this neat chain of command can appear threatened in interesting ways. In Varro's *De re rustica*, in certain circumstances, written texts can be used as a means of imparting knowledge to slaves: a farmer's calendar is to be inscribed on the wall of the villa; the slaves in charge of looking after sick livestock and sick slaves are to carry around written instructions, some of which are derived from the agricultural writings of Mago the Carthaginian.[9] For Varro, writing can be used as a means of asserting the value of a particular form of inherited knowledge over the practical experience of the slave.

The management of slave labour is an important theme in Roman writing on agriculture, and Varro's treatment of the topic is perhaps especially dehumanising.[10] Cato had already asserted that an old or a sick slave should be sold off like unwanted equipment or worn-out oxen (Cato, *Agr.* 2.7), a practice that Plutarch was to criticise as unnecessarily heartless (Plutarch, *Cat. Mai.* 5.1). In Book 1 of Varro's text, we find the famous characterisation of a slave as *instrumentum vocale*, a speaking tool (Varro, *Rust.* 1.17), and in Book 2, slaves are treated as just another type of livestock, who, with a bit of ingenuity, can be discussed under the same headings as farm animals (Varro, *Rust.* 2.1.25, 2.10). The conceptual effort involved in treating slaves as livestock is highlighted in the dialogue: Tremelius Scrofa has laid out his nine headings and nine categories of animals in a virtuoso display of organisational skill, when Atticus interjects that the categories of breeding and bearing young do not apply to mules or slaves. A discussion follows in which it is agreed that these categories can apply to slaves though not to mules, and so, somewhat bathetically, shearing and dairy products are to be thrown in as a substitute so that eighty-one subject headings can still be used (Varro, *Rust.* 2.1.25) – a self-aware joke, perhaps, about the limits or artificiality of strict organising systems.[11] In the event, Illyrian and Liburnian women are singled out as especially good at producing children when Cossinus and Varro discuss the matter later in the book (Varro, *Rust.* 2.10.8–9).

---

[8] The estate manager was usually, but not always, a slave: Columella for one repeats the advice of Publius Volusius, that it is better to let a distant estate to free tenants rather than give it over to a slave *vilicus* who cannot easily be monitored (Col. *Agr.* 1.7). On the *vilicus*, see Aubert 1994: 117–220; Carlsen 1995; Maróti 1976.

[9] Farmers' calendar: Varro, *Rust.* 1.36.21. Veterinarian/medical instructions: see below, n. 18.

[10] On slavery in the Roman agricultural writers, see Dumont 1986; Martin 1972.

[11] For parody in this passage, see Nelsestuen 2011: 334–5.

Despite this use of structure to assimilate slaves to animals, Varro's *De re rustica* is unusual in insisting that literacy is necessary not just for the *vilicus*, but for the head herdsman and the head shepherd and the slaves in charge of each of the different types of livestock. The *vilicus* in Varro's text is assumed to be able to read, but this was not always the case. Although Cato mentions leaving written instructions for the manager (Cato, *Agr.* 2.6), according to Columella, literacy was not always considered necessary or even desirable in a *vilicus*: Celsus had argued that it was undesirable, since an illiterate man cannot easily cheat on the accounts (Columella, *Rust.* 1.8.4). Literacy is not imagined to have such disruptive effects in Varro's dialogues, where reading allows the slave to understand and follow instructions, and the landowner to assert control from a distance. After an account of the divisions of the year and the tasks to be performed in each, Tremelius Scrofa finishes by saying that 'what I have said should be written down and displayed in the villa, especially so that the *vilicus* will know it' (Varro, *Rust.* 1.36). As Scrofa has been talking for nine long sections and used several different ways of dividing the year, a great deal of condensing would have to take place to produce a text that could be set up in the villa. It is likely that Varro has parapegmatic texts in mind here, which could take the form of lists and tables of astronomical, meteorological and calendrical information, and were sometimes displayed in public inscriptions.[12] A parapegmatic device hung among the messages and warnings for household slaves on the fictional wall of Trimalchio's villa.[13] A farming calendar on the wall of the villa could have rhetorical and symbolic functions, as well as instructional ones. It stands as a reminder to the *vilicus* of the year's work, and remains to teach him how to farm, in the absence of the landowner.

The *vilicus*' role is to act as a substitute for the absent owner and run the farm in his place. In agricultural writing, the *vilicus* can be seen to mirror the owner in various ways: knowledge, moral character, intelligence. The good *vilicus*, like the good owner, must have knowledge in order to have the authority to give orders (Columella, *Rust.* 1.8.3). Columella writes that the *vilicus* should be from the country and avoid the vices of the town in terms that recall standard rhetorical topoi, as well as Mago's injunction that

---

[12] 'Parapegmata' can also refer to ancient instruments which use pegs to track cyclical astronomical and meteorological phenomena; how these worked and their relationship to the written texts is debated: see Lehoux 2007; Taub 2003: 20–33. On farmers' calendars, see also Wenskus 1998.

[13] Trimalchio's parapegmata has different coloured pegs for lucky and unlucky days (Petron. *Sat.* 30, discussed by Lehoux 2007: 483).

the best way to be a good farmer is to not keep a house in the town.[14] Pliny says that the *vilicus* should be as close as possible to the owner in intelligence, but should be unaware of the fact; beyond this, Pliny tells us that he will not include a full account of the *vilicus*' duties because, he says, Cato has already fully covered the subject (Pliny, *HN* 18.36). The way Cato does this is surprising. He begins by setting out the *vilicus*' main responsibilities, which are to follow the instructions of the master in all things that must be done on the farm and in managing the household (Cato, *Agr.* 142). In this section, he uses third-person subjunctives, 'he should make sure', 'he should act' (*curet, faciat*), just as he had done earlier when advising on the duties of the watchman (Cato, *Agr.* 66–7). But when he turns to give instructions on how the *vilicus* should manage the work of the *vilica*, the female manager, Cato writes directly to the *vilicus*, in the same way as he has been addressing the reader throughout:

> *Vilicae quae sunt officia, curato faciat. Si eam tibi dederit dominus uxorem, ea esto contentus. Ea te metuat facito. Ne nimium luxuriosa siet.*

> Make sure that the *vilica* fulfils whatever her responsiblities are. If the master has given her to you as a wife, be satisfied with her. Take care that she is afraid of you. She should not be too extravagant. (Cato, *Agr.* 143)

Cato assimilates the landowning reader to the slave manager, by placing both in the position of readers who need to be instructed in the art of farming. The *vilicus* in his role as instructor of the *vilica* learns directly from the text, in the same way as the landowner has been implicitly learning how to instruct him.[15] Cato's suggestion that the *vilicus* might be among his readers is unexpected. As recent work on literacy in antiquity has emphasised, there are many different levels of literacy: the ability to understand a calendar, manage an inventory or keep accounts does not mean being able to read a literary text, even one as spare as Cato's.[16] The turn towards the *vilicus* as reader may be best understood in the context of Cato's closeness to oral forms of communication in his writing: it is not so much that Cato imagines the *vilicus* is reading, as that all his readers are placed in the role of listeners.[17]

In Varro's dialogues, slaves are also imagined as readers of an agricultural writer, though not of Varro himself. In Book 2, speakers repeatedly suggest

---

[14] For Mago's maxim that farmers should sell their town houses, see Pliny, *HN* 18.35 (Pliny thinks this is a step too far).
[15] On the *vilica* and her relationship to the *vilicus*, see Roth 2004.
[16] See, for instance, Johnson and Parker 2009, esp. articles by Woolf 2009 and Werner 2009.
[17] On orality in Cato, see Diederich 2007: 156–60.

that the slaves in charge of the various types of livestock, and of the slaves' health, should keep with them written instructions on how to treat various illnesses.[18] In the case of cattle, it is made explicit that these written instructions are extracts from the work of Mago the Carthaginian:

> *De sanitate sunt complura, quae exscripta de Magonis libris armentarium meum crebro ut aliquid legat curo.*
>
> There is a great deal of material on their health, which has been excerpted from Mago's books, and I make sure that my head herdsman reads some of this frequently. (Varro, *Rust.* 2.5.18)

Both the status of the readers and the nature of the source are interesting here. Each of Varro's speakers assumes that the different types of slave-herdsmen will be able to read the medical instructions, until the final time it is mentioned, when the idea is finally defended:

> *Quae ad valitudinem pertinent hominum ac pecoris et sine medico curari possunt, magistrum scripta habere oportet. Is enim sine litteris idoneus non est, quod rationes dominicas pecuarias conficere nequiquam recte potest.*
>
> The manager ought to have in writing everything that is relevant to the health of flocks and people and that can be treated without a doctor. Anyone who is illiterate is not suitable for the position, because he cannot possibly produce proper accounts of the livestock for the owner. (Varro, *Rust.* 2.10.10)

What Varro describes here seems unlikely to have been commonplace for slaves of this status.[19] Columella at least makes no mention of literacy in his description of the qualities – diligence, thrift, endurance – that make a good head herdsman or chief shepherd (Columella, *Rust.* 1.9.1–2). And in any case, keeping accounts is a different matter from using written texts in treating the sick.

There are practical reasons why writing down medical recipes and instructions is a good idea, in that they are difficult to remember and accuracy matters, but how written information was to be used, and who should use it, could be contentious. Lesley Dean-Jones has suggested that the traditional rhetoric against doctors began in the fourth century BCE precisely because the development of medical literature allowed charlatans to practise medicine based on their reading rather than on practical training.[20]

---

[18] Varro, *Rust.* 2.2.20 and *Rust.* 2.1.23 on sheep; *Rust.* 2.3.8 on goats; *Rust.* 2.5.18 on cattle; *Rust.* 2.7.16 on horses; *Rust.* 2.10.10 on treating slaves.
[19] Aubert 1994: 177–8; Harris 1989: 256–7.   [20] Dean-Jones 2003.

For Pliny, the simple remedies of the Italian countryside, administered by the *paterfamilias*, are an ideal of medical practice exemplified by Cato the Elder, who warned his son not to trust Greek doctors (Plin. *HN* 29.14). Cato is supposed to have treated his household using a range of homespun recipes, which he wrote down in a short collection (*commentarium*) arranged by *materia medica* (Plin. *HN* 29.15). In Pliny's somewhat reactionary rhetoric, professional Greek medicine, with its over-sophisticated and ever-changing theories, is unnecessary, if the simple remedies of Cato's (or Pliny's) texts are employed.[21] In the case of sick slaves, Varro's speaker makes clear that the written instructions are to be used in cases where a doctor is unnecessary (Varro, *Rust.* 2.10.10), and it is possible that the extracts (*exscripta*) from Mago that Varro had in mind were similar to the *commentarium* of Cato to which Pliny refers in his work. It is not possible to know what Mago's prescriptions looked like, or even if these extracts were a standard set, though the speaker at least takes no credit for compiling them himself.[22] The written text, in this case ultimately derived from a Carthaginian's work, allows a slave to stand in for the *paterfamilias* in taking care of sick slaves and animals.

In these instances in Varro's text, the authority of written knowledge is maintained in the face of slaves' practical experience. The slaves in charge of veterinarian and medical care get their knowledge from an authoritative set of written instructions, derived from the work of Mago the Carthaginian. They do not, as we might have assumed, draw on their own experience of working with animals or on advice they might hear from other slaves. Unlike Varro, whose knowledge comes from what he has read, what he has heard from experts and what he has learned from experience (Varro, *Rust.* 1.1.11), the slaves who are directly responsible for the work of managing the sick are written of as though they draw on texts only. The calendar of the year's work can be written up on the wall of the villa, on the instructions of the owner, so that the *vilicus* will know it. Unlike the practical knowledge that might pass from worker to worker, written knowledge is easily accessible to elite readers, who use this knowledge to instruct their managers. Extending this access to written material to workers might lead to uneasy moments, as when Cato's reader morphs suddenly into a *vilicus*, but the overall effect seems to be to shore up the authority of the written text as a means of acquiring expertise. The idea of a slave who learns from Mago might in one way

---

[21] For a discussion of how Pliny's emphasis on natural remedies and attitude to Greek doctors emerges from his approach to nature in the work as a whole, see Hahn 1991.

[22] Heurgon 1976: 452–3 suggests these extracts (*exscripta*) may have been a standard set; Skydsgaard 1968: 109 thinks they were a private selection.

represent an optimistic vision of the democratisation of knowledge, but it works to occlude the possibility that slaves who work the land could have autonomous access to forms of knowledge not captured by written texts – perhaps so that when a landowner like Columella asks his slaves' advice, he can do so in the security that he is really the expert.

Varro's Sources and the Use of Written Texts

This confident vision of written texts as a means of maintaining authority and acquiring expertise among elite landowners, of making cithara-owners into cithara-players, is not the only way that writing is presented in Varro's *De re rustica*, however. It is also possible to trace a more pessimistic impression of the ability of written texts to hold and transmit information on agriculture, even in the way that Mago is used here. Mago the Carthaginian is given great prominence in Varro's initial list of sources in the preface to Book 1, and Varro dwells on the process of translation and epitomisation involved in making this text available for a Roman audience:

> *Hos nobilitate Mago Carthaginiensis praeteriit, poenica lingua qui res dispersas comprendit libris XXIIX, quos Cassius Dionysius Uticensis vertit libris XX ac Graeca lingua Sextilio praetori misit; in quae volumina de Graecis libris eorum quos dixi adiecit non pauca et de Magonis dempsit instar librorum VIII. Hosce ipsos utiliter ad VI libros redegit Diophanes in Bithynia et misit Deiotaro regi.*

> Mago the Carthaginian is even more distinguished than the Greek writers. He covered these disparate subjects in 28 books, written in the Punic language. Cassius Dionysius of Utica produced a twenty-book edition in Greek, dedicated to the praetor Sextilius. In this edition, he used a great deal of material from the Greek writers that I mentioned, and took about eight books' worth of material from Mago. Diophanes of Bithynia produced a useful shorter edition of six books, dedicated to King Deiotarus. (Varro, *Rust.* 1.1.10)

For Varro, Mago is the most authoritative of non-Roman writers on agriculture.[23] According to Pliny, the senate ordered that his huge work on agriculture be translated into Latin after the sack of Carthage in 146 BC, despite the fact that Cato's *De agri cultura* had just been published; all other books from the library at Carthage were distributed among the

---

[23] On Mago, see Devillers and Krings 1996; Greene and Kehoe 1995; Fantar 1998; Heurgon 1976; Mahaffy 1889.

African provinces (Plin. *HN* 18.22–23, cf. Columella, *Rust.* 1.1.13). It has been suggested that this act of translation had political overtones – a symbolic annexation of Carthaginian expertise on agriculture, and perhaps also a deliberate snub to Cato.[24] There is something potentially bathetic in the idea that the ultimate fate of Mago the Carthaginian's great work on agriculture is to be read in extracts by a head herdsman with a sick cow on his hands. And yet it is also the logical end point of a long process of selective reading and fragmentation. Mago had gathered together the material that Greek authors dealt with separately, only to have it broken down into ever smaller texts by the readers who followed. Mago's work, for all his authority, is never read in its entirety.

The question of whether and how writing on agriculture is read by potential farmers surfaces again in Varro's discussion of his sources. In his first preface of the *De re rustica*, in which Varro promises the work to Fundania, he says that before he gives her an account of recent conversations he has had on agriculture, he will give her the names of Greek and Roman writers on agriculture whom she can consult for further information (Varro, *Rust.* 1.7). He goes on to provide a series of lists of Greek authors, beginning with the kings Hiero of Sicily and Attalus Philometor, then a list of philosophers, beginning with Democritus, Xenophon, Aristotle, Theophrastus and Archytas, followed by a list of philosophers with place names attached who are arranged alphabetically by first letter (Varro, *Rust.* 1.8), followed by a further alphabetical list of philosophers for whom he has no place of origin (Varro, *Rust.* 1.9). He then comments that all these writers wrote in prose, but that there are also Hesiod and Menecrates, who dealt with these subjects in poetry (Varro, *Rust.* 1.9), before we come to the paragraph about the work of Mago the Carthaginian and its various translations (Varro, *Rust.* 1.10). The different ordering criteria give the impression that these authors are being displayed, and that the lists represent a show of erudition as much as an actual reading list. Columella provides a very similar list of Greek authors, though in a simpler order, and it has been suggested that it is a conventional one, originating, perhaps, with Cassius Dionysius, the translator of Mago.[25] Unlike Columella, who goes on to provide a similar list of Roman authors (Columella, *Rust.* 1.1.12–14), Varro lets this rather flat list of Greek experts stand alone in the preface and leaves the Roman authors to be discussed by the speakers within the dialogue itself. K. D. White suggested that Varro may have chosen not to list the Roman authors because they might

---

[24] Devillers and Krings 1996: 495–7; Heurgon 1976.
[25] Columella, *Rust.* 1.1.6–10; Baldwin 1963; Martin 1985: 1967.

have looked rather unimpressive after the parade of Greeks, but, as Silke Diederich has argued, the effect of this omission is a strong contrast between Greek and Roman authorities on agriculture: where the Greeks remain static on the page, Roman authors are part of the discussion, and – in the case of Tremelius Scrofa and Varro himself – the writer speaks in the text, and gives his opinions in dialogue with others.[26]

The dialogue form of the text allows Varro to show us elite Romans sharing knowledge from various sources with one another. The speakers in the dialogue are presented as individuals with whom Varro has spoken and learned from. Not all of these speakers are real people: Varro presents a combination of actual contemporary Romans, like Atticus, Varro's father-in-law Fundanius and Tremelius Scrofa, and characters with appropriately punning names that play on the elements of agricultural knowledge that they discuss. Within the dialogue, the farms and farming practices of contemporary Romans are used as examples and evidence. Not only is Varro's own experience on his farms presented as a source of expertise, but the farms of the characters in the dialogue and a range of other famous Romans such as Lucullus and Hortensius are brought into play.[27] The appeals to the example of the farms of living Romans root the dialogue and the opinions of the characters in the contemporary reality outside the text, and create a sense of authenticity and immediacy in the text. And yet the fictional characters and their farms are written into the dialogue, and discussed on the same level as these real examples. In this playful literary context, it is possible to find layers of irony in the ways in which other examples of texts on agriculture are spoken about by the characters in Varro's book.

In the opening of the first dialogue, the characters fulfil Varro's authorial promise in the preface to demarcate carefully what is and what is not proper to the discussion of agriculture. This leads to a discussion of the merits of the two Sasernas, father and son, two of the first Romans to write on agriculture (Varro, *Rust.* 1.2.22–28). Tremelius Scrofa notes disapprovingly that they included a discussion of clay pits in their work and comments that just because it is a profitable use of the land does not mean that it counts as agriculture. This leads to a dispute – another character, Stolo, claims that Scrofa is criticising the Sasernas as a result of jealousy and is being small-minded in picking out something to criticise rather than praising the many

---

[26] Diederich 2007: 25–7; White 1973: 468.
[27] Clearly one reason they are introduced is as illustrations of contemporary luxury: Hortensius and Lucullus as owners of famous fishponds (Varro, *Rust.* 3.3.10, 3.17.5–9); Lucullus' aviary as a place designed for both pleasure and profit (Varro, *Rust.* 3.4.3); Hortensius' game preserve, where animals are trained to come when a horn is blown (Varro, *Rust.* 3.13.2).

excellent features of their work. He is then challenged to quote something from the Sasernas, by a character, Agrasius, who is so proud of his own knowledge of these authors that he thinks others will be less familiar with their works. Recipes for insecticide and depilatories are quoted, and – at Varro's instigation – a charm for foot pains for the benefit of Varro's father-in-law, Fundanius. This leads to playful criticism of Cato the Elder's decision to include recipes and healthcare prescriptions in his work on agriculture, and in particular his famous championing of the cabbage's healing properties. Like good symposiasts, the speakers engage with that literature through their playful and competitive quotations, though here the quotations are of obscure charms for curing sore feet, and advice on killing insects with wild cucumber, rather than anything more elevated.

Earlier Roman writers are introduced in a competitive context in the dialogue: they form the locus of a dispute on what is and what is not agriculture that is important to the characters in the dialogue and to Varro's own aims in the work. Varro's books on agriculture highlight the clarity of their organisational structure and the neatness of the system of categorisation employed. At the start of Book 2, for instance, the practice of animal husbandry is broken down into eighty-one separate divisions which the speakers proceed to fill in over the course of the rest of the book (Varro, *Rust.* 2.1.11–25). At the start of Book 1, this extended discussion of previous writers chooses ground on which Varro is something of a virtuoso. A better structure and clearer definition of agriculture is promised, and the earlier writers' work is gently mocked. But the multiple voices in the dialogue allow for the possibility that this critique of previous writers is based on partial or unfair readings of their work. The writer, Tremelius Scrofa, is accused of ignoring what is good about the Sasernas' work out of agonistic envy of his predecessors; the great Cato is reduced to some faintly comical advice about eating cabbage before getting drunk. And, as Fundanius points out, if the advice is good, whether about feet or about insects, it might not really matter if it does not properly belong to the study of agriculture, if the information provided is useful to the reader. Varro's dialogue highlights that readers have their own agendas, and their use of earlier writers, like the slaves' use of Mago, are shaped by their own circumstances and concerns. Learning how to farm from a text on agriculture requires critical engagement, because the text may contain problematic information or need to be broken into smaller, more useful, sections. The fun of the discussion, and the amusing information that is quoted, raises another spectre: the knowledge may remain on the page for many readers, something to be enjoyed and talked about rather than put into practice.

Writers in the Dialogue: Varro's Authorial Voice

The gap between writing and reading about agriculture and the practice of agriculture is also visible in the ways in which writers are presented as characters in Varro's text. It is possible to see a similar gentle mockery of both Tremelius Scrofa and of Varro himself in the way they are constructed as speakers in the dialogue. In the preface to Book 1, Varro says one of the roots of his knowledge is 'what he has heard from experts' (*a peritis*; Varro, *Rust.* 1.1.11), so when Tremelius Scrofa is introduced as '*peritissimus*', 'the greatest expert', the implication is that Scrofa is someone Varro has learned from listening to. Scrofa becomes the key interlocutor through which Varro structures and sets out knowledge on farming in the first two books of the dialogues; he is the main speaker in the first book, and an important voice in the second. Varro did know Scrofa; they had served together on Caesar's agrarian commission in 59 BC. But Varro's representation of Scrofa here has been interpreted both as tribute and as parody, not least because it is surprisingly difficult to reconstruct the views of the historical Scrofa on the basis of Varro's text.[28]

Varro makes no mention of Scrofa as a writer, but we know from Columella that he produced a work on agriculture, which Columella describes as eloquent, setting him just before Varro in the list of Roman writers on agriculture (Col. *Rust.* 1.1.12). Jacques Heurgon suggested that Varro's failure to mention Scrofa's writing might have been because Varro had used Scrofa more as an oral source; P. A. Brunt and R. Martin thought that Varro may not have been familiar with Scrofa's work at the point at which he wrote Book 1 of the *De re rustica*.[29] The relationship between Scrofa's writing and Varro's depiction of his views has been debated: the few opinions that Columella attributes to Scrofa do not tally easily with those that Varro puts in his mouth in the *De re rustica*, although recent work by G. Nelsestuen has shown that it is possible to reconcile them.[30] Varro's Scrofa is pedantic and, in some cases, out of step with contemporary thinking on agriculture. J. E. Skydsgaard found so many problems with Scrofa's views on agriculture, as Varro describes them, that he suggested that Varro might have had some sort of deliberately satiric intent, though he went on to discount the idea,

---

[28] Nelsestuen 2011 is the best introduction to our state of knowledge on Scrofa, and offers a new reading of the evidence. See also Kronenberg 2009: 76–80 on the playfulness of this parody, with discussion of alternate views.

[29] Brunt 1972: 307; Heurgon 1978: xliv; and Martin 1971: 242, on which, see Nelsestuen 2011: 327–38.

[30] For discussion, see: Kronenberg 2009: 77–81; Nelsestuen 2011: 337–47; Noè 1977.

since he could see 'no apparent reason for Varro to make his immediate predecessor a main character in order to parody him'.[31] More recently, this idea of parodic intent has taken hold. Leah Kronenberg has suggested that the depiction of Scrofa and his interlocutor Stolo is meant as a 'parody of academic debates in the Late Republic, debates which are best represented in serious guise by those in Ciceronian dialogue';[32] G. Nelsestuen also sees mild parody of pedantic intellectualism here, including a measure of self-parody.[33] It is possible that the difficulty in separating Scrofa's views from those of Varro and the depiction of Scrofa as a speaker rather than a writer may also be a sort of joke about the dialogue form: Scrofa plays Socrates to Varro's Plato (or Xenophon). In any case, Varro's depiction of Scrofa as a character in the dialogue allows him to gently satirise Scrofa the writer, at the same time as he acknowledges him as a leading authority on agriculture.

There is a similarly playful gap between Varro's self-presentation as an author in his prefaces and his construction of himself as a character in the dialogues. In the preface to Book 1 of the *De re rustica*, Varro claims his practical experience as a farmer as a source of knowledge, one of the roots of the text (Varro, *Rust.* 1.1.11). Varro is both writer and a character in what follows – he speaks, is spoken to and is spoken about by various characters, and his practical experience as a farmer is demonstrated, challenged and appealed to at various points by characters within the text. In the preface to Book 2, on the management of livestock, Varro dedicates the book to a new addressee who has a particular interest in cattle.

> *E quis quoniam de agri cultura librum Fundaniae uxori propter eius fundum feci, tibi, Niger Turrani noster, qui vehementer delectaris pecore, propterea quod te empturientem in campos Macros ad mercatum adducunt crebro pedes, quo facilius sumptibus multa poscentibus ministres, quod eo facilius faciam, quod et ipse pecuarias habui grandes, in Apulia oviarias et in Reatino equarias, de re pecuaria breviter ac summatim percurram ex sermonibus nostris collatis cum iis qui pecuarias habuerunt in Epiro magnas, tum cum piratico bello inter Delum et Siciliam Graeciae classibus praeessem.*

> I have produced a book on one of these subjects, agriculture, for my wife Fundania, because of her farm. So this one is for you, Turranius Niger, my friend – since you love cattle so much, and your feet are always dragging you to buy at the market Campus Macri – so that you can meet your many pressing expenses more easily. I can do this more easily, because I myself have large livestock holdings – sheep in Apulia and cattle near Reate. I will run through a quick summary of livestock farming, based on some

---

[31] Skydsgaard 1968: 10–37. [32] Kronenberg 2009: 76. [33] Nelsestuen 2011: 333–7.

> conversations with people who own large cattle herds in Epirus that I had during the war with the pirates when I was in command of the Greek fleets between Delos and Sicily. (Varro, *Rust.* 2.pr.6)

Here Varro puts his friend Turranius Niger in the same position as his wife, Fundania: just as she owns a farm, Turranius Niger owns cattle and so might be expected to take a particular interest in a book on livestock, whether because he needs instruction, or simply enjoys engaging with others' expertise. Varro claims a particular authority to speak on livestock on account of the large quantity of sheep and cattle he has on his estates, and his authority as a farmer is implicitly underpinned by his authority as a military commander.

The value of Varro's personal experience is complicated, however, by Varro's characterisation of himself within the dialogue, where he appears not only as a landowner, but also as someone who is called upon for his knowledge of philosophy. At the outset of the discussion, Varro is asked to speak on the history, prestige (*dignitas*) and science (*ars*) of farming. He agrees to talk about the first two, but hands the subject of practical farming over to Scrofa, who agrees to do it, but only if Varro shares his knowledge too, as Scrofa jokes that Varro is a champion livestock farmer in Epirus (*Epirotici pecuariae athletae*: Varro, *Rust.* 2.1.2). Varro, the character, agrees, with an aside to the reader in his authorial voice:

> *Cum accepissem condicionem et meae partes essent primae, non quo non ego pecuarias in Italia habeam, sed non omnes qui habent citharam sunt citharoedi: Igitur, inquam...*
>
> Since I accepted the proposition and mine was the first role – not that I do not have livestock in Italy myself, but not every cithara owner is a cithara player – I began, 'So... ' (Varro, *Rust.* 2.1.3)

Varro's allusion to drama draws the reader's attention to the dialogue as fiction, and to his own role-playing within it. He also critiques Scrofa's assumption that owning livestock implies expertise in livestock farming: for a moment, he opens up the possibility that he, like Fundania and Turranius Niger, may be in possession of a farm, but not necessarily the master of it. When he speaks, it is not the practical wisdom of experience that he shares with the group, but the historical and philosophical knowledge that comes from reading and scholarship. In an ostentatiously learned opening, Varro cites Thales, Zeno, Pythagoras, Aristotle and Dicaearchus in his first sentence, and goes on to present a traditional account of the evolution of agriculture from Golden Age acorn gathering to the domestication of animals

(Varro, *Rust.* 2.1.3).³⁴ Elsewhere too, Varro is called upon for his knowledge on a point at issue based on his scholarship: at 2.5.13, the speaker looks to Varro for an explanation of a peculiar way of knowing whether male or female has been conceived, because Varro has read Aristotle.

Varro's reading and scholarship are a key part of his characterisation in the dialogues, and one source of his authority among the other speakers. But philosophy can be seen to be an alternative, theoretical discourse that is antithetical to the practical knowledge necessary for agriculture. Although Varro lists a large number of philosophers in the preface to Book 1, within the *De re rustica*, Theophrastus' *De causis plantarum* is marked off as more suitable for philosophers than farmers, though still interesting and useful (Varro, *Rust.* 1.5.1–2). When Varro's personal experience is called upon by other participants in the dialogue, it is not always practical experience of farming. Speakers refer to the example of Varro's villa bought from M. Piso (Varro, *Rust.* 3.3.8); Varro's expertise on mules drawn from his large herd is invoked (Varro, *Rust.* 2.8.6), and his comments on the distance of the summer from the winter pasturage for his sheep (Varro, *Rust.* 2.2.9). But Varro's interventions as speaker in the dialogue also include anecdotes and mirabilia – his leporaria where the boars and deer eat together at the sound of a horn (Varro, *Rust.* 3.13.1); a sow he saw that was so big a mouse made a nest in it (Varro, *Rust.* 2.4.12); his observations of the unusual childrearing and sexual practices among the women of Illyricum (Varro, *Rust.* 2.10.8–9); and an anecdote about Hortensius' great attachment to his fish (Varro, *Rust.* 3.17.5–9). It is a whimsical array of knowledge, the sort of far-fetched or anecdotal material that often requires the citation of a written source. In his role as a character, Varro produces the sort of dubious aside that Saserna and Cato are criticised for elsewhere (Varro, *Rust.* 1.2.28).

Where Varro does speak most extensively from personal experience in the case of his lengthy description of his aviary in Book 3, it presents a further set of potential problems. Although Varro presents a critique of present-day luxury in his prefaces to Books 2 and 3, and although numerous characters critique examples of luxury in the dialogue, Varro's longest contribution from personal experience is on the subject of a huge aviary which he has built solely for pleasure.³⁵ Appius remarks that Varro's aviary surpasses that of Lucullus, who is renowned for his profligate, modern luxurious living: Varro as character criticises him for as much at 1.2.10. When

---

³⁴ On Varro's engagement with these philosophers here, see Diederich 2007: 340–52.
³⁵ Book 3's focus on profits and luxury products has long been considered problematic in the context of Varro's moral programme. For a good discussion, see Diederich 2007: 352–64.

Varro speaks within the dialogue, it is as a connoisseur-owner of this luxury aviary rather than as a practical, plain-spoken man of the earth.[36] Even more problematically, the aviary, as Varro describes it, defies attempts to reconstruct what it looked like or how it could have functioned.[37] The apparent impossibility or implausibility of what Varro describes has led scholars to find allegorical and satirical meanings in his description of his birdhouse: Carin Green and Leah Kronenberg have argued that Varro's *De re rustica* plays off Cicero's *De re publica*, and can be read as a comment on politics of the late republic and its political discourse.[38] Kronenberg suggests that Varro's aviary symbolises a contemplative life that is an alternative to engagement with the turbulent politics represented in the framing narrative of Book 3, set in the Villa Publica during an aedile election marred by electoral fraud.[39] When he comes to draw on it, Varro's practical knowledge exemplifies the contemporary excess criticised in the prefaces, and cannot easily be translated from the page to the real world. More fundamentally, if Varro's writing about agriculture represents a parody of pedantic scholarly discourse, then does this imply that writing about the practice of agriculture is an inherently foolish enterprise? Even the great work of Mago is subject to fragmentation, its ultimate reader a slave herdsman, while writers like the Sasernas and Cato are flawed and subject to the animus of their readers. And writers like Varro and Scrofa can also be unreliable voices when they speak in dialogue with their peers.

If Varro intended his dialogues to poke mild fun at the project of writing about agriculture, the point was apparently lost on his immediate successors, whose reading of his work is sometimes similar to the partisan and partial readings of Cato and the Sasernas dramatised by Varro. For Columella and Pliny, Varro is one of the great authorities on agriculture. He holds a special place alongside Cato as one of the pioneering scholars in a Roman tradition of writing about farming that included the two Sasernas, Tremelius Scrofa and Celsus.[40] In Columella's list of sources, he comes between Scrofa and Virgil, with the claim that he gave style to the subject (Columella, *Rust*. 1.1.14). When Pliny discusses the most prestigious writers on agriculture, the kings and generals who wrote about it, Cato the Elder and

---

[36] On Varro's complex commitment to the idea of traditional values and the *mos maiorum* in his contemporary political context, see Diederich 2007: 297–368. Kronenberg has suggested that one of the targets for Varro's satire in the *De re rustica* is hypocritical moralising about luxury: Kronenberg 2009: 99–107.

[37] On reconstructions, see Flach 2002: 341–52.

[38] Green 1997; Kronenberg 2009: 73–129.   [39] Kronenberg 2009: 116–29.

[40] See White 1973 on Roman agricultural writers.

Mago, he brushes past the philosophers and poets he has listed in Book 1 to single out Varro, who decided to write about agriculture when he was 80 years old (Plin. *HN* 18.23).

Neither Columella nor Pliny pays any overt attention to the fact that Varro is writing a dialogue, although Pliny's praise for Cato's terseness as opposed to the *sermones* of unnamed others may be aimed at Varro (Plin. *HN* 17.35).[41] Both these later readers of Varro ignore the narrative framework of the text and the conceit of different speakers; anything said by any character in the dialogue is attributed to Varro's authority. This in itself is not especially surprising: it is difficult to determine what different genres signified for readers within the field of technical writing, and it is only relatively recently that the question has been of concern in modern scholarship.[42] Pliny in particular can sometimes appear quite blind to context in his extraction of facts from other people's texts. Pliny attributes two peculiar facts to Varro's authority that are thrown up as asides by characters in the *De re rustica*: an olive tree becomes barren if a goat licks it, and it is best to cut your hair at the full moon.[43] Still, in the case of Xenophon's *Oeconomicus*, a Socratic dialogue, Columella does make a distinction between writer and speaker and cites Ischomachus rather than Xenophon, whereas Pliny refers only to Xenophon and Cicero, who produced a Latin translation.[44] For Columella, Varro's characters, even Tremelius Scrofa, remain literary devices on the page, whereas Ischomachus is viewed as a historical figure whose opinions are representative of a longer tradition: 'antiquity hands down precepts in the persona of Ischomachus' (Columella, *Rust.* 12.3.5).[45] In Columella's work, Varro is frequently linked with Cato as one of the *antiqui*, representatives of what used to be done in the old days, a role that Cato generally performs alone in Pliny's *Natural History*.[46] Writing about agriculture can

---

[41] *Plus dixit una significatione quam possit ulla copia sermonis enarrari*: 'He says more with one suggestive phrase than could be explained in any amount of discussion'. *Sermo* can be used to mean dialogue or literary conversation, though it could also have the meaning of loose, conversational style as opposed to Cato's condensed, axiomatic prose: *OLD s.v.*

[42] On the literary nature of technical literature, see, for instance: Asper 2007; Doody et al. 2012; Fuhrmann 1960; Horster and Reitz 2003; Kullmann et al. 1998; Meissner 1999; Pigeaud and Pigeaud 2000; Taub and Doody 2009.

[43] On goats and olive trees, Varro, *Rust.* 1.2.18 and Plin. *HN* 15.34; on hair cutting and the moon, Varro, *Rust.* 1.37.2 and Plin. *HN* 16.194.

[44] Col. *Rust.* 11.1.15, 12.3.5; Plin. *HN* 18.224, where the reference is to Cicero's translation.

[45] On Columella's use of his sources, see Baldwin 1963, and on his approach to the past, Noè 2001.

[46] Columella, *Rust.* 3.3.2: Cato and Varro on yields in the past; 3.9.3 as *antiqui*; 6 pr.7 on old Roman opinions; 8.8.9: Varro on profitability of pigeons contrasted with present day. Pliny uses Varro's aviary to comment on the intervening period at *HN* 17.50, but it is Cato who is most

mark time in the long tradition of Roman practice, its authority continuing to resonate, even when its information on practical matters has been superseded.[47]

Tradition, and the importance of that tradition, lends weight to Roman discourse on agriculture and the land. The ideal of Cincinnatus, the senator called from the plough to lead the army, could represent a poignant image of a particular lost model of Roman masculinity, long after it had become a hackneyed topos. In the late republic and early empire, the simplicity of their agricultural past was a recurrent theme in Roman literature, cut through with nostalgia and with irony.[48] As the author of the *Antiquitates* and *De lingua latina*, Varro was an author interested in the Roman past, and the origins of Roman words and traditions, demonstrated in the *De re rustica* in a recurring interest in old word forms and etymologies.[49] As Silke Diederich has shown, Varro's writing on agriculture is deeply concerned with the *mos maiorum* and its compatibility with the political and economic realities of his time.[50] Pliny makes the fact that kings once wrote about agriculture a sign of its importance and dignity (Plin. *HN* 18.22), but the idea that landowners must read books to learn how to become farmers could also be seen as symbolic of its changing place in the life of the Roman elite.

Writing on agriculture was an important means of asserting control over knowledge about the land, and allowing for its transmission to those members of Rome's cultural elite who could not call a spade a spade for want of having seen one. The weakness of writing as a substitute for lived experience is hinted at by Varro's playful dialogues, with their gentle mockery of elite readers and writers. The lived experience of slaves who worked the large estates of Roman landowners is not acknowledged, except in the imperative that the landowner must know more than the estate-manager. Written texts are a means of making this possible, in that the landowner is supposed to glean from them enough information to be able to instruct and assess the *vilicus* in his work. When the estate-manager and the head herdsman are shown reading in Varro's text, this can be seen as an assertion of the authority of what can be written down and known by elite readers over the

---

frequently used as a means of commenting on changes over time: for instance *HN* 14.44–6 on types of vines; 15.24 on olive oil; 17.115–16 on grafting.

[47] On developments in Roman technology, see Greene 2000; Rossiter 2007; Stoll 2005.
[48] On Varro's relationship to this tradition, see Diederich 2007: 364–5.
[49] On old forms of words and etymologies, see for example: Varro, *Rust* 1.2.1 (*aeditumus*), 1.2.14 (*villicus*), 1.7.10 (*prata*), 1.48.2 (*spica*); 1.50.1 (*messis*).
[50] Diederich 2007: 297–368.

practical expertise of those who work the land. But it can also stand as a reminder of the fragility of written texts, and the humble contexts in which knowledge about agriculture is ultimately put into action: the great work of Mago is dismantled through successive translations and recensions to find itself in excerpted form in the pocket of a slave, who uses its prescriptions to treat a sick cow. Or so Varro makes 'Vaccius' tell us, a character whose name puns on the Latin word for 'cow' (*vacca*). Varro's dialogues are highly literary in style. In the first line, he makes *otium* the context for his writing of them, and in the dramatic settings of the dialogues, conversation about agriculture represents a rest from political, military and religious responsibilities. Varro's writing on farming is also intended to be enjoyed, though the promise that it is practical could be part of its charm for readers who left the business of farming to the managers of their estates. Like farming itself in Varro's formulation, writing and knowledge about agriculture could produce pleasure as well as profit, useful for showing off to friends in the forum, whether or not it is ever put to use on the farm.

9 | The Limits of Enquiry in Imperial Greek Didactic Poetry

EMILY KNEEBONE[*]

Didactic Authority

Declarations of authority and expertise play a crucial role in proclaiming and reinforcing the instructive potential of didactic texts. In offering up a defined body of information on a specific topic, and in foregrounding the educative potential of their works, didactic authors necessarily claim for themselves a position of authority and place their own professed credentials under scrutiny. As Catherine Atherton observes, 'given its self-imposed task, at least the *appearance* of authority proves essential to the didactic enterprise, and even subversion of authority, or of expectations about authority... presupposes (expectations of) access to authority, perhaps authority which is unique or uniquely appropriate'.[1]

Didactic poetry, however, is a form of literature that operates at the – to us, relatively unfamiliar – juncture between epic poetry and technical prose writing, and each of these traditions brings with it established, but by no means always compatible, conventions about authority and expertise. Authors of ancient prose manuals and technical handbooks, as this volume itself outlines, offer up a wide range of authoritative strategies by which they (affect to) guarantee the validity of the information they transmit. Ancient prose authors' claims about their own proficiency often rest, for instance, on references to their personal experience, scholarly activity or professional credentials, to oral reports, eye-witness accounts or written sources (both poetic and prosaic) or on the citation of, or polemical engagement with, the views of laymen, rivals, experts and predecessors.

Some of these are strategies also to be discerned in Greek didactic poems, yet ancient didactic poets frequently orientate their authoritative claims less towards their own technical expertise in the subjects they treat, and more towards their status, and expertise, as poets. Remarkably few Hellenistic and imperial Greek didactic poets seem, or indeed claim, to have composed

---

[*] I am very grateful to Richard Hunter, Tim Whitmarsh, Jason König and Greg Woolf for their helpful comments on an earlier draft of this chapter.
[1] Atherton 1998: viii (emphasis in the original); cf. Dalzell 1996: 25.

their works from professional or practical experience in their topic, nor do they usually represent their own experience as more than mere embellishment of the material at hand. Furthermore, while the factual information in their poems is usually drawn from prose treatises, sometimes intermingled with poetic accounts, oral lore or personal experience, these sources are rarely differentiated or explicitly discussed. Instead, the claims made by didactic poets about the provenance of their information traditionally rest, at least at first glance, on their professed proximity to the gods and Muses, however we may choose to interpret that claim.

Hellenistic and imperial Greek didactic poets, then, were rarely professional experts in the technical fields they treated, and to demand fresh information, direct experiential authority or comprehensive practical instruction from these poems is often to misunderstand their poetic, didactic and rhetorical aspirations. At the same time, however, these didactic poets do frequently point to the value of the information their poetry disseminates, explicitly or implicitly emphasising their knowledge and mastery of the topic at hand, drawing attention to the scope and significance of their treatment and promoting its perceived value for the addressee or reader, whether in practical, theoretical, moral or aesthetic terms. This chapter explores some of the tensions and conventions at work in such poems, probing the gaps between technical and poetic expertise in Greek didactic poetry of the Roman imperial period and scrutinising the ways in which these didactic poets draw attention to the margins and the inadequacies of the information they present, and to the limitations of their own perceived power to instruct.

Greek didactic poetry enjoyed considerable success during the Roman imperial period, and Greek didactic poems are known to have been composed on an impressive range of topics. My analysis in this chapter focuses principally on three complete or near-complete Greek hexameter didactic poems from the second and early third centuries CE: Dionysius of Alexandria's *Periegesis of the Inhabited World*, composed under Hadrian and outlining the geography of the *oikoumene* in just under 1,200 verses; the five-book *Halieutica* by Oppian of Cilicia, addressed to Marcus Aurelius and Commodus and treating of fish and fishing at sea; and Ps.-Oppian of Apamea's four-book *Cynegetica*, addressed to Caracalla and describing wild animals and the methods by which they are hunted.[2] I will also glance briefly at

---

[2] Greek text is quoted from the editions of Brodersen 1994, Fajen 1999 and Papathomopoulos 2003 respectively. The *Cynegetica* may perhaps be incomplete, on which see Agosta 2009: 12–13; Hopkinson 1994: 197; cf. Bartley 2003: 20; Whitby 2007: 125–6.

Ps.-Manetho's *Apotelesmatica*, a compilation of astrological didactic poetry dating probably from the second and third centuries CE, and at the scant remains of the forty-two-book medical *Iatrica* of Marcellus of Side, composed in the second century CE. Imperial Greek didactic poems are seldom analysed side by side in modern scholarship, yet this chapter argues that, taken together, these poems point to a flourishing tradition of imperial Greek didactic poetry that self-consciously articulates a central and coherent series of questions about the nature and boundaries of authority, knowledge and expertise.

The Boundaries of Knowledge

The Greek didactic poets discussed in this chapter treat of the earth, the sea and the sky and their various occupants: Dionysius the Periegete describes the inhabited world, Oppian the sea and its fish, Ps.-Oppian terrestrial animals and Ps.-Manetho the stars. In promising to recount, order and organise information about realms as vast as the earth, sea and sky, however, these second- and third-century Greek didactic poets often place considerable emphasis on the magnitude of the task they undertake. Dionysius' *Periegesis* and Oppian's *Halieutica* in particular are suffused with a powerful awareness of the immensity of the natural phenomena they set out to depict. Dionysius is here concerned primarily with innumerable rivers and islands and with vast bodies of land and sea, Oppian with the boundless sea and the teeming multitudes of fish within it; both place weight upon the enormity of their subject matter, which they repeatedly describe as ἀπειρέσιος: boundless, infinite, vast or uncountable.[3]

---

[3] ἀπειρέσιος, ἀπείριτος, etc. are used in the *Periegesis* both of geographical features (*Perieg.* 4, 57, 119, 323, 430, 514, 550, 613, 616, 644, 659, 977, 1030, 1137) and their numerous inhabitants (*Perieg.* 165, 217); similarly, in the *Halieutica* both of the sea (*Hal.* 1.85, 1.601, 3.646) and the vast numbers of fish to be found within it (*Hal.* 1.796, 2.552, 571, 4.130, 4.496, 4.502, 4.683). Cf. the poem comprising Books 2, 3 and 6 of Ps.-Manetho's *Apotelesmatica*, where ἀπειρέσιος and ἀπείριτος are used of the infinite nature of the stars and heavens at *Ap.* 2.27, 2.29, 2.87, 3.231; suggestively, the word is not used in this way in the remainder of the collection. The *Cynegetica* uses such language more frequently of impressively powerful animals: thus *Cyn.* 2.530, 2.535, 3.40, 3.416 (although note *Cyn.* 2.517, 2.257, 2.272). Cf. also μυρίος at *Perieg.* 166, 1168; *Ap.* 2.3; *Hal.* 1.80, 1.775, 2.439; *Cyn.* 1.271, 1.166, 1.400, 2.83, 3.34. This language is in part rooted in Homeric and Hesiodic descriptions of the boundless earth and sea: see, e.g., Hom. *Il.* 1.350, 7.446, 20.58, 24.342, 24.545; *Od.* 1.98, 4.510, 5.46, 10.195, 15.79, 17.386, 17.418, 19.107; Hes. *Th.* 109, 187, 678, 878; *Op.* 160, 487. In the final stages of completing this chapter I have benefited greatly from the publication of Jane Lightfoot's commentary on the *Periegesis*; see now her discussion on immensity in Dionysius at Lightfoot 2014: 96–7, 133–4, on which cf. Jacob 1981: 41, 49–50.

These didactic poems, moreover, are to different degrees structured around a tension between the vast scope of their subject matter and the necessarily bounded nature of their individual poetic projects. Indeed, Dionysius' *Periegesis* is in many ways dominated by the contrast between its ambitious subject matter and its deliberately circumscribed form: this is a poem that aims to describe the entire *oikoumene* in little more than a thousand verses.[4] Yet while Dionysius foregrounds the size of the world his poem describes, he also proclaims his mastery of the topic at hand, emphasising his accuracy and repeatedly declaring that he could 'easily' describe certain geographical features in his poem (ῥέα, 345; ῥεῖα, 707; ῥηιδίως, 881), a claim which in turn informs the addressee's projected ability to grasp the topic easily (ῥέα, 280) thanks to Dionysius' descriptive powers.[5] Dionysius' 'easy' distillation of the whole world into such succinct poetic form itself asserts his self-proclaimed control of the factual material, the poetic medium and the pedagogical enterprise he undertakes.

For all this 'rhetoric of ease',[6] however, Dionysius also uses the very vastness of his subject matter to draw attention to the ways in which his ability to provide information is severely hampered; indeed, his claims of ease often mark precisely what he *could*, but in the end chooses not to, describe in the poem. Thus, while Dionysius declares himself perfectly able to speak of the more significant islands in the ocean, he also declares that there are countless (ἀπειρέσιαι, 613; ἀπείριτοι, 616) islands he would not find it easy to name (τῶν οὐ ῥηίδιόν μοι ἔνιο ἔμεν οὔνομα πασέων, 619). The poet's inability to describe these smaller islands indicates the unimaginably vast nature of his subject matter, yet it also foregrounds the process of the compilation and dissemination of knowledge. The gathering of information, as Dionysius seems to imply at 618, is hampered not only by the sheer quantity of islands, but also by the fact that so many of these smaller islands are unsuitable for exploration. Knowledge is here necessarily shaped by the environment of the observer. Thus, too, in the *Halieutica*, Oppian's didactic poem on fishing, it is largely at the boundaries of the sea – near the shore or the surface, where fish approach the sphere of human activity and knowledge – that men are able to examine the sea and mark out its attendant life forms.[7]

---

[4] See esp. Lightfoot 2014: 94–8 on the tension in the *Periegesis* between 'open' and 'closed' lists, between 'orderly control' and 'sketchiness'.
[5] Here echoing the language of Nicander's proems (on 'ease' in Nicander, see Clauss 2006 and Jacques 2007: lxxv); see Hunter 2004b: 223–4; Lightfoot 2014: 103–4, 108–9.
[6] The phrase is that of Hunter 2004b: 224.
[7] See, e.g., *Hal.* 1.363–5, 3.205–18, 3.636–48, 4.349–53.

The formless mass of the open sea, by contrast, is huge, bewildering and beyond the limits of enquiry.

The notion that the vastness of the cosmos should impose limits on mortal knowledge and on the didactic poet's enterprise finds an important Hellenistic precedent in Aratus' *Phaenomena*, a poem whose influence over later Greek didactic poetry was far-reaching. Aratus clearly delineates the scope of his own poetry by declaring himself uncomfortable discussing the planets (*Phaen.* 460–1), and outlines the boundaries of mortal enquiry by claiming first that nobody could describe the constellation of the Kneeler precisely (*Phaen.* 64–6) and second that the stars were so numerous and so individually alike that they were by necessity grouped as constellations instead of being assigned individual names; the stars below the Hare, moreover, are said to be too hazy even to be assigned a collective constellation (*Phaen.* 367–85).[8] The reader of Aratus' didactic poem, in other words, is repeatedly confronted with a (partially) visible reminder of mankind's limited cognitive and visual capacities.

Aratus opens the *Phaenomena*, furthermore, by discussing the two stable poles of the sky, remarking upon the invisibility of the south celestial pole in contrast to the north pole, which is fully visible to men: ἀλλ' ὁ μὲν οὐκ ἐπίοπτος, ὁ δ' ἀντίος ἐκ βορέαο| ὑψόθεν ὠκεανοῖο ('but one [of the poles] is not visible, while the opposite one in the north is high above the horizon', *Phaen.* 25–6). From the start of his poem, then, Aratus touches implicitly on the constraints of human vision: only one pole is visible to men at any one time, and the poet, quite naturally, presents the sky from the standpoint of a native of the northern hemisphere. The adjective ἐπίοπτος (*Phaen.* 25) is rare, and before the fourth century CE is attested only in Aratus, in discussions of the Aratean passage and in Oppian's *Halieutica*. Oppian, like Aratus, uses the word at the start of his poem: having opened the *Halieutica* with an expression of confidence about what he will describe, Oppian immediately undercuts this certainty by alluding to the difficulties of this enterprise, describing the attempts of men to explore and chart the sea: ἀίδηλον ἐπιπλώουσι θάλασσαν| τολμηρῇ κραδίῃ, κατὰ δ' ἔδρακον οὐκ ἐπίοπτα| βένθεα ('they sail over the unseen sea with daring heart, and they look down over its invisible depths', *Hal.* 1.9–11). In emphasising the vast and baffling nature of the sea in his proem, Oppian echoes not only Aratus' initial statement about what may and may not be observed of the sky, but also a sense of the

---

[8] Nor, fittingly, is the name of their namer recorded. On this passage, see further: Erren 1958; Kidd 1997: 318–19; Martin 1956: 55–60, 1998: II, 302–7; Pendergraft 1990; Semanoff 2006: 310–14.

magnitude of the realm under discussion, and, in this case, the sheer impenetrability of the sea. As the scholia explain *ad loc.*, Aratus uses the Ocean at *Phaen.* 26 to signify the horizon: for both Aratus and Oppian, therefore, the sea represents a vast and symbolic impediment to the observation of what lies beneath it.[9]

At several points in the *Periegesis*, Dionysius similarly focuses on the gulf between locally orientated, and thus severely limited, mortal knowledge, and inexpressibly vast natural phenomena. As critics have pointed out, the poem offers its reader, at least in part, a 'bird's-eye view' of the world, encompassing continents, oceans and islands at a single glance.[10] Yet on a number of occasions we are also told that men themselves have arbitrarily divided natural features in accordance with their own, necessarily small-scale, perceptions, and that humans tend to be concerned only with the immediate manifestations of far larger phenomena, each tribe focusing solely on that which comes into its limited purview.[11] Thus, when speaking of the Taurus mountains and the ἀπειρέσιοι rivers which flow from it, Dionysius readily admits to his inability to name them all (646: τίς ἂν πάντων ὄνομ' εἴποι), while simultaneously distancing himself from the very need to do so.[12] The men who live next to the streams may concern themselves with such local names, but Dionysius himself has bigger fish to fry.[13] Dionysius, that is, offers not an exhaustive catalogue of small-scale local details, but rather a global overview, covering in their entirety only the most important tribes that inhabit the area (νῦν γε μὲν ἔθνεα πάντα διίξομαι, ὅσσ' ἀρίδηλα ἐνναίει, 650–1), a project he represents as being overseen and directed by the Muses, who are addressed immediately afterwards (651). The tension here between comprehensiveness and selectivity lies at the heart of Dionysius' project, and his very brevity foregrounds that process of selectivity and prioritisation. The poet's perceived expertise at this point, we are to infer, consists primarily in sifting expertly through the data and choosing only the most salient features to convey to the reader.

Analogous claims about poetic selectivity are employed by Ps.-Oppian in the *Cynegetica* (early third century CE): we are told, for instance, that

---

[9] Cf. Bartley 2003: 31; Kidd 1997: 180.
[10] See esp. Jacob 1990: 23–8; Lightfoot 2014: 120–7, with further bibliography.
[11] Dionysius describes, for instance, the human division of the Earth into three continents, despite its structural unity (*Perieg.* 7–8), and speaks of the many names which mortals have given to the Ocean (*Perieg.* 27–8), on which see further below.
[12] Cf. Aratus at *Phaen.* 1036–7.
[13] Drawing here on Hes. *Theog.* 367–70, a parallel noted already by Eustathius 638 (Müller 1855–61: Vol. 2, 336); see Hunter 2004b: 224–6.

species of horses are as diverse as the countless tribes of men (ἵππων δ' αἰόλα φῦλ', ὅσσ' ἔθνεα μυρία φωτῶν), yet despite this breadth of subject matter (ἀλλ' ἔμπης) the poet will single out the most important types (*Cyn.* 1.166–9). Ps.-Oppian also devotes the end of Book 2 to listing those animals too insignificant to be discussed at length: Μοῦσα φίλη, βαιῶν οὔ μοι θέμις ἀμφὶς ἀείδειν·| οὐτιδανοὺς λίπε θῆρας, ὅσοις μὴ κάρτος ὀπηδεῖ ('Dear Muse, it is not proper for me to sing about trivial things; leave behind the worthless beasts that are not endowed with strength', *Cyn.* 2.570–1), referring to cats, squirrels, hedgehogs, dormice and other such negligible species. Judicious discrimination and a sense of what is and is not appropriate to the task at hand are ostentatiously placed at the centre of the didactic poet's project of sifting and selecting material to relay.

Ps.-Oppian's address to the Muse and advertisement of his own poetic selectivity in Book 2, moreover, in turn recalls the poet's generic *recusatio* in the proem of the first book of the *Cynegetica*, an extraordinary dialogue which may itself be characterised as a dramatisation of the tension between authorial selectivity and universalism. Artemis, entering into dialogue with the narrator, lists in this proem those topics – Bacchic, heroic, argonautic, martial and erotic – which the poet is not to treat, ordering him instead to sing only of hunting (*Cyn.* 1.20–40). Yet as Artemis' anthropomorphic language begins to reveal, the boundaries of Ps.-Oppian's subject turn out to be highly permeable, for we observe as the poem progresses that each of the goddess's 'forbidden' themes recurs later in this poem, this time recast onto the bestial plane: we hear of the loves, battles and quasi-epic feuds of wild beasts, of Bacchants turned into leopards and of the argonautic king Phineus metamorphosed into a mole.[14] The claim of careful selectivity promoted at the start of the *Cynegetica* thus pulls continually against the totalising impulse of didactic poetry, against the poet's profoundly incorporative view of the knowledge he is to transmit.

The self-contained Books 2, 3 and 6 of Ps.-Manetho's astronomical didactic *Apotelesmatica* (second century CE) likewise return frequently to the vastness of the poet's subject and to the impossibility of providing a comprehensive account of all the information to be gleaned from the stars. The movements and configurations of the stars, we are told, are countless and ineffable (ἄσπετ' ἀπείριτά τ'), and nobody could possibly comprehend, let alone relate, them all (πάντα μὲν οὖν οὐκ ἄν τις ἑῷ φράσσαιτ' ἐνὶ θυμῷ,| οὐδ' ἐνέποι); the poet will therefore declare only the material that has been

---

[14] On which see Asper 1997: 104–6; Bartley 2003: 172–3; Costanza 1991; Goldhill 2006: 151–4; Koster 1970: 155; Paschalis 2000: 218–21; Schmitt 1969: 47–57; Whitby 2007: 132–3.

selected by the gods (*Ap.* 3.230–3).¹⁵ Nor is the pedagogical onus placed solely on the didactic poet, for the reader too will be expected to apply the principles outlined in the poem to new and analogous contexts; the poet's task, as Ps.-Manetho has it, is not to cover every aspect of his proposed subject, a feat impossible for the mere mortal, but rather to provide enough illustrative information for the careful reader to draw their own inferences about comparable phenomena.¹⁶ As Richard Hunter notes of Hesiod and Aratus, '"didactic poetry" does not have to be comprehensive to be "didactic." It gives us examples, exemplary signs, to guide us as we move beyond the confines of the poem.'¹⁷

Although Hunter draws a broad contrast here with prose manuals, this is also a strategy to be found in a number of ancient prose treatises. Indeed, a tension between the desire for exhaustive coverage and the necessarily – if to different degrees – selective role of the author or compiler is a recurring theme in encyclopaedic prose literature more widely. Hilary Clark well observes that 'the encyclopedic text is marked by several important paradoxes: in seeking to totalize and eternalize knowledge, the writer finds the project shadowed by incompletion and obsolescence'.¹⁸ Pliny the Elder's encyclopaedic *Natural History*, to take but one example, is a text characterised both by the author's obsession with comprehensiveness and by his awareness of its ultimate impossibility. Indeed, as Pliny all but points out, the very mass of detail required for the comprehensive treatment of a particular topic may itself become an impediment to a synoptic, comprehensive view of the whole; in the words of Sorcha Carey, 'the paradox of [Pliny's] quest for totality is that, as Pliny himself admits, in order to catalogue the whole world, he has to leave things out'.¹⁹ Likewise, in his *Oneirocritica*, Artemidorus makes much of his comprehensive research and the encyclopaedic scope of his project, yet he too, as Daniel Harris-McCoy has shown, remains heavily aware of the necessarily incomplete nature of his project, thanks in particular to his emphasis upon the need to interpret dreams in light of

---

¹⁵ Cf. also *Ap.* 6.112–13, 6.222–6, 6.338–9.
¹⁶ See esp. *Ap.* 6.260–1: ἀμφί γε μὴν τεκέων γενεῆς ἅλις· ἢ γὰρ ἐχέφρων| ἀνὴρ καὶ ἀπὸ τῶνδε τὰ δὴ λίπον ὦκα φράσαιτο. For further discussion of the implied reader and/or addressee in Greek and Latin didactic poetry, see esp. Schiesaro *et al.* 1994; Volk 2002.
¹⁷ Hunter 2004a: 234; cf. Hunter 1995.
¹⁸ Clark 1992: 97. See also König and Woolf 2013: 7–11, esp. 8: 'it is in fact a standard feature of encyclopaedic rhetoric to undermine or throw doubts on its own claims to totality even as it makes them, to reveal the precariousness of encyclopaedic aspirations to comprehensiveness'; cf. also Harris-McCoy 2013 on the 'fragmentary encyclopaedia'.
¹⁹ Carey 2003: 22, referring here to Plin. *HN* 3.42; on Pliny's obsessive 'desire for totality', see further Carey 2003: 17–32.

individual and cultural contexts.[20] As in Ps.-Manetho's poem, that is, the attentive reader is expected to be able to extrapolate from and supplement the information provided in the text.[21]

Divine and Mortal Knowledge

In contrast to such prose treatises, however, each of the illustrative passages of didactic poetry we have just examined – Dionysius' appeal to the Muses (*Perieg.* 651), Ps.-Manetho's comment on divinely selected information (*Ap.* 3.233), Ps.-Oppian's dialogue with Artemis (*Cyn.* 1.17–40) and his later address to the Muse (*Cyn.* 2.570ff.) – represents the tension between immensity of subject matter and authorial (in)capacity as a consequence above all of the crucial distinction between divine and mortal capability, hence their recourse to superhuman agency. The limited nature of the information offered in each poem is depicted, that is, not as a function of the poet's own lack of technical or professional expertise in the field, but as an inevitable function of the human condition. This is a poetic topos rooted, at least in part, in the Homeric narrator's address to the Muses at the start of the Iliadic Catalogue of Ships, where the poet appeals to the Muses for information, declaring himself unable to name the multitude of Greeks at Troy, even had he ten tongues and mouths, an unbreakable tongue and a heart of bronze.[22] As Andrew Ford observes, 'the appearance of the Muses in the *Iliad*, book 2 ... is not simply a scene of instruction but also one of selection: it is the point at which both the immense Greek host and the ineffable oral tradition must be cut down to manageable, significant figures.'[23] This Iliadic passage establishes an important precedent for later Greek didactic poets first in the narrator's reliance on the Muses to supply information about an unimaginably vast topic, and second in the poet's emphasis upon the necessarily selective nature of his narrative enterprise: even with the help of the Muses, we observe, the Homeric narrator will go on to speak only of the Greek leaders, not the πληθύς of *Il.* 2.488.[24]

---

[20] Harris-McCoy 2013. [21] E.g. at Artem. *Oneir.* 1.21, 1.74, 2.25, 4.4.
[22] Cf. the Homeric expressions of narratorial incapacity at *Il.* 12.176, 17.260–1; *Od.* 3.113–17, 4.240–1, 7.241–2, 11.328–30, 11.517; see de Jong 2001: 102–3.
[23] Ford 1992: 79.
[24] The Catalogue was much imitated and discussed in antiquity, not least, presumably, in Apollodorus' commentary. See Cribiore 2001: 194–5 for the popularity of the Catalogue of Ships in ancient educational contexts; cf. also Cavavero 2002: 57–9; Ford 1992: 72–89. The use of the topos in Latin literature – see esp. Ennius, *Ann.* 469–70 (Skutsch 1985); Virgil, *Georg.* 2.43–4; Lucretius, as quoted in Servius' commentary on Virgil, *Georg.* 2.42 – is well traced in

Significantly, moreover, the influence of this precedent is evident in the fact that Dionysius, Oppian, Ps.-Oppian and Ps.-Manetho each make a further, and this time explicit, declaration of the distinction between mortal and divine knowledge, linked in each case to the issue of narrative selectivity and to the professed limits of the author's ability to provide information. Here, we witness imperial Greek didactic poets self-consciously situating their works within a long-standing tradition of distinctively epic knowledge-ordering strategies. Let us start with Dionysius. The *Periegesis*, as scholars have observed, is a poem in general much indebted to the Iliadic Catalogue of Ships, which functions, in effect, as a Homeric archetype for the presentation of a compact hexameter geographical catalogue.[25] At the end of the *Periegesis*, Dionysius concludes his catalogue of nations with the claim that, while he has now spoken about the most prominent tribes, many more remain beyond the remit of his poem, and must remain the subject of divine knowledge alone:

> τόσσοι μὲν κατὰ γαῖαν ὑπέρτατοι ἄνδρες ἔασιν·
> ἄλλοι δ' ἔνθα καὶ ἔνθα κατ' ἠπείρους ἀλόωνται
> μυρίοι, οὕς οὐκ ἄν τις ἀριφραδέως ἀγορεύσαι
> θνητὸς ἐών· μοῦνοι δὲ θεοὶ ῥέα πάντα δύνανται.
> αὐτοὶ γὰρ καὶ πρῶτα θεμείλια τορνώσαντο   (1170)
> καὶ βαθὺν οἶμον ἔδειξαν ἀμετρήτοιο θαλάσσης
> αὐτοὶ δ' ἔμπεδα πάντα βίῳ διετεκμήραντο,
> ἄστρα διακρίναντές, ἐκλήρωσαντο δ' ἑκάστῳ
> μοῖραν ἔχειν πόντοιο καὶ ἠπείροιο βαθείης.

> Such are the most notable men on the earth. But countless others wander here and there over the land, and these no mortal could describe clearly; only the gods can do everything easily. For they marked off the first foundations and made known the deep tract of the immeasurable sea, they marked out all steadfast things in life when they distinguished the stars, and to each man they assigned a share in the sea and the deep land. (*Perieg.* 1166–74)

Once again we are presented with the poet's portrayal of excluded information as an issue not of personal experience, but of universal human

---

Hinds 1998: 35–47; see also Thomas 1988: 163–4, who distinguishes helpfully between 'unwillingness' and 'incapacity'.

[25] It is, for instance, surely no coincidence that Dionysus' poem should stretch to 1,186 lines, recalling the 1,186 ships described in the Catalogue of Ships (for all that certain verses of the poem are usually deemed spurious). On the relationship between Dionysius' *Periegesis* and the Catalogue of Ships, see: Hunter 2004b: 225; Jacob 1981: 47–50; Lightfoot 2008.

capability (... οὓς οὐκ ἄν τις ἀριφραδέως ἀγορεύσαι| θνητὸς ἐών, 1168–9).²⁶ This is a phrase, and a sentiment, which finds repeated echoes in imperial Greek didactic: note, for instance, the parallel with Book 6 of Ps.-Manetho's *Apotelesmatica*, where the poet declares his own inability to describe the entirety of his chosen topic: ἀλλὰ γὰρ <u>οὔτις θνητὸς ἐὼν</u> πάσας κε δύναιτο| πρήξιας ἢ τέχνας εἰπεῖν, ὅσσας μερόπεσσιν| ἀστέρες ἐν σφετέροισιν ἐμοιρήσαντο δρόμοισιν ('But no mortal would be able to speak of all the affairs or professions that the stars in their courses have assigned to men', Ps.-Man. *Ap.* 6.541–3).²⁷ In Dionysius, however, the familiar tension between enormity and selectivity is presented in terms that themselves mirror the cosmic order: the world may be vast, deep or immeasurable, but each man has nevertheless been allotted his own defined share (μοῖρα) within it. The poem's task of marking out, dividing up and making sense of the world echoes the very actions of the gods in dividing the cosmos into its component shares.²⁸

We have already noted that Oppian's *Halieutica*, like the *Periegesis*, is informed by a tension between its ἀπειρέσιος subject matter and the task of comprehending and recounting such immensity. Yet while Dionysius waits until the conclusion of his poem to highlight this gulf between the world and its cataloguers, between mortal and immortal capabilities, Oppian sets this tension centre-stage, outlining the limitations of human knowledge in the proem to the first book of the *Halieutica*:

μυρία μὲν δὴ φῦλα καὶ ἄκριτα βένθεσι πόντου (80)
ἐμφέρεται πλώοντα· τὰ δ' οὔ κέ τις ἐξονομῆναι
ἀτρεκέως· οὐ γάρ τις ἐσίκετο τέρμα θαλάσσης,
ἀλλὰ τριηκοσίων ὀργυιῶν ἄχρι μάλιστα

---

²⁶ θνητὸς ἐών: cf. *Il.* 16.154 (Achilles' horse), 22.9 (Achilles himself); the former is parodied by Matro of Pitane fr. 1.52 Olson-Sens, on which cf. Olson and Sens 1999: 23, 58. Note too the contrast between divine and mortal ability at *Od.* 10.305–6 (*moly*) and 4.379 (divine knowledge). Cf. also Ibycus 282.23–6 *PMG*, similarly contrasting mortal and divine (in)capacity to recount information in full, on which see Lightfoot 2014: 504; Wilkinson 2013: 71–3. Khan 2004: 242–3 finds in Dionysius' language an echo of Cleanthes fr. 1.38; contrast Magnelli 2005: 107 n.11.
²⁷ Priority here is difficult to establish with certainty. Although the two poets were evidently near contemporaries, neither the *Periegesis* nor the poem comprising Books 2, 3 and 6 of the *Apotelesmatica* may be dated with precision: the *Periegesis*, as the acrostic at 513–32 implies, was probably composed under Hadrian (see esp. Amato 2003; Jacob 1991), perhaps in the 130s; Ps.-Manetho's description of his horoscope at *Ap.* 6.739–50, on the other hand, dates the poet's birth to 80 CE, for which see Garnett 1895; Neugebauer and Van Hoesen 1959: 92.
²⁸ Cf. Hunter 2004b: 226 on Dionysius' geographic diversity as 'a sign of divine care'; Khan 2004: 240, apropos of Dionysius' acrostics: 'the clearly marked order of Dionysius' poem ... mirrors the order of the *oikoumene*'.

ἄνερες ἴσασίν τε καὶ ἔδρακον ἀμφιτρίτην.
πολλὰ δ' (ἀπειρεσίη γὰρ ἀμετροβαθής τε θάλασσα)   (85)
κέκρυπται, τά κεν οὔ τις ἀείδελα μυθήσαιτο
θνητὸς ἐών· ὀλίγος δὲ νόος μερόπεσσι καὶ ἀλκή.
οὐ μὲν γὰρ γαίης πολυμήτορος ἔλπομαι ἅλμην
παυροτέρας ἀγέλας οὔτ' ἔθνεα μείονα φέρβειν.
ἀλλ' εἴτ' ἀμφήριστος ἐν ἀμφοτέρῃσι γενέθλη   (90)
εἴθ' ἑτέρη προβέβηκε, θεοὶ σάφα τεκμαίρονται,
ἡμεῖς δ' ἀνδρομέοισι νοήμασι μέτρα φέροιμεν.

Countless and indistinguishable are the swimming tribes contained in the depths of the sea; nobody could name them accurately. For nobody has reached the limit of the sea, but up to about 300 fathoms men know and have observed the sea. Yet many things remain hidden (for the sea is boundless and immeasurably deep), and no mortal could tell of these unseen things: for the mind and strength of mankind is slight. I think the sea does not nourish fewer herds or inferior tribes than bountiful mother earth. But whether the offspring are evenly matched on both sides, or whether one exceeds the other, the gods judge plainly, whereas we must make our measurements using human perception. (*Hal.* 1.80–92)

Oppian here lays out the epistemological core of the *Halieutica* and sets this huge physical space, and the countless creatures it contains, within a tradition of Greek hexameter poetry treating unimaginably vast topics. Dionysius, in particular, looms large in Oppian's portrayal of the unplumbed sea and its myriad inhabitants. Oppian's statement that μυρία μὲν δὴ φῦλα καὶ ἄκριτα βένθεσι πόντου | ἐμφέρεται πλώοντα (*Hal.* 1.80–2) evokes the proem to the *Periegesis*, where Dionysius had promised to treat the ἀνδρῶν ἄκριτα φῦλα by starting with the βαθύρροος Ocean (*Perieg.* 1–3).[29] Oppian's parenthetical statement of infinitude – ἀπειρεσίη γὰρ ἀμετροβαθής τε θάλασσα (*Hal.* 1.85) – is reminiscent not only of Dionysius' widespread obsession with ἀπειρέσιος geographical features, including the sea, but also of the βαθὺν οἶμον ... ἀμετρήτοιο θαλάσσης that Dionysius' gods revealed to mankind at the end of the poem (*Perieg.* 1171). And Oppian's claim of mortal incapacity – τά κεν οὔ τις ἀείδελα μυθήσαιτο| θνητὸς ἐών (*Hal.* 1.86–7) – pointedly echoes Dionysius' οὓς οὐκ ἄν τις ἀριφραδέως ἀγορεύσαι| θνητὸς ἐών (*Perieg.* 1168–9); the sentiment, as we shall see, will in turn be reworked by Ps.-Oppian in the *Cynegetica*.

---

[29] Cf. also Parmenides 28 B6.7 D-K; see Magnelli 2006: 243.

Measuring the World

The juxtaposition of these passages, I suggest, reveals not only imperial Greek didactic poets' evocation of archaic epic traditions in which the magnitude of one's subject is presented in divine or cosmic terms, in contrast to the limited capacity of the narrator *qua* mortal, but also a concern with intellectual authority and with the process of ordering and cataloguing the world itself. Oppian's emphasis on assessing and measuring the sea and its inhabitants (ἡμεῖς δ' ἀνδρομέοισι νοήμασι μέτρα φέροιμεν, *Hal*. 1.92) builds on his description at the start of the poem of both the achievements and the difficulties involved in this process: ἀίδηλον ἐπιπλώουσι θάλασσαν| τολμηρῇ κραδίῃ, κατὰ δ' ἔδρακον οὐκ ἐπίοπτα| βένθεα καὶ τέχνῃσιν ἁλὸς διὰ μέτρα δάσαντο| δαιμόνιοι ('men sail over the unseen sea with daring heart, and they look down over its invisible depths, and by their arts they have divided up the measures of the sea, marvellous beings', *Hal*. 1.9–12); the statement is preceded by a laudatory address to the emperor and followed by a description of the perils of sailing and the difficulties of catching fish, in contrast to the pleasures of life on land (*Hal*. 1.12–55). Oppian opens the *Halieutica*, in other words, by invoking the emperor's power and by referring to the professional expertise of the fishermen whose art (τέχνη, *Hal*. 1.7) he is to outline in his poem; that expertise, moreover, is described in terms of the measuring and mapping of the sea itself, an attempt to understand and impose order on an inherently treacherous and inscrutable body of water. In his repeated addresses to the emperor, Oppian represents knowledge about the natural world – both the information obtained by marine exploration and that disseminated by the poet himself – as intimately bound up with Roman imperial power, a tendency echoed and amplified by Ps.-Oppian, above all in the proem to the first book of the *Cynegetica*; both poems are addressed to the Roman emperor(s), and both poets foreground the relationship between their projects of ordering and cataloguing the world and the emperor's control over that world.[30]

---

[30] Similar patterns may be traced across a number of Imperial Greek didactic poems. Roman imperial power is more subtly evoked in Dionysius' *Periegesis*, although Hadrian is of course written into the fabric of the poem in the acrostic at 513–32; Marcellus of Side's extensive medical didactic poem (on which see further below) is declared, in an epigram perhaps authored by Marcellus himself, to have been acquired by Hadrian and Antoninus Pius for their imperial libraries. On the relationship between imperial power and the systematisation of knowledge in imperial texts, see esp. Potter 2004: 172–4; König and Whitmarsh 2007a.

What is more, Oppian's interest in this process of measurement, and its association, later in the proem, with a divinely all-encompassing view, evokes not only Dionysius' claims about the gods at the end of the *Periegesis*, but also Dionysius' own efforts to catalogue the world in his poem. For, as Dionysius declares, the poet may never himself have sailed at sea or seen the nations and places of which he speaks (*Perieg.* 707–14),

> ἀλλά με Μουσάων φορέει νόος, αἵτε δύνανται
> νόσφιν ἀλημοσύνης πολλὴν ἅλα μετρήσασθαι
> οὔρεά τ' ἤπειρόν τε καὶ αἰθερίων ὁδὸν ἄστρων.

but the mind of the Muses conveys me, they who, without roaming about, are able to measure much sea and mountains and mainland and the path of ethereal stars. (*Perieg.* 715–17)

In contrast to the frequent claims of ancient prose geographers to have traversed and viewed the regions which they describe, Dionysius disclaims personal experience in favour of the intellectual 'journey' he has undertaken courtesy of the Muses and the authoritative traditions they embody. As readers of the poem have long observed, Dionysius' Muses must here be understood to represent not only the omniscient, synoptic vision associated with the gods and Muses of archaic epic poetry, but also the traditions of scholastic endeavour associated with the Alexandrian Mouseion and the bodies of archived knowledge on which this poem is so evidently based.[31]

This section of the *Periegesis*, moreover, pointedly reframes Hesiod's statements in the *nautilia* of the *Works and Days* about his lack of practical experience in sailing and his privileged relationship with the Muses: δείξω δή τοι μέτρα πολυφλοίσβοιο θαλάσσης | οὔτε τι ναυτιλίης σεσοφισμένος οὔτε τι νηῶν ('I will show you the measures of the loud-roaring sea, although I have expertise neither in sailing nor in ships', *Op.* 648–9). As Hesiod goes on to imply, practical or professional knowledge of the field is not necessary for his purposes, for the Muses have 'taught' (ἐδίδαξαν, *Op.* 662) the poet to

---

[31] The Hesiodic and Hellenistic backdrop to this passage has been much discussed, as has Dionysius' use of the Muses to represent 'education' and intellectual or scholastic activity. See esp. Amato 2005: 89–101 and 314; Hunter 2004b: 228–9; Jacob 1990: 41–51 (and cf. 18–28); Lightfoot 2014: 107–8, 419–23; for the Hellenistic 'intellectualization of the Muses' and their 'role as goddesses who preside over education, scholarship and learning', see Bing 1988; Murray 2004: 386. On 'savoir livresque et autopsie' in geographical writings, see Marcotte 2000: 20–4. See also Bowie 2004: 178; Khan 2004: 235–6, 241. At *Hal.* 1.680 Oppian uses the Muses to suggest education and intellectual pursuit, on the broader resonances of which see Kneebone 2008.

sing even that of which he has no personal experience.³² The significance of this passage for the authoritative claims of later didactic poets, who frequently had little or no practical experience of their chosen subject matter, is immense.³³

As West observes of *Op.* 648, 'μέτρα is loosely used of the rules and formulae known to the expert';³⁴ Hesiod's claim here establishes space for an expertise available to him precisely because of his poetic authority, guaranteed by his proximity to the Muses and set in parallel to, and obviating the need for, the practical expertise of the sailor.³⁵ This notion of a specifically *poetic* expertise juxtaposed with the practical knowledge of the technical expert is further refracted not only in Dionysius' measuring/travelling Muses (the verb μετρήσασθαι here evoking both theoretical measurement and the physical traversal of that space),³⁶ but also in Oppian's use of the word μέτρα in the proem first of the practical experts who measure and sail over the sea (*Hal.* 1.11) and second of the poet's evaluation of his own ability to discuss his chosen subject matter (*Hal.* 1.92), the latter positioned pointedly after the poet's invocation to the gods and Muses (*Hal.* 1.73–9). In both cases, the juxtaposition of practical and poetic expertise underlies these poets' claims for the authoritative basis of their didactic poetry, and in both cases this poetry is framed as a fundamentally intellectual pursuit (νόος, *Perieg.* 715; νοήμα, *Hal.* 1.92). We might even detect a further

---

³² Cf. Hes. *Theog.* 22, where the Muses are said to have 'taught' (ἐδίδαξαν) the poet.
³³ This is not, of course, to imply that the authors of all ancient technical prose treatises necessarily wrote as professional specialists with personal experience in their chosen field (see, e.g., A. König and Doody, this volume, for discussion of the issue); however, the authorising strategies typically employed by ancient prose authors are, I suggest, rather different from those outlined above. The technical expertise (or otherwise) of didactic poets was evidently much discussed in antiquity, especially in relation to Aratus. See, e.g., Cic. *De or.* 1.69, apropos of Aratus and Nicander, for the implication that later didactic poets were widely acknowledged to write from a standpoint of poetic, rather than technical, expertise.
³⁴ West 1978 *ad loc.*
³⁵ For μέτρον in association with poetic σοφία, see also Solon fr. 13.51–2 W; Theognis 876; on Hesiodic authority in the *nautilia* and beyond, see: Hunter 2014: 52–8; Rosen 1990, esp. 101; Steiner 2005. On μέτρον and the intersection of practical and poetic expertise, see Dougherty 2001: 20–3; Ford 2002: 18; Purves 2010: 77–9. Cf. also the response of the Delphic oracle to Croesus' test of oracular authority at Hdt. 1.47.3: οἶδα δ᾽ ἐγὼ ψάμμου τ᾽ ἀριθμὸν καὶ μέτρα θαλάσσης ('I know the number of grains of sand and the measures of the sea'), on which see Kindt 2006: 36–7. As Purves 2010: 152 notes, 'the all-seeing knowledge of the Delphic oracle validates Herodotus' text in a way that is similar to the way in which the visual range of the Muses lends credibility to the voice of the epic poet'; note that Herodotus explicitly draws attention here to the fact that the oracle is delivered in hexameters.
³⁶ Cf. Amato 2005: 91, n. 108. The verb is also used by Oppian of Odysseus' traversal of the sea (*Hal.* 2.504) in a passage itself containing verbal echoes of Dionysius' portrayal of Odysseus; for Homeric parallels for this usage cf. *Od.* 3.179, 10.539, 12.428.

poetological claim in Oppian's μέτρα of *Hal.* 1.92: this, we understand, is to be information disseminated specifically in *metrical* form, poetry that activates the authorising force of the hexameter as a culturally privileged vehicle for knowledge.

Dionysius, as we have noted, explicitly advertises his purely intellectual engagement with his material; Oppian, on the other hand, leaves the issue of personal experience implicit, but certainly never suggests that he has firsthand knowledge of fish or fishing.[37] Rather, references to the fisherman's expertise are separated in the *Halieutica* from the poet's self-presentation, at times even distancing the poet from knowledge possessed by fishermen.[38] Thus, while effecting the transition from zoology (Books 1 to 2) to fishing (Books 3 to 5), the poet offers a list of the implements used by fishermen (3.72–91) but provides no practical information about their respective merits or suitability for different species or circumstances. Instead, Oppian observes that some fishermen prefer one method of catching fish, others another, but τῶν πάντων καὶ μέτρον ὅσον καὶ κόσμον ἑκάστου| ἀτρεκέως ἴσασιν, ὅσοι τάδε τεκταίνονται ('those men who devise all these things know accurately the due measure [μέτρον] and good order of each', 3.90–1). The poet lays no claim to detailed knowledge about the practicalities of fishing tackle, a topic which is subordinated to the fisherman's expert judgement about what is appropriate. Instead, Oppian in the proem to this book lays weight upon his specifically poetic authority and alleged proximity to the gods: it is, we are told, the gods who have established Oppian as ὑμνητήρ and source of delight to the emperor (3.7), and Hermes who is exhorted to guide and direct the poet's song (3.9–12). Indeed, the *Halieutica* is presented throughout as a poem closely associated with the gods and Muses: Oppian prays for his poetry to be divinely authorised and directed (1.73–9), songs (ἀοιδαί) at large are represented as the gifts of Apollo and the Muses (2.26) and the Muses are said to have crowned Oppian with the divine gift of song (4.7–10).[39]

Yet we must be careful here not to underplay the contrast between Dionysius and Oppian in their portrayal of the gulf between mortal and divine knowledge. For while Dionysius at the end of his poem refers to the

---

[37] Cf. Amato 2005: 97, n. 120; Bekker-Nielsen 2002: 30, 2006: 83; Rebuffat 2001: 29–33, although the latter needs some qualification.

[38] E.g. at *Hal.* 2.661–3; 4.300–1.

[39] This association with the divine is heightened by the poet's hymns and prayers to Poseidon and the sea gods at 1.73–9; to 'Father Zeus' at 1.409–20; to the gods at large at 2.38–42; to Zeus and the gods at 2.685–8; to Hermes at 3.9–28; to Eros at 4.11–39; indirectly, to Poseidon at 5.679–80.

gap between divine and human knowledge in order to highlight the gods' efforts to aid mankind, and while Dionysius' measuring Muses conduct him metaphorically across the world, Oppian is left at the start of his poem to understand the world with merely mortal understanding, for all his implied poetic proximity to the divine.[40] Many marine matters, as Oppian reports at 1.85–6, are hidden (κέκρυπται) from mortals and remain unseen or unknown (ἀείδελα), a state of ignorance which itself has a powerful didactic heritage.[41] Oppian's expression of mortal incapacity (1.80–92) is double the length of his address to the gods and Muse just before it (1.73–9), and the placement of his discussion of mortal knowledge and its limitations, providing the transition between his appeal to the gods and the start of his technical material (from 1.93 onwards), signals the troubled relationship between the two, a point magnified by the fact that his catalogue starts at the familiar shores and shallows before moving into the ἀμέτρητος ('unmeasured', 'immense' or 'immeasurable' 1.179) deep. Oppian's appeal to Poseidon and the sea to let him speak of ὑμετέρας ἀγέλας καὶ ἁλίτροφα φῦλα ('your shoals and sea-nourished tribes', 1.76) aids him little in attempting to resolve the subsequent question of whether the ἀγέλας οὔτ' ἔθνεα ('herds/shoals or tribes', 1.89) 'nourished' by the sea or the earth are greater. Instead, we are told, the gods know with certainty εἴτ' ἀμφήριστος ἐν ἀμφοτέρῃσι γενέθλη |εἴθ' ἑτέρη προβέβηκε ('whether the offspring are evenly matched on both sides, or whether one exceeds the other', 1.90–1), whereas in the mortal sphere all assessment must take place by human estimation (1.92).

There is no necessity, of course, for Oppian to raise and immediately fail to answer the question of whether the land or sea contains more species. Rather, the issue serves to exemplify the poem's didactic mode from the outset, instantiating its claims about the limitations of mortal endeavour. This is not, we are to understand, a poem which simply offers up the technical knowledge of an expert in fishing; rather, it is one which asks questions about the nature and provenance of knowledge itself. It is, moreover, worth stressing that Oppian nowhere proclaims his personal ignorance about the technical content of his poetry. Instead, while the poet repeatedly uses the

---

[40] Thus, while the ability to fish and to observe the sea is portrayed in Book 2 as a divine gift (*Hal.* 2.29–31), this does little to mitigate the uncertainty created at the start of the *Halieutica*; indeed, the statement is framed by the poet's professed ignorance as to the identity of the god who bestowed this gift.

[41] Cf. esp. Hes. *Op.* 42, where the gods keep the means of life 'hidden' from men (κρύψαντες γὰρ ἔχουσι θεοὶ βίον ἀνθρώποισιν) and Arat. *Phaen.* 768–72, where men do not yet know everything from Zeus, for much remains hidden (ἀλλ' ἔτι πολλὰ | κέκρυπται, 769–70), although it may be revealed by Zeus in signs.

εἴτε... εἴτε formulation to express his *aporia* in the face of alternative possibilities, this *aporia* is restricted to matters mythological and theological, with the sole exception of the question about marine and terrestrial species raised at the start of the poem.[42]

Oppian's references to topics on which mortals are deemed unable to pronounce, moreover, orientate his poetry towards theological, and above all hymnic, traditions of debate about the nature of knowledge, especially knowledge about the divine. It is, for instance, telling that Oppian's ἀμφήριστος... γενέθλη at 1.90 should recall in its phrasing the opening to Callimachus' *Hymn to Zeus*, where the poet questions whether Zeus should be hymned as Cretan or Arcadian: ἐν δοιῇ μάλα θυμός, ἐπεὶ γένος ἀμφήριστον ('my heart is greatly divided, for his birth is disputed', Call. *H*. 1.5). That this Callimachean verse was probably itself adapted from Antagoras of Rhodes' *Hymn to Eros*, a hymn which also offers various alternative genealogies for a god, is given further significance by Oppian's presentation of alternative genealogies for Eros in the extended 'hymn' to Eros in the proem to Book 4 of the *Halieutica* (4.11–39), there too introduced by the εἴτε... εἴτε formulation and a question about the god's disputed γενέθλη (*Hal.* 4.23-28).[43] In gesturing towards these claims of *aporia* in the face of partial, and merely mortal, understanding of the divine, Oppian situates his poetry within yet another tradition of culturally authorised hexameter discourse, this time the literary hymn, a form into which is woven a consciousness not only of the socially privileged role and expertise of the poet, but also of the mortal impossibility of fully comprehending the nature and magnitude of one's subject.

The boundaries between didactic and hymnic traditions had long been highly permeable: both Hesiod and Aratus open their didactic poems with a

---

[42] Thus, the poet remarks that it is impossible for a mortal to say (θνητῷ γὰρ ἀμήχανον ἐξονομῆναι, *Hal.* 1.411) whether Zeus is immanent or resides in the sky, and suggests in the last book that humans were either created by Prometheus or were born from the blood of the Titans. See *Hal.* 1.355-6 (human or divine invention of ships); 1.410-11 (immanence of Zeus); 2.35-7 (identity of the sea god); Eros: 4.23-8 (genealogy of Eros); 5.6-11 (genesis of humans). Oppian twice mentions fish whose reproductive mechanisms are unknown, yet these too are said to be beyond the limits of mortal knowledge (*Hal.* 1.593-4, 773-5). Ps.-Oppian, by contrast, only rarely expresses uncertainty, as e.g. over the breeding habits of rhinoceroses (*Cyn.* 2.561-5).

[43] On the philosophical and theological contexts for Callimachus' adaptation of Antagoras, see esp. Cuypers 2004; Stephens 2003: 79–82, 91. The discussion of Eros' genealogy in Plato's *Symposium* is surely also central to Oppian's account; my argument for hymnic precedents here rests less on any specific allusion to Antagoras than on the broader intellectual tradition of conveying *aporia* in the face of alternative genealogical accounts. Note too that human and divine capability is again contrasted in vv. 92–3 of Callimachus' *Hymn to Zeus*.

'Hymn to Zeus', and the 'technical' material in both poems may itself be read in terms of a meditation on Zeus' divine ordering of the world. Yet Oppian's evocation of hymnic traditions in man's inability to know the sea seems also to recall the hymnic echoes in the opening and closing verses of Dionysius' *Periegesis*, where the poet invokes not, as we might have expected, a named deity, but rather his geographical subject matter, and in particular the Ocean.[44] Likewise, at the start of his poem Dionysius employs the characteristic hymnic topos of polyonymy to depict the vast Ocean, again representing his subject matter in quasi-divine terms and drawing attention to the gulf between small-scale mortal perception and the immensity of the poem's subject: πάντη δ' ἀκαμάτου φέρεται σθένος Ὠκεανοῖο, | εἷς μὲν ἐών, πολλῆισι δ' ἐπωνυμίηισιν ἀρηρώς ('on every side is borne the strength of the inexhaustible Ocean, which, although it is one, is called by many names', *Perieg.* 27–8).[45] Oppian, as we have seen, alludes verbally throughout the proem of his first book to Dionysius' own proem and *envoi*; it seems highly likely that this consciousness should also be reflected in his quasi-hymnic depiction of the sea's vastness. Both the *Periegesis* and the *Halieutica*, then, are presented as divinely authorised literary texts, the knowledge disseminated in these poems purportedly guaranteed by the gods and Muses themselves. Yet this poetic association with the divine, I have argued, is at the same time intertwined in both texts with the poets' elevation of their technical subject matter itself to the level of quasi-hymnic discourse; neither poem opens with a hymn to a deity, but each defers its divine invocation until long after their chosen discipline has been introduced and explored (Dionysius' Muses are first invoked at 62–3, Oppian's deities at *Hal.* 1.73–9). While Dionysius' Muses are depicted as representatives of a scholastic or intellectual tradition of archived knowledge, however, and his gods beneficently mark out and divide up the world for humankind, the proem of the *Halieutica* instead highlights mankind's ultimate inability to explore

---

[44] See esp. *Perieg.* 1–3, 27–8, 1181–6. On hymnic elements in these lines, see Bowie 1990: 72; Effe 1977: 193; Hunter 2004b: 218; Lightfoot 2014: 89–90 and *ad loc.* Dionysius' use of hymnic formulations is further mediated through verbal allusion to (quasi-)hymnic literary precedents, esp. the opening verses of Apollonius' *Argonautica*, as well as Hesiod's *Theogony* and Callimachus' *Hymn to Delos*. As Lightfoot 2014 notes *ad loc.*, Dionysius' use of the word ὕμνος in the penultimate line of the poem may also be significant. For the 'didactic hymn' as an ancient tradition, see, e.g., Gordley 2011; for a contemporary Hadrianic 'hymn' to the Ocean, cf. the lyric poet Mesomedes' poem on the Adriatic (fr. 2, Heitsch, *Die griechischen Dichterfragmente der römischen Kaiserzeit*).

[45] On polyonymy as a characteristic hymnic strategy, see Bremer 1981: 194–5; Versnel 2011: 49–84 discusses 'dubitative formulas' and the use of the polyonomastic topos 'to articulate the inscrutable magnitude of the supreme divine power, which cannot be captured in one name' (50).

and comprehend the mysteries of the cosmos, for all the power of the gods.

Methods of Enquiry

We have so far observed the ways in which imperial Greek didactic poets foreground issues of comprehensiveness and knowledge in their poetry in order to situate their works within the intellectual traditions associated with long-standing and culturally authoritative (epic, didactic and hymnic) Greek hexameter poetry. The limited scope of the information provided in these didactic poems, as we have seen, is repeatedly presented by the poets as an inevitable function of the sheer size of the topics they treat, yet it also operates at times as a statement about the poet's expertly synoptic and selective role, about the privileged association of poetry with the divine, and about the prestigious literary heritage of the hexameter form. It is worth remembering, however, that this was by no means the only way for imperial Greek poets to construct and underscore their didactic authority. Let us turn, by way of contrast, to a rough contemporary of these second- and third-century didactic poets: Marcellus of Side, a doctor who authored a forty-two-book hexameter medicinal didactic *Iatrica* or *Chironides* under Hadrian and Antoninus Pius.[46] For Marcellus' poem affords us a glimpse into a different, and rarer, mode of authorial self-positioning: that of the didactic poet whom we know to have been a professional expert on his topic. The start of the surviving portion of Marcellus' poem outlines the terms of his enquiry:

> εὖ δὲ καὶ εἰναλίων ἐδάην φύσιν ἰήτειραν
> <σχ>ήμασι παντοίοισιν ἐμὸν νόον ἐξερεείνων,
> ὡς αὐτός τ' ἐνόησα καὶ ἄλλων μῦθον ἄκουσα·
> ὧν τοι ἐγὼ πληθὺν ἠδ' οὔνομα πᾶν ἀγορεύσω.
> βένθεα κητώεντα πολυσκοπέλοιο θαλάσσης   (5)
> ἰχθύες ἀμφινέμονται ἀπείριτοι ἀργινόεντες
> παμμέλανες περκνοί τε καὶ αἰόλον εἶδος ἔχοντες·
> φάγροι τε γλαῦκοί τε, πρέποντες βούφθαλμοί τε ...

> And I have learned well the medicinal nature of sea-creatures, searching my mind thoroughly in all sorts of ways, both as I myself have observed and material I have heard from others: I will tell you the multitude and

---

[46] See anon. *Anth. Pal.* 7.158.

all the names of these [creatures]. Around the monstrous depths of the sea with its many cliffs there range innumerable fish, bright, wholly black, dusky, and variegated in appearance: the sea-bream and the (?) dogfish and the prepon and the bogue... [47] (Marc. Sid. fr. 63.1–8)[48]

There follows a lengthy catalogue of fish. Marcellus claims in this poem to have integrated his professional experience with transmitted wisdom (ἄκουσα here presumably spanning both written sources and oral accounts), a statement which well advertises his status as both doctor and scholar. This kind of explicit methodological claim about the poem's sources is rare in hexameter didactic poetry, but is of course a recurrent feature of ancient technical prose writing. Yet Marcellus is not just a doctor but a poet, and his allusive language and erudite use of epithets engages with and manipulates a number of (didactic) epic conventions about his subject matter.[49]

In the first place, Marcellus' self-proclaimed project of describing and naming the whole multitude of fish (ὧν τοι ἐγὼ πληθὺν ἠδ᾽ οὔνομα πᾶν ἀγορεύσω) directly rebuts the rhetoric of the Iliadic Catalogue of Ships, where, as we have seen, the Homeric narrator had declared precisely the opposite: πληθὺν δ᾽ οὐκ ἂν ἐγὼ μυθήσομαι οὐδ᾽ ὀνομήνω (*Il.* 2.488). Marcellus' 'comprehensive' catalogue of fish is introduced, in other words, using language which evokes the very archetype of epic catalogue poetry, while simultaneously rejecting its predecessor's epistemological premise. What is more, with the observation that these creatures are ἀπείριτοι (a term whose epic and didactic resonance has been discussed above),[50] Marcellus conveys in his long list of fish names a sense of comprehensive mastery of his subject: he truly is to tell us the names of *all* these fish, evoking the epic immensity of his subject matter, but offering no indication that its vast scope offers any impediment to his ability to convey its contents exhaustively.

In contrast to Oppian's unknown and immeasurable sea, moreover, Marcellus' sea is πολυσκόπελος, a *hapax* that suggests not an unbounded

---

[47] Neither the γλαῦκος nor the πρέπων can be identified with certainty, although both recur in ancient natural historical accounts; the former refers to a kind of shark or dogfish. 101 verses of this poem survive. Bowie 1990: 69 n. 32 is surely right to doubt the suggestion of Effe 1977: 196–7 that this fragment belongs not to the *Iatrica*, but to a separate didactic poem on medicines obtained from fish.

[48] Text and numbering from Heitsch, *Die griechischen Dichterfragmente der römischen Kaiserzeit*.

[49] Nor is this solely a question of epithets: the word ἐδάην in v.1 features prominently in imperial didactic poets' descriptions of their didactic enterprises, and is used by poets of their knowledge of their subject matter at, e.g., Ps.-Man. *Ap.* 6.732; Opp. *Hal.* 4.635, 5.675, Ps.-Opp. *Cyn.* 2.561, 4.18; of fishermen's expert knowledge at Opp., *Hal.* 4.300; of knowledge imparted to the reader by his poem at Ps.-Man. *Ap.* 6.326. Cf. also Arat. *Phaen.* 376.

[50] Cf. Lightfoot 2014: 97, n. 48.

body of unknowable water, but a sea surrounded by promontories, headlands and lookout-places: this, we infer, is a realm whose inhabitants are eminently observable and available to be catalogued and used. The term evokes in particular the depiction of Scylla in the *Odyssey*, and especially the (instructional) account given by Circe to Odysseus in which she describes Scylla fishing from her cliff (σκόπελος). Circe here offers up, we might say, a miniature catalogue of fish-names as she describes Scylla fishing for dolphins and dogfish in the hope of finding a κῆτος (cf. Marcellus' βένθεα κητώεντα, 5) amongst the countless (μύρια) creatures of the sea (*Od.* 12.94–7). Eight of the nine σκόπελοι to be found in the Homeric epics, moreover, are used of Scylla's cliff: could Marcellus' *hapax* πολυσκόπελος even be read as an erudite joke about this very frequency?[51]

Nor does Marcellus' engagement with epic fish end here, for his use of the relatively unusual verb ἀμφινέμονται in the phrase ἰχθύες ἀμφινέμονται (6) surely picks up Aratus' description of the location of Pisces: τὸν δὲ μετὰ σκαίροντα δύ' Ἰχθύες ἀμφινέμονται| "Ἵππον (*Phaen.* 282–3). In each case, Marcellus recasts (didactic) epic discussions of fish in a new, scientific, context, expanding considerably on their scope and comprehensiveness: we are now presented not with two fish, nor with a list of three different types of fish, but with an immensely long list which highlights its own comprehensiveness and allusive epic engagement. In his evocation of an ἀπείριτος subject matter, his own declared professional experience and learned comprehensiveness, in his epic echoes and his pointed engagement with the Catalogue of Ships in *Iliad* 2, Marcellus draws repeated attention to epic and didactic topoi only to advertise his distance from them. This, we are to understand, is the didactic poem of a man who presents himself as both doctor and poet.

If Marcellus' poetry represents a point of self-conscious intersection between the 'epic' and 'scientific' didactic traditions, however, it remains an intersection orientated primarily towards the epic mode; here we may in turn contrast the traditions of hexameter didactic epic with the Hellenistic iambic geographical didactic tradition exemplified by Ps.-Scymnus (late second century BCE) and Dionysius son of Calliphon (probably first century BCE), both of whom ostentatiously align their poems with prosaic geographical traditions rather than with epic poetry.[52] Ps.-Scymnus, for instance, offers in relatively swift succession an indication of his poetry's

---

[51] Thus *Od.* 12.73, 80, 95, 101, 108, 220, 239, 430. Marcellus perhaps even alludes further to the central significance of πολυ- compounds in the *Odyssey* itself, in this field outdoing even his Homeric model.

[52] As well as with comic traditions, as Ps.-Scymnus states overtly from the first verse.

comprehensiveness (73–4), a long list of prose sources cited by name and at times differentiated according to perceived reliability (109–27) and a delineation of his methods, which he paints as the combination of ἱστορία, autopsy and travel (128–36). The poet's emphasis upon ἱστορία in particular (111, 132; cf. 44) aligns Ps.-Scymnus' self-proclaimed investigative methods with his depiction of prose historians such as Timaeus (214), Herodotus (565) and Ephorus (fr. 15b10 Marcotte).[53] Here the choice of iambics over hexameters clearly represents a marked generic choice and a self-conscious affiliation with a more prosaic, and 'scientifically' orientated, mode of didactic poetry.

Each of the Greek didactic poets we have considered may be seen to occupy a different position on a spectrum of generic affiliation differentiated less by subject matter than by choice of metre, and bolstered by implicit or explicit claims about methodology, comprehensiveness, personal experience and the citation of sources. At the far end of this didactic spectrum, however, lie technical prose works.[54] While no ancient treatise on fishing survives in more than exiguous fragments, we may learn much from the comparison of Dionysius' *Periegesis* with geographical prose works, or of Ps.-Oppian's *Cynegetica* with prose treatises on hunting.[55] Arrian's *Cynegeticus* (second century CE), for instance, well illustrates some of the authoritative conventions to be observed in technical prose treatises of the period. In contrast to the near-contemporary didactic poems this chapter has explored, Arrian explicitly represents the value of his treatise as lying in its supplementation of information provided by previous treatment of the topic; he does so, moreover, by naming and evaluating his predecessor's work throughout. From the outset, Arrian remarks not only upon the topics covered by Xenophon in his *Cynegeticus* (Arr. *Cyn.* 1.1–3), but also that Xenophon had made no mention of the Celtic gaze-hounds more recently introduced to the hunt, or of Scythian and Libyan horses, an omission which Arrian sets out to rectify at length (Arr. *Cyn.* 1.4); the first word of the work, indeed, is 'Xenophon', marking Arrian's explicit engagement with the

---

[53] See further Boshnakov 2004: 7–9; Effe 1977: 184–7; Hunter 2006, esp. 132–3; Korenjak 2003: 16–18; Marcotte 2000: 20–4.

[54] Which may themselves variously mark their engagement with poetic conventions; cf. e.g. Arr. *Cyn.* 35, on Homer, or the place of Hesiod in Xen. *Cyn.*, on which see now Hunter 2014: 59–64. Consideration of scientific prose authors' attitudes towards didactic poetry lies beyond the scope of this chapter, but offers fascinating insights into the role and reception of Greek didactic poetry in 'professional' contexts. Galen's self-professed relationship with the medical didactic poems of Andromachus and Damocrates, for instance, is well explored in Totelin 2012; Vogt 2005; von Staden 1998.

[55] A strategy traced across several ancient disciplines in Hutchinson 2009.

Cynegeticus of his predecessor.⁵⁶ Arrian, moreover, expressly frames this engagement with his predecessor in the same way that Xenophon himself, as Arrian observes (Arr. *Cyn.* 1.5), had set out to supplement Simon's treatise in his work on horsemanship (Xen. *Eq.* 1.1), thus adopting and surpassing Xenophon's own 'rhetoric of completeness'.⁵⁷ Arrian here overtly writes himself into an accumulative prose tradition of supplementation and correction, portraying his work as one in a series of prose manuals whose declared purpose was to fill the gaps in recorded material, implicitly aiming at eventual comprehensive coverage of the topic at hand.⁵⁸

We may note too that Arrian places considerable weight in his treatise on autopsy and personal experience, offering a lengthy description of the hunting hound he himself raised and several times referring to his own practices and observations while hunting.⁵⁹ Here we may again contrast the didactic tradition represented by Dionysius and Oppian, neither of whom presents himself as having any first-hand practical expertise in his subject; indeed, both distance themselves carefully from this kind of first-hand experience. This, however, is by no means to say that Greek imperial didactic poets could not also point to their personal engagement with their subject, merely that their ability to provide information was not usually deemed to depend upon this personal experience. Thus, while Ps.-Oppian never portrays himself as an expert in zoology or in hunting – indeed, the poet's conversation with Artemis at the start of the *Cynegetica* explores his potential to compose on any number of topics, from contemporary wars to erotic poetry, and he elsewhere mentions information that he has been unable to ascertain – autoptic experience is nevertheless afforded a privileged position in the *Cynegetica*. We hear, for instance, of Ps.-Oppian's personal sightings of Cappadocian horses, ostriches and a marvellous black lion; this last is especially marked in its differentiation between first- and second-hand material: ἔδρακον, οὐ πυθόμεν, κεῖνόν ποτε θῆρα δαφοινόν ('I once saw, and did not

---

⁵⁶ Arrian's attitude towards his predecessor is almost entirely approbatory, although note the disagreement with Xenophon over hares at Arr. *Cyn.* 16.6, which Arrian declares 'the only point on which I do not agree with my namesake'. On Arrian and Xenophon, see Hutchinson 2009: 205–6; Phillips and Willcock 1999; Stadter 1976; Swain 1996: 246–7; on Arrian's citational practice, see Stadter 1980: 57–8.

⁵⁷ Cf. Hunter 2004a: 234.

⁵⁸ Cf. Wietzke, this volume, on mathematical supplementation and correction.

⁵⁹ Arr. *Cyn.* 5.1–6; cf. 7.2 (dog), 16.5 (hares), 35.1 (sacrifice). Cf., for instance, the prefaces to the five books of Artemidorus' *Oneirocritica* (second to early third century CE), similarly filled with claims about the author's research methods, including travel, personal experience and the comprehensive gathering of earlier material on the subject; Artemidorus too repeatedly marks his engagement with, and supplementation of, previous treatment of the topic (even, in the preface to Book 4, including his own earlier books). Cf. Harris-McCoy 2012: 37, 40–1.

[just] learn about, that savage beast', *Cyn.* 3.46).⁶⁰ Autopsy is here presented as superior to mere book-learning, and worthy of explicit discussion for its own sake. In no case, however, does the poet claim to offer the reader new information based on such experience;⁶¹ rather, the poet's emphasis is on the visual world of imperial spectacle (the black lion, for instance, has been transported for the emperor's eyes), and on the relative reliability of different sources of information.

We have already noted the ways in which Dionysius and Oppian mark out the boundaries of their poetry in terms of mortal capability. Ps.-Oppian offers his own version of this topos in the proem of Book 4, a passage which effects the poem's transition from zoology to hunting:

> ἤθεα πολλὰ πέλει κλειτῆς πολυαρκέος ἄγρης, (10)
> ἄρμενα καὶ θήρεσσι καὶ ἔθνεσιν ἠδὲ χαράδραις,
> μυρία· τίς κεν ἅπαντα μιῇ φρενὶ χωρήσειεν
> εἰπέμεναι κατὰ μοῖραν ὑπ' εὐκελάδοισιν ἀοιδαῖς;
> τίς δ' ἂν πάντ' ἐσίδοι; τίς δ' ἂν τόσον ὠπήσαιτο
> θνητὸς ἐών; μοῦνοι δὲ θεοὶ ῥέα πάνθ' ὁρόωσιν. (15)
> αὐτὰρ ἐγὼν ἐρέω τά τ' ἐμοῖς ἴδον ὀφθαλμοῖσι,
> θήρην ἀγλαόδωρον ἐπιστείχων ξυλόχοισιν,
> ὅσσα τ' ἀπ' ἀνθρώπων ἐδάην, τοῖσιν τὰ μέμηλεν,
> αἰόλα παντοίης ἐρατῆς μυστήρια τέχνης.

> There are many kinds of glorious many-netted hunting, countless kinds, suited to the animals and species and ravines. Who could contain them all in their single mind and describe them all in order in melodious song? Who could look at them all? Which mortal could see so much? Only the gods see everything easily. But I myself will tell both of the things I saw with my own eyes while approaching in the thickets the hunt that bestows splendid gifts, and as much as I have learned from the men who concern themselves with these things, the quick-moving mysteries of all kinds of desirable skill… (*Cyn.* 4.10–19)

Ps.-Oppian here signals his relationship to both the *Periegesis* and the *Halieutica*. His claim of authorial incapacity (*Cyn.* 4.12–15) picks up on those of Oppian (τά κεν οὔ τις ἀείδελα μυθήσαιτο| θνητὸς ἐών· ὀλίγος δὲ νόος

---

⁶⁰ Horses: θαῦμα δὲ Καππαδόκεσσι μέγ' ἔδρακον ὠκυπόδεσσι ('I have seeen a great marvel in swift-footed Cappadocian [horses]', *Cyn.* 1.198); ostrich: ναὶ μὴν ἄλλο γένεθλον ἐμοῖς ἴδον ὀφθαλμοῖσιν| ἀμφίδυμον, μέγα θαῦμα, μετὰ στρουθοῖο κάμηλον ('moreover, I have seen another hybrid breed with my own eyes, a great marvel, the camel combined with the sparrow', *Cyn.* 3.482–3). Cf. Agosta 2009: 25.

⁶¹ The poet does, however, mention information he has gleaned from lion-keepers: *Cyn.* 3.53–5.

μερόπεσσι καὶ ἀλκή, *Hal.* 1.86–7) and Dionysius (οὕς οὐκ ἄν τις ἀριφραδέως ἀγορεῦσαι| θνητὸς ἐών· μοῦνοι δὲ θεοὶ ῥέα πάντα δύνανται, *Perieg.* 1168–9), while his μυρία (*Cyn.* 4.12) echoes Dionysius' μυρίοι (*Perieg.* 1168) and Oppian's μυρία (*Hal.* 1.80), all three in the same *sedes*. Yet whereas both Dionysius and Oppian had addressed the impossibility of describing the countless inhabitants of the land and sea respectively, Ps.-Oppian focuses rather on the various branches of a resolutely human activity: hunting. His focus is akin to that of Oppian, namely the realm of vision, but Ps.-Oppian, as ever in the *Cynegetica*, is concerned more with what may be seen than with what may not, and above all with the testimony of his own eyes and its implication for his poetry (thus *Cyn.* 4.13). This is a poem oriented towards human endeavour and the relationship between personal experience and theoretical knowledge rather than the gap between divine and human mortal understanding. Ps.-Oppian's statement that the gods see all things thus cedes directly to a discussion of his own experience and its role in his poetry, combining claims to autopsy, information obtained from experts and divinely conferred knowledge (*Cyn.* 4.16–22). The poet's synthesis of second-hand knowledge and personal experience is in turn echoed in his subsequent description of the impact of his poem on its addressee, Caracalla: the emperor, we hear, is to receive knowledge of how to pursue his quarry so that in his hunting he may be μακαριστὸς ὁμοῦ παλάμῃ καὶ ἀοιδῇ ('blessed in both deed and song at once', *Cyn.* 4.24). Nowhere, however, does Ps.-Oppian speak again of his own first-hand experience.

As we have seen, both Dionysius and Oppian use the vocabulary of measurement to juxtapose practical and poetic expertise in their chosen disciplines. Dionysius' use of the verb μετρέω in association with the Muses suggests both the theoretical geography in which he partakes and the physical experience of travelling and exploration (which he leaves to the authors of the works contained in the Mouseion), while the μέτρα to which Oppian frequently returns encompass the physical process of traversing and marking out the sea and its inhabitants, as well as the professional expertise of the fisherman, who knows which implements to use and how; these are in turn set alongside the poet's own measured assessment of his topic and his skill in turning this material into metrical form. In this light we may note that Ps.-Oppian, like Dionysius and Oppian, similarly depicts his own poetic activities in language which plays evocatively with the connection between his poetic composition and the technical discipline described in that poetry. For while Dionysius depicts himself embarking upon an intellectual 'journey' in which he is 'carried' by the Muses across the world his poetry describes, Ps.-Oppian 'wanders' and pursues the 'path' of his song along an untrodden track akin to that on which the hunter pursues his prey.[62] Thus, Artemis' initial instructions to the poet to sing of hunting rather than more familiar subjects – ἔγρεο, καὶ τρηχεῖαν ἐπιστείβωμεν ἀταρπόν,| τὴν μερόπων οὔπω τις ἑῆς ἐπάτησεν ἀοιδαῖς ('arise, and let us tread a rugged path that no mortal has yet trodden with his song', *Cyn.* 1.20–1) – are in turn echoed by the poet's own advice on hare-hunting at the end of the *Cynegetica*: ναὶ μὴν ἀτραπιτοῖο πολυστιβίην ἀλεείνειν| καὶ πάτον, ἐν δ' ἀρότοισι γεωμορίῃσι τ ἐλαύνειν ('moreover, avoid the much-trodden path and the beaten track, but pursue [hares] in the fields and the tilled lands', *Cyn.* 4.433–4).[63] These didactic poets may claim little practical experience in the subjects they treat, but their poems carefully manipulate the metaphorical connections to be drawn between poetic composition and practical expertise in the field.

Conclusion: The Limits of Enquiry

This chapter has examined a group of imperial Greek didactic poets who readily admit to the partial and selective nature of their information, who reflect actively on what can and cannot be treated in their poetry and who each comment on the discrepancy between the 'infinite' nature of their subject matter and the finite nature of their poetry. Their very virtue as poets is presented in terms of their synoptic view and their selection of salient information for the reader, while their poems are each in different ways structured by this tension between selectivity and the drive to catalogue and incorporate. Each of these poets, moreover, locates their work within a shared epistemological framework which traces its roots back to archaic epic poetry and which shifts the question of comprehensiveness away from issues of professional or technical competence and towards the idea of mortal incapacity to depict an immense, even boundless, topic. Their evocation of epic, didactic and hymnic traditions, and their layered interweaving of archaic and Hellenistic precedents, sets their work in a deeply poetic space imbued with the sense of a shared literary and intellectual heritage, yet one also devoted to describing and cataloguing a world now oriented in large part towards Roman imperial power.[64]

---

[62] Cf. e.g. *Cyn.* 1.368–9. On metapoetic journey imagery in didactic poetry, see Fowler 2000; Volk 2002, esp. 20–4.

[63] On the Callimachean resonances of Artemis' words, see esp. Asper 1997: 105–6; Costanza 1991; and the bibliography in n. 15 above.

[64] Any examination of the relationship between later Greek and Latin didactic poetry lies well beyond the scope of this chapter, although it is worth observing that themes such as authorial

We have, however, seen that alternative authoritative tactics were employed by other Hellenistic and imperial didactic poets, particularly those composing in the iambic didactic tradition, and that several of these poets did indeed put forward the kinds of claims to comprehensiveness, personal experience and the citation of named sources that we might more readily associate with technical prose treatises. Within the hexameter didactic tradition, on the other hand, Dionysius, Oppian and Ps.-Oppian rework similar topoi in very different ways: Dionysius ends his poem with a reflection on the beneficent activities of the gods and the relationship between bounded individuals and the cosmos at large; Oppian opens his poem with a stark reminder of human limitation and raises important methodological questions about the ways in which we view and interpret the world; Ps.-Oppian in turn reworks both Dionysius' and Oppian's declarations of mortal constraint in order to foreground the importance of sight and its impact on his own poetic enterprise. Together, therefore, these didactic poets present themselves as engaging in an ongoing debate about the nature and boundaries of knowledge, interrogating the complex relationship between practical and technical expertise, compilatory or synoptic scholarly endeavour and 'divinely authorised' poetic craftsmanship.

> selectivity, (lack of) comprehensiveness, technical and poetic expertise, the boundaries of knowledge and the appeal to divine authority recur throughout Latin didactic poetry. On, e.g., authority, selectivity and technical expertise in Virgil's *Georgics* in relation to Latin agricultural prose literature, see Doody 2007; Spurr 1986, esp. 166–9; Thomas 1987; on the boundaries of knowledge in the *Georgics*, see Schiesaro 1997. On the implications of Manilius' inconsistencies and lack of comprehensiveness, see Volk 2009: 116–26, 174–82; for the boundaries of the world and of knowledge in the *Astronomica*, see Kennedy 2011: 181–7. For the proliferation of the epic 'incapacity' topos of *Il.* 2.488–90 in Latin poetry, see Hinds 1998: 35–47, esp. in relation to Virgil, *Geo.* 2.42–4; Lucretius in Servius' commentary on Virgil, *Georg.* 2.42; Ov. *Ars Am.* 1.433–6.

10 Expertise, 'Character' and the 'Authority Effect' in the *Early Roman History* of Dionysius of Halicarnassus

NICOLAS WIATER*

Introduction: 'Literary Authority' vs 'Authority Effect'

The last two decades have witnessed a remarkable increase of scholarly interest in the works of Dionysius of Halicarnassus. Even though the most significant progress has been made on Dionysius' classicism,[1] scholars have also begun to explore Dionysius' historical work, his *Early Roman History*, or *Roman Antiquities* (*Antiquitates Romanae*), not only from a historical but also a literary perspective.[2] As part of this renewed interest in the *Antiquities*, Dionysius' authority as a historical narrator has also received some attention in two important studies by John Marincola and Clemence Schultze.[3]

John Marincola discusses Dionysius alongside and in comparison with other Greek historical authors in his *Authority and Tradition in Ancient Historiography*, while Clemence Schultze's 2000 article is devoted entirely to Dionysius' authority as a historical narrator. Both scholars are interested in what could be called authorial strategies of establishing an authoritative *persona*. In this process, Dionysius' expertise plays a key role: Dionysius repeatedly engages in a 'conspicuous display' of his comprehensive knowledge of the sources, thus demonstrating, as Marincola points out, 'both inclusivity... and superiority'.[4] This display of knowledge goes hand-in-hand with a demonstration of Dionysius' 'critical faculty and judicious

---

* I would like to thank Jason König and Greg Woolf for inviting me to contribute to this volume, and Jason König in particular for all his helpful remarks, comments and suggestions. Thanks are due also to Aiste Celkyte for her comments on an earlier draft of this chapter.
[1] See, e.g., de Jonge 2005a, 2005b, 2008; Hidber 1996, 2011; Porter 2006a, 2006b; Wiater 2011a, 2011b, forthcoming.
[2] The most important monographs are Delcourt 2005 and Gabba 1991; see further Fox's discussion of Dionysius' image of the regal period in Fox 1996 along with Fox 1993, 2001; other important studies include Fromentin 1988, 2006, 2010.
[3] Marincola 1997; Schultze 2000.
[4] Marincola 1997: 235; cf. Schultze 2000: 14, 21. De George 1985: 22 refers to this concept of authority as 'epistemic'.

scrutiny of what others have said'.[5] Moreover, Dionysius stresses his long-standing personal familiarity with both Rome and the Romans as well as his knowledge of Latin by giving a prominent place to Latin sources alongside Greek ones.[6] Both Marincola and Schultze have also pointed out how Dionysius evokes, and enters a dialogue with, his prominent Greek predecessors, especially Herodotus and Thucydides, but also Polybius, Theopompus and Timaeus.[7]

Marincola and Schultze approach authority from an author-centred perspective in which Dionysius' narratorial authority is virtually identical with the image he creates of himself as a narrator. This results clearly from Marincola's definition of authority as 'the rhetorical means by which the ancient historian claims the competence to narrate and explain the past, and simultaneously constructs a persona that the audience will find persuasive and believable';[8] and even though Schultze never explicitly defines the concept, the similarities of her methodology and results to Marincola's show that she operates on the same premises.

When exploring 'literary authority',[9] the author's self-presentation is certainly a key factor: authority in literature is inevitably related to the author, and Schultze and Marincola have contributed greatly to furthering our understanding of Dionysius' authority as a narrator. But if we take the connection of authority and persuasion seriously, which is (rightly) implied in Marincola's definition of the concept, I, at least, cannot help but wonder whether an author's display of expertise is really sufficient to convince readers to accept his own opinion as opposed to those of other, equally learned and, hence, authoritative, figures.[10]

I would argue that a full understanding of authority, just like a full understanding of persuasion in a political or court speech, has to explore the role

---

[5] Marincola 1997: 235; cf. Schultze 2000: 21: 'The claim to exhaustive... and meticulous reading is... the most important single element of Dionysius' claim to authority'.

[6] Marincola 1997: 123, 234, 245 (on the importance of the epichoric tradition); Schultze 2000: 11, 19–21 on Dionysius' claim to be particularly qualified to handle a Roman subject.

[7] Delcourt 2005 index s.vv. Polybe, Thucydide, Timée de Tauromenium; Marincola 1997: 123, 246, n. 147; Schultze 2000: 10, 11, 19–20. Both Delcourt and Schultze underestimate the complexity of Dionysius' engagement with Polybius, see Wiater 2011b: 194–8, cf. ibid. 143, n. 403, 183–4.

[8] Marincola 1997: 1.   [9] Marincola 1997: 1.

[10] It seems to me that the narrator's authority is particularly important when the reader is confronted with conflicting, (seemingly) equally convincing alternatives, when, to put it differently, the author expects readers to subscribe to his view only because of their assessment of his reliability and the trust they place in his judgement, that is, because it is Dionysius who is subscribing to it. I have therefore focused on such passages in my analysis of Dionysius' 'literary authority'.

10 Expertise, 'Character' and the 'Authority Effect' in Dionysius

of the author and his 'authoritative *persona*' as part of the larger communicative relationship, the *rapport*, which the speaker, or author, establishes with his addressees. We can then conceive of authority as the result of a two-sided process in which the speaker/author is involved as much as the recipient. Such an approach receives support from Dionysius' own description of how he expects his *Early Roman History* to interact with his addressees: in a strikingly physical image, Dionysius describes how his narrative will 'extract' (ἐξελέσθαι) his contemporaries' erroneous opinions about the Romans and replace them (ἀντικατασκευάσαι) with the true ones (τὰς ἀληθεῖς): the *Antiquities* will, quite literally, re-configure his readers' minds (τῆς διανοίας) (1.5.1). At the same time, however, Dionysius is keen to stress that this operation is the result of a negotiation: the readers' intellectual 'enlightenment' regarding the Romans' ethnic origins is the 'promise' (ὑπισχνοῦμαι) that his narrative holds and will be contingent on a proper 'demonstration' (ἐπιδεῖξαι) (1.5.1). Dionysius himself conceives of the process of 'education', or rather, re-education, that his work is designed to bring about as a dialectics between an overt demonstration of his skills and expertise and a negotiation with the reader.

Giving the reader a more central role in our concept of 'literary authority' seems important for two reasons. To begin with, an author's influence on his readers, including his ability to make them accept his claim to authority and, consequently, adopt his point of view, is much more elusive than the establishment of an authority-relationship in real-life interactions. Neither Marincola nor Schultze seem to pay enough attention to this difference. But in contrast to real life,[11] authority in literature cannot be imposed: an author cannot pull rank. Moreover, authority in real-life encounters often depends to a large extent on the physical presence of and immediate interaction with the person who is acknowledged as an authority. The presence of an author in and through his text is much less immediate and therefore gives much greater weight to the recipient's side of the process. Much more than in extra-textual reality, 'literary authority' is the result of a negotiation that cannot rest on an author's demonstration or assertion of 'epistemic authority' alone.[12]

Furthermore, in the case of the *Early Roman History* in particular we have to take into account that Dionysius is not writing for uneducated 'lay readers' who read his work only for pleasure and are simply ready to accept his superior knowledge. As Dionysius explains at 1.8.3, his primary addressees are highly educated specialist readers who are either active politicians or

---

[11] Cf. De George 1985: 15. [12] On 'epistemic authority', see n. 4 above.

experts in political theory and philosophy.[13] Not all of them might have been equally familiar with all the historical and antiquarian material that Dionysius has studied, but it is probably safe to assume that many had more than just a cursory knowledge at least of the more prominent Greek and Roman historical works to which Dionysius refers. Such 'expert readers' are likely to have seen themselves as Dionysius' intellectual equals and we can assume that they were more than prepared to evaluate critically or even challenge Dionysius' interpretation of Roman history. Yet, as de George points out, the acceptance of someone's authority always implies the acceptance of a 'superior/inferior relation'.[14] Assertions of knowledge and expertise as such, then, are unlikely to have a strong effect on such readers unless Dionysius has prepared their acceptance by creating a reading situation that instils a sense of inferiority and inability to deal with the material competently even in confident, highly educated readers.

I therefore propose a different approach to authority in Dionysius' *Early Roman History*. From Marincola and Schultze I retain the emphasis on Dionysius' demonstration of skills and expertise, his 'epistemic authority'. But these alone are not sufficient to create 'literary authority'. Rather than a static and 'top-down' phenomenon based exclusively on the author's self-presentation, we should see 'literary authority' as a form of interaction between author and reader in which the author's self-presentation is necessarily complemented by positioning the reader towards the text in such a way that he or she is willing to accept his authority.[15] Put differently, 'literary authority', as I understand it, includes 'epistemic authority', the display of superior knowledge and expertise, but sees the latter as only one part of a larger, complex relationship between author and reader:[16] 'literary authority' is not simply a characteristic of the author, but the result of the specific design of the text. In order to take this peculiar relational, or 'interactive', nature of authority in literature into account, and to avoid confusion

---

[13] καὶ τοῖς περὶ τοὺς πολιτικοὺς διατρίβουσι λόγους καὶ τοῖς περὶ τὴν φιλόσοφον ἐσπουδακόσι θεωρίαν; Dionysius acknowledges that some people might just want to read his work for amusement (καὶ εἴ τισιν ἀοχλήτου δεήσει διαγωγῆς ἐν ἱστορικοῖς ἀναγνώσμασιν), but the pleasure resulting from his narrative is merely a by-product of the variety of material which he covers and by no means the main purpose of his work. On the interpretation of this passage, see Nicolas Wiater, 'ΕΞ ΑΠΑΣΗΣ ΙΔΕΑΣ ΜΙΚΤΟΝ: Dionysius of Halicarnassus on the Design of his Historical Work' (in preparation); meanwhile, cf. Fromentin 1993. Dionysius clearly envisages 'upper-class Roman' (Bowersock 1965: 131; cf. Luraghi 2003: 281, 283) as well as educated Greek readers (Bowersock 1965: 131; Delcourt 2005: 65–9; Schultze 1986: 138–9).

[14] De George 1985: 15.

[15] For a discussion of passages illustrating this, see the third section below.

[16] Cf. De George 1985: 14–16 on authority as a 'relational quality'.

between the different concepts of 'authority', 'epistemic authority' and 'literary authority', I propose to use the term 'authority effect'.[17]

The remainder of this chapter will flesh out these considerations. The following section is centred on passages representative of 'literary authority' in Marincola's and Schultze's sense, that is, passages featuring an overt demonstration of superior knowledge and skills on Dionysius' part, but this section considers these passages in the light of author-reader interaction. My aim is to show that a proper understanding of authority in literature, the 'authority effect', is possible only if both of these factors, Dionysius' self-presentation and the reader's involvement in the narrative, are taken into account. One can even, I will argue, go so far as to say that dialogic interaction with the reader becomes itself a form of authorial control.

The reader's interaction with the text will play a central role also in the third section. The main focus of the discussion in this section, however, will be on aspects of Dionysius' self-presentation whose crucial importance to the creation of the 'authority effect' has not yet been recognised because Marincola and Schultze identify Dionysius' authority exclusively with his 'academic qualifications'.[18] The discussion will show, by contrast, that other, more 'irrational' elements of Dionysius' self-presentation, his 'character', are of equal importance for the creation of the 'authority effect':[19] the qualities of character that Dionysius encourages his readers to recognise in him, for example, his moral integrity, piety and political views and even his emotional involvement with the historical actors, contribute just as much to the reader's acceptance of Dionysius' authority as the demonstrations of his superior knowledge.

In order to put the 'authority effect' in the *Antiquities* into a larger perspective, I have endeavoured to compare and contrast Dionysius' practice with that of other historical writers. I should emphasise that the purpose of these comparisons is to give some indications about the ways in which Dionysius' practice might be distinctive or, indeed, similar to that of other historians. I do hope that they might stimulate further thought about the role of the reader in constructions of 'literary authority' in general, but they are not designed to be a comprehensive attempt to place Dionysius in the

---

[17] This term is loosely inspired by Barthes' famous notion of 'reality effect' ('l'effet du réel').
[18] This anachronism is deliberate, as Marincola's and Schultze's studies alike seem tacitly to presuppose the modern academic system in which 'credibility' depends, at least in theory, exclusively on the author's education and demonstrated professional competence, whereas an author's moral or religious views are explicitly excluded as irrelevant to the validity of his or her statements.
[19] On the importance of 'trustworthiness' and 'character' in the creation of authority, see De George 1985: 40.

tradition of ancient historical writing. As important and desirable as such a study would be, it goes far beyond what this chapter can, and wants to, achieve. Sporadic though they are, those side-glances on other historical writers will be sufficient to demonstrate, I hope, that compared to authors like Herodotus and Diodorus, Dionysius is distinctive in his almost obsessive concern with controlling his reader's views and opinions. On the other hand, there seem to be interesting parallels between Dionysius' literary practice and Nicole Loraux's reading of Thucydides' 'héroïsme de l'intellect' and the 'submission' to his interpretation of the Peloponnesian War that he expects, or, rather, demands, from his readers.[20]

'Let Every Reader Judge As He Thinks Proper': Reader Participation and the 'Authority Effect' in the *Antiquities*

The first passage I will discuss is Dionysius' version of Aeneas' flight from Troy to Italy (1.45.4–49). A brief summary of these rather convoluted chapters will be useful for the subsequent discussion.

Dionysius begins his account by stressing the great variety of available versions (1.45.4): 'concerning the arrival of Aeneas in Italy, since some historians have been ignorant of it (ἠγνόηται) and others have related it in a different manner (διαπεφώνηται), I wish to give more than a cursory account, having compared the histories of those writers, both Greek and Roman, who are the best accredited (τάς τε τῶν Ἑλλήνων καὶ τὰς Ῥωμαίων τῶν μάλιστα πιστευομένων ἱστορίας παραβαλών)'.[21] The following two chapters, however, give a surprisingly straightforward account of Aeneas' departure from Troy without signs of doubt or disagreement. It is only at the beginning of 1.48 that the reader learns that he has just read the version presented by Hellanicus, one of the 'ancient historians' (τῶν παλαιῶν συγγραφέων), deemed by Dionysius to be the 'most credible' one (πιστότατος): 'There are different accounts given of the same events by some others, which I look upon as less probable (ἧττον... πιθανούς) than this. But let every reader judge as he thinks proper' (κρινέτω δὲ ὡς ἕκαστος τῶν ἀκουόντων βούλεται, 1.48.1).[22]

1.48.2 to 1.49.2 then present the readers with several divergent accounts which Dionysius partly summarises and partly introduces through verbatim passages. Chapter 48 contains different versions of Aeneas' departure itself and the events leading up to it, beginning with Sophocles' *Laocoon* (1.48.2),

---

[20] Loraux 1986.   [21] Translations of the *Antiquities* are from Cary's Loeb, often adapted.
[22] Cf. Fox 1996: 56–7, 78–80.

followed by the version of Menecrates of Xanthus (1.48.3). The chapter concludes with a brief summary of other versions, the authors of which remain unnamed (1.48.4):

> Others say that he chanced to be tarrying at that time at the station where the Trojan ships lay; and others that he had been sent with a force into Phrygia by Priam upon some military expedition. Some give a more fabulous (μυθωδεστέραν) account of his departure. But let the case stand according as each man can convince himself (ἐχέτω δ' ὅπῃ τις αὐτὸν πείθει).

In the following chapter 49, Dionysius enumerates different versions of the events after the departure, which, he points out, 'creat[e] still greater difficulty for most historians' (ἔτι πλείω παρέχει τοῖς πολλοῖς τὴν ἀπορίαν, 1.49.1). Dionysius mentions Cephalon of Gergis and Hegesippus of Mecyberna (Chalcidice), Ariaethus (or Ariathus, possibly from Tegea), as well as the Arcadian poet Agathyllus (1.49.1–2).

Aeneas' arrival in Italy, by contrast, is beyond doubt, as it is attested by Roman sources and customs to which Dionysius can refer for confirmation (1.49.3). The subsequent chapters therefore inform the reader about monuments testifying to Aeneas' route and conclude with Dionysius' affirmation, at 53.4 to 53.5, that these monuments prove – against 'some historians' (1.53.4) – that: (1) Aeneas did, in fact, come to Italy; (2) this Aeneas was the son of Aphrodite and Anchises, and not some other Aeneas, for example, the son of Aeneas' son Ascanius, as held by several historians; and (3) Aeneas remained in Italy and did not return to Troy, as was claimed by some on the basis of what according to Dionysius (ὡς μὲν ἐγὼ εἰκάζω) is a misinterpretation of the famous passage *Iliad* 20.307–8, which predicts the long-lasting reign of Aeneas and his descendants (1.49.5).[23]

Dionysius' treatment of the journey of Aeneas demonstrates the extent to which his authority is based on an interplay between (demonstration of his) expertise and dialogic interaction with the reader. Dionysius twice overtly

---

[23] Cf. 7.72.3, where Homer is introduced as 'worthiest of credence' as the source for 'the purest, most typically Greek practice' (Schultze 2000: 36). Here and elsewhere Dionysius bolsters his claim to authority by aligning himself, his methods and views, with 'the poet'. These passages cannot be discussed here. It is worth noting, however, that they point to another difficulty with Marincola's conception of 'literary authority' which explicitly excludes 'authority in the sense of an established political, religious, or social power, something external to the history itself' (Marincola 1997: 1, n. 1). By associating his position with Homer as well as famous predecessors, Dionysius is, in fact, creating authority *within* the text through association with the repute of Homer and other authors *outside* the text, in the culture and society within which his text is written and read. Factors 'external to the history itself' thus turn out to be of crucial importance to an author's authority within the narrative.

leaves the decision between the different versions up to the reader, and at first glance, the different alternative versions to choose from confirm the impression of the reader's freedom of choice. At a closer look, however, the situation is much more complicated. To begin with, Dionysius prejudices his readers from the start by contrasting the straightforward and unproblematic narrative from Hellanicus with the confusing chaos of the alternative versions accumulated in the following chapters. Notice in particular the wealth of foreign place names, only a few of which would presumably have meant much to his readers, which are crammed into Dionysius' summary of divergent accounts of Aeneas' journey along with gratuitous further information on the places themselves, as well as a battery of names of authors that already in Dionysius' time must have been known only to readers with a particular antiquarian interest (1.49.1–2):

> For some, after they have brought him as far as Thrace, say he died there; of this number are Cephalon of Gergis and Hegesippus, who wrote concerning Pallenê, both of them ancient and reputable men. Others make him leave Thrace and take him to Arcadia, and say that he lived in the Arcadian Orchomenus, in a place which, though situated inland, yet by reason of marshes and a river, is called Nesos or 'Island'; and they add that the town called Capyae was built by Aeneas and the Trojans and took its name from Capys the Troan. This is the account given by various other writers and by Ariaethus, the author of *Arcadica*. And there are some who have the story that he came, indeed, to Arcadia and yet that his death did not occur there, but in Italy; this is stated by many others and especially by Agathyllus of Arcadia, the poet.

Dionysius does nothing to help the reader assess the reliability of any of these authors; quite to the contrary, by calling Cephalon of Gergis and Hegesippus 'ancient and reputable men', just as he had characterised Hellanicus, whose version he accepts as 'most creditable', as an 'ancient' historian, Dionysius insinuates that their accounts cannot be so easily dismissed.[24] Moreover, in order to assess the value of these alternative routes, readers would not only have to have an extensive and detailed geographical knowledge of the Mediterranean, they would also have to figure out how these individual bits of information fit in with and compare to the version of Hellanicus, which is adopted by Dionysius. We cannot know whether many of his readers would have been familiar with the latter before reading it a few chapters earlier; but the fact that Dionysius deems it necessary to identify

---

[24] Cf. Schultze 2000: 33 on the connection felt between the antiquity of a source and its importance and reliability.

Hellanicus as the author of his preferred version suggests that he does not presuppose that most of his readers would be able to recognise it on their own. How much easier, then, simply to trust Dionysius' judgement and subscribe to the version he has identified as the 'most credible' one.

The comparative μυθωδεστέραν (1.48.4, cited above) further fosters such an attitude by making readers unsure about their ability to 'judge for themselves', because it implies a decrease in credibility from Sophocles' version to that of the anonymous authors. But it is not immediately patent, nor does Dionysius explain, why some of the less trustworthy (ἧττον ... πιθανούς, 1.48.2) versions should be more trustworthy than others, and we should not simply assume that this would have been any more obvious to Dionysius' readers than it is to us.[25] The purpose of this statement is not to help the reader make a decision; if anything it complicates matters because it implies that it is easy to misjudge how unreliable a particular version really is. A reader accepting any of these unreliable versions could, at best, 'convince himself' (αὐτὸν πείθει), but who would like to risk exposing his own ignorance by falling for one such μυθώδης account? Clearly, one cannot simply 'let the case stand' according to 'what each man convinces himself of' (1.48.4).

The *tour de force* of alternative versions of the events confronts the reader with a veritable jungle of different opinions which even the historians find difficult to sort out: reading through this confusing multitude of divergent versions readers will find themselves subject to the same kind of ἀπορία which Dionysius attests to the historical authors (1.49.1). I would suggest that this effect is deliberate. The enumeration of the different versions thus has two, closely interrelated purposes: on the one hand, it demonstrates Dionysius' expertise by proving his initial statement that he has 'compared the histories of those writers, both Greek and Roman, who are the best accredited' (1.45.4); on the other, it is designed to diminish the reader's confidence in his own ability to extract a meaningful interpretation from the confusing, multifarious and arcane material.

The confrontation with the chaotic state of the tradition demonstrates effectively what kind of Herculean labour lies behind Dionysius' choice of Hellanicus' version as the 'most credible' one and that Dionysius is far better equipped to make this choice than the reader or, indeed, most historians.

---

[25] Cf. Schultze 2000: 24, who points out that many of the authors cited by Dionysius 'although they must have been better known to educated readers of Dionysius' own day... will still have been relatively obscure'. She considers the citation of 'such recherché material' as part of Dionysius' attempt to 'bolster the impression of [his] erudition'; this view is compatible with my interpretation.

The paradoxical effect of the 'choice' offered to the reader is that the very multitude of options, the *embarras de choix*, leads him to realise that he lacks the competence to make a decision and happily acknowledges Dionysius' authority to make the choice for him. Dionysius' authority is therefore as much an effect of the demonstration of his expertise as of the effect of the reading process on the recipient.

It must have (and for the modern reader still does) come as a relief when the account reaches more stable ground at 1.49.3. And Dionysius makes sure that the reader knows that he owes this feeling of secure orientation in the past to him and his expertise. Several times he stresses his personal knowledge of the topography and customs of different parts of the Mediterranean in order to confirm the information provided by the sources.[26] He informs his reader, for example, of the promontory Cinaethion, named after Cinaethus, one of Aeneas' companions (1.50.2), and of the temple of Aphrodite in Leucas 'which stands to-day on the little island between Dioryctus and the city' and is 'called the temple of Aphroditê Aeneias' (1.50.4). He also mentions bronze mixing bowls, offerings of the Trojans at Dodona, 'some of which are still in existence and by their inscriptions, which are very ancient, show by whom they were given' – as elsewhere in the *Antiquities*, the phrasing here clearly implies autopsy and is designed to demonstrate the secure foundations of Dionysius' account.[27]

Whether or not Dionysius really visited these places or had personal knowledge of the customs he mentions is difficult to know and is perhaps more likely for Roman buildings, customs and institutions than for the sacrificial rites on Samothrace. The question is of secondary importance for understanding how authority works in the *Antiquities*. The decisive point is that Dionysius presents himself as having personal knowledge of the places and customs that he mentions – who would be able to refute him? More importantly, Dionysius' repeated affirmations of the secure foundations of his knowledge form an effective contrast with the loss of the reader's sense of orientation in the jungle of different versions in the previous chapters: as soon as Dionysius takes over the lead from the reader again, the narrative becomes stable and the confusion disappears.

---

[26] 1.49.3: 'evidences of it are to be seen in the ceremonies observed by them both in their sacrifices and festivals, as well as in the Sibyl's utterances, in the Pythian oracles, and in many other things, which none ought to disdain as invented for the sake of embellishment. Among the Greeks, also, many distinct monuments remain to this day on the coasts where they landed and among the people with whom they tarried when detained by unfavourable weather. In mentioning these, though they are numerous, I shall be as brief as possible.'

[27] Cf. Fromentin 1988: 320–1; on autopsy in the *Antiquities*, see Andrén 1960; Wiater 2011b: 199–201.

Dionysius has asserted his claim to authority-through-expertise by demonstrating to his readers what the narrative would be like if they were in charge of it. By concluding this section with his criticism of other historians' incompetent interpretation of *Iliad* 20.307–8, which led to their erroneous narratives of the past (ἀπάτη, 1.53.5), Dionysius further bolsters this 'authority effect' by demonstrating his competence as an interpreter of Homer.[28] The message to the reader is clear: when dealing with the difficult early Roman history (χαλεπούς, 1.8.1), there are pitfalls everywhere that even many specialists have not managed to avoid, and it takes a whole set of different skills, ranging from knowledge of the Latin language and Roman customs via geographical knowledge and skills in interpretation of Homeric poetry, to handle the material properly.

Dionysius employs a similar narrative strategy to create the 'authority effect' when discussing the exact date of the foundation of Rome.[29] Here, too, Dionysius leads the reader into a thicket of conflicting interpretations – summarising first Greek (1.72), then Roman authors (1.73). The most interesting passage for the present enquiry is 1.73.4. Having explained that some authors relate not just one, but two foundations of Rome (διττὰς εἶναι τῆς Ῥώμης κτίσεις, 1.73.3 – the reader was confronted with a similar problem regarding different Aeneases at 1.53.4), Dionysius remarks:

> if anyone desires (εἰ δέ τις... βουλήσεται) to look into the remoter past, even a third Rome will be found, more ancient than these, one that was founded before Aeneas and the Trojans came into Italy. This is related by no ordinary or modern historian (οὐ τῶν ἐπιτυχόντων τις οὐδὲ νέων συγγραφεύς), but by Antiochus of Syracuse, whom I have mentioned before.[30]

The issue of potentially different foundations of Rome is never discussed, let alone resolved, by Dionysius. The purpose of this remark is less to help establish the date of the foundation of the city than to convey to the readers the chaotic state of information on the subject. As at 1.49.5 (above), the difficulty of making an informed decision in this matter is underlined by

---

[28] Cf. n. 23 above.
[29] For a good discussion of this passage, see Schultze 2012: 120–5, rightly criticising P. M. Martin's reading of the passage. My own interpretation is compatible with Schultze's view that 'the recurrent foundations of Rome have an importance for Dionysius... not in that he buys into the notion of successive Romes receding back into the mists of time [as Martin claimed, N.W.], but in that it enables him to concretize the nature of disputes that lie within the plupast even – or, perhaps, *especially* – of a city so great as Rome' (125, emphasis in the original).
[30] Cf. 1.12.3, 22.5, 35.1. Schultze 2012: 123 convincingly reads the direct quotation from Antiochus' work at 73.4 'as a move to assert the historian's command over a contested plupast by a decision to impose closure'.

the emphasis on the widely accepted reliability of Antiochus which would have misled most readers to accept his version. Even more significant is the phrasing 'if anyone desires'. Βουλήσεται ostensibly invites the reader to pursue yet another, even more complicated line of enquiry. But as in Dionysius' discussion of the journeys of Aeneas, the reader is meant to be discouraged by the very invitation: βουλήσεται makes it perfectly clear that any further enquiry, and the complications arising from it, would be entirely at the reader's discretion and, hence, the reader's own responsibility – much better to leave this to an expert like Dionysius who can sift through all the material and choose the most reliable one on the reader's behalf.

The purpose of Dionysius' long enumeration of alternative versions is not to enable readers to form their own judgement, nor simply to disclose the material Dionysius has been working with in the interest of transparency. These enumerations have a calculated narrative effect: they confront the reader with his helplessness and lack of competence vis-à-vis the abundance of conflicting traditions and prepare Dionysius' intervention which will create order out of chaos and provide the guidance the reader (now knows he) needs. The way in which Dionysius provides this guidance further enlarges the distance between himself and his readers: the confusion about the precise date of the foundation of Rome, it turns out, can be resolved only if the historian is not only in command of the relevant Greek and Roman traditions, but also a specialist in chronology.[31] Drawing on his expertise with the chronology of Eratosthenes as testified to by another work of his, the *Chronoi*, in which he has scrutinised Eratosthenes' standards and confirmed their validity (1.74.2–4),[32] Dionysius will settle, under his readers' eyes, once and for all a centuries-old controversy about the time of Rome's foundation.

Again, Dionysius' authority is clearly not based exclusively on this demonstration of his 'competence' and expertise. Dionysius stresses that he, unlike Polybius, who based his dating 'without further examination upon the single tablet preserved by the high priests, the only one of its kind', was 'determined to set forth the reasons that had appealed to me, so that all might examine them who so desired' (1.74.3). Dionysius is here employing the familiar method of achieving authority and credibility through discrediting a famous predecessor. But he is doing much more than that. He bases his claim to greater expertise, and hence superiority over Polybius,

---

[31] Cf. Schultze 2000: 38; ibid. 39 on 'chronological exactitude' as a 'prime requisite for a historian'. See the detailed discussion of Dionysius and Roman chronology in Schultze 1995.
[32] On this work, very little of which survives, see Schultze 1995: 193.

not simply on one piece of arcane evidence accessible only to himself, but specifically on the fact that he reached his conclusions in an 'open' process witnessed by his readers. That shows, first, that the strong interaction with the reader in the passages discussed above is a deliberate narrative strategy, and, second, that Dionysius employs this narrative strategy in order to create what I have called the 'authority effect': it is because his readers have witnessed how Dionysius manages and makes sense of the complicated material with skills far beyond their own, that they are prepared to accept his word over that of his famous predecessor.

This passage is so important for understanding Dionysius' 'authority' because the strategy employed by Polybius, the claim that he had access to special or even secret sources, is precisely one of the 'rhetorical strategies' identified by Marincola as creating 'literary authority'.[33] The fact that Dionysius explicitly rejects using this effective rhetorical strategy further supports the observations made in the first part of the chapter, that 'authority', at least in the *Antiquities*, is not identical with a simple demonstration of the author's exclusive knowledge. On the contrary, Dionysius creates an 'authority effect' by stressing that his material is accessible to everybody and presenting his enquiry as a process which appears to be open to everybody, but through this very openness makes (most) readers aware of their lack of expertise while demonstrating Dionysius' own.

This might be an appropriate point to cast an, albeit quick and superficial, glance at Herodotus and Thucydides which, I hope, will provide some interesting differences and similarities. I should reiterate that the purpose of the following paragraphs is merely to offer some ideas on how the 'authority effect' in Dionysius can be situated within the historiographical tradition, not to provide an in-depth discussion of authority in Herodotus and Thucydides.

As is well known, Herodotus frequently reports alternative versions of the same events.[34] That this is part of his historical method is clear from the well-known passage at 7.152.3, 'I must tell what is said, but I am not at all bound to believe it, and this comment of mine holds about my whole *History*',[35] which follows the report of three different versions of the Argives' attitude during the war. This passage is programmatic not only for Herodotus' habit of listing alternative versions. It also shows that he does not use his expertise to demonstrate his superiority to the reader in order

---

[33] Marincola 1997: 107–9.
[34] See Lateiner 1989: 76–90, with a full list of such instances; cf. Asheri *et al.* 2007: 20–1.
[35] Translations from Herodotus are taken from Herodotus, *The History*, tr. David Grene, Chicago 1987, adapted, when necessary.

to influence his choice of one of them. On the contrary, Herodotus and his reader seem to meet eye-to-eye in their consideration of the different versions, both equally entitled to their respective opinion.

Rather than trying to impose his view on the reader, Herodotus seems to invite his reader to follow his example and pick the version he finds most plausible, or, as Emily Baragwanath and Mathieu de Bakker attractively suggest, 'to consider a wider sweep of history and different perspectives'.[36] And even when Herodotus does express an overt judgement about an alternative version, as, for example, at 8.119.1 and 120.1,[37] the careful phrasing (οὐδαμῶς ἔμοιγε πιστός; λέγοντες ἔμοιγε οὐδαμῶς πιστά – emphasis added) seems to be designed to leave room for the opposite view so as to neither influence the reader's choice nor to offend the beliefs of his informants. As Lateiner put it, Herodotus 'is conserving accounts, not imposing interpretations of even a rationalized, tested "best version"'.[38] Leaving the choice to the reader thus seems to be as integral to Herodotus' historical method as the attempt to limit and control this choice is to Dionysius'.

This view seems to receive further support from those passages in which Herodotus gives no indication of his own preference at all, as at 5.44 to 5.45, which report the respective versions of the people of Sybaris and Croton of the same event, including the 'evidence' (μαρτύρια) produced by each side. Herodotus concludes with a phrase that is familiar from Dionysius, but has diametrically opposite implications: 'That is the evidence that each side musters. Whichever one finds convincing should be given preference' (καὶ πάρεστι, ὁκοτέροισί τις πείθεται αὐτῶν, τούτοισι προσχωρέειν, 5.45.2). Herodotus is clearly demonstrating his 'epistemic authority' by giving the detailed summary of the story and evidence presented by either side.[39] But unlike Dionysius, he does not use this situation to make the reader feel his own inferiority and inadequacy in dealing with the material: Herodotus rather brings the reader up to speed so he can make his choice based on the same knowledge and material of which Herodotus disposes. He is

---

[36] Baragwanath and de Bakker 2012a: 27. Baragwanath 2008 argues that the presentation of alternative motivations is designed to draw the reader into 'beginning a process of weighing up alternatives' and thus actively to involve them in making sense of the variegated and often contradictory historical material, the 'process of evaluation' (241).

[37] Cf. 4.11.1: ἔστι δὲ καὶ ἄλλος λόγος ἔχων ὧδε, τῷ μάλιστα λεγομένῳ αὐτός πρόσκειμαι.

[38] Lateiner 1989: 79; cf. the qualifications in Baragwanath and de Bakker 2012a: 42–3.

[39] Cf. Vivienne Gray's (2012: 189) observation that these 'formulae renounce the omniscience of the poets to construct a more persuasive truth that admits to fractured and incomplete knowledge... only to assert what he does know against what he does not'. I would only take issue with the implication that this is the exclusive ('only') function of such statements.

diminishing the gap between himself and the reader where Dionysius seeks to widen it; the choice he offers to his audience is a real one.[40]

Why an apparently similar procedure in Herodotus and Dionysius has such diametrically opposite implications is difficult to determine. I would suggest that the categorically different relationship between Herodotus and Dionysius and their respective sources plays an important role: Herodotus' historical project, his ἱστορίη, is defined by the collection and (critical) redaction of information from predominantly oral sources;[41] this information constitutes the raw material which Herodotus transforms into his narrative, and much of it is collected and made accessible in this form for the first time in his work. When Dionysius was writing his *Early Roman History*, in stark contrast, his main concern was not the collection of new material: most, if not all, of the relevant information was already accessible in a wide range of Greek and Roman historical and antiquarian works. The main challenge faced by Dionysius was to master the sheer volume of pre-existing information and create a new interpretation out of already known material that had a chance to distinguish itself within a fiercely contested market of 'true' narratives about the Romans.[42]

The authors on which Dionysius drew for his own account are not simply his sources in the same sense in which his informants were for Herodotus; they are also his competitors. While for Herodotus, there was usually a difference between his sources and his historiographical predecessors from which he needed to distinguish himself, these were identical for Dionysius. Dionysius therefore had a much greater need to establish (and, that is, convince his readers of) his superiority and greater authority vis-à-vis

---

[40] Cf. 6.14.1, 137, 8.84.2.

[41] On Herodotus' sources, see Rengakos 2011a: 365–6; on ἱστορίη ('enquiry'), see Asheri *et al.* 2007: 7–8, 15–23.

[42] See the long list of predecessors at 1.6. The cornerstones of Dionysius' claim to originality are (1) his thesis, fundamental to his interpretation of early Rome, of the Greekness of the Romans (1.5.2) (cf. Fromentin 1998: xxxi–xxxiv; Delcourt 2005: 81–115 traces the intellectual tradition behind this idea. Dionysius was apparently anticipated by Heracleides Ponticus (*FGrH* 840 F 23) and, possibly, Acilius (Cornell 2013: 7 F7, comm.), but he is unique in the comprehensiveness and systematic implementation of this theory (113–15)); (2) his emphasis on the origins and beginnings of Rome as key to the understanding of Roman power in the present (1.4; Wiater 2011b: 171–98); (3) the claim that his work provides the first detailed narrative of this period and, therefore, the first adequate representation of its importance in (Greek) literature (1.5.4, 6.2; cf. Wiater 2011b: 192–3); and (4) the exhaustive wealth of different kinds of information (external wars and internal strife, constitutions, and customs and way of life of the early Romans) united in his work, combined with its innovative 'design' (σχῆμα) (1.8.1–3, note ἅπαντας, ὁπόσας, πάσας; cf. n. 13 above). On different 'true' narratives about the Romans, see 1.4–5 with Wiater 2011b: 100–6, 185–91.

a well-established historical tradition than Herodotus. This pressure of the competition, along with the prospect of the prestige of having produced *the* definitive 'true' account about the Romans, might explain why Dionysius was so keen on controlling his readers' choices, while Herodotus was prepared to enable them to make their own.

On the other hand, it is perhaps not surprising to find that there seem to be some interesting similarities between Dionysius' practice and the relationship established by Thucydides between himself and his recipients; after all, Thucydides, too, was acutely aware of, and strongly positioned himself with regard to, his predecessors, including poets and the (in)famous λογογράφοι (1.21.1).[43] In an important article, Nicole Loraux has shown that Thucydides all but identifies his interpretation of the Peloponnesian War with the war itself, constantly inducing the reader to accept his narrative as the final and definitive version of events and seeking to discourage any further questions and investigation.[44] Thucydides achieves this effect, according to Loraux, by establishing himself, mainly through his chapters on historical method, as the only competent narrator of the war and, moreover, the only one whose intellectual capacities can manage the complexities of the historical material. Loraux calls this Thucydides' demonstration of his 'héroïsme de l'intellect', and time and again he reminds his readers of the 'Herculean labour', the πόνος, that he has undertaken in collecting, assessing and ordering the historical material on their behalf.[45] The reader, on the other hand, is not supposed to engage independently with the raw material; on the contrary, Thucydides, as is well known, all but denies access to his sources and allows us to interact only with the narrative he has constructed on their basis. Instead, Thucydides makes the reader re-trace his own thoughts about the material (rather than encouraging him to form an alternative opinion) and re-enact his own choices: 'il suffit pour chacun de prendre à son propre compte le discours de l'historien, ce qui revient à en occuper répétitivement toutes les positions de pensée'.[46] The reader thus finds himself in a position of 'complete submission to the historian' ('une entière soumission à l'historien').[47]

There are some interesting parallels between my view of the 'authority effect' in Dionysius' *Early Roman History* and Nicole Loraux's reading of Thucydides' relationship with his reader. Like Thucydides, Dionysius leaves his readers in no doubt about the extraordinary time, effort and competence

---

[43] Cf. Rengakos 2011b: 404.   [44] Loraux 1986: 142, 153, 159.
[45] Loraux 1986: 154 ('héroïsme de l'intellect'), 155 (πόνος, citing 1.20.3; 22.3).
[46] Loraux 1986: 157; cf. 158: 'mimétiquement, le lecteur refera tout le parcours de Thucydides'.
[47] Loraux 1986: 157.

necessary to master the difficult, enormous and variegated material. As I have argued above, Dionysius, too, makes the reader re-enact his, Dionysius', intellectual processes with the purpose of creating a feeling of inadequacy and inferiority that stops the reader from attempting his own enquiries and producing an alternative narrative on his own. Loraux's concepts of the demonstration of the 'héroïsme intellectuel' and Thucydides demanding unconditional acceptance of his own account are equally useful for describing the 'authority effect' in Dionysius; in analogy to her statement that 'le présent appartient à l'historien', one could say that Dionysius makes the same claim for the Roman past.

The decisive difference between Thucydides and Dionysius is their authorial presence in the text. Loraux rightly emphasises that Thucydides withdraws from the narrative to create a reading situation in which the facts seem to 'speak for themselves': it is only through subtle reminders throughout the narrative that he creates and maintains the reader's recognition of the extraordinary intellectual achievement that made such a supposedly unfiltered and un-mediated account possible.[48] Dionysius, in stark contrast, creates a strong authorial presence in his text.[49] Where Thucydides hides his 'intellectual heroism' behind the text and allows it to surface only at strategically important moments, Dionysius almost indulges in a display of the 'research protocol' and the process of which his *Antiquities* is the result.[50] As the passages discussed in this section have shown, Dionysius makes sure that his reader never forgets about him and his achievements; on the contrary, Dionysius encourages the direct comparison between himself and his recipients in order for the latter to realise that they can never be equals.[51]

---

[48] Loraux 1986: 142 ('l'effacement du discours'), 149–52, esp. 150: 'Thucydide attend de son lecteur... qu'il admire l'écriture en acte et qu'il oublie qu'elle est un acte. Qu'il sache que l'œuvre est un résultat, mais qu'il ne demande pas à en savoir plus sur la recherche qui l'a produite'; 152: 'Thucydide donne juste assez d'indications pour que le lecteur zélé puisse reconstruire ce que fut le temps de l'investigation'.

[49] In that respect Dionysius very much stands in the tradition of Hellenistic historiography, especially Polybius (cf. n. 7 above), despite his well-known negative judgement on the *Histories*. But in contrast to Dionysius, Polybius generally combines his criticism of predecessors with a clear and unambiguous statement of what he regards as the one and only true version, as, e.g., at 3.6–10.6 or 20.1–5. While Herodotus enables the reader's choice and Dionysius and Thucydides seek to minimise and control it, Polybius attempts clearly to eliminate alternatives from the start. Note, however, that the polemic which is characteristic of Polybius (Walbank 1962), Timaeus and other Hellenistic historians (Baron 2013:113–37) is almost entirely absent in Dionysius (Schultze 2012: 124).

[50] I borrow the term 'research protocol' from Loraux 1986: 151.

[51] Note that Loraux's reading of Thucydides also provides support to the general premise informing the present discussion, that the overt demonstration of knowledge and expertise is

For Truth and Justice: Authority, Morals and Belief

As suggested in the first section of this chapter, expertise and competence are not the only factors of importance for creating the 'authority effect' in the *Antiquities*. From the very beginning of his work, Dionysius combines the claim to special knowledge with a claim to a particular 'moral expertise' which informs his work and which is an essential aspect of his relationship with his reader.[52] In this section, I will first explain the importance of the moral values of author and reader alike for Dionysius' historical project and then explore how Dionysius draws on this moral component of his work for the creation of the 'authority effect'.

As a historian, Dionysius says at 1.6.5, he is not only devoted to truth (τῆς ἀληθείας), but also to justice (τοῦ δικαίου). The latter term not only refers to the moral contents of his work, in which the assessment of the character and actions of the historical actors plays a prominent role.[53] Dionysius also regards his work, both its presentation and its contents, as a direct manifestation of his own character (1.1.3):

> those who base historical works upon deeds inglorious or evil or unworthy of serious study, either because they crave to come to the knowledge of men and to get a name of some sort or other, or because they desire to display the wealth of their rhetoric, are neither admired by posterity for their fame nor praised for their eloquence; rather, they leave this opinion in the minds of all who take up their histories, that they themselves admired lives which were of a piece with the writings they published, since it is a just and a general opinion that a man's words are the images of his mind.[54]

---

not sufficient for, let alone identical with, the creation of 'literary authority': in fact, Thucydides' authority, his towering presence as the only representative of the true account of the Peloponnesian War (Loraux 1986: 153), is due almost exclusively to the particular reading experience created by his account and the way it positions the reader towards the narrative, while the overt display of knowledge, skills and expertise, Thucydides' 'epistemic authority', is reduced to a minimum.

[52] This aspect of the self-presentation of historical writers is still underexplored. For an interesting discussion of the role of morals and character in Galen's self-presentation and construction of authority, see Barton 1994b: 139, 143–7. I am grateful to Jason König for pointing me to Barton's study. See also Trapp in this volume.

[53] Dionysius shares his interest in morals and character with Hellenistic historiography; see, e.g., Pownall 1998 and 2004; Hobden and Tuplin 2012. Dionysius' *Antiquities* attempt to create a historical work whose moral impact is on a par with that of Isocrates' speeches, an 'Isocratean historiography', as it were; see Wiater 2011b: 149–54. Cf. Wiater 2011b: 65–77 on the role of Isocrates and classical Greek values in Dionysian classicism; Dionysius' entire concept of historiography is strongly informed by his notion of classical Greek moral-cum-political values, see Wiater 2011b: 130–65.

[54] Not simply 'the style', but 'the work is the man'. On this passage, see Wiater 2011b: 75–6 with n. 223 (with further literature).

Dionysius endows his historical project, to give the early Roman past the representation it deserves and thus enable a fairer judgement of the Romans, their character and the legitimacy of their rule, with a moral component: by telling the truth about the Romans Dionysius is, quite literally, doing them justice and demonstrating his own commitment to truth and justice.[55] At the same time, Dionysius is setting himself apart from historians who (he claims) presented a deliberately distorted image of the Romans in order to 'humour barbarian kings (βασιλεῦσι βαρβάροις) who detested Rome's supremacy, – princes to whom they were ever servilely devoted (δουλεύοντες) and with whom they associated as flatterers (τὰ καθ' ἡδονὰς ὁμιλοῦντες), – by presenting them with "histories" which were neither just nor true (οὔτε δικαίας οὔτε ἀληθεῖς)' (1.4.3).[56] Dionysius is thus connecting the accuracy and credibility of his work and, by implication, his authority as a historical narrator, with core elements of Greek identity: these Greek authors have had to give up two key values traditionally claimed by the Greeks, freedom and justice: in order to slander the Romans, one has to become a barbarian. By writing the first (or so he claims) true history of *Roman* culture, politics, customs and achievements, Dionysius, in stark contrast, presents himself as defending precisely those *Greek* values.[57]

Dionysius' emphasis on the importance of the moral implications of his work for the reader's acceptance of his authority as a historical writer is not confined to his, the author's, qualities of character. His moral-cum-true narrative can be appreciated only by readers whose character is as morally sound, and whose interest in the truth as genuine and unbiased, as that of Dionysius. The *Antiquities*, Dionysius says at 1.6.5, is for 'all good men and toward all who take pleasure in the contemplation of great and noble deeds' (ἅπαντας ... τοὺς ἀγαθοὺς καὶ φιλοθεώρους τῶν καλῶν ἔργων καὶ μεγάλων); only an 'uncivilised' (ἀγρίως) and intrinsically 'hostile' (δυσμενῶς) reader would reject Dionysius' conclusions after reading his work (1.5.2). Just as Dionysius links the readers' recognition of his authority in factual matters

---

[55] Esp. 1.6.3: 'I have determined not to pass over a noble period of history which the older writers left untouched, a period, moreover, the accurate [ἀκριβῶς] portrayal of which will lead to the following most excellent and just [δικαιότατα] results'; cf. 1.5.2: 'I shall omit nothing worthy of being recorded in history, to the end that I may instil in the minds of those who shall then be informed of the truth [τὴν ἀλήθειαν] the proper [ἃ προσήκει] conception of this city'; 1.5.4.

[56] The association of Greek historians enslaved to the pleasures of barbarian kings and the lack of 'justice' and 'truth' in their narratives evokes the classical Hellene-Barbarian antithesis which is fundamental to Dionysius' classicist ideology as well as his interpretation of the Romans as Greeks; see Hidber 1996: 25–30; Wiater 2011b: 66–7, 93–100, 155–8, 187–8, 2011a.

[57] Note also the contrast between the 'flattery' of the barbarised Greeks (χαριζόμενοι, 1.4.3) and Dionysius' rejection of flattery as a possible motive for embarking on his historical project (οὐχὶ κολακείας χάριν, 1.6.5).

with the recognition of their own lack of competence in these questions, he links the acceptance of his interpretation of the Romans with the question of his readers' moral integrity: readers are given the choice to accept or reject his conclusions, but those opting for the latter reveal not only their flawed characters, but also their lack of education, their 'uncivilised' nature that is impenetrable even to the scholarly and morally soundest explanations. The historians criticised by Dionysius for giving up crucial aspects of their Greekness to please barbarian kings (above) thus become a serious warning to readers who do not subscribe to Dionysius' view: they, too, run the risk of disavowing those moral-cum-political standards that define them as Greeks in the first place.

This moral aspect of the 'authority effect' remains important throughout Dionysius' narrative. Repeatedly, Dionysius presents himself as an expert in Greek and Roman morals, both past and present. At 2.27.1, for example, he praises the Roman father's absolute control over his son, including his right to sell his son up to three times for profit – 'a thing', he adds in an aside, 'which anyone who has been educated (τραφείς) in the lax manners of the Greeks (ὑπὸ τοῖς Ἑλληνικοῖς ἤθεσι τοῖς ἐκλελυμένοις) may wonder at above all things and look upon as harsh and tyrannical'. The reader's acceptance of the greatness of early Roman austerity here goes hand-in-hand with the recognition of the moral inferiority of their own tradition and upbringing. Passages such as this one must have presented a particular challenge for Dionysius' Greek readers, especially because he does not attempt to convince his readers by way of rational argument or a rhetorical strategy of persuasion. Dionysius expects them to accept his judgement merely because as an authority in Greek and Roman morals, he simply knows better.

Dionysius' claim to have exclusive knowledge of the Romans is an important part of his claim to authority.[58] At 1.6.5, Dionysius expresses his gratitude to Rome for the 'education' (παιδεία) he received there during his stay in the city (διατρίψας ἐν αὐτῇ) since his arrival in Italy 'at the very time that Augustus Caesar put an end to the civil war [c. 30 BCE]' (1.7.2).[59] The latter statement in particular is more than merely a piece of biographical information. Dionysius is programmatically associating the beginning of his career as a historical writer with the beginning of the 'new era' of Augustus' reign: not only has Dionysius been there from the start; as the reader knows from 1.3.4, where Dionysius names the consuls of 7 BCE, he has been with the Romans for over twenty years, while his gratitude for the Roman παιδεία makes it clear that he was more than a passive observer; he was deeply

---

[58] Cf. n. 6 above.   [59] Cf. Wiater 2011a: 85–7.

10 Expertise, 'Character' and the 'Authority Effect' in Dionysius

shaped himself by those two decades of intensive and first-hand contact with the Romans.

We need to read passages like 2.27.1 (cited above) in conjunction with these statements in the proem in order to appreciate their significance. Dionysius' positive judgement of the early Romans' morals, and his concomitant criticism of the 'lax' morals of the Greeks, are designed to demonstrate the extent to which his knowledge of and contact with the Romans have influenced his own character; at the same time, they signal to his Greek readers that they are not even aware of the defects and shortcomings of their own education, because unlike Dionysius, they have not profited from the salutary acquaintance with the (early) Roman character. Dionysius is thus presenting himself as a Greek whose additional, Roman παιδεία enables him to assess the moral value of Greek customs and upbringing with an objectivity which many of his Greek readers are incapable of. The provocation implied in his harsh judgement of the 'lax manners' of his fellow Greeks is therefore an important part of creating the 'authority effect'. It demonstrates the depth of Dionysius' acquaintance with the Roman character as well as his independence as a researcher who has overcome the limited perspective of his original upbringing and achieved an objectivity inaccessible to his readers and other Greek historians.

At the same time, however, Dionysius' expertise alone is not sufficient to create the 'authority effect'. Dionysius also puts his readers and their own moral integrity to the test: if they, too, regarded the father's rights over his son as 'harsh and tyrannical', they will have to realise that they, too, lack a proper moral perspective because of their 'lax' upbringing. Dionysius' text thus offers his readers a choice: will they persist in their distorted point of view or attempt to liberate themselves from it, just as Dionysius achieved moral integrity through his interaction with the Romans? The intensive and detailed contact with Dionysius' true and morally uncompromised narrative of early Rome provides his Greek readers with the same opportunity for correcting their moral standards as the long-term stay in Rome provided to Dionysius himself. It is in passages such as this one that the purpose of the *Antiquities* as a process of re-education, as Dionysius defines it at 1.5.1,[60] is most apparent: readers can either accept Dionysius' authority in matters of morality and customs, subscribe to his point of view and correct the flaws of their own upbringing, or they can (stubbornly) reject Dionysius' interpretation and thus reveal their moral shortcomings, as well as lack of education.

---

[60] See p. 233 above.

The complex of morals, reader-response and authorial self-image produce an effect of narratorial authority also in those passages in which Dionysius discusses the role of the gods in history.[61] These are particularly important for the present enquiry because by definition, questions of belief cannot be settled through rational argument. Already in antiquity, the credibility of divine intervention in history was a matter of intense debate, and opinions ranged from utter rejection (e.g. by the Academics) to approval (Dionysius), along with several attempts to compromise, for example, by relegating divine intervention to the pre-historical, 'mythical' period only (Varro), rationalising the 'myths' (Diodorus), including myths for educational purposes despite doubts about their authenticity (Strabo, Diodorus), or distinguishing different degrees of truth (Strabo).[62] The narrator's authority therefore played a crucial role in the reader's acceptance of a version of the past that endorses the notion of divine influence on the course of history.

At 2.68.1, Dionysius addresses the question of the credibility of reports of the direct intervention (ἐπιφάνειαν) of the goddess Hestia in Roman life. Dionysius marks this question as controversial: 'these things, however incredible (παράδοξα) they may be, have been believed by the Romans and their historians have related much about them'. Dionysius then distinguishes between two groups of readers (2.68.2): 'those who practise atheistic philosophies (τὰς ἀθέους... φιλοσοφίας),[63] – if, indeed, their theories deserve the name of philosophy, –' and 'ridicule all the manifestations of the gods (ἁπάσας διασύροντες τὰς ἐπιφανείας τῶν θεῶν) which have taken place among either the Greeks or barbarians', will 'also laugh these reports to scorn (εἰς γέλωτα πολὺν ἄξουσι τὰς ἱστορίας) and attribute them to human imposture'. Those, by contrast, 'who do not absolve the gods from the care of human affairs, but, after looking deeply into history (διὰ πολλῆς ἐληλυθότες ἱστορίας), hold that they are favourable to the good and hostile to the wicked, will not regard even these manifestations as incredible (ἀπίστους)'.

The readers' attitude towards the concrete historical situation – why should they accept the 'incredible' alternative, that the early Romans were in direct contact with Hestia, rather than the rational one? – and the more fundamental question of whether or not the gods play a role in history, is contingent on their assessment of Dionysius' credibility as a historical narrator, his authority. Rather than Dionysius' professional competence, however,

---

[61] The most comprehensive (but often merely descriptive) study of religion and the gods in Dionysius is Mora 1995; Sautel 2010 discusses the epiphany of the Dioscuri at *Ant. Rom.* 6.13.
[62] See Gabba 1984: 860–3; Wiseman 2002.   [63] Presumably, the Epicureans (Gabba 1984: 859).

it is his moral and religious credibility that is the essential factor in this process.

Dionysius' reader is already aware of Dionysius' own reverence for the gods (his εὐσέβεια) from passages such as 1.67.4, where Dionysius practises self-censorship in order not to violate the sacredness of the *Penates*.[64] Dionysius thus presents his handling of the historical material as an expression of piety: the self-imposed limits of his narrative replicate the limited access to the actual sanctuary, and Dionysius strongly distinguishes himself from historians whose works are not informed by the same piety.[65]

In the same vein, Dionysius phrases the rational alternative in the above passage, that any alleged manifestation of Hestia is just 'human imposture', as a transgression committed by the adherents of 'atheistic philosophies'. Apart from questioning the legitimacy of philosophical belief systems that deny divine interest in human affairs, Dionysius' distinctly negative representation of their views (διασύροντες, εἰς γέλωτα πολὺν ἄξοντες) clearly marks them as acts of blasphemy bound to offend the (allegedly nonexistent) gods.

Dionysius counters these 'blasphemous' attitudes by emphasising historical evidence. The phrase 'all the manifestations of the gods which have taken place among either the Greeks or barbarians' implies a rebuttal of the 'philosophers'' arrogant confidence: does not the sheer number of these manifestations and the fact that they have been attested all over the world, by Greeks and non-Greeks alike, suggest that there is more to them than sheer 'human imposture'? The same point is taken up more overtly in διὰ πολλῆς ἐληλυθότες ἱστορίας ('those who have gone through a lot of history'): it is history itself that teaches us that divine interference is common everywhere and at all times.[66] The 'atheistic philosophers' thus seem not only blasphemous, but ignorant: their lack of competence in assessing the relevance of the historical evidence is paired with their lack of moral integrity and piety. It is difficult not to refer Dionysius' subsequent remark, that the gods are

---

[64] 'For my part, I believe that in the case of those things which it is not lawful for all to see I ought neither to hear about them from those who do see them nor to describe them; and I am indignant with every one else, too, who presumes to inquire into or to know more than what is permitted by law'.

[65] Sautel 2010: 385 convincingly reads Dionysius' positive account of the epiphany of the Dioscuri as a reply to a more sceptical view of the event, such as the one held by Cotta in Cicero, *De Natura Deorum* 3.11–13 (note the ironic, mocking tone of his statement).

[66] The same argument is used by Quintus at Cic. *Div.* 1.12; cf. Wiseman 2002: 340–1. Cf. Sautel 2010: 385 on 6.13.4: Dionysius sets out to confirm the annalistic record of the epiphany of the Dioscuri by collecting 'many proofs' (πολλὰ σημεῖα), namely monuments, rituals and customs, which he himself has witnessed (*autopsy*).

'favourable to the good and hostile to the wicked', to the 'atheistic' philosophers, who should be wary of divine vengeance for their attitude, and the good, who, like Dionysius, revere the gods and know their history and can be sure of divine favour, respectively.

It is, therefore, not because of Dionysius' professional competence (nor the quality of his arguments, for there are none) that readers are expected to subscribe to his position. Instead, Dionysius stresses the religious dimensions of his and his readers' involvement with the past: denying the authenticity of divine manifestations in early Rome is more than an intellectual exercise, a 'philosophy' (if, indeed, it deserves that name). It is about making a fundamental decision about your attitude towards the gods in general and accepting the potentially serious consequences of your decision: 'Those who argued that myths had no place in history were not just defining a genre; they were putting into practice a controversial belief about the nature of the gods.'[67] Dionysius has already demonstrated his piety, which informs the range of his enquiry, and accepted divine intervention as the better alternative to rational explanation. He now invites his readers to side with him or to count themselves among the blasphemous and wicked – the choice is theirs, or is it?[68]

At this point, it might be instructive to compare the way in which Dionysius handles the divine in his text, and how he uses it to influence his reader's attitude, with a striking passage from the *Bibliotheke* of his near-contemporary Diodorus. At 11.13.1, Diodorus reports that after the battle at Artemision, in which the Greeks were unable to defeat the large fleet of the Persians (12.6), a storm arose which was 'in the process of destroying many of the [Persian] ships anchored outside the harbour'.[69] The Greeks interpret

---

[67] Wiseman 2002: 348; cf. ibid. 353. Sautel 2010: 386 detects a moral message ('une leçon édifiante, celle de la piété envers les dieux') in Dionysius' account of the epiphany of the Dioscuri at Lake Regillus.

[68] This strategy is even more obvious at 8.56.1: 'It would be in harmony with a formal history and in the interest of correcting those who think that the gods are neither pleased with the honours they receive from men nor displeased with impious and unjust actions, to make known the epiphany of the goddess at that time, not once, but twice, as it is recorded in the books of the pontiffs, to the end that by those who are more scrupulous about preserving the opinions concerning the god which they have received from their ancestors such belief may be maintained firm and undisturbed by misgivings, and that those who, despising the customs of their forefathers, hold that the gods have no power over man's reason, may, preferably, retract their opinion, or, if they are incurable, that they may become still more odious to the gods and more wretched'; cf. Wiseman 2002: 346–7.

[69] Translations of Diodorus are from the Loeb translation of Oldfather *et al.* 1933–67, mostly adapted.

(δοκεῖν) this as divine intervention in their favour, 'so that the Greek force become a match for them and strong enough to offer battle', but both their assessment of the situation and their interpretation of the event as a sign of divine support are sorely mistaken (οὐ μὴν ἀλλά): the damage is much less severe than they had assumed and the Persians remain, in fact, in possession of their entire fleet (ἁπάσαις ταῖς ναυσίν).

It is remarkable how Diodorus initially leads the reader to share the Greeks' mistaken interpretation of the situation – a process initiated by the ambiguous διέφθειρεν[70] and (seemingly) borne out by the following statement that 'the Greeks grew ever more bold, whereas the barbarians became ever more timorous before the conflicts which faced them' – and then debunks our own error along with that of the Athenians: the Athenians' belief (shared, presumably, by the reader) that as at previous occasions, the gods were supporting them in battle, is so thoroughly disavowed as to seem silly in the first place.[71] Based on the preceding discussion, it is difficult to believe that Dionysius, had he described the event, would not have tried to keep the possibility of some sort of divine intervention open. He might even have thought that the narrative trick through which Diodorus so effectively exposes this belief as a total failure to grasp the real nature of the situation, might itself constitute a dangerous act of disrespect for the gods and their role in history.

A sustained comparison of Dionysius and Diodorus would be a worthwhile undertaking, but is beyond the scope and possibilities of this chapter. Returning to Dionysius, I will now conclude my discussion by considering yet another passage in which authority is bound up with religious belief as well as Dionysius' relationship with 'philosophers'. At 1.77, Dionysius addresses the question of whether Rhea Silvia, the mother of Romulus and Remus, was raped by a man disguised as a god (one of her suitors or even Amulius himself, 1.77.1) or, in fact, by 'the divinity of the place', i.e. the god Ares, as 'most writers relate' (77.2):

---

[70] Both Oldfather and Green 2006 misidentify διέφθειρεν as an aorist, when, in fact, it is a conative imperfect ('was in the process of/about to'; see Kühner and Gerth 1904: 2.1, pp. 140–2). The aorist ('destroyed') does not make sense in light of the following οὐ μὴν ἀλλ' ἀναλαβόντες ἑαυτοὺς ἐκ τῆς ναυαγίας ἁπάσαις ταῖς ναυσὶν ἀνήχθησαν ἐπὶ τοὺς πολεμίους, 'but no, they [the Persians] recovered themselves from the shipwreck and set sail with all their ships against the enemy'. Green's translation 'with their whole [surviving] fleet' significantly distorts the sense: the whole point of the passage is the strong contrast (οὐ μὴν ἀλλά) between what the Greeks *believe* is happening (and their interpretation of this event as divine support) and the actual situation, namely that the storm did, in fact, not bring about the destruction of any ships and reduce the enemy's fleet at all: it was merely 'about to'.

[71] Cf. 13.12.6, where he, however indirectly, blames Nicias' δεισιδαιμονία for the Athenians' catastrophic losses in Sicily.

> This is not a proper place to consider what opinion we ought to entertain of such tales, whether we should scorn them as instances of human frailty (ἀνθρωπίνων ῥᾳδιουργημάτων) attributed to the gods, – since a god is incapable of any action that is unworthy (ἀνάξιον) of his incorruptible and blessed nature (τῆς ἀφθάρτου καὶ μακαρίας φύσεως), – or whether we should admit even these stories, upon the supposition that all the substance of the universe is mixed, and that between the race of gods and that of men some third order of being exists which is that of the daemons, who, uniting sometimes with human beings and sometimes with the gods, beget, it is said, the fabled race of heroes. The philosophers have said enough about these matters. (1.77.3)

While Wiseman is right that 'Dionysius does not endorse the first (or either) of these alternatives',[72] one wonders whether it really does not matter to him which of these his reader subscribes to. We have seen above that Dionysius is concerned with the religious propriety of both his narrative (which is also a reflection of his own character) and his readers' attitudes towards it.[73] The assumption that Rhea would have been raped by the god Ares himself obviously goes against Dionysius' sensitivity to religious decorum; and even though he does not reject this alternative outright, his phrasing does imply that accepting it involves attributing actions to the gods that are 'unworthy' of their nature and might be taken by them as an act of blasphemy and therefore potentially expose the reader (as well as the historian himself) to divine retribution. Moreover, not unlike an 'Alexandrian footnote',[74] the mention of the philosophers might be an allusion to Plato's famous criticism of improper narratives about the gods (*Resp.* 377e6–83a5);[75] Roman readers might also recall Varro's 'obsession', his criticism of the *theologia fabularis*, the 'tales unworthy of the gods' typical of poetry and drama.[76]

But it is also clear that Dionysius does not want to dismiss divine involvement completely:[77] introducing the 'mythical' (μυθολογοῦσι, 77.2) version

---

[72] Wiseman 2002: 343.
[73] Cf. the discussion of 1.6.5 in the second section of the chapter above. [74] Ross 1975: 78.
[75] Cf. Fromentin 1988: 323 with n. 23. [76] Wiseman 2002: 335 (characterising Varro).
[77] Fromentin 1988: 322, wrongly takes ἐπὶ τὸ τῇ ἀληθείᾳ ἐοικὸς μᾶλλον (1.79.1) to refer to the rationalising version of the rape. But Dionysius is not referring to the different versions of the rape specifically, but to the entire tradition about Rhea's pregnancy and how it became known to Amulius. His point is that all historians agree on the basic 'facts' of these events and that disagreements concern details only, some of which are more credible (ἐπὶ τὸ τῇ ἀληθείᾳ ἐοικὸς μᾶλλον) than others (ἐπὶ τὸ μυθωδέστερον). Nothing in the text allows us to refer either of these to specific elements of the rape episode; on the contrary, as the above discussion has shown, Dionysius does not exclude divine involvement in this event as such; he only rejects the idea that it was the god Ares who raped Rhea. But that does not mean that he subscribes to the rationalising version that the father of Romulus and Remus was Amulius.

10 Expertise, 'Character' and the 'Authority Effect' in Dionysius

as the consensus of the majority at the very least prompts readers to be careful before rejecting it outright. Romulus and Remus were too special to be regarded simply as the product of the rape of their mother by some random suitor or a scoundrel like King Amulius.

Again, attention to the role of the reader can further our understanding of this passage. Dionysius is, no doubt, showing off his comprehensive knowledge of the historical tradition as well as his competence in religious matters, but he also uses them to involve the reader in his narrative. His expertise enables him to identify and point out the historical problems posed by the two different versions: the first one is trivial and unlikely, the second one blasphemous. In so doing, Dionysius first creates an *aporia* for the reader to struggle with and then draws on his philosophical expertise to help them solve it. He offers his recipients an alternative that allows them to retain the element of divine intervention without offending religious decorum: not Ares himself raped Rhea Silvia, thus compromising his divine nature, but a *heros*, a supernatural creature (τὸ δαιμόνων φῦλος) which partakes in the human and divine spheres alike.[78] And he supports his theory by referring the reader to the philosophical tradition which has sufficiently explored the nature of these divine beings.[79]

As in the previous examples, the hierarchy between Dionysius and the reader, the 'authority effect', is the result of the combination of Dionysius' display of his expertise and his interaction with the reader. Dionysius demonstrates that his expertise enables him to perceive problems that his readers might not have noticed – and are not prepared to deal with – and, at the same time, to find a solution to these problems. Dionysius does not tell the reader that he should subscribe to his theory, that Rhea was raped by a *daimon*; he leaves him a choice. But unless the reader knows as much about the nature of the *kosmos* (77.3) and the different divine creatures in it as Dionysius, unless, that is, the reader is familiar with all those philosophical discussions that Dionysius refers him to, but does *not* share his piety and religious sensitivity, he will probably find it wiser to regard Dionysius as the authority in the field to whose opinion he should subscribe.

---

[78] Note how Dionysius prepares this compromise already when introducing the second version by speaking of 'the divinity' (τοῦ δαίμονος, 77.2), rather than 'the god', 'to whom the place was consecrated'. Cf. Mora 1995: 180: 'L'importanza dei *daimones* nella terminologia Dionisiana porta a pensare che di fatto Dionigi risolva positivamente la questione [of their existence]'; for a collection of passages where Dionysius uses *daimon*, see ibid. 70–2.

[79] Gabba 1984: 860 therefore understands the dynamics of the passage better than Wiseman (see n. 72) when he states that by informing the reader of the philosophical theories about 'heroes' or *daimones*, 'di fatto le riconosce'.

Conclusion

This chapter had two interrelated aims. It sought to propose a different way of approaching authority in literature and to illustrate this approach by way of an analysis of key passages from Dionysius' *Early Roman History*. The core assumption underlying my discussion of Dionysius' authority is that authority in texts should not be equated with the author's overt demonstration of his competence and expertise. These are important aspects of 'literary authority', to be sure, but they should be seen as elements of a larger pattern of interaction between author and reader. Authority is created by, and is part of, the reading experience. It is constituted by the interplay of the author's self-presentation and the ways in which he positions his readers towards his narrative and, consequently, himself as the narrator. I therefore suggested speaking of a text's 'authority effect', rather than focusing exclusively on the narrator's 'authorial *persona*'. Moreover, our understanding of authority in literature would profit, I think, from a broader concept of which elements of the author's self-presentation contribute to the 'authority effect': while previous scholars have almost exclusively focused on Dionysius' display of his 'academic qualifications', 'irrational' factors such as the display of his morals, respect for the gods and piety, in short, his 'character', should be regarded as equally important.

The second and third sections aimed to support these considerations by discussing the 'authority effect' in Dionysius' *Early Roman History*. While both sections aimed to demonstrate the importance of author-reader interaction for the 'authority effect', the second section focused on passages in which Dionysius' display of his expertise is paramount, whereas the third section was centred on passages in which Dionysius' 'character' is more prominent. In both cases, the discussion confirmed the importance of the reader's involvement as a complement to authorial self-presentation: when Dionysius demonstrates his expertise through a display of his knowledge, he also creates a reading situation that makes the reader, quite literally, experience his own inability to deal with the complex and confusing material, thus creating a strong contrast with Dionysius' comprehensive and masterful command of the evidence. He offers the reader the opportunity to form his own opinion about the historical evidence only to make him realise that he lacks the competence and expertise to do so.

To a certain extent, this relationship of superiority that Dionysius establishes with his reader and that is designed to bring about his unconditional acceptance of Dionysius' version of the events, is similar to the 'submission of the reader' that Nicole Loraux has identified as typical of Thucydides' narrative. Thucydides, however, intervenes in his narrative much more subtly than Dionysius, who writes himself into his own text as a strongly present, controlling force. An important contrast to Dionysius' practice is Herodotus who often is genuinely unconcerned about presenting the reader with 'plural pasts', leaving them to choose whichever version they regard as most plausible or even to forego their choice.

The discussion in the third section explored the importance of Dionysius' display of moral values and piety for the 'authority effect'. In these instances, Dionysius tries to urge the reader to accept his interpretation of the past on account of his 'moral', rather than his 'epistemic', authority. This is particularly visible in Dionysius' attitude towards the controversial question of divine intervention in history. Dionysius fully endorses the view that gods influence the course of events, he even defines the opposite position as blasphemous. Dionysius' narrative thus becomes a statement of his own piety and an expression of his expertise in morals and religion. Again, Dionysius' self-presentation is supplemented by a specific kind of reader involvement: ostensibly, Dionysius leaves it up to the reader to take his judgement on the events for granted (accept his authority) or to reject his view. But he also makes it unmistakably clear that a rejection of his version of events would constitute an act of impiety that might anger the gods, as well as reveal the readers' flawed character and ignorance. As with the examples discussed in the second section, it is the reading experience that steers readers towards accepting Dionysius' claim to represent the 'true' narrative because of his qualities of character in the first place: in the end, Dionysius' narrative makes the reader feel that accepting his claims to superior knowledge and expertise and, hence, subscribing to his view, is the much safer alternative.

11 | The Authority of Galen's Witnesses

DARYN LEHOUX

ἀμείνω δὲ τῶν παραδειγμάτων ἐστίν ὧν αὐτόπται γεγόναμεν

Those things of which we are eyewitnesses are better than paradigmatic examples.

(Galen, *MM* K9.608)

Experience is the Highest Court

To anyone who has read much Galen, it will seem more than a little strange for me to say that Galen doesn't repeat himself often. But in fact he doesn't – at least, not when it comes to case studies. If we comb through Susan Mattern's list of known Galenic case studies (all 358 of them),[1] we find only thirteen instances Galen likely uses more than once (sometimes details are just vague enough in a repetition to make us less than positive that the case is the same in both stories), plus a tight-knit string of five that shows up in both the *De simplicium medicamentorum* and the *Subfiguratio empirica*. Add to these about six or seven other cases that are uncertain, and we see that the vast majority of Galen's case studies are one-off affairs. So when, on reading Galen, one thinks one recognises a patient or their illness and treatment from another story, one perks up one's ears a little.

One pair of cases with significant overlap turns up in the *Prognosis* and in *On Antecedent Causes* (extant only in mediaeval Latin translation). Here are the two versions:

> καὶ τό γε σοῦ παρόντος γενόμενον ἡνίκα περὶ φλεβοτομίας ἐσκέπτοντο τῶν ἐν Ῥώμῃ πρωτευόντων ἔνιοι διὰ τῶν ὑπομνημάτων ὧν ἐποίησα δείκνυται... ἐμοὶ δὲ τά τ' ἄλλα διασκεψαμένῳ πάντα καὶ τὸ κατὰ τὸ δεξιὸν μέρος τῆς ῥινὸς ἄχρι τοῦ μήλου θεασαμένῳ τὴν τέως ἀμυδρὰν ἐρυθρότητα πολὺ δή τι νῦν ἐμφανέστερον γενομένην, ἐπίδοξον ὅσον οὔπω κατὰ τὸν δεξιὸν μυκτῆρα τὸ τὴν αἱμορραγίαν ἔσεσθαι σαφῶς ἐφαίνετο. καί τινι τῶν παρόντων οἰκετῶν τοῦ κάμνοντος ἠρέμα διαλεχθείς, ἀγγεῖον ἔχειν ἕτοιμον ὑπὸ

---

[1] Mattern 2008: App. B. Her list is near-exhaustive.

τὴν ἐφεστρίδα τῶν ἐπιτηδείων δέξασθαι τὸ αἷμα, κἄπειτα φθεγξάμενος εἰς ἐπήκοον πάντων ἰατρῶν, ἐὰν βραχύτατον προσμείνωσι, θεάσεσθαι τὸν ἄνθρωπον ἐκ δεξιοῦ μυκτῆρος αἱμορραγοῦντα. τῶν δὲ γελασάντων ... ἅμα τ' οὖν ἔλεγον ταῦτα καὶ τὸν δάκτυλον ἡμαγμένον ὁ κάμνων ἐξείλκυσεν, ὅ τ' οἰκέτης προσδραμὼν ὑπέθηκεν αὐτῷ τὸ ἀγγεῖον, ἐφ' ᾧ μεγίστης κραυγῆς, ὡς οἶσθα, γενομένης οἱ μὲν ἰατροὶ πάντες ἔφυγον.

What happened in your presence when some of the leading doctors in Rome were considering phlebotomy is discussed in my case-studies... I examined everything closely, including the right side of the (patient's) nose extending to the cheek bone, where I saw the previously faint redness becoming much more visible. It then became clearer than ever that the haemorrhage would come from his right nostril. So I spoke quietly to one of the patient's slaves there, and told him to have ready under his cloak a bowl suitable for catching blood, and then I announced in the hearing of all the doctors that, if they would only wait a moment, they would see the man bleeding from the right nostril. They laughed... But just as I was saying this, the patient drew out his bloodstained finger, at which the slave rushed forward and held the bowl underneath. As you know, there was a great shout and all the doctors fled.[2]

nec etiam oblivisceretur eorum quae nuper facta sunt... ac etiam quoniam dubitantibus illis primo et timentibus et ignorare confitentibus ad quid finiret egritudo iuvenis qui desipiebat, ridens ego: 'emorragiam patieris parum post ex naso,' dixi, 'et totaliter liberaberis ab egritudine,' illis vero deridentibus et talia divinare dicentibus esse impossibile repente dextra ei naris effluxit sanguinem, et quoniam sudorem predixi.

[The rhetor Menander] had not forgotten what happened recently... that the other doctors had at first been in a state of doubt and apprehension, agreeing that they had no idea how the disease of the young man who was delirious would turn out, and how I, laughing, had said: 'in a short time you will suffer a haemorrhage from the nose, then you will recover from the disease completely', and how, while the other doctors were laughing at this and saying it was impossible to divine such things, blood suddenly flowed from his right nostril and he broke into a sweat, as I had predicted.[3]

---

[2] Galen, *Praen.* 13.1–10. The translation of the *Prognosis* throughout this chapter is or follows Nutton 1979. For a list of the abbreviations in use for Galen's voluminous works, see Mattern 2008: App. A, and/or Hankinson 2008b: App. 2 (not all works have completely standardised abbreviations and there are differences between these two lists, alas); this chapter follows Hankinson's list, in line with the rest of the volume.

[3] Galen, *CP* 2.16–3.17. The translation of the *Antecedent Causes* throughout this chapter is or follows Hankinson 1998.

The parallels are striking, although there are some divergent details that may indicate that we have here two different patients. In any case, for my purposes a strict identity is not really as important as the significant thematic similarities in how Galen presents these case studies, his relationship to the other doctors in each, and, most importantly, his use of corroborating witnesses to bolster the authority and veracity of the claims. In both instances the other doctors laugh at Galen, and in the *Antecedent Causes* version Galen also laughs at the other doctors.[4] What the doctors are deriding in both cases is Galen's presumption that he can make a prediction so very precisely, and in both cases they are proved spectacularly wrong. The extreme implausibility of Galen's success is highlighted by the scornful use of the verb 'to divine' in place of 'to predict' in the Latin passage. We see him similarly accused of sorcery and divination in the first of these two passages. Indeed, the accusation of sorcery by 'other doctors' is a favourite motif of Galen's, one he loves to repeat whenever he talks about his incredible success at prediction.[5]

What is particularly interesting in the present cases, though, is Galen's use of named eyewitnesses to buttress the plausibility of each of the accounts. Not content to rely on the authority of his own testimony (which would hardly be disinterested), Galen has recourse to naming for his reader at least one individual who was actually present at each event in order to both lend authority and give life to his version of the events. In the *Antecedent Causes* passage, he mentions his witness, Menander, as having 'not forgotten' the event. He also spends some time bolstering Menander's credentials before presenting his eyewitness testimony (a practice borrowed from the courtroom).[6] We are explicitly told that Menander is a rhetor, and that this gives him solid qualifications when it comes to adjudicating arguments. In particular, he 'is not ignorant that experience is the highest court' (*experientia est maximum iudicatorium*). So here we see a double rhetorical game being played by Galen: his witness is most reliable, and in particular the reliability of the witness is couched in terms of Menander's awareness of just exactly the epistemological point Galen needs to make in underscoring the

---

[4] For what it is worth, the Latin translator of the *Antecedent Causes*, Niccolo da Reggio, uses a slightly different word for the scornful laughter of the other doctors than he does for Galen's own laughter (*derideo*, the root of our 'deride', versus *rideo*).

[5] See, e.g., *Praen.* 1.7, 1.8, 7.9 (includes Galen laughing), 7.13 (includes Boethus laughing at Galen only to be shown up in the end), 10.17, 10.18 (has the other doctors laughing derisively: καταγελῶντες). Compare also his *Hipp. Prog.* K18b.2, K18b.11; *Hipp.Off.Med.* K18b.715; *CAM* K1.292 (more laughing); *Hipp.Epid.* K17a.250 (discussing the case of Eudemus also found in the *Prog.*); *Di.Dec.* K9.833; *Ut.Diss.* K2.906; *HNH* K15.57; *Loc.Aff.* K8.362. On Galen's use of the motifs of divination and sorcery, see Hankinson 2005a; Van Nuffelen 2014.

[6] For this use in Roman scientific texts, see Lehoux 2012: 77–105.

reliability of witnessing in general: Menander knows that seeing is believing, even if in this instance the reader's own seeing is through Menander as proxy.

In the *Prognosis* case, on the other hand, the witness in whose presence the story unfolds is none other than Galen's dedicatee for that work, Epigenes, who is repeatedly addressed with the simple second-person pronoun, 'you' (σύ). From a rhetorical point of view, the use of the dedicatee as a witness seems an even more effective strategy than naming a third party such as Menander, at least if we assume the reader to take the conventions around dedications and their texts even partly at face value. Perhaps no sophisticated reader of Galen would completely miss the transparently manipulative use of 'you will remember the event', directed at an addressee, but there is still something compelling about the pretensions surrounding the involvement of the dedicatee that strengthens the veracity of the reported witnessing. This is because there is an implied assent on the part of the dedicatee-witness – one that comes from the assumption that he is, just as we are, a reader of the text – which then corroborates the report offered by Galen. Galen is effectively signalling to the reader his confidence that the dedicatee, as a fellow reader himself, will see nothing in the story to which he should object. Where Menander in the *Antecedent Causes* is said to have witnessed an event, we hear the account of that witnessing only from Galen, Menander himself appearing in some ways as just another character in the story who has no voice of his own in the narrative, and who, for all we know, may never even see the account Galen offers. In the *Prognosis*, on the other hand, Epigenes is addressed directly, 'you remember how ... ' and the rhetorical move serves to hint very strongly that Epigenes agrees with the version Galen is presenting. Perhaps this is the reason why, in the other case studies in the *Antecedent Causes* and in the *Prognosis*, Galen tries as often as possible to have his dedicatees present at one after another of his cures, his diagnoses, and his ever-grand and stunning Sherlockian revelations.[7]

What is doubly interesting, though, is that, effective though it may be, Galen avails himself of the dedicatee as witness in relatively few of his works. Having combed through his uses of pronouns and verbal endings for the second person in the *TLG*, I have found that Galen only addresses three

---

[7] This phenomenon in the *Prognosis* and its epistemological/rhetorical import has been noted by Mattern 2008: 83–4 and Wenskus 2010: 88–9. In many ways, the present chapter is an expansion of Mattern's original idea. On the use of the second person in other ancient scientific texts, see Hine 2009; Keen 1985; cf. also Gilmartin 1975; Gibson 1997; on the uses of first-person verbs and pronouns in Galen, see Nutton 2009; in other ancient scientific and medical texts, see Lehoux 2013; Hine 2009.

kinds of people in the second-person singular. Two are very common, the third quite a bit rarer. The common uses are when Galen wants to give instruction to his reader or to call on their knowledge in some way ('if you cut here you will see... '), or when he is feigning a dialogue, as he so often does, with some real or imagined opponent, and so 'arguing with' an Erasistratus or a Chrysippus directly ('you say that... but in fact...), or even an entirely unnamed adversary ('you [who] say... '). The third use, the one that we see in the *Prognosis* and *Antecedent Causes*, where the dedicatee is called on as a witness ('you yourself saw this') is significantly rarer in Galen (it occurs, so far as I can find, only in these two works, in *On Not Grieving*, in the *Method of Medicine* and in a small handful of passages in four other works).[8] It will thus repay some attention as an epistemological strategy for bolstering Galen's own authority vis-à-vis his case histories.

Galen begins his little book on antecedent causes with a par-for-the-course mention of how his dedicatee, Gorgias, asked him to write up his ideas on the subject. We see this kind of thing all the time in ancient dedications, but with Galen it always seems to take on a different flavour. Part of it is just the way Galen's massive personality shines through in virtually every work he writes. We might call him supremely combatant (or, closely related: pathologically insecure) and thus explain his apparent inability to resist a jab at anyone he feels he has bested or has been slighted by, or else his compulsive need to brag about the important people who are 'amazed' by his work. A delightful line from the *Antecedent Causes* sums up how one often feels when Galen gets so worked up: *unde, et si non vis, exempla alia sustine* (6.58) ('please put up with some more examples, even if you don't want to'). But let's for the moment pretend that his combativeness and triumphalism is, rather than a distasteful personality quirk, instead a deliberate rhetorical strategy – something he is doing intentionally for effect rather than something he can't help himself from doing.

So Galen begins the *Antecedent Causes* by addressing Gorgias, who, we learn, asked him for a treatise on the subject of antecedent causes. But the fact that he asked Galen for this treatise is not, as it turns out, the first thing we learn about Gorgias. Here is what Galen says on introducing him:

---

[8] These are: *MMG* K11.125.4, 128.9–11, and *Ord.Lib.Prop.* K19.43.15, all very similar to passages in the *MM* that I shall discuss below, *Aff.Dig.* K5.43.15, very similar to passages in the *De indolentia*, discussed below, and *Ven.Art.Diss.* K2.779.3. Mattern catalogues the *Praen.* and *MM* instances on p. 234, nn. 37–8, but not the *Antecedent Causes* instances. She also lists two instances of invocation of the addressee's experience from one other work, the *Loc.Aff.* (her cases # 132 and 133), but in these cases the dedicatee is simply said to know something distantly related to the events, not to have actually witnessed the events himself, and so I do not discuss them here.

> tu autem, o Gorgia, licet deriseris de huiusmodi medicis et maxime nuper videns quemdam spretum et valde redargutum ab ipsis operibus artis, tamen rogasti nos submemorationes scribere tibi de procatarticis causis.
>
> And although you have laughed at doctors of this sort, Gorgias, having recently seen one confounded, indeed refuted, by the actual results of this art, none the less you have asked me to write you a résumé of matters concerning antecedent causes. (Galen, *CP* 1.6)

And so before we get to the usual 'you asked me to write...' that is so common in ancient dedications, we first are made aware of Gorgias as a witness to a refutation by Galen of the kinds of sophistical physicians who either doubt or over-complexify the idea that some kinds of causes ('antecedent' ones) can act irregularly, not affecting every patient every time, or not affecting every patient in the same way every time.

Having thus introduced his reader to the dedicatee Gorgias, Galen turns immediately to his first mention of 'our friend', the witness Menander. But it is not quite yet Menander-as-witness-to-the-bleeding-nose-incident to whom we are introduced. Perhaps a little curiously, we are instead first introduced to Menander as a patient himself. Thus, we meet Menander in the company of Gorgias, himself acting as a witness to Menander's case. Galen tells the reader that he and Gorgias had experienced just the subject matter of his present book, and that this happened 'yesterday', *pridie* (probably used loosely to mean 'very recently', but even still, a remarkably vivid time frame for a case study, even for Galen).[9] Menander had, it seems, come down with a fever after watching a contest at the stadium under (unspecified) weather conditions to which he and the whole audience had been exposed. Galen and Gorgias then go to see him (... *visitantes nos* ...) and in the presence of other doctors, Galen orders (*iussimus*) Menander to undergo a course of treatment to which the other doctors object, and a heated debate ensues as a result of which, Galen tells us, Gorgias requested the present treatise be written.[10] That Gorgias himself was present is highlighted several times in the passage,[11] even if one might be led to some confusion by Galen's liberal

---

[9] *CP* 2.11. That it cannot be meant literally is shown by the time frame of Galen's whole account, which continues the story 'on the second day', *secundo die*, where we should expect *hodie* ('today') if *pridie* had been used literally to refer to yesterday. (I suspect this *pridie* is picking up on the *nuper* at 1.6). See also Hankinson 1998: 155–6.
[10] On the competitive nature of Roman medicine, see Mattern 2008 and 2013.
[11] In addition to the explicit (but not necessarily definitive) use of *nos visitantes*, we also see a well-timed *ut nosti* ('as you know ... '), as well as Gorgias' very strongly implied presence at the ensuing debate among the doctors and perhaps also his inclusion under the *recessimus simul* at the end of the debate.

interspersion of the 'royal we' with instances of what must be the proper first-person plural 'we'.

And so we get a complex interleaving of witnesses in these passages. Gorgias is said at the outset to have laughed at certain doctors, presumably in the case of Menander. Menander's case is then introduced, but before Galen can finish the story, he interrupts it to tell us why Menander trusted him so much. Menander had, in the event, previously seen three successful treatments by Galen of similar cases, the third (and most detailed) of which was the man with the bleeding nose that started us off on this investigation. Menander, then, having this basis for trust in Galen, serves to bring the complex narrative back into Gorgias' presence by asking Galen, in light of all this, what he, Menander, should do for his own treatment. So the dedicatee serves as witness to a case whose very dynamics (and in particular Menander's confident faith in Galen in the face of the other doctors' contrary advice) depend on three previous cases, themselves witnessed by the patient who is now being witnessed by Gorgias. Gorgias thus lends authority to the Menander story, and Menander lends authority to the nosebleed story, *but so does Gorgias himself* by virtue of his having witnessed Menander in the act of trusting Galen, which 'witnessed *fides*' is clearly intended to corroborate the events we are told Menander had seen.

In a similar way, in virtually every other case in the *Antecedent Causes*, Galen is careful to stand Gorgias somewhere in the background as a witness. So for the young man who became desiccated by wrestling while oiled up, the story begins with *nosti autem et alium adolescentem* ('and you know another young man'), a point reiterated during the story: *nosti igitur* ('thus you know').[12] That Gorgias is not just being reminded of a story he has heard, but of a story he actively participated in, is carefully underscored towards the end of the tale:

> *quin etiam memor es qualiter tibi largiens, quando iam melius se habebat, non dedi cibum ei semel ante immissionem paroxysmi, ut videres qualiter exsolueretur et infrigidaretur et prope in impulsualitatem veniret.*
>
> You remember too how, as a favour to you, after he had recovered somewhat, I withheld food from him on one occasion before the paroxysm, so you could see how he grew cold and his pulse virtually stopped. (Galen, *CP* 3.29)

Gorgias is here intervening in the treatment to propose an experiment. It is not just Gorgias' memory that Galen wants to parade before his reader,

---

[12] Galen, *CP* 3.22, 3.26.

but Gorgias' active participation in what amounts to a potentially dangerous demonstration of theoretical principle.

Likewise, in the *Prognosis*, a wonderful collection of case histories if ever there was one, Galen again and again invokes his dedicatee Epigenes as witness. Galen begins by telling us how ignorant he was at first of the wickedness of Rome's doctors. He discovered this rot at the core when he treated the philosopher Eudemus: ὡς ἐπίστασαι σὺ ὁ μάλιστα παρὼν τῇ νόσῳ πάσῃ μέχρι τέλους ἀπ' ἀρχῆς Εὐδήμου τοῦ περιπατητικοῦ φιλοσόφου ('as you well know, since you were present from beginning to end throughout the whole illness of Eudemus the peripatetic philosopher') (Galen, *Praen.* 2.1). As with Gorgias in the earlier example, Epigenes-the-witness is here invoked vividly as a participant in the theoretical debate: ἐγὼ μὲν οὖν ἐρομένῳ σοι, μέμνησαι γὰρ πάντως καὶ τοῦτο ... ἔφην ('When I was asked by you (and you remember this completely), I said') (Galen, *Praen.* 2.9).

In many another case in the *Prognosis*, Galen likewise tries never to enter a room unaccompanied by his dedicatee. Just like the *nosti* invoked in the *Antecedent Causes*, we again get a constant invocation of the witness in the background of each carefully painted scene. ὡς οἶσθα ('as you know') or οἶσθα σύ ('you know') is invoked repeatedly throughout the work,[13] with other telltale phrases salt-and-peppered in besides: σοῦ παρόντος ('you being present', 2.15); γινώσκεις and γινώσκοντί σοι ('you know', 2.16, 5.13, 3.1); ἤκουσας αὐτὸς σύ ('you yourself heard', 3.17); ὑπὸ σοῦ τε παρακληθείς ('I was summoned by you', 5.2); σὺ δέ, ὦ Ἐπίγενες, ἔγνως ('you know, Epigenes', 10.17).

There are a few cases, though, where Epigenes is not present. The first cluster of these occurs in chapters 5 to 7, where Galen recounts to us three incidents. One is a story about a dissection he was performing for an audience, where a philosopher by the name of Alexander caused Galen to storm out of the room in a snit when Alexander insisted that Galen's demonstration would be meaningless until he could first prove that the evidence of the senses could be trusted. The second incident is a remarkable case study, where Galen performs a controlled experiment to diagnose a woman whose illness stems from a secret love for a dancer named Pylades, and the third a similar case where a slave was ill from worry about some missing money. In these extended stories, Galen clearly misses the presence of Epigenes, insofar as he catches himself immediately after their telling to try and re-leverage Epigenes' authority. Epigenes now becomes a new kind of witness, one attesting to events he did not himself see: τούτων μὲν οὖν, ὦ Ἐπίγενες,

---

[13] E.g. at 2.15, 2.21, 2.22, 5.9, 9.7, 11.8, 14.1.

αὐτὸν ἔχω μάρτυρά σε, πολλῶν δ' ἄλλων ἑτέρους ἁπάντων τῶν πραχθέντων μοι κατὰ τὴν πρώτην ἐπιδημίαν, ἤκουες δ' αὐτὸς παρ' αὐτῶν τῶν ὑπ' ἐμοῦ θεραπευθέντων ('I have you to corroborate these tales, Epigenes, and other people can attest all the many other achievements of my first stay, and you yourself heard of them from those who were treated by me') (Galen, *Praen.* 8.1). Epigenes may not have been present, but he is called on as a second-hand witness, one who can apparently verify that he heard these same accounts from others or from the patients themselves.

Another sign that Galen has been missing the presence of his dedicatee in these three cases is the way in which Galen very subtly tries to hide Epigenes' absence in the case of the dissection and the argument with Alexander. Where he cannot bring himself to say that Epigenes remembers particular events in the room with Alexander, or that Epigenes himself heard or saw something in the thick of the action, Galen tries very hard to hide this fact. Instead, what he does is to have Epigenes remember a host of circumstantial details around the event, while quietly being absent for the event itself. So Epigenes is said in the middle of the story to 'know' Alexander's key personality flaw, his φιλονεικία, a love of quarrelling (Galen, *Praen.* 5.13, 5.16). In this way, Galen allows himself to insert another (otherwise routine-looking) γιγνώσκεις ('you know'), but this one pointed only peripherally at the actual scene under discussion. Galen thus uses Epigenes' knowledge of Alexander's quarrelsomeness to paper over the fact that, on closer inspection, Epigenes seems not to have actually witnessed *this particular quarrel* of Alexander's. Similarly, Epigenes is said to know the patron of the event, Flavius Boethus, that he was an ex-consul, and that he was a lover of virtue and learning (φιλόκαλός τε καὶ φιλομαθής) (Galen, *Praen.* 5.9). In this statement a double strategy is at play: Epigenes is being suggestively inserted into the narrative with as much directness as Galen dares, and at the same time the *actual* witness to the events, indeed their very sponsor, Boethus, is treated just as we saw Menander handled earlier: with an opening paean to his virtue and reliability. Galen also brings Epigenes as close to the heart of the affair as possible by 'reminding' him how the whole thing began (ἀναμνήσω σε πρότερον ὅθεν ἤρξατο, 5.7) – a phrase just fuzzy enough that Epigenes himself would read it as, presumably, an innocent call on his assent to something he knows by hearsay, but one worded so as to subtly lead the reader's eye away from Epigenes' absence. Galen's purpose in reminding Epigenes of the event is, he says, in case Epigenes should want to re-tell the story to someone worthy. And notice here another subtle ploy: Galen is setting the story up as one valuable enough to be worth re-telling, and also pointing out that it is the sort of story only suitable to be shared τινι τῶν ἀξίων κοινωνίας τοιούτων λόγων ('with someone among those worthy of participation in such debates') (5.7).

After this, Epigenes again becomes less of an explicit presence for a while as Galen begins to frequent rooms higher and higher up the social ladder, closer and closer to the imperial bedchamber itself. Nevertheless, Epigenes seems to have access to the emperor's words (or at least reports of them) when he is invoked at 11.8 as witness to Marcus Aurelius' claim that Galen was 'first among the physicians'. Finally, though, Galen does again explicitly interject Epigenes back into the preceding series of stories at 12.11, where he says αὐτὸς γὰρ σύ, ὦ Ἐπίγενες φίλτατε, τὰ γεγραμμένα μοι περὶ τούτων ἁπάντων ὧν ἐθεάσω με προλέγοντα γινώσκεις ἐπιδεδειγμένα πρὸς Ἱπποκράτους εἰρῆσθαι ('You yourself know, my dear Epigenes, that I have said that the things I have written about all these predictions that you have seen me make, had been already demonstrated by Hippocrates') (Galen, *Praen*. 12.11). 'All the predictions that you have seen me make', as though Epigenes had been at his side all along. This claim is doubly reinforced as Galen very carefully and more explicitly brings Epigenes back into the rooms of the sick as he closes the book (and it will be worth noting that Galen uses the explicit presence of Epigenes as a framing device at both ends of the work, with all the attempts to implicitly assert his authority as an indirect witness occurring in the middle, a pattern we will see him repeat in other works).

Thus, in book 13, when Galen brings up the nosebleed case that overlaps so much with that in the *Antecedent Causes*, he once again bodily brings Epigenes back into the room (remember that Gorgias had been absent in the *Antecedent Causes* version), saying that these events happened in front of Epigenes himself (13.1). Galen describes the event as τό γε παρόντος γενόμενον ('what happened in your presence'),[14] which he follows up a little later in the story with a ὡς οἶσθα ('as you know').

The *De indolentia*

In another, recently discovered treatise, the *De indolentia* (*On Not Grieving*), Galen avails himself of a similar framing device to that in the *Prognosis*,

---

[14] One early edition (Basil, 1538) and Niccolo's fourteenth-century Latin translation both add an explicit word for 'you' (σοῦ, *te*) to this sentence, although given both the grammar and the wider context, where Galen has just been talking of Epigenes' witnessing in very explicit terms (αὐτὸς γὰρ σύ... ὧν ἐθεάσω... γινώσκεις, 'for you know about what you saw'), it is clear that Epigenes must be the implied subject of παρόντος.

calling on his dedicatee (or in this case, his correspondent) as witness most prominently at the beginning and end of the work. Given the very different subject matter of the *De indolentia*, however, the kinds of witnessing Galen ascribes to his correspondent play out slightly differently.

Galen writes this short text, as he himself tells us, in response to a letter he had received from an acquaintance. The short treatise takes the form of his epistolary response. Unsurprisingly, Galen frequently refers to his correspondent as 'you'. Many of the turns Galen makes to his correspondent are for our purposes not particularly meaningful, more stylistic than authoritative.[15] But in light of the present analysis, it is hard not to notice that in the opening and closing pages of the book, Galen's calls to the dedicatee take on a different tone, one much closer to the epistemological witnessing we have seen in our other works.

The basic premise of the *De indolentia* is that Galen's correspondent wants to know how it is that Galen never grieves, even when suffering what look to be unbearable losses. The particular crisis that prompted the exchange of letters that produced the *De indolentia* is Galen's loss, in the great fire of 192 AD, of many or most of his books – some irreplaceable – as well as a great store of expensive and near-unobtainable drugs, one-of-a-kind pharmacological recipes, a quantity of gold, and a large number of specialised bronze medical instruments, many designed in wax by Galen's own hand. In the face of this tremendous loss, we are told that Galen showed no grief. As Galen goes on to describe the magnitude of his losses, manuscript upon manuscript, rare ingredient after rare ingredient, and as it becomes clear just how much of his personal and professional wealth is gone forever, it becomes hard to believe that anyone could take such a devastating blow with equanimity (this is made all the more clear in the picture Galen paints of the grammarian Philides, who had a complete breakdown at his own losses in the same fire). And it is this disbelief Galen seeks to offset in the reader with his very first invocations of his correspondent. In only the second sentence of the text, Galen tells the story of the time his correspondent saw Galen lose most of his servants to a plague, but still Galen maintained a cool detachment:

---

[15] E.g. ἴσως ἂν οὖν φήσεις ('perhaps you will say') (13.3); πολλάκις ἤκουσας παρ' ἐμοῦ λεγόμενον ('you have often heard it said by me') (14.7–8); πέπεισαι δ' οἶμαι καὶ αὐτός ('you also believe, I think') (18.1); σὺ γινώσκεις αὐτός, ὡς ἂν ἐξ ἀρχῆς συναναστραφεὶς καὶ συμπαιδευθεὶς ἡμῖν ('you know these things yourself, having been raised and educated with me from the beginning') (16.18–19); and there is even a τὸ γάρ τοι δεινότατον... λέληθέ σε ('you do not know the worst part') (5.16). Similar instances can be found at 6.16, 8.11, 8.14, 9.15, 11.7–9 and 18.19–20.

παρὼν μὲν αὐτὸς ἔφης ἑωρακέναι κατά τινα τοῦ πολυχρονίου λοιμοῦ
μεγάλην ἐμβολὴν ἀπολέσαντά με τοσούτους οἰκέτας οὓς σχεδὸν εἶχον ἐν
τῇ Ῥωμαίων πόλει, ἀκηκοέναι δὲ καὶ πρόσθεν ἤδη μοι γεγονέναι τι τοιοῦτον εἰς χρήματά τε τρίς που καὶ τετράκις ἁδραῖς ζημίαις περιπεσόντα· ἔφης
αὐτὸς ἑωρακέναι με μηδὲ ἐπὶ βραχὺ κινηθέντα.

You said that you yourself were present and saw when a great attack of a long-lasting plague killed almost all the servants that I had in the city of the Romans, and you had heard before this that something of the sort had already befallen me three or even four times, with my estate being overturned by terrible losses. You said that you yourself saw me moved not even in the slightest. (*De indolentia* 2.5–12)[16]

And so we have a precedent for Galen's calmness in the face of adversity and great financial loss. Notice how Galen emphasises his correspondent as a witness: Galen does not simply say that the correspondent saw the event, but that he has verbally acknowledged having done so ('you said that you saw'). That the correspondent volunteered this testimony – apparently spontaneously and likely in the very letter to which Galen is responding, as we shall see – serves to underscore for the reader just how remarkable the witnessed event must have been. So too, Galen's wording of the correspondent's account includes emphases that are, strictly speaking, grammatically unnecessary: the correspondent did not just say that he saw the event, but that *being present* (παρών), he *himself* (αὐτός) saw it. Notice also how Galen repeats the assertion of witnessing at the end of the passage, again with an emphatic αὐτός, again in the form of voluntary, spontaneous testimony offered by his correspondent. Indeed, this use of the correspondent's own testimony to his witnessing, in the form of some variant on 'you said that you saw' repeats again and again in the text: 'you said that you yourself know' (2.16); 'you said you were amazed' (3.3); 'you said you know' (4.1); 'you say you have never seen me grieving' (21.16–17).

One such passage where the correspondent is said to have volunteered information occurs at a key point, when Galen describes his immediate reaction to the news of his losses in the fire, which is really the moment of truth for the whole narrative, the moment where even a philosopher of mettle might forget himself in the shock of it all. What is most interesting is that Galen acknowledges that the correspondent himself was *not* a witness at this moment. Nevertheless, he still ties his friend as closely as possible to the

---

[16] Citation of this text is not yet, to my knowledge, standardised. My references are to page and line numbers in Boudon-Millot and Jouanna's 2010 Budé edition. Translations of this text are my own throughout.

act of witnessing, in this instance by using a proxy (and here we may have the best evidence yet that Galen's narrative is factually constrained by the historical presence or absence of his addressee as reader): ὁπόσα μὲν καίρια καὶ αὐτὸς ἔφης ἐπίστασθαι, πεπύσθαι δέ τινος τῶν ἀγγελλόντων σοι μηδὲν νῦν ἀνιαθῆναί με φαιδρόν τε καὶ τὰ συνήθη πράττοντα καθάπερ ἔμπροσθεν ('You said you yourself knew how serious this was, and that you had learned from one of your messengers that I did not grieve, but was cheerful and went about my business as before') (*De indolentia* 2.15–3.2). The correspondent may not have been present, but he at least heard it from one of his own messengers, who, Galen leads us to believe, was himself a direct witness. And here again Galen vivifies the correspondent's role by leveraging his voluntary testimony ('you said you knew and had learned').

Finally, Galen underscores the magnitude of his loss and the powerful significance of his equanimity by bringing in the correspondent's reaction to Galen's response to the fire as a kind of measuring-stick. In the first such passage, he says: ἔφης ... θαυμάζειν ... ὅτι ... ἀλύπως ὤφθην φέρων[17] ('You said that ... you were amazed ... that I was seen bearing myself without grief') (*De indolentia* 3.3–5). And then a little later,

> τὸ γὰρ μηδὲ τῶν τοιούτων ἁπάντων ἁπτομένων ἀνιαθῆναί με θαυμασιώτερον ἐδόκει σοι καὶ πάνυ μοι τοῦτ' ἐφαίνου γράψαι ἀληθῶς· ἐν γὰρ τῇ Καμπανίᾳ πυθόμενος καὶ αὐτὰ διεφθάρθαι, πάνυ ῥᾳδίως ἤνεγκα τὸ πρᾶγμα, μήτε βραχὺ κινηθείς.
>
> It seemed most amazing to you that I did not lament being beset by all these things, and you appear to have written correctly: for I was in Campania and, [being told] these things were destroyed, I bore the matter very easily, being moved not in the slightest. (*De indolentia* 5.5–9)

Here we see Galen finally make overt reference to the fact that his correspondent actually reported what he knew of the events to Galen in the letter to which Galen is responding. Again, it is the correspondent's amazement that is being called on as a measure for the scale of Galen's equanimity.

The *De methodo medendi*

Turning back to medical texts, we find dedicatees being frequently called on as witnesses in one of Galen's larger and more influential works, *Method of Medicine*. In the *MM*, he outlines in detail his therapies for a wide range

---

[17] Here the main verb, ἔφης, carries over from the previous sentence.

of ailments, as well as the reasoning behind many of his treatments. For good measure, he also sprinkles liberal criticisms of his opponents' practices throughout, although there is a more pronounced polemicist tone in the first half of the book than the second. The work is a large one and, as he tells us himself, it was written in two parts, the first six books dedicated to a certain Hieron, and the last eight books dedicated to someone called Eugenianus. The last half of the work, that to Eugenianus, is said by Galen to have been taken up 'a long time' after the first half had been completed and all but abandoned.[18]

The tone of the text is once again quite different from those we have so far been examining. Rather than a straightforward polemic such as *On Antecedent Causes* or an autobiographical collection of competitive case studies such as the *Prognosis*, the *Method of Medicine* is much closer to something like a teaching text. Imperative verbs, verbal adjectives in -τέος ('it is necessary to ... '), and instructional asides such as 'direct your attention to the following' or 'as you know from [some previous demonstration or text]' are common throughout. In this long work, Galen explicitly addresses his dedicatees over fifty times – it is perhaps telling in this context that more than half of all instances of the verb ἐθεάσω ('you saw') in the entire Galenic corpus are found in this one work. A few of these instances are for our purposes relatively innocuous, pointing out that Hieron or Eugenianus is familiar with some condition or other, or else familiar with some writing of Galen's. The vast majority of his references to his dedicatees, however, are more pointed, and fall for the most part into two broad categories. First, in both halves of the work the dedicatee is called on frequently to back Galen up on his criticisms of other doctors. Second – and this applies almost exclusively to Eugenianus (which is to say it predominates in the second half of the text) – Galen calls on his dedicatee as witness to the effectiveness of a specific treatment, often in the context of a particular case study, but also commonly in a more general formulation ('you have often seen me cure ... ')

That 'other doctors' are idiots is a frequent theme in Galen. In the *De methodo medendi* Galen repeatedly underscores this claim by calling on his dedicatee for corroboration. In a number of instances Galen's strategy is very general, simply citing his dedicatee's agreement that the majority of (unnamed) 'other doctors' are fools, as at K10.76 where he says: ἔνιοι δ' οὐ

---

[18] Johnston and Horsley 2011 speculate that there may have been something on the order of twenty years between the two halves, which is certainly possible, although we really have no idea.

μόνον οὐκ ἀπέδειξαν, ἀλλ' οὐδὲ παρὰ τῶν ἀποδεικνύντων μανθάνουσι· καὶ τό γε πλεῖστον γένος, ὡς οἶσθα, τῶν νῦν ἐπιπολαζόντων ἰατρῶν, ἔστι τοιοῦτον ('Some not only do not give proofs but also do not learn from those who do give proofs, and the greatest class [of these], as you know, is that of the currently fashionable doctors').[19] Calling on the dedicatee's agreement in his general derision of other doctors recurs frequently in both halves of the *MM*. Thus, the dedicatee is said to 'know' that other doctors are unable to explain their reasoning (K10.34); mistake terminological differences for real differences (K10.44); shun demonstration (again) (K10.113); desire to appear wise rather than be wise (K10.114); foolishly change from one drug to another without justification (K10.169); falsely claim to be 'methodical' (K10.204); harm patients through fasting (K10.542); do not want to be seen to be learning anything new (K10.560); are unreasonably afraid of phlebotomy and administering cold drinks (K10.627); cause hectic fevers by improper timing of food (K10.692); and cause inflammation through ignorance (K10.782).

What these general instances all share is the fact that the other doctors who come in for such criticisms are unnamed, and the faults they have are described very broadly, not clearly applying to any one specific instance of failure. Similar to these are a host of positive claims made by Galen, saying that his dedicatees have 'often' witnessed him curing people in the past, where he names or describes a condition but does not cite any specific instance or occasion of witnessing in particular. So he says of Hieron (and it is worth noting that this is the only instance in all of the six books dedicated to him that he invokes Hieron as a witness to such a general treatment claim),

> ἐθεάσω δὲ καὶ ὀφθαλμῶν ὀδύνας σφοδροτάτας ἰασαμένους ἡμᾶς ἢ λουτροῖς ἢ οἴνου πόσεσιν ἢ πυρίαις ἢ φλεβοτομίαις ἢ καθάρσεσιν, ἐφ' ὧν οὐδὲν ἄλλο ἔχουσιν οἱ πολλοὶ τῶν ἰατρῶν ἢ ταυτὶ τὰ δι' ὀπίου καὶ μανδραγόρου καὶ ὑοσκυάμου συντιθέμενα φάρμακα, μεγίστην λώβην ὀφθαλμῶν ... καὶ πολλοὺς οἶσθα μετὰ τὰς τοιαύτας χρήσεις τῶν φαρμάκων, ἐπειδὰν ἀμετρότερον προσαχθῇ, μηκέτ' ἐπανελθόντας εἰς τὸ κατὰ φύσιν.

> You have also seen me cure the severest pains of the eyes either with baths, draughts of wine, heating, phlebotomy, or purges, whereas the majority of doctors have nothing else for this than those drugs made from poppy juice, mandrake, and henbane which are the greatest destruction to the eyes ... And you know that after the use of these drugs, when applied too liberally, they never return to their natural state. (*MM* K10.171)

---

[19] All translations from this text are my own.

Again, no specific case is cited, only the general claim that Hieron 'has seen' Galen cure patients with the condition in question (with the implication, given the list of different remedies employed, that he has seen it often). By contrast to the isolation of this one incident in Hieron's half of the work, Galen calls on Eugenianus in such general cases frequently: as having seen him cure 'many' diseases caused by imbalance of hot, cold, wet and dry in some body part or other (K10.465), as having seen him cure 'some' patients with poor stomachs in a single day by administration of cold water (K10.467), and as having 'often seen' (πολλάκις ἐθεάσω) Galen give these same patients foods cooled with snow (K10.468). Galen reiterates this point later when he says that Eugenianus has often seen him (πολλάκις ἐθεάσω) administer cold water (K10.622). Eugenianus has often seen Galen (πολλάκις ἐθεάσω, again) cure people with diseases of excess through the use of fasting (K10.689) and to have always seen (ἐθεάσω ἀεί) Galen correctly predict death in patients with such diseases when fever intervened (K10.690). Just as with the patients suffering from poor stomachs, a single day sufficed for Galen to treat some patients with inflammation of the eyes 'as you saw', he says to Eugenianus (K10.902). Finally, Eugenianus is said to 'know' that Galen has kept many people of a hot and dry constitution healthy by maintaining their regimen to ensure free-flowing pores (K10.551), and because of his marvellous and conspicuous successes, Galen was, as Eugenianus is yet again said to know, called a wonder-worker (K10.683 and K10.684).

In many other instances, however, we come much closer to the kinds of uses to which Galen put his dedicatees in the *Prognosis*, *Antecedent Causes* and the *De indolentia*, which is to say he calls on them as witnesses to specific, historical events in order to lend authority and credibility. Thus, on the foolishness of individual doctors, Hieron is said to 'surely remember' (μέμνησαι δήπου) one doctor who treated a dirty wound with 'the usual green drug, having mixed it with honey', but to no effect.[20] In a similar vein, Galen says to Hieron 'surely you know how one of the stupid Methodists' failed to understand how the seasons of the year work to heat or cool bodies.[21]

But again, compared to Eugenianus' half of the *MM*, there is a relative dearth of specific cases for which Hieron is called on as witness (and notice, too, how vague both of the above instances really are). In the latter half of the work, however, we see Eugenianus present for a number of much more richly described cases, signalling what must be a deliberate shift in

---

[20] *MM* K10.202. On the importance of the qualification 'surely' in these instances, see below.
[21] οἶσθα δήπου κἀνταῦθα τῶν ἀναισθήτων Μεθοδικῶν… (*MM* K10.213).

rhetorical strategy.[22] Thus, at K10.550, we find a string of case studies for which Eugenianus is said to have been present:

> διὰ τοῦτο οὖν ἡμᾶς ἐθεάσω καὶ τὸν οἰνάνθῃ χρώμενον ἐν τῷ βαλανείῳ κωλύσαντας χρῆσθαι, καθ' ὃν ἔφησε χρόνον ἀνωμαλίας αἰσθάνεσθαι. καὶ τὸν ὡσαύτος ἐκείνῳ μετ' ἐλαίου βραχέος οἴνῳ πολλῷ, καὶ τὸν τῷ σχινίνῳ δὲ χρώμενον ἡνίκα πυκνώσεως ᾔσθετο, καὶ τοῦτον, ὡς οἶσθα, τοῦ λοιποῦ χρῆσθαι διεκωλύσαμεν. εἴρξαμεν δὲ καὶ τὸν εἰληθεροῦντα καθ' ἑκάστην ἡμέραν καὶ τὸν τῇ κόνει χρώμενον. ἑτέρῳ δ' εἰς ἀνωμαλίαν ἀφικομένῳ πυρετώδη τὸ σθοδρὸν τῆς τρίψεως ἐπανεῖναι προσετάξαμεν. ἄλλον δ', ὡς οἶσθα, ἐθεράπευσα θαυμάσαντα πῶς ἑνὶ λουτρῷ μετὰ τρίψεως... ἐχρῆτο δὲ κἀκεῖνος ἐλαίῳ... ὃ καλοῦσιν Σπανόν. ἀφελόντες οὖν τοῦτο... εὐθέως ἀπεφήναμεν ὑγιῆ.

Because of this you saw us prevent the man who was using grape salve in the bathhouse from using it at a time when he said he felt an irregularity. And similarly, the man using that salve made from a little oil in a lot of wine, and the one using mastic salve when he noticed hardening – I prevented this man, as you know, from using it in the future. I also stopped the man who sunbathed every day and the one who used ashes. And I commanded for another man who had come violently into irregular fever to lay off of massage. Another, as you know, was amazed when I treated him with one bath and a massage. For he was using an oil... the one they call Spanish. Taking this away... we immediately proved him healthy. (*MM* K10.550–1)

What clearly connects all these case studies is Galen's intervention to stop each patient from using some treatment or other to which they were habitually turning. What also seems to connect these studies, but on closer inspection does not quite do so, is the presence of our dedicatee. In fact, however, Eugenianus disappears for the cases in the middle, being said explicitly to have been present only for the first and last treatments in addition to one other near the beginning, as follows:

> man who used grape salve: 'you saw us prevent him'
> man who used salve from oil and wine: no comment
> man who used mastic salve: 'you know we prevented him'
> man who sunbathed: no comment
> man who used ashes: no comment
> feverish man who used massage: no comment
> man who used Spanish oil: 'you know I treated'

---

[22] By contrast, Johnston and Horsley 2011: xcvii–xcviii see a stylistic continuity across the two halves, although they do note that case histories are more common in the second half, and they wonder if Galen's voice is not also more authoritative in the second half.

In a sense, this set of cases is a fast-paced microcosm of the structure we have already seen in the *Prognosis* and the *De indolentia*, where Galen frames each treatise with a series of dedicatee testimonials at the start and end, only to have the eyewitness silently disappear without comment for a string of cases in the middle. One may be tempted to see in the present, quickly described set nothing more than omission by concision but, given the precedent of the other texts, where the implications are more deliberately muddied for rhetorical effect, one wonders whether Galen is not calling on his witness in the middle cases of this passage more by rhetorical implication than he is in historic fact.

But as case studies go, all of these so far have been relatively vague. We know nothing at all about the patients: were they young or old? Bilious or phegmatic? For the purposes of simply pointing out Galen's success at medicine, or perhaps for warning future doctors away from certain treatments, such vagueness is perhaps acceptable, but for learning *why* a particular treatment did or did not work in a particular case, it is less effective. By contrast, Galen goes into considerably more detail in other case studies in the *MM*, deliberately offering them, for the purposes of teaching, as παραδείγματα, 'paradigmatic examples'.[23] One such case sees Galen self-consciously reflect on the epistemological status of paradigmatic examples versus eye-witnessing. The differences between these, as Galen handles them, turn out to be less clear in deed than they seem at first to be in word. He begins the discussion by calling to Eugenianus' mind a case that he had been present for: καὶ δὴ παραδείγματος ἕνεκα ἀναμνήσω σε δυοῖν νεανίσκοιν οὓς ἐθεάσω μεθ' ἡμῶν ('For the sake of a paradigmatic example, let me remind you of two youths whom you saw with us') (*MM* K10.608). Galen then goes on to describe their individual physical conditions and constitutions, and how the fevers that they came down with differed. He then breaks the narrative briefly to reflect on the pedagogical and epistemological significance of the use of paradigmatic examples versus witnessing:

> ὁποίαν δ' ἑκατέρῳ τὴν ἴασιν ἐποιησάμεθα καιρὸς ἂν εἴη λέγειν, ἐπειδὴ μάλιστα μὲν χρὴ γυμνάζεσθαι τοὺς μανθάνοντας ὁτιοῦν ἐπὶ παραδειγμάτων· οὐ γὰρ ἀρκοῦσιν αἱ καθόλου μέθοδοι πρὸς τὴν ἀκριβῆ γνῶσιν. ἀμείνω δὲ τῶν παραδειγμάτων ἐστὶν ὧν αὐτόπται γεγόναμεν· ὡς εἴ γε πάντες οἱ διδάσκειν ἢ γράφειν ὁτιοῦν ἐπιχειροῦντες ἔργοις ἐπεδείκνυντο πρότερον αὐτά, παντάπασιν ἂν ὀλίγ' ἄττα ψευδῶς ἦν λεγόμενα.

---

[23] One could legitimately render the Greek word into English as its simple cognate 'paradigms', but that word has come to have an often restricted meaning among modern historians and philosophers of science and so I will avoid it here. See Kuhn 1962.

> This is a good time to say what sort of cure we brought about in each case, since it is especially necessary for students to practice on paradigmatic examples, because general methods are not sufficient for a precise inquiry. But those things of which we are eyewitnesses are better than paradigmatic examples, since if all those who teach or write anything whatsoever had, using their hands, previously proved those things with actions, there would be a lot fewer false statements. (*MM* K10.608–9)

Here again, the context is clearly pedagogical, and there are three teaching tools being discussed. On the one hand, there are what Galen calls 'general methods', which I take to refer to broad constructs for understanding constitutions or how certain medicines interact with certain conditions, for example. At the next level of specificity are the παραδείγματα, the paradigmatic examples, which refer to case studies passed down from teacher to student or culled from medical texts, that are taken as examples that put the general methods to the test. These at least nominally real-world examples would then complement, complicate or challenge the student's theoretical understanding. But so far this is all effectively book-learning. What Galen really lauds in this passage are cases that the student has actually seen himself, what we might call 'real'-real-world examples.

Having extolled the virtues of getting one's hands dirty through personal experience, Galen then proceeds to rail against doctors who try to raise their reputations through social visits rather than doing what Galen has always done, which is to tirelessly practise medicine. Galen then gets down to the business of describing in minute detail the two youths whom Eugenianus had seen with him. Over the course of six Kuhn pages Galen goes into great detail about what exactly happened on each day of each patient's fever, and how he treated them (one common element is the surprising and presumably dangerous induction of fainting, λειποθυμία, through phlebotomy). Curiously, and contrary to his usual practice in such cases, Galen does not belabour the presence of Eugenianus throughout this long passage. In fact, it is only in the opening sentence to this section, 'let me remind you of two youths whom you saw with us', that his dedicatee gets any mention at all until much later. Instead, happy with having established Eugenianus' presence with this two-pronged stake at the outset (*reminding* him of *what he saw*),[24] Galen shifts his rhetorical focus for the next six pages to emphasise very minutely how he cured these two patients through employing what

---

[24] Galen uses a similar move with a case study at *MM* K10.856–7 and at K10.909 he asks Eugenianus to remember the case of the philosopher Theagenes the Cynic, although it is unclear there whether Eugenianus was an actual witness or if this was merely a famous incident.

he had learned from his many experiences of continuous fevers. He completes the case studies by saying that his opponents cannot 'offer any experience or any true argument' for why they do not follow his method of treatment in such cases.[25] This pairing of experience and true *logos* turns out to be a foreshadowing, for having given us an account of these experiences, he now completes the picture by offering us five more pages outlining the theoretical reasons why his treatment by phlebotomy works for continuous fevers, before finally reminding us again of Eugenianus' presence, saying 'as you saw me doing throughout' (ὡς ἐθεάσω διαπαντὸς ἡμᾶς ποιοῦντας) (*MM* K10.619).

And so we find Eugenianus pinning the stories down at both ends, with the stories ostensibly serving to prove a point about the importance of experience over the use of paradigmatic examples in teaching. At the same time, though, whatever good Galen's own experience was meant to be for his own practice as set out in these cases, for the reader these cases are really no more than two new *paradeigmata*, two examples to complement our theoretical understanding and our own experience. But in the very high levels of detail and in the compound witnessing (Galen's and Eugenianus'), could Galen be trying to epistemologically strengthen these case studies, moving them into a cleverly crafted grey zone between common paradigmatic examples and the reader's own witnessing?

Possibly. He clearly does recognise (or hope) that his cases, particularly in a text with such a strong pedagogical focus as the *De methodo medendi*, will serve as *paradeigmata* for future students. And just as in this passage, there is a case in book 8 of the *MM*, where, 'reminding' Eugenianus of yet another case (ἀναμνήσω σε ...), Galen says of the patient that 'this man will suffice, for the sake of clarity, as a paradigmatic example for the argument' (ἀρκέσει γὰρ ἕνεκα σαφηνείας οὗτος οἷον παράδειγμά τι τοῦ λόγου γενέσθαι) (*MM* K10.536). Note that he does not explicitly say that Eugenianus witnessed this other case personally, although Galen does go out of his way to craftily imply something of the sort, preceding it half a page or so earlier with 'you saw, I am sure, some people who were sick in this way' (ἐθεάσω δὲ δήπου ... ἐνίους οὕτω νοσήσαντας), and interjecting three further references in the course of the story to what Eugenianus 'knows' about it (οἶσθα twice and ἐπίστασαι once). But Eugenianus' testimony for this particular case is, strictly speaking, weaker than the previous, not only through Galen's continued insistence on repeatedly modifying the assertions with the adverb δήπου, meaning something like 'surely', 'perhaps' or 'I presume', but also by

---

[25] οὔτ' ἐμπειρίαν οὐδεμίαν οὔτε λόγον ἀληθῆ προστησάμενοι, *MM* K10.616.

Galen's switching of the verb to 'you know' in place of the earlier, stronger, 'you saw'.

A particularly interesting example of such cautious weakening is used for Hieron in book II of the *MM*, where Galen addresses him as follows:

> παραγέγονας γὰρ δὴ μυριάκις αὐτοῖς ὡς ὑπὸ σκορόδων ὄντως καὶ κρομ-μύων τῶν ἐλέγχων ἀναγκαζομένοις δακρύειν· οἶσθα δὲ δήπουθεν ὡς καὶ πολλοὶ πολλάκις ἡμῖν ἐξωμολογήσαντο καταμόνας αἰσθάνεσθαι μὲν ἤδη τῶν κατὰ τὴν σφετέραν αἵρεσιν ἀτόπων.

> You have been present ten thousand times when (other doctors) have been forced to tears by refutations just as they really would be by garlic and onions, and surely you know that many have often confessed to us in private that they now see the inconsistencies of their own sect. (*MM* K10.114)

Here we see simultaneously the strengths and weaknesses of the generalised witness claim laid bare. Hieron is called on to agree to what is strictly speaking impossible, with a clearly exaggerated μυριάκις, 'ten thousand times'. Obviously the number is not meant to be taken literally, but at the same time its force lies in the very scale of the exaggeration. In almost the same breath, however, when Galen comes to finish the story with the claim that the other doctors are forced to tears and to secretly recant their doctrinaire commitments, he once again weakens his witness's engagement with a modifier, δήπουθεν, again to say 'of course', 'surely' or even 'perhaps ... you know'[26] (which in the company of the adverbial καταμόνας, 'in private', should likely lead us to read the ἡμῖν, 'to us', as just another royal we, rather than an assertion that Hieron was actually present with Galen to hear the purported confessions). Still, Galen does try to bring Hieron on board as corroboration of the dicier part of the story, hoping, no doubt, to buttress for his reader what would otherwise be a suspicious-sounding claim that his opponents are actually willing to secretly confess to him the weaknesses of their own sects while they continue to maintain (as he goes on to tell us) those same doctrines in public. Curiously, this use of δήπου/δήπουθεν to soften an ostensible witness claim occurs fairly frequently in the *MM*, but otherwise rarely in the Galenic corpus.[27]

---

[26] Compare also another instance on K10.114.
[27] See *MM* K10.114 (2x), 171, 202, 203, 204, 213, 539, 540, 785, 800. The only two other instances I can find of οἶσθα δήπου in Galen occur at *PHP* K5.3.6, where it is rhetorically (and probably sarcastically) addressed to the long-dead philosopher Chrysippus to show up his shortcomings in anatomy, and at *Nat.Fac.* K2.12, where it is addressed to the reader: 'Surely you know (the differences distinguished by) taste, smell, and sight'. The other instances I have found of a second-person singular verb modified by δήπου occur in (non-testimonial) instructional contexts in the *AA* (K2.608, 623), *UP* K3.548, 816, *Loc.Aff.* K8.364, *Cur.Rat.Ven.Sect.* K11.281, *Comp.Med.Loc.* K12.918 and the *Trem.Palp.* K7.608, as well as a line directed at Aristotle at

In some ways, the qualifications that litter this instance show up the larger contrast between Galen's generally weaker use of witnessing in the Hieron half of the *MM* versus the Eugenianus half. For as it turns out, we have an instance of Eugenianus witnessing a similar confession to this, and one can't help but notice how all the qualifications suddenly drop away, how Galen instead makes a much more explicit claim to a specific event, which he then goes on to tie to the more general claim about doctors recanting and gnashing their teeth. Look at how much more direct and effective this passage is than the previous:

> ἔνιοι δὲ, ὡς ἤκουσάς ποτέ τινος ἐξ αὐτῶν ὁμολογήσαντος ἡμῖν ἰδίᾳ τἀληθῆ μετὰ τοῦ δακρύειν, οὐδὲ τῶν ἐπιτηδείων εὐπορῆσαί φασιν, ἐὰν οἱ νομίζοντες αὐτοὺς ἐπίστασθαί τι θεάσωνται μανθάνοντας. ἀλλὰ σὺ μὲν ἅπαξ ἤκουσας τοῦτο, θεοὺς δ' ἐγὼ σύμπαντας ἐπόμνυμι πολλοὺς ἐμοὶ πολλάκις ὡμολογηκέναι μόνῳ ταῦτα μετὰ τοῦ δακρύειν.

> Some, as you once heard one of them in tears admitting the truth to us in private, say they would not have much in the way of qualifications if those who thought them to be knowledgeable saw them learning something. Although you heard this once, I swear by all the gods that many of them often admitted these things to me with tears in private. (*MM* K10.560)

The contrast with Hieron could not be stronger: where in Hieron's case, Galen makes do with a lukewarm 'surely you know that often' (because you have seen? because you have heard me relate? because you *may have* heard me relate at some point?), with Eugenianus we get a specific instance, explicitly cited and doubly reinforced. 'You once heard one of them', and 'you heard this once' are strong claims to a particular instance of witnessing, and the verb ἤκουσας, used twice, is considerably more pointed than the vaguer 'you know', never mind the temporal markers Galen uses to signal the specificity and uniqueness of the occasion (ποτέ, ἅπαξ). As if this were not enough, Galen goes on to try and underscore his insistence that these events happened not just this one time, but involved 'many' doctors, 'often' (πολλοὺς πολλάκις), which he secures for good measure with an oath, 'by all the gods'. In light of all this, the insistence that the doctors admitted their errors 'with tears' – a claim he makes twice to Eugenianus and once to Hieron – becomes a curious detail. In the Hieron case the claim rings about as true as the clearly exaggerated μυριάκις, 'ten thousand times', and his comparison of their tears to those from onions and garlic may be charming, but

---

Sem. 1.5.20 and one to a fictional opponent at *Aff.Dig.* K5.11. See also *Thras.* K5.877, 879; *Diff.Puls.* K8.680. The closest we come to the uses we find in the *MM* is at *Aff.Dig.* K5.41: γινώσκεις δὲ δήπου καὶ σὺ τοὺς παῖδας, οἷς μὲν ἂν ἡσθῶσι, ταῦτα μιμουμένους, 'you know, perhaps, how children will imitate the things they like'.

it is hardly convincing (how strong was ancient garlic?). With Eugenianus, then, one wonders how Galen expects his readers to understand the other doctors' crying. Surely it is meant metaphorically? Curious if so, given the wider verisimilitude he is striving for in the other details of the story.

Conclusions

Having begun with a pairing of case studies that had much in common in terms of the use of the addressee as authoritative witness, we end with a contrast that shows up, I think, Galen's careful handling of individual witnesses in different situations, particularly with regard to their proximity to the action. As in the *Prognosis* and the *De indolentia*, Galen tries in the *MM* to bring his witnesses bodily into the room whenever possible. But when that is not possible, he tries as often as not to leverage their authority by hinting that they were there, or calling on their knowledge of specific details in, or surrounding, the situation.

This care with which Galen frames his calls on his addressees as witnesses, subtly distinguishing witnessing from hearsay while at the same time blurring the lines as best he can, shows, I think, that Galen wants his contemporary readers (some of whom will be familiar with some of the events, presumably) to take the conventions around dedication and correspondence at face value. His liberal use of the addressee as witness in most of those works where he employs the strategy shows that he wants to try and maximise the epistemological effect of the dedicatee's authority throughout. For all that this would appear to be an effective strategy – particularly in situations where there may well have been competing narratives in circulation – it seems that Galen does not avail himself of it all that often in his massive oeuvre. But in those works where he does use it, by and large he does so with vigour and with frequency. The structural aspects we have seen of his use of this rhetorical device, where he shows a tendency to frame narratives with dedicatee-witnessed events at the beginnings and ends of important or surprising passages or even of whole works, while often (but not always) letting them slide into the background in the middle, may be a way of muddying the waters, allowing him to 'remind' a dedicatee of a series of events that he saw some part of, while hoping to lend the authority of eye-witnessing to the whole.

12 | Anatomy and *Aporia* in Galen's *On the Construction of Fetuses*

RALPH M. ROSEN

'So an account of the soul as constructor of the parts of the body is problematic from any perspective.'

(Galen, *Foet.Form.* 98.3 N)

Epistemological Tensions in Greek Medicine

Medical knowledge was dispensed in a variety of ways in Greek antiquity, each with its own claims to efficacy and superiority. The early Hippocratic authors give us a taste of this variety, if only indirectly through their polemical attempts to stake out their own intellectual and professional territory[1] against a host of antagonists and detractors. In its broadest contours, the basis of their defence lay in a curious, often confusing, blend of inductive and deductive logic about the material world, empiricism and dogmatic theorising.[2] Hippocratic treatises, for example, often invoke their commitment to an *akribeia* ('precision') founded on the evidence of the senses, and a desire to turn the fruits of their investigations into theoretical principles that would be useful for doctors confronting unknown diseases in the future. Their rivals, as they tell us, could offer only theories and practices with no rational grounding, no satisfactory syllogisms to account for cause and effect in sick and healthy bodies.

Galen shared many aspects of this Hippocratic approach to scientific knowledge six centuries later, and his repeated insistence on logical precision in the theory and practice of medicine became a cornerstone of his self-fashioning as a formidable authority on both medical and moral truth. As with Hippocratic writers, Galen's own claims to authority often rested on his assertions that other medical writers (contemporary and historical) did not properly understand logical inference or the proper use of empirical

---

[1] On Hippocratic polemic in general, see: Ducatillon 1977: esp. 89–143; Jouanna 1999: 181–209; and Nutton 2013: 64–71.
[2] Lloyd 1979: 126–68, 1991: 112–20 (on Greek science generally); Jouanna 1999: 245–58; Mann 2008.

evidence,[3] and, indeed, many of his treatises were composed, as he tells us himself in *On My Own Books*, specifically to respond to what we might call 'bad science'.[4] We are more accustomed to the brash, self-confident Galen than to a more humble version of the man,[5] but there were a few subjects which, he was not ashamed to admit, stumped even his rational powers. The most perplexing cluster of scientific and philosophical conundra for Galen concerned the nature of the psyche. He reflects on his *aporia* about the soul in his late, retrospective work, *On My Opinions*, frustrated by his inability to prove its materiality or immateriality, and resigned to a position of agnosticism.[6] For the most part, however, the problem of the soul's materiality was not an obstacle to Galen's agenda as scientific phenomenologist and practical healer. He had plenty to say about the complex interaction between psychological states and physiology without needing to pinpoint the physicality of the soul.[7] But this particular epistemological problem did highlight the limitations of Galen's empiricism and his fondness for description and demonstration. If the soul cannot be seen or touched, a variety of questions about causation cannot be answered: without knowing *what* the soul is made of, for example, exactly *how* can we say it animates a body? Even teleological questions (*why* do animate beings have souls to begin with?) are less easy to answer when we have no sensible evidence of the soul's 'thereness'.[8] When Galen chooses to confront these epistemological limitations in his

---

[3] Barnes 1993, 2003; Hankinson 1991a: xxii–xxxiii, 2008: 165–78; Lloyd 1996a.

[4] E.g. *Lib.Prop.* 1.7–12 on Galen's vehement disagreements with one Martialius, an Erasistratean anatomist: 'and so because of him I wrote six books rather zealously on Hippocrates' anatomy and three on Erasistratus'. See in general: Lloyd 2008: 37–48; Rosen 2010.

[5] Both sides of Galen's professional persona are summed up at Nutton 2013: 234; see also Hankinson 2008a: 22–4 and Mattern 2013: 112.

[6] E.g. *Prop.Plac.* 7.1 (Boudon-Millot and Pietrobelli 2005): ἐπειδὴ τὴν τῆς ψυχῆς οὐσίαν ἀγνοεῖν ὁμολογῶ μοι γινώσκειν (τε οὐ)δ' ἡ θνητή τίς ἐστιν ἢ ἀθάνατος... ('Since I admit that I do know what the substance of the soul is, and whether it is mortal or immortal...') See also Nutton 1999: 161 on this section of *Prop.Plac.*: 'Although Galen frequently declares that questions on the mortality and corporeality of the soul are irrelevant to practical medicine, he keeps returning to them, especially in this treatise. His agnosticism ... enables Galen to withdraw gracefully from any discussion once his position is threatened, and it contrasts him with those who believe strongly in one position or another ... '

[7] See in general, Donini 2008: 185–7; more specifically, see Galen's treatises *The diagnosis and treatment of the affections and errors peculiar to each person's soul, The Capacities of the soul depend on the mixtures of the body* (= *QAM*), now available in Singer 2013, with his remarks at 32–3, 340, n. 19, 346–54 and 358, with n. 63. These two works often seem to be in tension with each other (each with rather different emphases in the debate on how body and soul influence each other), but neither requires a definitive position on the composition of the soul. On Galen's various attempts to locate and define the nature of the soul, see Singer 2013: 33.

[8] Galen does on occasion come close to identifying a materiality of the soul, such as in *De placitis Hippocratis et Platonis* (= *PHP*) 7.3.19–22, where he discusses the *pneuma* of the brain as 'the first instrument of the soul towards all sensations', and implies that he has at least considered

writings, his characteristically strident rhetoric of authority is at least temporarily muted, and allows for a more nuanced, realistic assessment of what his scientific method can and cannot achieve.

Methodology and Epistemology in Galen's *On the Construction of Fetuses*

One place where Galen addresses at some length the problem of the soul in the context of a discussion of scientific method is his work *On the Construction of Fetuses*[9] (Περὶ κυουμένων διαπλάσεως / *De foetuum formatione*).[10] On the face of it, Galen's discussion of the soul here is substantively not much different from passages in other works where he also dilates on the impossibility of knowing the soul's origins and materiality.[11] But what makes this work unusual, if not unique, among the others is that its very subject – the creation of a new human body – demands a direct confrontation with the central questions about the soul in a way that the others do not.[12] Since new human beings must at some point be 'animated' independently of their parents if they are to survive after birth, this implies a moment of 'ensoulment' when the fetus, and then infant, takes on an identity that is as much immaterial and impalpable (as manifest in character or behaviour, for example) as material and perceptible (as in bodily form, qualities or characteristics). No discussion of embryology, ancient or modern, can fully avoid grappling with this moment between becoming and being, and Galen felt duty-bound to register his *aporia* on the matter honestly and not, as we shall see, without some hint of pathos. Indeed, it is striking how Galen calls attention to this *aporia* in *Foet.Form.*, and its interaction with his more characteristic

---

(if only to reject) the idea that this *pneuma* might also be the 'substance [ousia] of the soul'. See also Donini 2008: 184–6.

[9] On the difficulty of translating διάπλασις adequately into English, see Singer 1997: 421, who opts for 'construction' on the grounds that it conveys the sense of structure and process inherent in the Greek word. Nickel's (2001) German 'Ausformung' seems as versatile as Singer's 'construction'. See further on the title, Nickel 2001: 43. Galen refers to this treatise by name in *Prop.Plac.* 3 (Boudon-Millot and Pietrobelli 2005) 174, 16 (the Greek text was unavailable before 2005) as Περὶ διαπλάσεως ἐμβρύων.

[10] References to this work throughout are taken from Nickel 2001, and indicated by Nickel's page and line number + 'N'. Nickel's edition includes the numeration of Kühn's nineteenth-century edition in the margins.

[11] See above n. 7.

[12] Debru 1991: 37 notes that the subject matter of *Foet.Form.* elicits from Galen more examples of 'degrees of plausibility' than most of his other treatises: 'Un traité comme le *De foetuum formatione*, où la doctrine dépasse difficilement le stade de l'hypothèse en raison de la difficulté de fournir des preuves expérimentales, en offre plus exemples que beaucoup d'autres'.

self-confidence about his contributions to scientific truth makes it particularly worthy of close consideration. What we will find, as I will argue, is that *Foet.Form.*, despite its authoritative stance on the need for anatomical knowledge in resolving the fundamental questions of embryological development, ends up an almost elegiac rumination on the very nature of 'scientific' (ἐπιστημονικός) evidence and method. This, too, is hardly unfamiliar territory for Galen in many of his other works,[13] but here the specific need to address the developmental mechanisms of the embryo and the teleology of its ensoulment allow for a particularly revealing discussion of the epistemological problems that such topics involve. *Foet.Form.*, as we shall see, shows Galen often struggling to reconcile two conflicting epistemological perspectives – on the one hand, a self-confident materialist who believes unwaveringly that the combination of sense-perception and logical thinking can lead to genuine scientific progress; on the other, a scientist who realises that the very nature of the questions he has set for himself in this work demands a certain measure of agnosticism that no amount of observation or anatomical dissection can overcome.

As is the case with many of Galen's works, *Foet.Form.* is as much about method as content. The treatise is in one sense 'about' the formation of the fetus, but it was inspired, as he makes clear, by his frustration with the pronouncements of certain philosophers about specific aspects of fetal development. It is obvious enough, perhaps, why a Greek philosopher would be interested in the generation of new human beings, but in the material world this involves biological processes, hidden within the body, and so largely inaccessible to non-specialists. As Galen notes in the opening paragraph, even 'those who dissect carefully' are still ignorant on many matters, so it is even worse when people (i.e. philosophers, Stoics especially)[14] simply rely on their own assumptions (*hyponoias*) without recourse to 'the things that are made apparent by dissection'. But as Gill has pointed out, Galen's stated objectives in this treatise seem somewhat at odds with his ability to accomplish them. Not only is embryology a particularly difficult scientific field to investigate empirically, but Galen seems to highlight this difficulty himself when he discusses in chapter 3 his change of view on the question of

---

[13] See Hankinson 1991b, 1994, 2008a and especially 2009, for accounts of Galen's idea of proper science and scientific method. Hankinson 2009 discusses at length Galen's sophisticated engagement with the epistemological tensions that good science entails, especially when it came to arguing for imperceptible truths from the perceptible. See esp. 223–5 on Galen's synthetic use of the earlier systems of logic (primarily Platonic, Aristotelian and Stoic).

[14] See Gill 2010: 125–8 for Galen's objections to Stoic embryology, specifically its claim that the heart is formed in the embryo earlier than the brain or liver and controls the subsequent formation of other organs.

whether the heart or liver formed first in the fetus.[15] The fact that he had so much trouble settling on the priority of the liver in this debate – a position he ultimately defends here,[16] but largely through indirect, inferential arguments – certainly tempers to some degree, at least, the criticism he levels against philosophers for their poor observational methods. One might even wonder why Galen would choose to take such a strong stand on the importance of anatomy in a treatise on *embryology*, a field of study in which systematic anatomy is so impractical in the first place. The basic technical limitations facing an embryological scientist in antiquity may be obvious enough, but at a more conceptual level the transformation from simple, inert material to complex and animated beings even today is difficult to account for in strictly anatomical terms. It is one thing to observe and describe the organs of a fully developed human body, and from there to draw inferences about their early formation,[17] but quite another to account for that moment when the human body actually comes into being and then develops into an autonomous creature.

So much of *Foet.Form.* dramatises this methodological tension, especially in its pursuit of the argument against the idea that the heart formed first in the embryo and was then responsible for the formation of the other organs. This question was a matter of some concern to Galen, since it was related to other debates, most notably played out in *PHP*, about the physiological and psychological roles of the liver and heart in the developed human body and the life it leads. But we may well ask how much relevance Galen's claims to anatomical expertise actually has to his attempts to sort out the roles of these two organs in the processes of fetal development. Part of the problem in answering this question is that the aims of the treatise itself are not especially clear, and its structure far from straightforward. It is only gradually that the reader begins to understand that Galen's ultimate interests seem to be limited to two interrelated topics – the question of the priority of the heart or liver in the construction of the embryo, and the roles of each organ in its development. Various ancillary questions arise along the way, as we shall see, such as the role of 'nature' in designing the body and driving its initial development, and pervasive throughout is the assumption that anatomical knowledge and sensory evidence are crucial for any attempts to answer them.

---

[15] *De Foet.Form.* 66, 19–24 N. See also Nickel 1989: 80–3 and Gill 2010: 127, with n. 152.
[16] See Gill 2010: 127 on the evolution of Galen's thinking about this issue as it can be traced across his other works.
[17] As Nickel and Gill suspect was Galen's basic *modus operandi* in *Foet.Form.*; see Nickel 1989: 82 and Gill 2010: 128.

By the time we get to the last section of the work, chapter 6, however, Galen's train of thought has led him to a different epistemological realm, increasingly distant from the phenomenology of embryo construction, its material causes and order of events. Instead, the treatise closes with extended musings about a host of far grander teleological questions, such as the question of animation – the role of the soul in the movement of muscles and how they coordinate with one another – and even the substance of the soul itself. On first glance, this strikingly aporetic ending seems somewhat discontinuous with the chapters which precede it, not only in subject matter, but also in tone. As I would like to show, however, what appears on the surface to be an 'honest' concession of intellectual limitation serves, paradoxically, to strengthen the rhetoric of authority that pervades the work up to this point. Indeed, it is precisely the expertise in anatomy that Galen (in his mind, at least) has demonstrated in the work which gives him the degree of comfort necessary to admit his limitations. No one else (he would say) can claim to know more about the anatomy of the embryo than what he has put forth in this treatise, and no one could possibly adduce anything anatomical to explain the impossible questions he brings up at the end. He is representing here, then, the definitive state of the discipline – if *he* cannot work out these questions, with all the anatomical knowledge he does know, then no one else can.

This stance makes the aporetic tone of the final chapter (6), with its probing teleological – even theological – questions, seem quite remarkable, especially given how long Galen sustains this mode. Chapter 6 takes up nearly a third of the treatise and its prolonged shift in focus has a slightly disorienting effect. One might, of course, simply conclude that Galen was simply writing or dictating quickly, free-associating and not worrying much about the ultimate coherence of the work.[18] But even if this were the case, it would hardly in itself diminish the significance of the topics he turns to in the last chapter. As with many scientists, it is the most impenetrable questions that

---

[18] The last sentences of chapter 5 read as if Galen is drawing his treatise to a close. At 90.17 N., he says, 'perhaps everything [the argument to that point] has been stretched out more than the topic warrants...' (ἀλλὰ ταῦτα μὲν ἴσως ἐπὶ πλεῖον ἢ κατὰ τὰ προκείμενα νῦν ἐκτέταται πάντα...). He then summarises his earlier complaint about the arrogance of those, ignorant of anatomy, who maintain that Nature formed the heart first in the embryo, and that this organ is responsible for the formation of the other organs. The opening of chapter 6 begins as if it will make a new start: 'And so if we pass on then to the central topic in this matter, we will show that they have not thought it worthwhile to investigate properly the research of doctors...' (μεταβάντες οὖν ἐπὶ τὸ μάλιστα προκείμενον ἐν τῇδε τῇ πραγματείᾳ δείξομεν αὐτοὺς μηδὲ τὰ πρὸς τῶν ἰατρῶν ἐζητημένα καλῶς ἀξιώσαντας ζητῆσαι).

ultimately attract him most, the ones that remain after he has laid out the area over which he *can* claim expertise.

Empiricism and Inference in Galen's Embryology

One reason why embryology is a particularly fraught area of enquiry is that it is very difficult to approach it strictly phenomenologically. Inevitably, as Galen himself finds, as soon as one lays out what can be known about the construction of the embryo from anatomical investigation, questions of *causation* arise. The embryo begins as an undifferentiated material quantity that develops into a complex creature with highly differentiated parts and systems, according to a predictable teleology. Even ancient anatomy, as Galen is at pains to show, can chart the teleological vector of the developing fetus – certainly in its broadest contours – with some accuracy. But how and why does it all happen? Drilling down to the level of morphogenesis, Galen could observe, for example, a certain order of events and expose, as he does in the treatise, the ignorance of ill-informed philosophers.[19] But, again, one wants to know what accounts for that particular sequence of events, what force or material is actually responsible for the differentiation of organs. The targets of Galen's scorn have their own ideas about causation which Galen disapproves of, but in trying to explain his disapproval he enters a different epistemological realm from where he began the treatise, a realm where anatomical knowledge is only of moderate, if any, use. In arguing, for example, against the proposition that the heart (or the liver)[20] was one of the first things to form in the embryo (πρῶτα γιγνόμενα, 86.15 N), Galen was able to note, from observation, that veins and arteries had to be in place prior to the formation of organs in order to supply them with nutrition.[21]

---

[19] On Galen's 'critical empiricism', see Boylan 1984: 110–12.

[20] Although much of the treatise, as noted earlier, is concerned to argue that the liver formed first of all the organs in the developing embryo, in this passage Galen focuses on what preceded even the formation of these primary organs. As he says, at first the formation of both the liver and the heart 'obviously requires a bloodlike substance for their development, which must necessarily come through the vessels from the uterus' (δεῖται γὰρ δή που προφανῶς εἰς τὴν γένεσιν αἱματικῆς οὐσίας, ἣν διὰ τῶν ἀγγείων ἐκ τῆς μήτρας ἀναγκαῖόν ἐστιν ἀφικνεῖσθαι, 86.16–17 N). Nickel 2001: 147 notes that this does not mean that Galen is saying that the heart and liver formed at the same time, second after the formation of blood vessels; Galen is not, in other words, contradicting his main tenet that the liver and heart formed in succession to one another. See also Gill 2010: 132–3.

[21] His conclusion comes at the end of a long passage in which he analogises (for the second time in this treatise; cf. 68–72 N) the vascular systems of the embryo to the roots and trunk of a tree:

This is a point he makes several times in the work, first at 60–2 N, where he carefully describes how veins and arteries from the uterus coalesce to form an entrance into the *chorion* of the embryo, much as the roots of a plant come together to form a trunk (60.2–15 N). His main point here, articulated at 60.15–25 N, is that actual organs cannot form until the vessels are in place, since 'there is no other material to sustain it apart from what is supplied from the animal which bears it... no part of it [is] capable of formation without this bloodlike substance'. This much accounts for the mechanism by which developing organs can be nourished, and they are conclusions available from anatomical observation of human and animal fetuses.

Galen's curiosity does not end here, however, and he also brings up the difficult question of how the entire process is initiated. Why do the vessels form in the first place, and where does the actual material that makes up the organs come from? These are not questions that can be answered by direct observation, so Galen can only offer what seems to him a 'reasonable' inference, namely that the seed itself (σπέρμα) is responsible for the initial formation of the blood vessels, and then for sustaining them with blood drawn from the uterus. In the midst of Galen's self-confident anatomical bravura, we should note the rhetoric of uncertainty along the way. It is an 'easy inference', he says at 659 (62.9–10 N), to suppose the blood vessels owe their initial formation to the substance of the seed (ἐκ τῆς τοῦ σπέρματος οὐσίας εὔλογόν ἐστι τὴν πρώτην γένεσιν ἐσχηκέναι), and 'surely likely' that the seed nourishes them and, subsequently, the organs (εἰκὸς δήπου τὸ διαπλάσαν αὐτὸ σπέρμα τροφὴν αὐτοῖς ἐκπορίζειν, 62.15 N – emphasis added). Here, and throughout the treatise,[22] Galen imagines the earliest stages of embryonic development as analogous to the growth of a plant seedling, with its roots, trunk and branches, so it is easy enough to understand how he might see the human seed operating analogously to a plant's seed. Indeed, he calls this analogy the 'foundation' (θεμέλια, 62.23 N) of the arguments he will be laying out in what follows, and he opens the next chapter (chapter 3) by announcing that he will consider 'how it is likely that

> the veins outside the embryo leading into the chorion are like the roots of a tree, while the veins inside the embryo, coming in through the naval, and then differentiating into the organs, are like its trunk. These are anatomically observable phenomena; his conclusion that the organs would need a nutritive supply system before they could form relies on the validity of his plant analogy.

[22] E.g. 58.23 N, 66.25 N, 68–70 N, 74.13 N, 72.8 N. See also Galen's theoretical justification for the plant-seed-human sperm analogy at *Sem.* 1.9.18 = 94.24–96.21 and 2.4.32–6 = 178.10–15 (De Lacy 1992). For the Hippocratic and Aristotelian background to the analogy, see De Lacy's note on *Sem.* 94.11–14 (De Lacy 1992: 218). See also Nickel 2001: 121–2.

the embryo is entirely formed by the power of the seed', and reaffirming his commitment to observational anatomical research (... ὅπως εἰκός ἐστι τὸ κυούμενον ὑπὸ τῆς κατὰ τὸ σπέρμα δυνάμεως ἅπαν ἐφεξῆς διαπλασθῆναι, τὴν ἀρχὴν τῆς εὑρέσεως αὖθις ἀπὸ τῶν κατὰ τὰς ἀνατομὰς ὁρωμένων ποιησάμενοι). While the statement seems at first as self-assured as before, it once again juxtaposes a creeping sense of uncertainty (εἰκός) with confidence that an empirical approach will demystify the earliest development of the fetus. The part that immediately follows launches into an impressively detailed description of the juncture of vessels at the navel of the embryo, and how from there the various organs begin to take shape (62–4N). But in fact he leaves behind any discussion of the 'power of the seed', presumably because this cannot be one of the ὁρώμενα he focuses on in this section. Nothing in Galen's anatomical observations of the embryo's vessels and organs, no matter how meticulous, can actually address the ultimate questions of causation that also interest him here. But the amount of detail he offers from his own study, largely of aborted human fetuses and animal dissection (64.25–66.18 N), distracts readers for the moment from questions of *how* and *why* an embryo can form from the 'planting' of a human seed in the uterus in the first place.

At 68–70 N, in the course of arguing that the liver forms before the heart in the developing embryo, Galen pauses to offer more detail on the plant analogy, since he is intrigued by the fact that, like plants, human embryos seem to have only nutritive needs in the earliest stages. They have no need, in other words, of an independent heart or pulse, nor of brain functions (ἅτε γὰρ ἀμίκτου τε καὶ μόνης οὔσης αὐτῆς, ὡς ἂν μήτε τὸ θυμοειδὲς ἐχόντων μήτε τὸ λογιστικόν, 21–3 N) so, he says, there is hope that the management (διοίκησις) of plants will turn out to be 'pure and untainted' (εἰλικρινῆ τε καὶ ἀνόθευτον). He proceeds to suggest that this is, indeed, the case, and at 667 he says directly that the 'embryo has the same management as that of plants in the first phase of formation'. In its earliest stages the embryo must be managed by the seed, which manages its nutrition and directs its gradual development. Galen's notion of the 'power of the seed', then, does little more than state a point from which the developmental processes begin. Clearly something happens when one places a seed in warm, moist soil, as Galen describes in some detail in 3.15–17 (68–9 N), just as something happens when the human seed is implanted in the uterus. This 'something' is the *dynamis* Galen keeps referring to, although it is only *inferred* from our ability to see that somehow the processes are, as he says, 'managed' (διοίκησις, 70.13 N) – they follow a predictable pattern, an order of events. Dissection and observation may show us the results of these processes, but cannot, in fact, tell us anything definitive about the logic or physiological mechanism behind them.[23]

The Limits of Anatomical Observation

When observation is itself inconclusive, the problems become even more intractable. Galen had to confront this challenge to his observational powers when trying to determine whether the heart, liver or brain formed first in the young embryo. It was a critical question for him to know the order of their formation for a number of philosophical reasons,[24] but, as he notes at 64.25–66.18 N, it is impossible to recognise articulated organs in the earliest phases of development (he invokes Hippocrates' famous description of an aborted six-day-old fetus which mentions a round red formation)[25] and by the time the fetus is thirty days old, the three organs are all recognisable, so it is impossible to tell which formed first. As he concludes at 66.11–12 N: 'it is not... possible to discover the exact time when the heart takes on the first stage of its construction... as long as the embryo is unarticulated' (οὐ μὴν ὁπηνίκα γε πρώτην ἡ καρδία τὴν ἀρχὴν τῆς διαπλάσεως ἔχει, δυνατὸν εὑρεῖν... ἔστ' ἂν ἀδιάρθρωτον ᾖ τὸ κύημα).

The candour of this passage about the limits of anatomy in solving these specific problems is rather disarming in the face of the explicit polemic of the treatise as a brief for the necessity of anatomical observation. When Galen admits in the ensuing section (66.19–32 N) that he had changed his mind on the question of the priority of the liver and heart in the formation of the fetus,[26] the epistemological conundra he was wrestling with

---

[23] See Gill 2010: 133–5 on Galen's criticism of the Stoic-Aristotelian notion of the heart as the organ responsible for the formation and management of the embryo, and his attempt to locate these functions in the logos of the seed.

[24] Chief among these being the hierarchy of functions ascribed to each organ, and his ongoing dispute with (most of) the Stoics on the role of the heart and brain in the psyche (i.e. which organ should be properly considered the *hegemonikon*, the rational 'ruling part', and which the seat of the passions). See Donini 2008; Gill 2010: 58–62; Tieleman 1996: 38–60. Gill notes, however, that Galen worries less in *Foet.Form.* about 'advanced or complex psychological faculties, such as being the seat of appetite or anger and other emotions' than in other works, such as *PHP*, simply because they 'are not relevant to the undeveloped state of the embryo'.

[25] Despite his observation that an unarticulated embryo can tell us nothing about the exact chronology of the formation of the organs, this does not stop him from stating that 'the thing which Hippocrates says appears to be round, and red inside the chorion of a six-day embryo, has to be the liver in an unarticulated and undeveloped state (ἀδιάρθρωτον ἔτι καὶ ἀδιάπλαστον)' (66.5–7 N).

[26] For details of Galen's change of opinion on the formation of the three main organs (with references to his earlier treatment of the question), see Gill 2010: 127–8.

12 Anatomy and Aporia in Galen's On the Construction of Fetuses

become clear. His grounds for writing in an earlier treatise (*De semine*) that the heart (and liver) formed early around the same time were, again, entirely inferential – he says there that he was 'led to this calculation ... ' (ἐπὶ τὸν λογισμὸν τοῦτον ἀγόμενος) by the observation that the heart was so important for a fully developed human being. Now, swayed by the *communis opinio* (also somewhat unusual for Galen to admit) that the διοίκησις of the embryo has no *need* of a heart early on, since in this period it behaves more like a plant, he has convinced himself that the liver must form first. As Gill notes, 'Galen's admissions of these variations in his own view ... might lead us to be rather sceptical about the validity of Galen's objectives in this essay', especially since 'his own account remains largely conjectural and is only indirectly supported by anatomical observation of fully formed animals'. It is unlikely that Galen would be oblivious to such an objection, and he would presumably reply that he never claimed that anatomy alone could solve *all* the questions raised by the mysteries of the developing embryo anyway. He would doubtless point out, moreover, that the λογισμός that led him to these positions consisted of inferences drawn initially from empirical observation (e.g. his argument that the embryo does not need a pumping heart early on, but it *does* need a source of nutrition, which the liver would provide). The rest of the passage, in fact, returns quickly to anatomical arguments as a means of corroborating his conclusions about the priority of the liver over the heart: 'I [then] sought [to figure out] *how* this [the formation of the heart *after* the liver] happens' (ἐζήτουν δ', ὅπως γίγνεται τοῦτο, 3.8 = 66.28 N, emphasis added). Investigating *how* the heart could be formed from the liver, in other words, involves accurate observation of the stages of articulation, as he proceeds to show. Any trace of the diffidence we saw in 3.5–9 (64.25–66.18 N) is immediately abandoned in 3.9ff. (= 66.19ff. N), as Galen launches into a μακρὸς λόγος based on his anatomical research. He effectively relaunches here the self-confident rhetoric with which he opened the treatise, censuring philosophers in particular for their ignorance of animal dissection 'skilfully undertaken on living creatures to expose the parts deep within ... ' (τὰ κατὰ τὴν ἐπὶ τῶν ζώντων ἀνατομὴν ἐγχειρουμένην τεχνικῶς εἰς γύμνωσιν τῶν ἐν τῷ βάθει μορίων, 3.12 = 68.2–3 N).

In the ensuing 'long account' of the formation of the heart (68–76 N), Galen takes clear delight in showing off his anatomical expertise, but his observations are systematic and precise, and supply him with evidence to support his conclusions about the formation of the organs. He has, for example, carefully observed the formation of what we would call the superior and inferior *venae cavae*, major veins returning blood to the heart from upper and lower parts of the body respectively,[27] and from this material he thought it made sense for the heart to form once the embryo outgrew its initial plant-like state. Once again, Galen seems confident enough about an order of events, but somewhat more reserved about how it all happens. He has observed that there comes a point when the embryo needs a pulse, but he can only attribute this to its inherent 'management' (διοίκησις):

> ... ὥστε τὸ κυούμενον οὐ μόνον ὡς φυτὸν ἔτι τὴν διοίκησιν ἔχειν, ἀλλ' ἤδη καὶ ὡς ζῷον, ὁποῖα ζῷα χῆμαί τ' εἰσὶ καὶ κήρυκες καὶ πίνναι καὶ ὄστρεα καὶ λοπάδες, ἤτοι γ' ὀλιγίστης, ἢ οὐδ' ὅλως δεόμενον κινήσεως σφυγμικῆς.

> ... with the result that the embryo's management is no longer merely plant-like, but now like that of an animal – animals such as clams, trumpet-shells, pen shells, oysters, or whelks, which really require very little movement of the pulse, if any at all. (74.13–16 N, tr. Singer 1997, emphasis added)

As this chapter (3) draws to a close, Galen relies on his preceding anatomical *tour de force* to summarise forcefully his conclusions: the liver comes first, followed by the heart; the liver is closer to the mother's womb than the heart, and the brain is set even further away than the first two. But it is noteworthy, too, that the dramatic rhetoric of this section relies again on mostly inferential reasoning in its attempt to explain the teleology behind the διοίκησις of the embryo. Consider his remarks about the position of the developing brain in the embryo:

> ... ὡς ἂν καὶ τῆς κατασκευῆς αὐτοῦ γενησομένης ὑστέρας, ὅτι μηδὲ χρῄζει τι τὸ κυούμενον ζῷον ἐγκεφάλου διὰ τὸ μήθ' ὁρᾶν αὐτὸ δεῖσθαι, μήτ' ἀκούειν, μήτε γεύεσθαι, μήτ' ὀσφραίνεσθαι, καθάπερ οὐδὲ τοῖς κώλοις ἐνεργεῖν, οὐδ' ὅλως ἑτέραν τινὰ προαιρετικὴν ἐνέργειαν ἢ τὴν τῆς ἁφῆς αἴσθησιν ἔχειν, ἢ φαντασίαν, ἢ λογισμόν, ἢ μνήμην. ὕστερον οὖν ποτε κατὰ τρίτην τάξιν χρόνου ἐγκέφαλός τε καὶ τὰ κατὰ τὸ πρόσωπον ἅπαντα διεπλάσθη, καθ' ὃν καιρὸν ἤδη καὶ τὰ κῶλα διηρθρώθη, καὶ πᾶν εἴ τι πρόσθεν εἴρηται μόριον εἰς τελειότητα τῆς ἑαυτοῦ κατασκευῆς ἀφίκετο.

> That part [the brain] is constructed later, *as the embryo has no use for a brain*: it requires neither the faculty of sight, nor that of hearing, nor that of taste, nor that of smell; nor does it need to work with the limbs, nor *to enjoy the use of any other voluntary function, nor the sense of touch, nor imagination, ratiocination, or memory*. Therefore, the brain and everything about the face, is constructed later on – third in chronological order. At this time too the limbs are articulated, and all the parts mentioned above which

---

[27] On Galen's vascular anatomy in this section, see Nickel 2001: 128–30.

conduce to the perfection of its constitution. (76.13–20 N, tr. Singer 1997, emphasis added)

In the opening sentence of this passage, the first clause ('that part [the brain] is constructed later... ') is something Galen could have observed, but the second ('the embryo has no use for a brain') can only be assumed *because* we cannot observe any evidence to the contrary. Galen's train of thought here is intuitive enough – if all the organs of sense have not fully formed yet, it is reasonable to assume that the embryo cannot use them. The assumption that the developing embryo has no need of 'imagination, ratiocination, or memory' is perhaps less obvious, though still understandable. But none of these assumptions will lead logically to the conclusion, as Galen would have it, that the brain, and indeed the entire head, must be formed 'third in chronological order'.[28] Galen still hungers to know the secrets of the embryo's 'management', but anatomy can only give him phenomenological data, not reasons, first causes or teleology.

As we have already seen, Galen was not unaware that his approach had its epistemological limitations, but at least his assumptions and premises relied on sensible *phainomena*, not, as in the case of his opponents, arguments that proceed from things that cannot be seen (*adela*).[29] The cumulative rhetorical effect of his vigorous and repeated insistence in the first five chapters on the importance of empirical evidence for the scientific study of embryology is to leave an impression of incontestable authority. For Galen, this will not seem especially surprising – one might even call this his standard rhetorical *modus operandi*.[30] What makes this stance especially interesting in the case of this treatise, however, is how it contrasts to the palpable *aporia* of the final section (chapter 6), where, as it seems, Galen has reserved discussion of all the most fascinating, but most difficult, questions of causation and teleological intention. Indeed, this chapter sometimes feels like a compendium of questions, particularly all those remaining after one has gone as far as one can go with anatomical work.

---

[28] See also Galen's further remarks in the subsequent paragraph about how little the embryo needs a brain, 78.2–5 N.

[29] In 80.3–6 N, Galen again complains of those 'who omit to begin their demonstrations from the specific nature of the subject under discussion, and attempt proofs which take non-evident assumptions (*adêla*) as their starting-points' (tr. Singer 1997). He proceeds to censure the Stoics for weighing in on these topics without knowing anything about anatomy, 'which seems strikingly useless for the purpose of reaching their goal...'. See Nickel 2001: 139 for discussion and similar references elsewhere in Galen (e.g. *PHP* 2.2.5, 104.6–15, De Lacy 2005).

[30] For a subtle discussion of Galen's self-fashioning rhetoric, see von Staden 2009.

If we examine the opening of chapter 6, it becomes apparent that Galen is looking to take a slightly different turn with his topic:

> Μεταβάντες οὖν ἐπὶ τὸ μάλιστα προκείμενον ἐν τῇδε τῇ πραγματείᾳ δείξομεν αὐτοὺς μηδὲ τὰ πρὸς τῶν ἰατρῶν ἐζητημένα καλῶς ἀξιώσαντας ζητῆσαι, ἀλλ' οἰομένους, ἐὰν εἴπωσιν ὑπὸ τῆς φύσεως διαπλάττεσθαι τὸ κυούμενον, εἰρηκέναι τι <u>πλέον ὀνόματος ἅπασι συνήθους</u>. οὐδεὶς οὖν οὕτως ἐστὶν ἐνεὸς ὡς μὴ νοεῖν εἶναί τινα τῆς τοῦ κυουμένου γενέσεως αἰτίαν, <u>ἣν ὀνομάζομεν ἅπαντες φύσιν ἀγνοοῦντες αὐτῆς τὴν οὐσίαν</u>. ἐγὼ δ' ὥσπερ ἔδειξα τὴν κατασκευὴν τοῦ σώματος ἡμῶν <u>ἄκραν ἐνδείκνυσθαι τοῦ ποιήσαντος</u> αὐτὸ <u>σοφίαν</u> τε ἅμα καὶ <u>δύναμιν</u>, οὕτως εὔχομαι δεῖξαί μοι τοὺς φιλοσόφους τὸν διαπλάσαντα, <u>πότερον θεός τίς ἐστι σοφὸς καὶ δυνατὸς ὡς</u> ἐννοῆσαι μὲν πρότερον, ὁποῖον ἑκάστου ζῴου προσήκει κατασκευασθῆναι τὸ σῶμα, δεύτερον δὲ τὴν δύναμιν αὐτοῦ, καθὰ προὔθετο κατεσκεύασαι, ἢ <u>ψυχή τις ἑτέρα παρὰ τὴν τοῦ θεοῦ</u>·

> If, then, we turn to the principal subject of this treatise, it will become clear, not only that they have not seen fit to make a proper enquiry into the researches of doctors, but also that they believe that in describing the embryo as constructed by Nature they are making *some utterance which amounts to more than a commonplace*. Surely no one is so stupid that he won't think that there is such a thing as the cause of formation of the embryo, *and that we all call this cause 'Nature', without knowing its substance*. Now, I have shown that the structure of our bodies *manifests to an incredible degree the intelligence and power of the one who made it*; and so I would hope that the philosophers could show me the identity of the maker. Is it a wise and powerful god, who has considered in advance how each animal's body should be constructed? And second, what is this power of his, by which he carried out the design he envisaged? *Is it some other soul apart from that of the god?* (90–1 N, tr. Singer 1997, modified, emphasis added)

Galen begins the paragraph with customary bravura. Getting right to the heart of the matter, he levels two charges against his philosopher-opponents – they ignore the empirical research of doctors, and then think it is sufficient simply to declare that the 'fetus was formed by Nature' – a statement that, as Galen says, is unobjectionable, but in itself says little. The first objection has been made many times since the beginning of the treatise, but the second opens up a line of enquiry that Galen has repressed until this point. *Who* or *what* is in fact responsible for the formation of the embryo? As we have noted earlier, this is not a question that anatomy can answer, and the best Galen has been able to offer so far has been his empirical data as a foundation for reasonable speculation. But here he admits that no one, in fact, knows what *physis* actually is, materially speaking (ἀγνοοῦντες αὐτῆς τὴν οὐσίαν), so simply declaring *that* nature is the 'cause' of the fetus says nothing other than that people have agreed to use the term *physis* as a name for this cause (ἣν ὀνομάζομεν ἅπαντες φύσιν). The next sentence reads as something of a challenge: Galen has done *his* part by offering evidence in his anatomical writings of the 'amazing intelligence' and 'power' of the 'creator' (τοῦ ποιήσαντος [σῶμα]), so now it is up to the philosophers to tell *him* something about what this creator actually is (ὥσπερ ἔδειξα ... οὕτως εὔχομαι δεῖξαί μοι τοὺς φιλοσόφους). Galen wants answers to the most fundamental questions: is the creator a god? Does it have a plan in advance for the construction of bodies, or does all this happen by some kind of ψυχή acting apart from a god?[31]

Galen's train of thought in the ensuing section, which draws the treatise to a close, is not at all straightforward, and seems to reflect his agitation at the complexity of the questions he has taken on. Throughout he seems intent on acting as if his expertise in anatomy has given him the upper hand in any discussion of such questions. It is anatomy, after all, which has convinced him, as he says at 92.6 N, of the 'intelligence and power' of one who made human bodies, and also that this intelligence has seen to it that the fetus develops 'towards an excellent end' (εἰς χρηστὸν τέλος, 92.15–16 N). But *how* this process actually works, and *why*, still seems to bother him, and he feels the need to explore some of the questions that his position might raise, even though this exploration ends up exposing a considerable amount of intellectual frustration. What we find in these concluding sections is an increasingly aporetic rhetoric, and a strikingly frank admission of intellectual defeat as he attempts to confront a nexus of issues that arise from asking the simple question, 'what is the cause (*aition*) of the construction of the fetus?'

The fundamental challenge in answering this question, as Galen lays it out, is deciding whether the 'cause' of the fetus is random or intentional and by design. This turns out to be a far more complicated question than might first appear, since Galen's discussion shows that discovering the cause of the formation of the fetus has consequences as well for the fully formed human body throughout its life. That is, if we are asking why the parts of the fetus develop as they do from the moment of conception and who is responsible for this, we are by implication asking how we can account

---

[31] Nickel 2001: 152 *ad loc.* notes the peculiarity of Galen's question, ... ψυχή τις ἑτέρα παρὰ τὴν τοῦ θεοῦ ('finden sich bei Galen keine Parallelen'), but explains that Galen seems to have in mind here the kind of cosmic soul that that he mentions more explicitly at 104.25–6.1 N ('die über den ganzen Kosmos hin ausgebreitete Seele').

for all movement in the body once the organs are formed and the body moves through time. The purpose (*telos*) of the formation of organs, in other words, anticipates the purpose of their functioning in the body after formation. This explains why in this section Galen continually uses examples drawn from his anatomical research to speculate about what is happening as the fetus develops. If we could properly account for voluntary and involuntary movement of muscles, for example, we would make some progress in understanding how and why these muscles came to be formed as they were in the fetus, since that initial formation had to come about with some purpose – a purpose which is only inchoate in the fetus, but fully realised in the body after birth and during its life.

Galen had little patience for the Epicurean view that the fetus forms without some sort of reason or skill to guide it, so convinced was he by his anatomical research that it must develop purposively towards an 'excellent end' (… κατά τινα κίνησιν ἄλογόν τε καὶ ἄτεχνον εἰς χρηστὸν τέλος ἀφικνεῖσθαι τὴν διάπλασιν τῶν ἐμβρύων). He took more seriously, however, the idea that some power (Nature, the gods) constructed the parts of the fetus and set the process of development in motion, but played no role itself after getting things underway. Galen likens this process to mechanical devices in a theatre, which operate on their own after the designer sets the first device in motion. The 'mechanical device' analogy has its attractions, it seems, because it allows for a knowing and skilled engineer at the beginning, but does not require rational oversight of every movement after the mechanism has been set in motion. At 92.25 N, however, he registers scepticism that such a mechanism would be able to produce an organism as complex as the body without making a mistake (a mechanism that involves 'such a great number of movements transferred from one to another'). What bothers Galen as well about this model is the idea that an 'irrational substance' (οὐσία ἄλογος) could be responsible for the succession of movements in a manner he describes as 'skillful' (τὴν ἀκολουθίαν … τεχνικήν). The reason people hold such a view, he claims, is because none of them has ever actually investigated how the parts of the body function (τί δὲ τοῦτ' ἔστι τὸ κατὰ τὰς ἐνεργείας τῶν μορίων, 6.6 = 94.5 N). Here again, we see Galen parading his authority as an anatomist and implying that a proper knowledge of anatomy would repudiate the 'mechanical device' analogy. He even ends this paragraph by stating that he will adduce two examples (the functioning of the hand and the tongue) 'for the sake of clarity' (ἕνεκα … σαφηνείας). In fact, what these examples end up revealing is Galen's intense *aporia*, not any kind of clear repudiation of the various accounts of the purpose and functioning of parts that he examines.

As we will see, Galen has carefully crafted this section to establish that, even though he will not be able to answer satisfactorily the questions that he has taken up in this passage, there is a kind of authority even in his *aporia* since only he has the expertise to know what epistemological limitations they reveal.

The argument is somewhat convoluted, but proceeds as follows: in the case of the hand and the tongue, a person can move each part according to his or her desire without knowing which muscles are responsible. The tongue is a particularly complex organ, and Galen even notes that professional anatomists do not fully understand the exact role of all the muscles involved in its movement (94.18–19 N). But, as with the hand, we somehow know how to move it, sometimes forming wholly new sounds and words in imitation of things we hear with no knowledge of how we are able to do so. While some anatomists explained this phenomenon by positing that each muscle is an animal, Galen finds it incredible that there would be so many individual 'animals' operating this way in a body (96.4 N). We see Galen in this section further working through his own intellectual struggles, entertaining and then abandoning various theories about voluntary and involuntary movement of body parts (96.14–8.6 N). In the end he judges that question impenetrable (τῶν ἀπορωτάτων ἐστίν, 96.29 N), and remains unsatisfied by any of the various theories he had heard from others to account for both the construction of the body *and* the use of its individual parts. As he concludes, the very attempt to account for the soul as constructor of the body is inherently fraught, if not impossible (ἄπορος οὖν ὁ περὶ τῆς διαπλασάσης τὰ μόρια ψυχῆς λόγος, 98.3 N), and the rest of the treatise becomes, in fact, an extended meditation on this *aporia*.

Epistemological *Aporia* and the Heroics of Intellectual Defeat

Galen reiterates his certainty that bodies have been created by a 'very intelligent and powerful craftsman' (σοφωτάτου τε καὶ δυνατωτάτου δημιουργοῦ, 98.25 N) – he cannot otherwise begin to explain the incredible complexity of the human body, and thinks that anyone approaching the matter with an unbiased mind (ἐλευθέρᾳ γνώμῃ, 98.5 N) would agree.[32] Galen may be comfortable enough with this conclusion, but he also realises that simply assuming *that* a 'craftsman' must be responsible for the creation of bodies

---

[32] On Galen's 'demiurge' in the context of 'theology', see Frede 2003; for an overview of Galen's notion of a demiurge within other ancient teleological debates, see Sedley 2007: 239–344.

does not get us very far in addressing the most important questions about the origins and teleology of human life – he still wanted to know *who* the craftsman is, how he goes about constructing and energising bodies, and even why. Galen's general interest here in bridging the metaphysical and the medical is familiar enough from his other works, but the protracted description of his attempt in 98.19–100.13 to find some satisfying, scientific answers to his questions is here unusually fraught and idiosyncratic. A closer examination of his rhetorical strategy in this passage shows how Galen attempts to maintain complete intellectual superiority over all others even as he concedes utter *aporia* in the face of these most fundamental questions. Taking the question of the nature of the craftsman to a group of philosophers, Galen writes:

> καὶ μαθητήν γε ἐμαυτὸν ὑποβαλὼν ἑνὶ τῷ πρώτῳ, καθ' ὃν ἐν γεωμετρίᾳ τρόπον ἀποδείξεις ἠκηκόειν, οὕτως ἤλπιζον ἀκούσεσθαι καὶ παρ' ἐκείνου· γνοὺς δ' αὐτὸν μὴ ὅτι γραμμικὰς ἀποδείξεις, ἀλλὰ μηδὲ ῥητορικὰς πίστεις λέγοντα, μετῆλθον ἐφ' ἕτερον, ὃς καὶ αὐτὸς ἐξ ἰδίων ὑποθέσεων ἐναντία τῷ πρόσθεν ἀπεφαίνετο, καὶ τρίτην γε καὶ τετάρτην πειραθεὶς οὐδενός, ὡς ἔφην, ἄμεμπτον ἀπόδειξιν ἤκουσα. λυπηθεὶς οὖν ἐπὶ τούτῳ μεγάλως, ἐζήτησα μέχρι κατ' ἐμαυτὸν εὑρεῖν τινα λόγον ἰσχυρὸν ἐπὶ ταῖς τῶν ζῴων κατασκευαῖς· εἶθ' εὕρισκον οὐδένα. τοῦτό γε αὐτὸ διὰ τοῦδε τοῦ γράμματος ὁμολογῶ, παρακαλῶν τοὺς περὶ ταῦτα δεινοὺς τῶν φιλοσόφων ζητήσαντας, εἴ τι σοφὸν εὑρίσκοιεν, ἀφθόνως ἡμῖν αὐτοὺς κοινωνῆσαι.

And so I presented myself as a student [to one of them] first of all, in the hope of hearing from him proofs of the same sort as I had learned in geometry. But when I realised that, so far from producing geometric-style proofs, he could not even utter rhetorical probabilities, I moved to another; he too began from his own personal assumptions, proceeding to prove the opposite to the previous philosopher. I tried a third, too, and a fourth; and from none of them, as I have said, did I hear a flawless demonstration. Much grieved at this, I sought on my own resources to find a watertight argument regarding the making of animals. But I found none. I admit this fact in the present treatise; and I call upon the best philosophers engaged on this matter, if they find some clever solution, to share it with us without jealousy. (100.3–13 N, tr. Singer 1997, modified)

Galen's narrative of his consultation with four philosophers has a veneer of humility about it – he offers himself as a *student* (μαθητήν), after all, hoping, he says, to hear from them logical demonstrations with the robustness of a geometrical proof, and he repeats his desire two more times in the next sentence (γραμμικὰς ἀποδείξεις... ἄμεμπτον ἀπόδειξιν). It is difficult to avoid the sense, however, that Galen is being disingenuous and ironic in portraying himself here as the eager student waiting for enlightenment at the feet of the philosophers.³³ The philosophers come across in his account as utterly incompetent and useless in these debates, with nothing whatsoever to contribute to enquiries of this sort. They cannot even offer 'rhetorical likelihoods' (ῥητορικὰς πίστεις), he says, and instead rely on 'personal theories/assumptions' (ἰδίων ὑποθέσεων). Galen is looking for something, in short, that philosophers just cannot offer, and, given his own vast philosophical training,³⁴ it is hard to believe he would not have known this before he approached them. Indeed, one wonders if he has simply fabricated the entire scenario for dramatic effect.³⁵ The image of Galen 'highly distressed' at his inability to get good answers from the philosophers (λυπηθεὶς οὖν ἐπὶ τούτῳ μεγάλως) has a ring of sarcasm about it, and sets the stage for him to present himself as going it alone as a medical man.

The overall effect of this passage, then, is shrewdly strategic: Galen can openly admit his *aporia* on the subject – he tried to come up with a 'solid argument' (λόγον ἰσχυρόν) on the construction of bodies on his own (μέχρι κατ᾽ ἐμαυτὸν), but found nothing – while still presenting himself as better equipped than anyone else for further pursuit of the problem. His parting shot to the philosophers comes in the form of a challenge (παρακαλῶν), an entreaty to them to share without jealousy (ἀφθόνως) any clever insight

---

³³ Plato, *Apol.* 21b–3b may lie in the background of this anecdote. Socrates there recounts his attempt to disprove (and then to explain when his attempt fails) the Delphic pronouncement that there was no man wiser than himself. He begins by questioning Athenians reputed to be *sophoi*, assuming he can find a wiser person among them, but moves on to poets and artisans when he cannot. I thank Christopher Gill for this reference. On Socrates as a role model for Galen in other contexts, see Rosen 2008.

³⁴ See, e.g., Barnes 2003; Frede 2003: 73–81; Hankinson 2003 and 2008a.

³⁵ See Nickel's comments about the philosophical background to this passage (2001: 161): 'Wie zuverlässig Galen das philosophische Lehrgut referiert hat, ist fraglich. Zumindest können in der wiedergegebenen Lehre, da sie als Meinungsäusserung mehrerer Philosophen ausgewiesen ist ... divergierende Einzelheiten unterschiedlicher Herkunft zusammengefasst worden sein. Dass sich Galen bei seinen Anspielungen auf das ihm vorgegebene Material dessen gedanklichen Kontext bis ins einzelne klargemacht hätte, ist unwahrscheinlich.' There are striking parallels also with another second-century figure, Lucian, whose *Icaromenippus* begins with the philosophical *aporia* (cf. 10.11) of its narrator Menippus. Menippus describes in ch. 5 how he 'selected the best of the philosophers' to teach him how to converse about astronomy and cosmology (ἠξίουν μετεωρολέσχης τε διδάσκεσθαι καὶ τὴν τῶν ὅλων διακόσμησιν καταμαθεῖν). Instead, however, he complains that they threw him into even greater *aporia* (εἰς μείζους ἀπορίας φέροντες ἐνέβαλον) with their conflicting and contradictory opinions. Branham 1989: 15–16, notes the Pyrrhonian sceptical background to this scenario, but points out that 'unlike the Pyrrhonists, Menippus persists in his desire to find something that truly can be taken seriously ... and refuses to rest content with the contradictory picture he has found' (16). See also Branham 1989: 224–5, n. 10.

they may discover on the matter. The message is slightly oblique, but clear enough: philosophers cannot help us with these questions, and even if they could, their petty squabbles might well keep them from sharing their knowledge. The challenge, therefore, is rather empty, since Galen knows that if any progress is to be made, it can be made only by someone who actually understands anatomy and biology.

Galen drives this point home implicitly in the topic he introduces immediately after his challenge to the philosophers, for he returns here to the kinds of questions raised by the movement of the tongue and the formation of words, that is, how a person can form sounds with the tongue without actually knowing what muscles and nerves are involved. These are not questions that a philosopher would know how to handle, and the abrupt turn in his discussion from the metaphysical abstraction of a cosmic demiurge to a child forming the words like σμύρνα with his tongue seems to highlight Galen's intention to parade his expertise. It is the scientific study of anatomy, as he makes clear, that allows him to ask the proper questions about the soul, a process that begins with observation of the material world through the senses, not *hypotheses* or unsupported conjectures. So it is the mention of a child's tongue that leads Galen to formulate his quandary over whether the divine craftsman who made the parts of the body ('whoever he is ... ') remains in the parts after their construction or somehow exists apart from them and animates them in some other way. Either option is full of problems, as Galen realised, and the end of this section (100.29 N) finds him quite at his wits' end. Unable to accept either that body parts might have individual animating souls or that a single 'common soul' (μία κοινή) manages all of them, he admits his *aporia* (εἰς ἀπορίαν ἔρχομαι) in a passage that lays bare Galen's struggle quite openly:

> εἰς ἀπορίαν ἔρχομαι μηδ' ἄχρι δυνατῆς ἐπινοίας, μήτι γε βεβαίας γνώσεως εὑρίσκων τι περὶ τοῦ διαπλάσαντος ἡμᾶς τεχνίτου. καὶ γὰρ ὅταν ἀκούσω τινῶν φιλοσόφων λεγόντων τὴν ὕλην ἔμψυχον οὖσαν ἐξ αἰῶνος ἀποβλέπουσαν πρὸς τὰς ἰδέας ἑαυτὴν κοσμεῖν, ἔτι καὶ μᾶλλον ἐννοῶ μίαν εἶναι δεῖν ψυχὴν τήν τε διαπλάσασαν ἡμᾶς καὶ τὴν νῦν χρωμένην ἑκάστῳ τῶν μορίων.

> I reach an impasse, unable to discover anything about the artificer who constructs us even in terms of probable conception, let alone a firm understanding. When I hear some philosophers assert that matter has been endowed with soul from eternity, and that by contemplation of the Ideas it forms or adorns itself, I realise all the more strongly that there must be only one soul, which both constructs us and continues to employ each of the parts. (100.23–9 N, tr. Singer 1997)

12 Anatomy and Aporia in Galen's On the Construction of Fetuses

As Galen makes clear, he is sympathetic to the basic Platonic notion that there is an eternal cosmic soul which models itself and its constructions (in this case, human bodies) on the Forms,[36] but reiterates his inability to square this with a body that can move its parts without 'knowing' what it is doing.

Indeed, in the section immediately following (100.30–2.9 N), Galen offers the study of anatomy, however incomplete and limited, as the only available strategy to combat the soul's ignorance of how its own parts work. In the case of his recent example of speech-formation, it must be left to the anatomist to pick up where the soul fails in its knowledge. The writing is subtle here, but it can be no accident that Galen counterposes an elaborate description of how sound is produced from the larynx, nose and mouth to the soul's own ignorance of such things. If we pay attention to Galen's phrasing in this section, we can witness him transforming his general *aporia* into an almost heroic endeavour.[37] One might well ask, after all: if the *soul*, consorting as it does with Platonic Forms, cannot know how the body actually works, how can humans go about discovering how it does? Anatomists like Galen try, nevertheless, and they can make *some* progress, at least, in understanding the movement of body parts. Confronted with the soul's ignorance of the movement of parts (ἡ ἄγνοια τῆς διοικούσης ἡμᾶς ψυχῆς τῶν ὑπηρετούντων ταῖς ὁρμαῖς αὐτῆς μορίων – emphasis added), Galen notes that anatomists have at least made *some* progress in describing how sound is produced: οἱ γὰρ ἐπιχειρήσαντες εἰπεῖν περὶ τῶν στοιχειωδῶν φωνῶν ἄχρι τοσούτου προῆλθον, ὡς ἀποφήνασθαι ... but after his long description of the physiology of speech, he must again admit to the limitations of such research – not even the best experts have been able to say, for example, exactly which muscles are involved (οὐδὲ τοῖς ἐπὶ πλεῖστον ἀνατομικῆς ἐμπειρίας ἀφικομένοις εὕρηνταί πω βεβαίως). By the end of this section, then, Galen has left a twofold impression on his readers, that anatomical research is our only hope for understanding physiological mechanisms, but that there are still great limitations to what it can solve.

Galen's purpose in stressing these limitations, however, seems not to end the treatise on a note of epistemological abjection, but rather to contrast

---

[36] On the Platonic, and middle-Platonic, background to this passage, see Nickel 2001: 159–61; with Moraux 1981: 104.

[37] See Hankinson 2009: 241–2 on Galen's self-presentation ('highly coloured and evidently self-serving; he is the adored hero of his own narrative'), but also his caution that 'the mere fact that a text is rhetorical does not entail that it contains no truth'. Beagon 2013 discusses a strand of heroism that suffuses Pliny the Elder's conception, a century earlier, of his own task in writing his *HN*. Galen's 'heroism' is more focused than Pliny's (and Galen does not explicitly use such a term of his endeavour), but his attitude of self-congratulation is similar.

his own intellectual honesty about these limitations with the ignorance of others who make pronouncements about the nature of the soul and the construction of the fetus in the absence of proper knowledge of anatomy. The final passages oscillate between his attacks on such people and confessions of his own inability to solve the questions everyone wants to answer. Galen's complaint is directed against those who bypass anatomy altogether (οἱ μηδὲν τούτων μήθ' εὑρόντες μήθ' ὅλως ζητήσαντες, 101.10 N) and argue (in the next sentence) from premises that cannot be *known* by the senses nor *demonstrated* by reason (... τὴν πρώτην εὐθέως ὑπόθεσιν ἄγνωστον μὲν αἰσθήσει, λόγῳ δ' ἀνεύρετον ὑποτίθενται – emphasis added). He seems, of course, to have philosophers primarily in mind here, as he returns to the gripes he voiced earlier in the treatise about those who insist that the heart was the first organ to form in the fetus and that ('therefore') the three parts of a Platonic soul reside therein. Galen adduces arguments from empirical observation to counter this kind of unsubstantiated speculation, but supplementing these with rational *inferences* drawn from his observations: (λόγῳ τε σκοπουμένους καὶ τοῖς ἐξ ἀνατομῆς φαινομένοις, 102.23–4 N, emphasis added).

From this moment of self-confidence and self-righteousness, Galen draws the treatise to a close by returning to a confessional mode. He is troubled again in 6.29–30 (102.27–4.14 N) by the question of the soul's substance (οὐσία) – does it come from the seed of each parent? Is it immaterial (ἀσώματον)? – and in response to the various explanations available at the time, he again professes *aporia*. None of them, he claims, has been 'scientifically demonstrated' (οὐδεμίαν... δόξαν ἀποδεδειγμένην ἐπιστημονικῶς), and because he cannot establish the οὐσία of the soul, he also cannot establish the 'cause' of the construction of the fetus: ἐγὼ μὲν οὖν ἀπορεῖν ὁμολογῶ περὶ τοῦ διαπλάσαντος αἰτίου τὸ ἔμβρυον (emphasis added in both). Galen's confidence in the power and intelligence of the constructing soul – really the only thing he feels sure of – confuses him when he tries to account for the more mundane aspects of the embryological development, specifically the way in which it seems to form like the seed of a plant, automatically and without any apparent reason or intelligence.[38] He rejects one contemporary Platonic argument that a soul extensive throughout the universe (τὴν δι' ὅλου κόσμου ψυχὴν ἐκτεταμένην) constructs embryos, because this would imply a soul that constructed low, disagreeable animals, such as worms and spiders as well as humans, and such a notion, Galen says, would

---

[38] On the Platonic and Aristotelian background of the 'vegetative soul' that Galen here refers to, see Nickel 2001: 167–9.

be 'practically impious' (πλησίον ἀσεβείας). In the end, what he claims to know 'about the cause that constructs living creatures' (περὶ τῆς διαπλαττούσης αἰτίας τὰ ζῶα) is quite limited and not especially useful – that it 'comes about from a great deal of skill and intelligence'. He draws more useful conclusions from anatomical observation about the interaction of body parts in the developing fetus, but, as he says, these occur *after* the body has been formed (μετὰ τὸ διαπλασθῆναι), and have little to do with the 'causes' or reasons for their formation.

Galen leaves us, then, with a mixed message of *aporia* and self-confidence, and a rhetorical strategy intended to privilege the latter over the former. On the one hand, nearly all of the grand questions of causation and psychic animation raised by the mysteries of animal reproduction cannot be answered. On the other, the ability to admit such *aporia* openly is presented as far more desirable – and honorable – than falsely claiming any certainty on such matters. Throughout this treatise, but especially in the last chapter (6), Galen continually stresses the impossibility of knowing certain fundamental things about embryogenesis, but as we have seen, this ignorance is always countered by the implication that if *anyone* could make some genuine scientific progress in these matters, he would be the one to do so. At one and the same time, then, Galen can both lament the limitations of human knowledge in the face of the deepest philosophical questions, and claim an intellectual superiority based on a full command of the contemporary methodological landscape. Galen may not have been able to offer an exact account of the cosmic or physiological processes at work *before* an embryo began to form, but the way forward could only begin with an accurate understanding of the processes we actually *can* observe once its development is set in motion. By the treatise's end, therefore, Galen is able to take the moral and intellectual high ground: he avoids the hybris of making unsubstantiated claims about the soul,[39] and shows how without the kind of anatomical knowledge he has demonstrated here and in other works, one has no business entering the fray.[40]

---

[39] As Galen is careful to stress in the penultimate sentence of the treatise (106.7–10 N), where he notes that, although he has been able to demonstrate the 'three causes of motion' (brain, heart and liver) that manage the body after the fetus forms, and throughout a body's life, he has never dared to comment on the substance of the soul (οὐσίαν ψυχῆς ἀποφήνασθαι μηδαμόθι τολμήσας). See further, Nickel 2001: *ad loc.*, 169–73, esp. 172.

[40] I am most grateful to Christopher Gill for his helpful comments on an earlier draft of this chapter.

13 | Varro the Roman Cynic: The Destruction of Religious Authority in the *Antiquitates rerum divinarum*

LEAH KRONENBERG*

Varro ubique expugnator religionis ('Varro is everywhere a destroyer of religion')

Servius *ad Aeneid* 11.787

Introduction

Our knowledge of Varro's sixteen books of the *Antiquitates rerum divinarum* (*ARD*) is largely dependent upon the quotations and summaries of the work found in Augustine's *De civitate Dei*. This situation is usually deplored by scholars, who caution of the distortions caused by the Augustinian lens, which is presumed to over-emphasise the contradictions and ridiculous elements in Varro's work.[1] Without discounting the great challenges involved in making sense of Varro's text based on the few selections of it found in Augustine and other Christian authors, this chapter will consider the possibility that Augustine's emphasis on the comic elements in Varro's depiction of Roman religion is not just a result of his hostile perspective, but is made possible by the satirical nature of the original text. After all, while Augustine is selective in his quotation of Varro, that does not mean he is inaccurate, and, despite his polemics, most find him a fairly reliable source for Varro's work.[2] I will argue that in the *ARD*, Varro takes on the pose of an expert in Roman religion and philosophy, but in the course of the work, he satirically dismantles his own authority through inconsistent and hypocritical attempts to meld philosophical approaches to religion with support for Roman cult.

---

* Many thanks to Jason König and Greg Woolf for helpful comments on this chapter. Thanks also to the members of the Workshop on Ancient Scientific, Technical and Medical Writing (Berlin, March 2013) and Jim Zetzel and Katharina Volk's Varro Seminar (Columbia University, April 2014) for useful feedback on the ideas in this chapter. I am grateful to Curtis Dozier, Duncan MacRae and Katharina Volk for providing access to unpublished material on Varro. Carin Green also generously shared unpublished material, and her work on Varro has been an inspiration; I dedicate this chapter to her memory.

[1] E.g. O'Daly 1999: 106. [2] E.g. Hagendahl 1967: 697; Momigliano 1984 in Ando 2003: 152.

My approach to the *ARD* is obviously quite different from traditional ones, which at one time treated it as a handbook for a religious revival in Rome. While modern readers tend to be more sceptical of this practical purpose of the *ARD*,[3] most recent formulations of its aims are not too different from the two purposes attributed to it by nineteenth-century scholars, which Jocelyn nicely sums up: 'One to promote a restoration of ancient belief and practice, the other to provide a more sophisticated intellectual basis for pious behaviour.'[4] Building on the work of Wissowa, Jocelyn was the first to question seriously the religious or philosophical goal of Varro's work, but he then had the difficult task of explaining just what kind of work it was. He decided it was scholarship for scholarship's sake and calls Varro's tone at times one of 'mildly cynical detachment' or 'tongue in cheek'.[5]

My goal is to push Jocelyn's view of the work further in the direction of cynicism, particularly of the Menippean variety. After all, if Varro's work were merely interested in antiquity for its own sake, then why would he intermingle fact and fiction in his description of Roman cult, as several scholars have suspected?[6] To explain Varro's errors or inventions, Jocelyn falls back on the traditional view of Varro as a victim of his own systematising obsessions.[7] In addition, most readers treat Varro's 150 books of *Saturae Menippeae* as an aberration in his oeuvre – the product of his early days and sharply distinguished from his serious scholarly endeavours. If readers do find common ground between the satirist and the scholar, it is in Varro's patriotic and moralising attitude or in the urbane sort of wit that has been allowed to works like the *De re rustica*.

However, the view of Varro's Menippean persona as a traditional Roman moralist has been called into question by Relihan's important study of ancient Menippean satire.[8] According to Relihan, Menippean satire, including Varro's version of it, is a subversive tradition of literature, which, far from promoting clichéd moralising, in fact parodies moralisers, philosophers, dogmatic systems and anyone with pretensions to knowledge or truth.[9] In addition, this parody can include self-parody on the part of the author: 'Varro's massive erudition, displayed in abstruse vocabulary, technical lists, etymologies, learned allusion, and philosophical argumentation, may be serious in the *Antiquitates* or the *De lingua Latina* but is turned to

---

[3] E.g. Powell 1996: 63–4. [4] Jocelyn 1982: 199. [5] Jocelyn 1982: 180 and 200, respectively.
[6] E.g. Jocelyn 1982: 199–200; Powell 1996: 64; Wissowa 1904: 304–26, 1921. Rüpke 2012 allows for the possibility of 'self-amused playing around' (182).
[7] Jocelyn 1982: 200. [8] Relihan 1993.
[9] See Relihan 1993: 12–36 for a general definition of Menippean satire. On Varro, see Relihan 1993: 49–74.

self-parody in his *Menippeans*'.[10] But what if Varro's *Antiquitates*, instead of representing 'straight' scholarship, display a similar parody of expert knowledge and traditional morality? Without denying that the genre of the *ARD* is ostensibly different from that of the *Menippeae*, I would still contend that a Menippean spirit can shape its overall tone.[11] As Relihan notes, 'Menippean satires are often constructed in their broadest outlines as parodies of other genres of literature or types of discourse'.[12]

While most Romans did seem to take Varro's 'serious' works quite seriously, it is also possible that Varro's lengthy works were admired more than they were read or understood.[13] Cicero's praise of Varro's *Antiquitates* in *Academica* 1.9 is frequently quoted, but the wider context in which he praises Varro – a context which includes Cicero's ulterior motive of wanting to earn a complimentary dedication from Varro in return – as well as his lack of engagement with Varro's work in any of his own theological dialogues – potentially qualifies the praise.[14] In addition, the generally serious reception of Varro's *ARD* is not proof of its serious intent (or rather, is not proof of its non-parodic nature; satire and parody can, of course, have very

---

[10] Relihan 1993: 30.

[11] Indeed, the very fact that Tertullian calls Varro 'the Roman Cynic' and 'the Roman Diogenes' when he introduces the Stoic-sounding '300 Jupiters without heads' (Tert. *Ad nat.* 1.10.43; *Apol.* 14.9) underscores the similarities between Varro's persona in the *ARD* and in the *Menippeae*, as does the uncertainty among scholars about which of these works contained this fragment: Cèbe 1999: 2118–19 argues for its place in the *Menippeae*; Cardauns 1976: 1.87 places it in his Appendix to Book XIV as a possible fragment of the *ARD*. If this fragment is from the *Menippeae*, it is certainly not the only instance of thematic overlap between the two works. Cardauns 1976: 2.228 comments on the similarities of *ARD* fr. 230–34 to *Sat. Men.* fr. 92 Astbury. On the many philosophical themes in the *Menippeae*, see also Lehmann 1997: 263–98; Mras 1914; Mosca 1937; Sigsbef 1976.

[12] Relihan 1993: 25. Cf. also Bosman 2012: 793: 'The Cynics were among the great literary innovators of antiquity. They put the Cynic stamp on philosophical genres such as dialogues, *politeiai*, symposia and epistles, incorporated and redefined *chreiai* and *gnomai* for their own purposes, and promoted extraliterary genres such as wills and diaries to satirical instruments'.

[13] As Cameron 2011: 620 argues, 'Not the least of the reasons Augustine turned to Varro was (I suggest) precisely that *nobody* was actually reading him anymore' (emphasis in the original). See also Vessey 2014: 268. Even contemporaries may not have read or relied on him: cf. Edwards 1996: 50: 'Varro's concern with preserving even the smallest and most puzzling details of Roman religious practice was one we need not assume to have been shared by many educated Romans. Even Livy, in whose history religion has a vital place, did not necessarily consult Varro's work directly'; Rawson 1985: 220: 'By the time of Q. Tubero, let alone Livy, Varro had made a mass of antiquarian material accessible; the refusal of the annalists to use it is dramatic'. In addition, scholars struggle to account for the relatively few citations of Varro in the Augustan antiquarian Verrius Flaccus; see Glinister 2007; Lhommé 2007.

[14] On the tense relationship between Varro and Cicero and Cicero's long campaign to achieve a dedication in a work of Varro, see Kronenberg 2009: 88–9 (with notes for earlier bibliography; see also Baraz 2012: 80–6, 207–9; Corbeill 2013: 13–15; Wiseman 2009: 107–29). I will discuss Cicero's relationship to Varro's theological writings further in the next section.

serious intent). Indeed, as theorists of parody frequently note, many works intended as parody are not interpreted as such by their readers.[15] Perhaps it took a poet like Ovid, himself a master of irony and generic creativity, to appreciate and respond to the underlying wit of Varro's *ARD*. While a study of Ovid's reception of Varro in the *Fasti* is beyond the scope of this chapter, I would simply note that many read the *Fasti* as displaying precisely the blend of learning and satire that I suggest is in Varro's *ARD*.[16]

A Menippean reading of the *ARD* helps to make sense of the many oddities in Varro's text and particularly of the tensions it displays between the conflicting authority of traditional Roman religion and Greek philosophy. Modern readers have defended Varro's desire 'to have things both ways' by citing the similarity of Varro's position to that of other late Republican figures, such as Cotta, the pontifex and Academic spokesman from Cicero's *De natura deorum* (*Nat. D.*), who supports Roman cult even as he maintains a philosophically sceptical position about the gods and their existence.[17] Indeed, Varro's stance in the *ARD* has been made emblematic of 'the capacity of educated Greeks and Romans of the post-classical era to entertain different kinds of assent and criteria of judgement in different contexts, in ways that strike the modern observer as mutually contradictory'.[18] There has been a trend in recent studies of late Republican religion to take more seriously such attempts to meld support for state cult with rationalising

---

[15] Morson 1989: 67 emphasises that 'mistaking an intended parody for its object... is especially likely to occur when the audience is remote from, and so unaware of, the original utterance' – a caution that is certainly pertinent for modern readers of Varro's *ARD* (and even for readers in late antiquity). Highet 1962: 72 even suggests that 'the best material parodies are those which might, by the unwary, be accepted as genuine work of the original author or style parodied'.

[16] For Varro's influence on the *Fasti*, see esp. Baier 1997: 165–74; Cardauns 1976: 2.126; Green 2002; Merkel 1841: cvi–ccxlvii. For a reading that allows for satire and irony in the *Fasti*, see Fantham 2002. While there is no exact parallel from antiquity for the type of work I am claiming the *ARD* to be (outside of Varro's own works), I might compare Plato's *Menexenus* (or even more central works such as the *Republic* and *Timaeus*) or Lucian's *How to Write History* as prose works not obviously marked as satiric, but which are susceptible to such a reading. In addition, Frischer's 1991 reading of Horace's *Ars poetica* as a parody of Peripatetic poetics (and whose speaker is a foolish pedant) and Wycislo's 2001 reading of Seneca's *De ira* as a parody of judicial *responsa* are relevant. Not surprisingly, all of these ancient authors influenced or were influenced by Menippean satire. For a Menippean reading of Varro's *De re rustica*, see Kronenberg 2009: 76–129, which also briefly makes a case for a satiric reading of his *De philosophia* and *De lingua Latina* (86–7).

[17] E.g. Cardauns 1976: 2.246; Powell 1996: 62, whose phrase I use above. Rosenblitt 2011 shows that a contemporary of Varro could take a critical view of a figure like Cotta: she argues that Sallust's *Historiae*, by focusing on Cotta's 'cynical misuse of a Roman religious rite of deep gravity' (398), counters Cicero's presentation of Cotta in the *Nat. D.*

[18] Feeney 1998: 14, explaining Veyne 1983's concept of 'brain-balkanisation', which he compares to Varro's 'three theologies' (15–16).

theologies and not simply to view them as symptoms of the decline of Roman religion.[19] But what if we have been too hesitant to allow any absurdity to Varro's various inconsistencies and contradictory impulses towards philosophy and cult and have missed some of the subtle ways in which Varro's approach differs from other late Republican manifestations of religious thought? The subtle differences open a space for irony and parody and for the possibility that Varro is commenting in a satirical fashion on the tensions in the late Republic between traditional state cult and philosophical approaches to theology – instead of simply exemplifying them.

Rationalising Religion in the Late Republic

Many excellent studies have detailed the general trends towards rationalism and systematisation in late Republican intellectual thought, and several have focused specifically on the development of new, rationalising approaches to religion.[20] Indeed, Beard, with reference to Cicero's *De divinatione*, has called this a period 'when "religion", as an activity and a subject, became clearly defined out of the traditional, undifferentiated, politico-religious amalgam of Roman public life'.[21] Rüpke slightly modifies Beard's thesis by placing the *De divinatione* in a broader process of rationalisation, which began earlier and included antiquarian as well as philosophical approaches to religion, though he also distinguishes antiquarian approaches to religion from rationalising approaches informed by Greek philosophy.[22] One of the difficulties in distinguishing antiquarian works on religion from rationalising ones informed by Greek philosophy is that we have no fully surviving examples of the former. However, their titles and fragments suggest works devoted to recording the origins and rules of narrowly defined religious topics, such as Veranius' *De verbis pontificalibus*, Granius Flaccus' *De indigitamentis* or the numerous works on augury attested in our sources.[23]

---

[19] Beard *et al.* 1998: 144–66 is representative of this more nuanced approach to late Republican religion. On the 'decline' narrative in general (and Varro's role in creating it), see Rives 2010: 247–51.

[20] E.g. Beard 1986; Brunt 1989; MacRae 2013a and 2013b (which has some critique of the 'meta-narrative of rationalization'); Moatti 1997 (esp. 173–86); Momigliano 1984; North 1986; Rawson 1985: 298–316; Rüpke 2012 (esp. 144–219). See also Volk in this volume.

[21] Beard 1986: 46. [22] Rüpke 2012: 186–7.

[23] On these works, see MacRae 2013a: 49–102; Rawson 1985: 302–3. Only Nigidius Figulus comes close to Varro in covering a wide range of religious topics, from augury, to haruspicy, to the gods, though he does not do so all in one work as Varro does. On Nigidius' works, see Rawson 1985: 309–12; Volk in this volume.

While our evidence for antiquarian works is meagre, Cicero is helpful in distinguishing antiquarian works from philosophical ones, since several times he defines his own philosophical project in opposition to antiquarian or technical writing. In the *De legibus*, he contrasts his interest in abstract, universal questions about law with the trivial topics handled by the jurisconsults (1.14, 2.47), and in the *Nat. D.*, the Academic spokesman Cotta ridicules 'those who scrutinise secret and obscure writings' (3.42) in order to reveal multiple Hercules and Jupiters.[24] Elsewhere, Cotta calls similar groups of scholars *genealogi antiqui* ('ancient genealogists', 3.44), *ii qui theologi nominantur* ('those who are called theologians', 3.53) and *antiqui historici* ('antiquarian researchers', 3.55). While Cotta's viewpoint need not be identical to Cicero's, Cicero's basic agreement with it is demonstrated by the type of theological writing that he produced – the *De legibus*, *De natura deorum*, *De divinatione* and *De fato*, all of which eschew antiquarian detail in favour of larger, philosophical debates about the nature of Roman religious laws, the gods, divination and fate.[25]

Momigliano sharply distinguishes Cicero's *Nat. D.* from Varro's *ARD* and questions whether Cicero even read Varro's work in its entirety based on his neglect of it in his own theological writings.[26] Momigliano goes too far, however, when he denies any 'respect for theism in general and for Roman traditional religion in particular' in Cicero's work.[27] In fact, the *Nat. D.* is quite respectful of traditional Roman religion and the *mos maiorum*, a trait shared by all of Cicero's writings on religion. While there has been intense debate and analysis of the potential changes in Cicero's positions on traditional Roman cult in his theological works,[28] Lehoux rightly emphasises some unifying themes that cut across all of Cicero's theological writings: the importance of genuine (and not feigned) *pietas* towards the gods, adherence to the *mos maiorum* and traditional *religio*, which is different from *superstitio*, and the desire to ground this genuine religious feeling in rational argument and knowledge of nature.[29] While it is true that Cicero ultimately suspends judgement like a good Academic in both the *Nat. D.* and the

---

[24] Walsh 1998: 198 (*ad* 3.42) notes, 'This may be a reference to the antiquarian researches of Varro... Varro claimed to have identified forty-three bearers of the name Hercules (so Servius on Virgil, *Aen.* 8.564)'.
[25] On Cicero's demarcation of his philosophical work from Varro's antiquarian endeavours at the beginning of the *Academica*, see Feeney 1998: 16–17.
[26] Momigliano 1984, in Ando 2003: 153–4.
[27] Momigliano 1984, in Ando 2003: 157. Koch 2003: 320 presents a similar view to Momigliano's.
[28] Lehoux 2012: 34–5 summarises the different approaches; see also Beard 2012: 37–9; Santangelo 2013: 10–23; Volk in this volume.
[29] See Lehoux 2012: 37–41 and Cic. *Leg.* 2.11–23; *Nat. D.* 1.3–4; *Div.* 2.148–9.

De divinatione (he makes no such pretence in the *De legibus*, in which his Stoic arguments are presented dogmatically),[30] he still distinguishes his position in the *Nat. D.* from that of the Academic spokesperson Cotta, who is content to have faith in traditional religion without a rational reason for doing so (*Nat. D.* 3.6). In contrast, Cicero prefers the Stoic Balbus' attempt to unite a philosophical understanding of the world with respect for Roman cult and custom (*Nat. D.* 2.71; 3.40). Lehoux compares Cicero's desire to ground traditional religion in natural philosophy with Varro's in the *ARD*,[31] but he also cautions, 'We should not ... overstate the similarities. Where Cicero's *religio* is rooted in a knowledge of nature that is possessed by wise men in the present, Varro's grounding seems to be historical, in that the rites, deities, and institutions of Roman religion preserve an ancient knowledge of the world that was allegorically encoded by wise men in the mists of time.'[32]

What both Momigliano and Lehoux are reacting to in different ways is the unique, hybrid genre that is Varro's *ARD*. Varro's work is both similar to Cicero's and vastly different. It is neither a straight work of antiquarianism, collecting arcane rules and rituals on a specific religious topic, nor is it a strictly philosophical work – and yet it contains aspects of each. Rüpke highlights the hybrid nature of Varro's genre by subtitling his chapter on Varro 'Crossing Antiquarianism and Philosophy' and emphasising that Varro 'legitimated the Roman wish to cling to *mos maiorum*, tradition, within the universalistic framework of Greek philosophy'.[33] However, as we have seen, Cicero had that basic goal, as well. The major difference between Varro's and Cicero's approaches to uniting the *mos maiorum* and traditional Roman religion with Greek philosophy is that Cicero never worked out how precisely to connect the two and does not focus on the details of Roman *religio*. Indeed, as has been frequently noted, a major flaw or oddity of his *De legibus* is that Cicero never really explains how the religious laws detailed in Book 2 are in fact based on the Stoic theory of natural law set out in Book 1.[34]

Balbus the Stoic similarly glosses over the details in his grounding of the traditional worship of the gods in Stoic philosophy in the *Nat. D.* After setting forth various etymological and allegorical ways of understanding popular mythical conceptions of the gods, he simply states that we should spurn such myths, but understand their origins and still worship 'the gods with the names that custom has given them' (*Nat. D.* 2.71). He follows this injunction with a comment on the general importance of worshipping the

---

[30] Cf. *Leg.* 1.39.   [31] Lehoux 2012: 39.   [32] Lehoux 2012: 39–40.   [33] Rüpke 2012: 180.
[34] E.g. Brunt 1989: 198; Dyck 2004: 323 (*ad Leg.* 2.23).

gods with a pure mind and separating *religio* from *superstitio* (*Nat. D.* 2.71). He traces the latter to the practice of people who spend all day praying 'that their children would outlive (*superstites*) them', while he connects the former to those 'who diligently examined and, as it were, retraced (*relegerent*) all the things which pertained to the worship of the gods' (*Nat. D.* 2.72). However, this vague distinction is obviously not a useful guide in reality, and Balbus' justification for supporting traditional Roman *religio* remains underdeveloped.

My argument is that Varro exploited the weak spots in these Ciceronian and Stoic-inspired attempts to meld Greek philosophy and traditional Roman religion by inventing a civic theology that might be considered a *reductio ad absurdum* of the Stoic position.[35] He revels in all the 'contingent details' that Cicero omitted[36] and puts the *relegere* ('retracing') firmly in *religio*, but the result is not a clear separation of *superstitio* from *religio* or a sustainable philosophical position. Instead, Varro's civic theology is inconsistent and self-undermining and (intentionally) plays into the hands of the critics of Stoic theology. Indeed, Cotta's Academic critique of Balbus' Stoic theology is helpful in underscoring the satiric elements of Varro's theology since, as we will see, Varro specifically emphasises and exaggerates the very areas that Cotta criticises.[37] Thus, instead of explaining away the inconsistencies and oddities of Varro's text by referencing the distorting Augustinian lens, Varro's obsessive pedantry or the rationalising tendencies of his age, I would like to consider the possibility that they are the product of a satirist, who revelled in exposing the limitations of human knowledge and in ridiculing the confident expertise of those who disregard their own ignorance, hypocrisies and selfish motives.[38]

---

[35] It is not a specific concern of mine to locate the exact source of the Stoic-sounding views displayed in the *ARD* or to distinguish possible Antiochan influence. I view Varro's satire as broadly directed against the general rationalising trend associated with Stoic-inspired thinkers, though at times he takes jabs at other types of rationalisers, such as Academic sceptics, or cynical politicians, who manipulate cult for their own ends. For an attempt to sort out the Antiochan elements of Varro's theology, see Blank 2012: 262–79; Van Nuffelen 2010: 170–4. For the Stoic influences on Varro's tripartite theology, see Cardauns 2001: 54–9; Dihle 1996; Lehmann 1997: 193–225; Lieberg 1973; Rüpke 2012: 172–85.

[36] Cf. Dyck 2004: 243: '[Cicero] is not concerned with providing exhaustive lists of gods, priesthoods, or festivals; indeed in a set of laws purporting to be "by nature" such contingent details should be reduced to a minimum'.

[37] For my purposes, the fact that the *Nat. D.*, begun in the summer of 45 BC, was published after the *ARD* (published some time between 63 and 45 BC, though usually placed in 47 or 46 BC; see Tarver 1996) is unimportant since the argumentation that Cotta uses is based on traditional Academic critique of Stoic theology.

[38] As such, Julius Caesar, the self-serving pontifex maximus, politician and pedant (as his *De analogia* suggests), makes a perfect dedicatee for such a work.

The Pious Impiety of Varro's Work

Augustine famously attributes to Varro a pious purpose in writing the *ARD* (fr. 2a):

> *Se timere ne pereant (sic dei), non incursu hostili, sed civium neglegentia, de qua illos velut ruina liberari a se (dicit) et in memoria bonorum per eius modi libros recondi atque servari utiliore cura, quam Metellus de incendio sacra Vestalia et Aeneas de Troiano excidio penates liberasse praedicatur.*

> He fears lest (the gods) perish, not by enemy incursion, but from the neglect of the citizens. From this neglect (he says) the gods are freed by him as if from downfall, and, through books of this kind, they are stored away and preserved in the memory of good men by a service more useful than that of Metellus, who is said to have rescued the sacred objects of Vesta from the fire and of Aeneas, who rescued the *Penates* from the destruction of Troy.[39]

Augustine, however, has his doubts about Varro's accomplishment and observes that if Varro 'were an assailant or destroyer (*oppugnator esset atque destructor*) of the so-called divine things which he wrote about, and spoke of things pertaining not to *religio*, but *superstitio*, I can't imagine he would have included in his work so many ridiculous, contemptible, and detestable things' (August. *De civ. D.* 6.2). Augustine's observation picks up on one of the primary ways in which Varro satirically undermines the authority of his persona: Varro highlights the destructive effect that rationalising approaches to religion have on *religio* even as he uses those same approaches to try to preserve it.

While Augustine's demonstration of how Varro's attempt to save Roman *religio* unintentionally (or pseudo-unintentionally)[40] destroys it may be coming from a Christian perspective, it is quite similar to the rhetoric of non-Christian philosophical schools in the first century BC, all of which accused their opponents' philosophies of actually destroying Roman religion instead of preserving it as they claimed.[41] As usual, Cicero is our best evidence. In the *Nat. D.*, Cotta comments on the very religious language of Epicurus' books about sanctity and piety towards the gods, which were

---

[39] I am using the Latin text of Cardauns 1976; translations are my own.

[40] Augustine allows that Varro may have secretly had doubts about Roman religion, but, due to societal pressure, could only subtly reveal those doubts under the guise of praising religion (*De civ. D.* 6.2).

[41] In addition, Servius had a similar reaction to Varro: though he considers Varro an expert in theology (*ad Aen.* 10.175), he also calls him a 'destroyer of *religio*' (*expugnator religionis, ad Aen.* 11.787) in conjunction with his rationalising demystification of a fire-walking ritual.

written as if by a *pontifex maximus*, and not by a man who in fact 'destroyed completely all *religio*' (1.115). Cotta makes similar criticisms of Balbus' theological arguments and concludes that even though the Stoics are aware of how important it is to spurn the myths about the gods 'in order for *religio* not to be thrown into disorder' (*ne perturbentur religiones*, 3.60), their approaches actually 'confirm the myths through interpretation' (3.60) (and thus, implicitly, throw *religio* into disorder).

Of course, Balbus, like Varro, professes to be saving *religio* from neglect. In particular, Varro's concern over the *civium neglegentia* ('the neglect of the citizens') parallels Balbus' similar complaint about the 'discipline of augury' being lost 'due to the *neglect* of the nobility (*neglegentia nobilitatis*)' (2.9, emphasis added).[42] Balbus argues that he is in a struggle with Cotta to defend the altars, hearths, temples and shrines of the gods, and the city walls, which Cotta as a pontiff should defend (3.94). In turn, Cotta makes clear that he *does* support traditional Roman religion as a *pontifex maximus* even if he cannot find a philosophical reason to as an Academic (3.5–6), and that he *wants* his arguments to be refuted (3.95). In the *De divinatione*, Cicero similarly defends Cotta's approach by limiting Cotta's goal to the refutation of Stoic arguments and not the 'destruction of *religio*' (1.8). Indeed, Cicero is very conscious of the fact that philosophical arguments about *religio* may appear destructive of it to the wrong audience. An unplaced fragment from Cicero, usually assigned to Book 3 of the *Nat. D.*, voices the recognition that rationalising approaches to *religio* could be dangerous for the public to know and could destroy *religio* (Lactant. *Div. inst.* 2.3.2).

Varro strongly associates this view, that philosophical knowledge of the gods is dangerous for the people to know, with his own persona in the *ARD*: in fr. 21, he relates that 'there are many things that are true, which it is not useful for the people to know, and it is even useful for people to believe in things although they are false'.[43] Yet, as we will see, Varro's defence of civic theology in the *ARD* depends on these 'dangerous' rationalising

---

[42] Cicero's character in the *De legibus* similarly bemoans the fading away of the true discipline of augury 'from age and neglect' (*vetustate et neglegentia*) (2.33), and Quintus Tullius Cicero complains in the *De divinatione* about the loss of many auguries through 'the neglect of the augural college' (*neglegentia collegi*, 1.28). These Stoic voices do not bemoan a general neglect of the gods, but only the specific art of augury. Varro's version, then, ramps up the threat to religion and creates a sense of crisis that his rationalising antiquarian work proposes to solve. On the element of 'invented crisis' in late Republican discussions of religion, cf. Beard et al. 1998: 117–26; Bendlin 2000: 133–5; Wardle 2006: 171.

[43] Varro also associates the view that 'some philosophical ideas are harmful for the people to know' (*De civ. D.* 4.27) with the pontifex Scaevola (whose tripartite theology Varro adopts in the *ARD*) in his *Logistoricus, Curio: De cultu deorum*.

techniques even as his purposes are purportedly practical and geared towards Roman citizens, not philosophers, whose controversies he 'removed from the forum, that is from the people, and kept locked up within the walls of their schools' (*De civ. D.* 6.5). Varro differs, then, from the other rationalisers we have examined by emphasising the danger of philosophical approaches to *religio* at the same time that he applies this sort of approach in the most comprehensive way possible to *religio* in the guise of creating a practical guide for his fellow citizens.[44]

Varro's focus on the danger of philosophical interpretations of *religio* and his symbolic eviction of philosophers from the forum evoke not just the tensions of the mid-first century BC, but also the cultural tensions of the previous century, which saw the eviction of philosophers and rhetors in 161 BC and the burning of Numa's books in 181 BC.[45] The latter event particularly captured Varro's imagination since it appears several times in his writing,[46] and I would argue that Varro consciously models the *ARD* on the dangerous, secret books of Numa. Varro's fullest version of the story is found in the *Curio: De cultu deorum* (fr. 3, Cardauns 1960): Varro tells of a 'certain Terentius' whose ox-driver was ploughing near the tomb of Numa Pompilius and dug up 'his books, in which the reasons for his religious institutions were written down' (*ubi sacrorum institutorum scriptae erant causae*). Terentius took the books to the praetor, who examined the beginning of them and referred the matter to the senate. 'When the leading senators had read certain explanations for why each element of cult had been established' (*cum primores quasdam causas legissent, cur quidque in sacris fuerit institutum*), they agreed with the dead Numa (i.e. that the books should not be read) and ordered the praetor to burn them. There were many variations on the story of Numa's books, but most versions label Numa's dangerous books Pythagorean.[47] Varro's version in the *Curio* is (almost) unique in explaining the content of the dangerous books instead as containing the '*causae* of sacred institutions'.[48] This is precisely the sort of language that Cicero

---

[44] E.g. fr. 3: 'Varro boasts that he benefitted his citizens with a great service, because he not only names the gods which it is necessary that the Romans worship, but also tells them what each one is in charge of'.

[45] Cf. Cardauns 1976: 2.149 (*ad* fr. 21).

[46] It appears in the *Curio: De cultu deorum* (August. *De civ. D.* 7.34), *Antiquitates rerum humanarum* (Plin. *HN* 13.87), and (probably) in one of the *Menippeae* (*Manius*). On the latter, see Cardauns 1960: 28; Cèbe 1985: 1167–9 (*ad* fr. 249); Krenkel 2002: 2.450–2 (*ad* fr. 255). Indeed, the story of Numa's books is perfectly at home in the *Menippeae* – perhaps more so than in history – as historians have struggled to find an explanation of the event. Briscoe 2008: 483 surveys the different theories and concludes, 'We must simply suspend judgement'.

[47] For analysis of the different versions, see Briscoe 2008: 480–3; Rosenberger 2003; Willi 1998.

[48] Ps.-Aurelius Victor, *De viris illustribus* 3.1 also labels the contents of the books *sacrorum causae*, but he could be dependent on Varro.

uses to describe the books of Varro in the *Academica* when he praises him for having 'revealed the causes' (*causas aperuisti*) of all human and divine things (1.9). Varro's version is also different from some in having 'a certain Terentius' (as opposed to Petilius) find the books – a name which is obviously significant for Marcus Terentius Varro.

Augustine himself draws a parallel between Numa's books and Varro's when he suggests that Varro's books should have been burned along with Numa's if Numa's books had simply contained natural explanations of religious rites like hydromancy (*De civ. D.* 7.35).[49] While Augustine ultimately assumes a difference between the books based on their differing receptions in Roman culture, Cardauns agrees that Varro saw a similarity between the two – 'namely a philosophical foundation of cult in the terms of (essentially Stoic) natural theology'.[50] Indeed, Varro in many ways makes himself into a latter-day Numa in the *ARD*. Livy describes Numa as the 'most learned man in all of divine and human law' (*consultissimus vir... omnis divini atque humani iuri*, 1.18.1.), and of course Varro's *Antiquitates rerum humanarum et divinarum* implicitly make the same claim. The difference is that while Numa kept his 'pious fraud' a secret when he invented Roman religion based on a fictitious meeting with Egeria (Livy 1.19.4–5) and buried his books, Varro reveals all the destructive secrets of Roman *religio*.[51]

In addition to Numa's books, Varro alludes to another historical/mythical model to frame his paradoxical purpose of saving *religio* through dangerous and even impious revelation: in his preface, he compares his service in rescuing the gods from neglect to the actions of Metellus, the famous *pontifex maximus*, and Aeneas in saving the *sacra Vestalia* and the *Penates* from destruction. Beard, North and Price call this a 'baroque (and grossly self-flattering) comparison',[52] and, indeed, its hyperbolic language fits most comfortably in a satiric work.[53] Yet, there is more than just satiric hyperbole in this comparison: the actions of Aeneas and Metellus activate another paradoxical model of preservation of *religio* through the (dangerous) revelation of secrets. After all, the sacred objects that Aeneas and Metellus bring

---

[49] Augustine hopes Numa's books contained conversations with demons via hydromancy, but he clearly is unsure of the exact nature of the *causae* since he leaves open whether they told of the passions of demons or contained an euhemerising theory about dead men becoming gods (7.35).
[50] Cardauns 1960: 27.
[51] On Varro's conscious parallelism with Numa, see also Peglau 2003: 148–56.
[52] Beard *et al.* 1998: 118.
[53] Cicero may frequently boast in his prefaces (and letters to Varro; e.g. *Fam.* 9.2.5) of the great service to the state his philosophical works can provide (see Baraz 2012), but only Varro compares his scholarly output to the benefits provided by the Sibyl (in *Rust.* 1.3), Aeneas and Metellus.

into the light – the *Penates* and the *sacra Vestalia* – are meant to remain hidden within the temple. Metellus' actions are particularly ambiguous in the literary record. While some accounts focus solely on the piety of Metellus' action (e.g. Cic. *Scaur.* 48; Val. Max. 1.4.5), others mention the blindness that resulted from the fire destroying the temple of Vesta (Plin. *HN* 7.139–41) and even make his blindness a punishment for seeing what he should not see (Sen. *Controv.* 4.2). Ovid's account does not mention blindness, but does underscore the paradoxical impiety of Metellus' pious action: "'Forgive me, sacred things', he said. "A man, I enter what should not be approached by man. If it is a crime, let the punishment for the crime redound on me'" (*Fasti* 6.449–51).[54]

As a latter-day Metellus, Varro brings with him a mixture of piety and impiety. Just as the sacred objects of Vesta are meant to be hidden deep within her shrine, so the secrets of Roman religion are meant to be kept secret – whether hidden in the *libri reconditi* ('secret books') of the priests and augurs, or buried with Numa.[55] Cicero has Cotta speak disparagingly of antiquarians 'who scrutinise secret, obscure books' (*qui interiores scrutantur et reconditas litteras*, *Nat. D.* 3.42, emphasis added), and Cicero himself disclaims a desire to scrutinise any books of the augurs that are secret in the *De domo sua* (*venio ad augures, quorum ego libros, si qui sunt reconditi, non scrutor*, 39, emphasis added), or to seem too curious about the esoteric details of pontifical law, *religio* or sacred rites (*Dom.* 121). Yet, these are the esoteric details that fill Varro's work, and, despite the controversial nature of such knowledge, he paradoxically boasts that he writes and investigates (*perscrutari*) these things in order to get the people to worship more, instead of less (fr. 12). Varro may be a latter-day Numa, but his books are not hidden; instead, he preserves the gods in his books (*per eius modi libros recondi*, fr. 2a, emphasis added) in order to reveal the dangerous secrets of Roman religion.[56]

---

[54] Cf. Williams 1991: 193: 'This mixture of the pious and impious is similarly apparent in Metellus' response to the conflagration, for in his actions he too proves to be *sanctus* and *sceleratus*, *profanus* and *pius* at one and the same time – a more complex, paradoxical figure than the paragon of heroic virtue paraded in schoolboy legend.' As scholars have noted, Ovid even gives Metellus' actions connotations of sexual violation: e.g. Littlewood 2006: 137 (*ad* 6.437–8); Newlands 1995: 138.

[55] On the importance of secrecy in Roman religion, see Savage 1945. On the *libri reconditi*, see Linderski 1985; MacRae 2013a: 185–6 (with further bibliography in his notes). While scholars have debated exactly how secret these books really were and how, if at all, they differed from antiquarian literature, all that matters for my argument is the veneer or rhetoric of secrecy that surrounds the religious knowledge they contain.

[56] Thus, Augustine appropriately characterises Varro's work as revealing the secrets of the mysteries (*De civ. D.* 7.5), even as Varro claims there is a good reason that the Greeks kept their

The Useless Utility of Civic Theology

In his exposition of Rome's civic theology, Varro takes on the pose of yet another *pontifex maximus*, namely Q. Mucius Scaevola (cos. 95 BC). Augustine tells us (*De civ. D.* 6.5) that Varro expresses in the *ARD* 'three kinds of theology' (*tria genera theologiae*), the mythical (*mythicon*), the physical (*physicon*) and the civil (*civile*) – the same tripartite construction that Varro attributes to Q. Mucius Scaevola in the *Curio* (*De civ. D.* 4.27).[57] While there has been much debate about the origins of this schematic division of theologies, I agree with Rüpke that it is a Varronian invention.[58] Of course, Varro did not invent it out of thin air, and it is closely based on the Stoic division between true philosophical accounts of the gods and the error-ridden mythical accounts, which nonetheless encapsulate some primitive truths that can be discovered through allegory and etymology.[59] However, despite the Stoic support for the worship of the popular gods in cult (*Nat. D.* 2.71), they did not develop a 'civic theology' per se or attempt to rationalise and defend in detail all the aspects of traditional cult – they focused primarily on reconciling the mythical and the physical theologies, to use Varro's terms.[60] Varro's civic theology, the only theology that he labels with a Latin term instead of Greek, is an invention designed to press this weak spot in Stoic theory to its breaking point. He develops a full-blown civic theology that attempts to unite *ratio* with *religio*,[61] but it abounds

---

mysteries behind closed doors (fr. 21). Varro quite literally reveals the secrets of the mysteries in fr. 206, when he rationalises the rites of the Samothracians – but his entire work might be considered an impious revelation of religious mysteries.

[57] On the attribution of Scaevola's tripartite theology to the *Curio* instead of the *ARD*, see Cardauns 1960: 34 (*ad fr.* 5). Like many scholars, I assume that Varro's Scaevola was not really the author of the tripartite Stoic-influenced theology attributed to his character.

[58] Rüpke 2012: 172–3: 'Godo Lieberg has tried to argue that, prior to Varro, a Greek doxographer of the late second century B.C.E. had already developed the concept of a *theologia tripertita*, or threefold theology. This thesis has found wide approval, but the term that was coined to describe the concept (and therefore his conceptual history) is not to be found in any ancient source – Tertullian comes closest in speaking of a *tripertita dispositio* – nor, frankly is there any reason to disclaim Marcus Terentius Varro's authorship of the concept. It is this encyclopedist and Roman magistrate, who lived from 116 to 127, to whom the sources give credit, even if we can put the concept also into the mouth of the slightly older *pontifex* P. [sic] Mucius Scaevola.' Cardauns 2001: 57 acknowledges that the source of Varro's tripartite theology is unknown, but thinks there must be a Greek source due to the Greek terms he uses. As Augustine emphasises, however, Varro uses a Latin term for his civic theology. See n. 35 above for further bibliography on Varro's tripartite theology.

[59] E.g. *Nat. D.* 2.60–72.   [60] See Algra 2009: 247.

[61] Cf. the formulation of Van Nuffelen 2010: '*ARD* does not only propose to inventorize the *theologia civilis*, but to show that it is formed in accordance with the philosophical truth' (186).

with the absurdities and inconsistencies that result from a philosophy that simultaneously supports traditional cult while disparaging every aspect of its attendant world-view.[62] These inconsistencies manifest themselves in Varro's sharply vacillating attitude towards the value of civic *religio*, which at times is presented as a fictive, corrupted sham, and at times as a reflection of true philosophical theology and a positive force in its own right.

Varro strongly associates the current state of civic *religio* with corruption by noting the greater purity of Rome's religion in Numa's times, when gods were worshipped without images, and the 'error' that Tarquinius Priscus added to Roman religion when he 'took away fear' (*metum dempsisse*) of the gods by introducing their images (fr. 18). Later, in Book 16, Varro is more positive about the images of the gods, which he treats as a tool well designed by the ancients so that people could 'approach the mysteries of the doctrine' (*adissent doctrinae mysteria*) and gain insight into the 'true gods' (*deos veros*) (fr. 225). Yet, even in its purest state, Varro presents civic *religio* as fictive: he explains in Book 1 that he 'wrote about human matters before divine, because states existed first, and then divine institutions were created by them', and 'just as the painter is prior to a painting' so are the 'states prior to things instituted by states' (fr. 5).

Varro sharply distinguishes his work from a work like the *Nat. D.* when he notes that he would have written about the gods first if he were writing 'on the whole nature of the gods' (*de omni natura deorum*, fr. 5),[63] and from the *De legibus* when he states that if he were 'founding a new state, he would rather have consecrated the gods and their names according to the rule of nature but that now, since he lives among an old people, he ought to uphold the history of the names and cognomens received from the ancients' (fr. 12). Cicero, in contrast, focused in the *De legibus* on the natural law that existed prior to states and which was relevant beyond the boundaries of Rome (1.19, 2.8). Varro, then, shows the same reverence for natural law that Cicero shows in the *De legibus* and that Balbus shows in the *Nat. D.*, but he couples it with a sixteen-book work detailing every aspect of civic theology. Thus, Varro constructs his position once again in a paradoxical manner: he is a civic theologian who recognises the inherent vacuity of civic cult and yet devotes all of his energy to explaining and supporting it in a manner

---

[62] While Roman religion did not have an explicitly formulated 'world-view' or set of doctrines or beliefs, I support the recent trend among scholars to allow for some concept of belief in Roman religion (e.g. Ando 2003: 143, 2008: ix–xvii; Bendlin 2000; Feeney 1998: 9–46; King 2003; Mackey 2009; Rives 2007: 47–50; Tatum 1999).

[63] As Augustine points out, Varro does end up writing quite a bit about divine nature in Books 14–16 (*De civ. D.* 6.4).

that goes far beyond the vague support and rationalisation of traditional cult voiced by Balbus or even Cicero.

Varro underscores the paradoxical pose of his persona when he outlines the great benefit he is providing his fellow citizens by carefully delineating which gods to appeal to for particular problems in human life. He uses the following analogy to support his point (fr. 3):

> *Non modo bene vivere, sed vivere omnino neminem posse, si ignoret quisnam sit faber, quis pistor, quis tector, a quo quid utensile petere possit, quem adiutorem adsumere, quem ducem, quem doctorem; eo modo nulli dubium esse... ita esse utilem cognitionem deorum, si sciatur quam quisque deus vim et facultatem ac potestatem cuiusque rei habeat. Ex eo enim poterimus... scire quem cuiusque causa deum advocare atque invocare debeamus, ne faciamus, ut mimi solent, et optemus a Libero aquam, a Lymphis vinum.*

> Not only is no one able *to live well*, but no one is able to live at all, if he does not know who is a blacksmith, who is a baker, who is a plasterer, from whom he can get something useful, whom to use as a helper, whom as a leader, whom as a teacher; in this way no one can doubt that *knowledge of the gods* is useful, if he knows what force, skill and power each god has over each thing. For from that knowledge, we will be able to know for what purpose to summon and invoke each god, lest we behave like comic actors and desire water from Liber and wine from the water nymphs (emphasis added).

Varro seems to support here a very traditional conception of Roman cult; yet, Stoic paradox and parody is still lurking beneath the surface, since his formulation of the underlying 'ethics' of traditional cult is brought into direct contradiction with Stoic ethics. For the Stoics, the only way to 'live well' is to live a virtuous life.[64] But according to Varro's presentation of the purpose of Roman *religio*, living well is defined not by virtue, but by knowing where to find a good blacksmith when you need one – or the equivalent divinity. Varro's formulation reduces the powers of the Roman gods to pedestrian, practical spheres instead of rationalising them as manifestations of true Stoic divinity.

Varro's formulation also reduces the goal of 'knowledge of the gods' (*cognitio deorum*) to a purely utilitarian and pedestrian one that contrasts

---

[64] E.g. Cic. *Paradoxa* 1.15: *profecto nihil est aliud bene et beate vivere, nisi honeste et recte vivere* ('surely the only way to live well and to live blessedly is to live morally and justly'). Cf. also Sen. *Ep.* 70.4: *non vivere bonum est, sed bene vivere* ('it is not good to live, but to live well'). On Varro's parody of Stoic Paradoxes in his *Menippeae*, see Sigsbef 1976.

starkly with Stoic accounts of how man gains true *cognitio deorum*. For example, Balbus describes how humans gain 'knowledge of the gods' through the contemplation of the heavens (*Nat. D.* 2.153):

> *Quae contuens animus accedit ad <u>cognitionem deorum</u>, e qua oritur pietas, cui coniuncta iustitia est reliquaeque virtutes, e quibus vita beata existit par et similis deorum, nulla alia re nisi immortalitate, quae nihil ad <u>bene vivendum</u> pertinet, cedens caelestibus.*
>
> The mind, observing these things, approaches *knowledge of the gods*, from which arises piety, which is closely connected to justice and the remaining virtues, from which a blessed life arises equal and similar to that of the gods, and in no way inferior to the life of the gods except in immortality, which has no bearing on *living well* (emphasis added).

For Balbus, knowledge of the gods yields piety, justice and a blessed life. For Varro, such knowledge allows one to avoid looking like a comic fool and yields concrete material benefits from the gods ... or does it? Elsewhere Varro expresses traditional Stoic views about the gods, when he notes that 'these alone seem to have discovered what god is who have believed that he is the soul which is governing the world with his motion and reason' (fr. 13), and when he expresses the belief that 'true gods' do not want or demand traditional sacrifices and respond with neither grace nor punishment if they are offered (fr. 22). Varro's claim at the end of fragment 3, that he does not want Romans to end up looking like the absurd comedians who ask for the wrong thing from the wrong god on stage, might hint that he himself is staging a comedy of sorts for his reader, one in which the Romans and their gods are the object of cynic derision.[65]

Varro's description of Roman divinities in Books 14 to 16 continues to parody civic theology and Stoic rationalisations of the gods. Once again, Cotta's critique of Balbus provides an instructive comparison: Cotta is critical of all the deified abstractions which the Stoic rationalising approach permits and suggests that such an approach opens the door to any sort of abstraction that we could imagine (*Nat. D.* 3.47). He particularly underscores the problematic deification of bad things such as Fever and Bad Fortune (3.63) and notes the difficulty of dismissing the worship of

---

[65] Elsewhere, too, Varro uses the language of mime and comedy in his work: e.g. Augustine notes that Varro depicts the 'feasting gods' (*epulones... deos*) as 'parasites of Jupiter' (*parasitos Iovis*) and suggests that a mime writer would have been trying to get a laugh from the audience if he had depicted them as such instead of attempting to make a favourable comment about the gods (*De civ. D.* 6.7).

Egyptian gods like Serapis and Isis if Stoicism can rationalise all the other gods (3.47).[66] While Balbus attempts to repudiate the myths of the gods which display their 'desires, sorrows and anger' (2.70) and the deification of vices (2.61), Varro's divinities exaggerate all the problems that Cotta found with Balbus' rationalising approach: as a latter-day Numa, he vastly inflates the pantheon with abstractions,[67] particularly with deities related to sex and other 'vices'.

For instance, Varro introduces a plethora of divinities associated with nuptial rites, all of which are only attested in Varro (with the exception of Venus and Priapus) (fr. 144–55; *De civ. D.* 6.9).[68] Every element of the marriage ritual has a divinity attached to it, including Virginensis, who undoes the bride's girdle, Subigus, who puts the bride underneath the husband, Prema, who keeps the bride lying still beneath her husband, and Pertunda, who oversees the penetration of the bride. Conception and Pregnancy bring with them their own entourage of deities, with a seemingly endless string of gods devoted to semen (*De civ. D.* 7.3). Varro also introduces a trio of otherwise unattested money gods (fr. 193), namely Pecunia, to whom one can pray for money, and Aesculanus and his son Argentinus, to whom one can pray for bronze and silver money, respectively.[69] Scholars are as uncomfortable with this trio of gods as the Church Fathers, but still have not concluded

---

[66] The deification of vices is a sticking point for the Stoics: such deification does not reflect the Stoic veneration of *ratio*, is not a sign of a providentially designed world that is 'all for the best', and the 'primitive truth' housed in these deities is one they would like to ignore. Perhaps for these reasons, Cicero forbids the deification of vices in his religious laws in the *De legibus* (2.19, 28) and cites as examples the altars to Fever and Bad Fortune (2.28). He concludes that 'if names must be invented', they should be positive ones like Safety (*Salus*), Honour (*Honor*), Wealth (*Ops*) and Victory (*Victoria*) (2.28).

[67] Cf. fr. 37 (*addidit*) *Numa tot deos et tot deas*, 'Numa added so many gods and so many goddesses'. On comparisons between Varro and Numa, see n. 51 above.

[68] Varro's 'Sondergötter', to use Usener's 1896 term for gods with a limited sphere of action in human life, have been the topic of much debate among scholars of religion, who have struggled to decide if these gods correspond to anything in actual cult, in popular belief, or in the pontifical books. For recent surveys of the various approaches to these gods, see Elm 2003; Perfigli 2004: 183–217. Wissowa was the first to cast profound doubt on the historical value of many of Varro's gods (see Wissowa 1904: 304–26, 1921). Despite Wissowa's critique, many scholars still rely on Varro's list of deities in their constructions of Roman religion, and even those who attribute his gods to antiquarian scholarship rather than cult practice (e.g. Scheid 1996) have not concluded that Varro invented them for a satiric purpose. Cardauns 1976: 2.240 resists the notion that Varro actually fabricated names and Rüpke 2012: 255, n. 58 suggests that 'Wissowa's (1904) fundamental criticism of the value of the names ... need[s] to be rejected in principle (not in every detail)'; however, as noted previously, Rüpke does not 'deny the possibility of self-amused playing around' (182).

[69] Varro associates Jupiter with Pecunia, as well, because Jupiter possesses everything (*De civ. D.* 7.12; fr. 238) – an association that particularly scandalises Augustine.

that they are meant to be satiric.[70] Yet, personification of abstractions like Pecunia are at home in satire and comedy: Horace makes reference to *regina Pecunia* (*Epist.* 1.6.37), Juvenal cynically comments on the fact that divinities like Pax and Fides are ignored while everyone worships Pecunia, even though she has no temple (1.112–16), and Plautus creates a list of pleasure-deities including Amor, Voluptas, Venus, Venustas, Gaudium, Iocus, Ludus, Sermo and Sauvisaviatio (*Bacch.* 115–16). In response to this list, the slave Lydus questions why Pistoclerus would associate with such harmful deities and rightly asks 'is there really a god Sauvisaviatio?' (*Bacch.* 120). But Pistoclerus responds with scorn and makes fun of Lydus' ignorance about the names of the gods. Varro's persona has the same comic arrogance of Pistoclerus and his divinities the same reality as Plautus' gods: they mix fact and fantasy in a satirical commentary on Roman morality and religion, in which Stoic vices dominate the pantheon and human beings struggle to manage risk and terrors in their life through *religio*.[71]

Varro's allegorising presentation of the 'select gods' (or the gods that have been given particular prominence in Roman religion) in Book 16 creates further parody of the Stoic approach. Cotta had criticised the Stoics for turning 'mad men into philosophers' by rationalising myths like Saturn's castration of Caelus (3.62), but Varro goes far beyond this standard rationalisation by also defending barbarian practices of human sacrifice: he explains that Saturn is the god of seed and sowing, and so the Carthaginians sacrifice children to Saturn and the Gauls sacrifice adults to him 'because of all seeds, the human is the best kind' (fr. 244). Varro had given voice previously to the standard moralising condemnation of the Bacchanalia (fr. 93), but he then reveals the truth of Cotta's critique when he rationalises the height of *superstitio*, namely the barbarian rites of human sacrifice.

---

[70] Or rather, it has not been suggested that *Varro* could be the author of the satire: so ingrained is the notion of Varro as a serious pedant that even though Cardauns allows that Varro may have found some of this information in a writer of comedy or satire, he assumes Varro must have mistakenly taken the information at face value instead of creating the satire himself; see Cardauns 1976: 2.217 (*ad* fr. 193).

[71] As Rüpke 2012 notes, Varro's divinities are 'organized around risky human action' (182). Indeed, in light of the non-providential world reflected by Varro's divinities (and emphasised in *De civ. D.* 6.9), it is pertinent to note that before Varro, Scaevola appeared in theological literature not as an exponent of a tripartite theology, but as proof of the indifference of the gods towards human beings and particularly good human beings: the *pontifex maximus* was notoriously slaughtered in front of the statue of Vesta in 82 BC during the violence between Sulla and Marius, and Cotta cites this event, among others, as an argument *against* Balbus' providential world (*Nat. D.* 3.80). The death of Scaevola is mentioned frequently enough (e.g. Cic. *Rosc. Am.* 33; *De or.* 3.3.10) that it seems likely any reference to him would bring to mind the unjust and violent death of a good man (and priest).

The Inexpert Expert

Varro's exposition is not just a parody of Roman religion and Stoic rationalisations of it; in taking on the tripartite theology of Scaevola, Varro parodies the sort of 'pedantic expert' persona that is frequently a target of Menippean satire.[72] Q. Mucius Scaevola was a jurisconsult obsessed with the classification and systematisation of Roman law. In the *De legibus*, Cicero criticises the Scaevolae (P. Mucius Scaevola and his son Q. Mucius Scaevola the Pontifex), and jurisconsults in general, for breaking a single area of knowledge into an infinite number of divisions, whether to obfuscate intentionally, to make their knowledge look more difficult or simply because they are bad teachers (*Leg.* 2.47). Of course, the Stoics love to break their discourse into divisions, as well, as Balbus shows when he starts his exposition by setting forth a fourfold division of the question (*Nat. D.* 2.3). Thus, in giving Q. Mucius Scaevola a Stoic pedigree, Varro intensifies Scaevola's philosophical and pedantic credentials and then takes them on himself in the ARD.[73]

Just as Varro's pedantic farmers lose control of their art of farming in the *De re rustica* by bungling their categories and giving impractical information (and misinformation) about farming,[74] so Varro's pedantic persona in the ARD creates divisions that make little practical sense for the teaching of Roman *religio*, have nothing to do with cult practice and are full of blunders or inventions.[75] Augustine is particularly perturbed by the lack of clear rationale for the category of the 'select gods', many of whom were already covered among the 'certain gods' and have lesser functions than some of the obscure 'certain gods' (*De civ. D.* 7.2–3). He was also bothered

---

[72] Cf. the disparaging reference to *multiplex scientia* ('encyclopaedic knowledge') in the *Agatho* (*Sat. Men.* fr. 6, Astbury 2002) and the many *Menippeae* involving the vain debates of quarrelling philosophers (usually Stoics and Epicureans) such as *Andabatae*, *Armorum iudicium*, *Caprinum proelium*, *Logomachia* and *Sciamachia*. For bibliography on these satires, see n. 11 above.

[73] Q. Mucius Scaevola the Augur was known to have studied with Panaetius, so perhaps Varro's character is a composite of these two different Scaevolae.

[74] On the pedantic farmers in the *De re rustica*, see Kronenberg 2009: 77–85; Nelsestuen 2011: 333–7. See also Doody in this volume.

[75] As Jocelyn 1982: 199–200 summarises: 'Wissowa and others demonstrated the differences between the highly practical *libri sacerdotum populi Romani* and what Augustine and others got from Varro's *Antiquitates rerum divinarum*. The names of many of the *di certi* turned out on close examination to be epithets of other deities or misinterpretations of obsolete words in Varro's source material or to belong to deities with quite different functions in actual Roman life. Many functions proved to be based on no more than Varro's own etymological guesses.' See also n. 68 above.

by Varro's inconsistent and contradictory attempts to relate the gods to the natural world (*De civ. D.* 7.28), and his wavering certainty about the information he relates (*De civ. D.* 7.17). Indeed, at times, Varro takes on the pose of a sceptic who is unwilling to state anything with certainty, particularly in Book 15 about the 'uncertain gods' (e.g. fr. 204 and fr. 228). Yet, in the very same book, he dogmatically explains the Samothracian mysteries through physical allegory (one of their statues signifies the sky, one the earth and one the Platonic ideas) and even promises to send the Samothracians a written exposition of their mysteries since they are ignorant of their meaning (fr. 206). We learn from Augustine that Varro contradicts his own allegorical explanations of these gods in the next book on the 'select gods' (7.28).[76]

I would also explain Varro's inconsistent and contradictory etymologies as part of this intellectual parody. Many scholars have pointed to the clashes between his etymological explanations in the *ARD* and *De lingua Latina* (or other works),[77] but even within the *ARD*, Varro offers conflicting explanations.[78] While Cardauns takes such inconsistencies as evidence of Varro's method of not limiting himself to one explanation,[79] there is a difference between explicitly offering several explanations at one time and contradicting oneself in different works or in different parts of one work without acknowledgement.

A fragment from Gellius, who is so often a useful guide to Varro, helps us to further characterise Varro's pedantic persona and etymological inconsistencies.[80] Gellius (1.18.1–2) notes that in Book 14 of the *ARD* (fr. 89), Varro takes Aelius Stilo to task for providing false etymologies (*causas falsas*) and not realising that certain Latin words are derived from Greek words no longer in use. However, Gellius also notes that in the very same book, Varro makes the precise mistake that he faults Aelius for in his own faulty explanation of the Latin word for 'thief' (1.18.3). Gellius is disturbed by Varro's hypocritical blunder and struggles to explain it, but

---

[76] Varro provides a different allegorical explanation of the Samothracian deities in *Ling.* 5.57–8, as well.

[77] On the differences between the etymological explanations of the *ARD* and those of the *De lingua Latina* and other works, see Cardauns 1976: 2.162, 2.207, 2.209, 2.192; Rüpke 2012: 156–9.

[78] E.g., in fr. 146, Iugatinus is linked to the yoking of man and wife (*coniunguntur*) and in fr. 164, to mountain 'ridges' (*iuga*).

[79] E.g. Cardauns 1976: 2.207 (*ad* fr. 122/123).

[80] On Gellius as a useful guide to Varro, see Relihan 1993: 54–9. On the influence of Varro's *Menippeae*, and particularly Varro's parody of intellectual pedantry, on Gellius, see Keulen 2009: 46–51, 58–65.

concludes that it is not his job to pass judgement on a man of surpassing learning like Varro (1.18.6). Yet, surely if Gellius noticed this inconsistency, Varro could have too, and I would suggest that Varro included it intentionally as part of his parody of pedantic intellectualism, which combined critique of others with ignorance of one's own faults.[81]

Indeed, Varro's self-parodic persona in the *ARD* might be best characterised by the type of arrogant expertise that Cicero disclaims in *De domo sua* 33:

> *Quid est enim aut tam adrogans quam de religione, de rebus divinis, caerimoniis, sacris pontificum conlegium docere conari, aut tam stultum quam, si quis quid in vestris libris invenerit, id narrare vobis, aut tam curiosum quam ea scire velle de quibus maiores nostri vos solos et consuli et scire voluerunt?*
>
> For what could be so arrogant as to try to teach the college of pontiffs about *religio*, about divine things, sacred rites or rituals, or so stupid as to tell you of things that have been found in your own books, or so officious as to wish to know those things which our ancestors wanted you alone to be consulted on and to know.

However, unlike Cicero's arrogant pedant, Varro's persona also has a Menippean (or even Socratic) edge that allows the reader the see how the confident teachings of his work fall apart. Thus, while Feeney argues that 'Varro's categories of arrangement... enabled him to dominate any body of knowledge, and it was this form of mastery that made him able to speak with expertise and authority',[82] I would suggest that it is precisely the authority of this form of pedantic mastery that Varro is undercutting in his satirical work on Roman religion.

Conclusion

Our knowledge of the *ARD* will always be partial and my satirical reading of the fragments open to the charge that I am falling prey to Augustine's selective (and polemical) citation of Varro's work. Indeed, it is difficult enough to prove the existence of subtle irony and satire in a fully surviving work, let alone one that is fragmentary, preserved through the lens of a hostile author

---

[81] I also imagine that some of Varro's etymological explanations were meant to be funny, such as the suggestion that Diana is a virgin goddess because roads do not bear children (fr. 276), or the similar suggestion that Vestal virgins are virgins because nothing is born from fire (fr. 282).

[82] Feeney 1998: 140.

and *sui generis* among surviving works from antiquity. In light of these difficulties, my modest hope is that this study preserves the possibility of another reading of the fragments, one which brings the *ARD* in line with the cynic wit of the *Menippean Satires*. Certainly, Varro must have taken scholarship seriously in some sense and enjoyed collecting human knowledge in order to write such massive compendia of it. However, his few scholarly works that survive, whole or in part, are more complex than simple compilations of knowledge, and perhaps Varro has been taken to exemplify a type of scholarship that he instead was commenting upon in a second-order (and satiric) fashion. If the Late Republic was a time when the elite's 'authority passes to specialists who can master increasingly complex and technical fields of knowledge,'[83] then Varro's works reveal the nagging resistance of the world to being mastered by reason and controlled by the experts.

[83] Wallace-Hadrill 1997: 21.

14 | Signs, Seers and Senators: Divinatory Expertise in Cicero and Nigidius Figulus

KATHARINA VOLK*

A Letter, A Death and Two Philosophical Treatises

In the late summer of 46 BC, Cicero wrote a letter to his friend and sometime political ally P. Nigidius Figulus.[1] Like Cicero, Nigidius had sided with Pompey in the Civil War, but unlike his friend, he had not obtained a pardon from Caesar and found himself in exile. Cicero exhorts Nigidius to keep up his spirits and promises to do everything in his power to have him recalled to Rome. However, his vague claims that Caesar seems favourably inclined to Nigidius and that the dictator's intimates, too, are pressing the exile's case do not inspire much confidence, and the letter as a whole is characterised by Cicero's despair at the present situation and survivor's guilt vis-à-vis his less fortunate Pompeian friends. And indeed, all efforts to bring about Nigidius' return were in vain: as Jerome informs us, he died in exile in 45 BC (*Chron.* 156.25–6, Helm 1956).

Unlike his famous correspondent, Nigidius remains little known even to professional classicists, which is a pity since, among the remarkable figures of the Late Republic, he is one of the most original and unusual.[2] While his undistinguished political career, characterised throughout by a staunch optimate stance, did not proceed beyond the praetorship (58 BC), Nigidius' fame as an intellectual rivalled that of his contemporary Varro, and his scholarly output (now surviving only in fragments) is as prolific in scope as it is quirky in content.[3] In addition to grammatical and scientific works, Nigidius is credited with treatises on religion and, in particular, divination;

---

* My thanks go to Jason König and Greg Woolf for inviting me to contribute to this volume, to Fritz Graf, Sarah Johnston and Jim Zetzel for commenting on an earlier draft and to Duncan MacRae and David Sedley for giving me access to unpublished material.
[1] *Fam.* 4.13 = 225 Shackleton Bailey 1977. Shackleton Bailey tentatively dates the letter to August 46.
[2] Generally on Nigidius, see Della Casa 1962; Kroll 1937; Rawson 1985: Index s.v.; and Volk 2016.
[3] On Nigidius' political career, see Broughton 1951–86: 2.190, 193, n. 5, 194, 239, 245 and 3.147; Della Casa 1962: 22–36 discusses his optimate stance. For Nigidius' reputation as an intellectual, see, e.g., Gell. 4.9.1, 4.16.1, 13.10.4, 13.26.1 and 17.7.4. The fragments of his works are collected in Swoboda 1889 and Liuzzi 1983.

Cicero describes him as an *acer inuestigator et diligens earum rerum quae a natura inuolutae uidentur* ('a keen and enthusiastic investigator of those things that appear to be concealed by nature', *Tim.* 1). This interest in the occult seems to have gone beyond the purely academic: Nigidius apparently was a practising astrologer, and in his death notice, Jerome intriguingly labels him *Pythagoricus et magus*.[4]

Some time after Nigidius' death, Cicero paid homage to his friend by featuring him as one of the speakers in his *Timaeus*, a dialogue on natural philosophy that has come down to us in mutilated form. While the first bit of text contains a eulogy of Nigidius and the setting up of the dialogue, the remainder of the work as we have it is a partial translation of Timaeus' speech from Plato's dialogue of the same name. It is generally assumed that in Cicero's version, Nigidius played Timaeus, as it were, uttering the Latinised Platonic text as all or part of his contribution to the discussion; what remains unclear is what transpired in the rest of the dialogue or whether Cicero in fact wrote anything beyond his translation from Plato.[5] In spite of these uncertainties, the presentation of Nigidius as a kind of Roman Timaeus was surely intended to be a tribute to a man as learned and well initiated into the secrets of the cosmos as his Greek counterpart.

At about the same time as he worked on the *Timaeus*, Cicero composed his *De diuinatione*, a dialogue in two books between himself and his brother Quintus on the validity of divination.[6] Relying on Stoic and Peripatetic arguments, Quintus in the first book affirms that it is possible to predict the future from signs provided by the gods, a stance that comes under vicious attack in Book 2, where Marcus uses the strategies of Academic Scepticism to disprove the existence of any kind of divination. *De diuinatione* does not mention Nigidius, but it has been suggested that by treating a topic so dear to the heart of his recently deceased friend, Cicero engages in a dialogue with the work of his colleague on an issue of common interest.[7]

---

[4] *Chron.* 156.25–6, Helm 1956. Jerome's designation may go back to Suetonius' *De uiris illustribus*; see Della Casa 1962: 9–36 and Musial 2001: 344–5. On Nigidius' Pythagoreanism (attested also in Cic. *Tim.* 1; Schol. Bob. Cic. *Vat.* 14), see now Volk 2016 and the literature cited there, esp. Musial 2001. His astrological activities will be discussed in the final section of this chapter.

[5] See now Sedley 2013 for a reconstruction of Cicero's aims in the work.

[6] The relative chronology of the *Timaeus* and *De diuinatione* is unclear. *Tim.* 1 provides two *termini post quem* for the former, Cicero's *Academica* (June 45) and the death of Nigidius (some time in 45), but the work could have been written any time between 45 and Cicero's death. The composition of *De diuinatione* took place in late 45 to early 44, straddling the Ides of March.

[7] See Engels 2007: 149; Hirzel 1895: 1.538, n. 1; Pease 1920–23: 12; and Timpanaro 1988: lxxx–lxxxi.

As has often been pointed out, a prominent aspect of the intellectual flourishing of the Late Republic was the definition of topics of study that had not previously been considered distinct realms of theory or practice.[8] This mental 'differentiation', to use the term of Beard *et al.* 1998, manifested itself in the development of separate discourses on such fields as rhetoric, law, grammar, literary history and religion, among others, and in the creation of massive bodies of works dedicated to collecting and interpreting the knowledge associated with each area.[9] One of these new topics of investigation was divination and its various subdisciplines: Nigidius, for one, published on private augury, haruspicy, dream interpretation, astrology and thunder omens,[10] and there was a fashion for specialised writings on augury (including a *De auguriis* by Cicero, an augur himself) and the *disciplina Etrusca*.[11] Cicero, however, seems to have been the first Roman writer to consider all predictive practices together as one phenomenon, indeed the first to use the Latin noun *diuinatio* to cover everything from augury to astrology to the omens contained in chance utterances.[12] This is not necessarily to say that the Romans did not have some overarching notion of 'divination', only that it had never before been made a separate topic of intellectual enquiry.[13]

As we will see in the following section, *diuinatio* is a slippery concept in Cicero, and modern scholars, too, find it difficult to define what exactly divination is. At the most basic level, divinatory practices aim at gaining, in the words of Johnston 2008: 3, 'knowledge of what humans would not otherwise know'; in the context of Ancient Greece and Rome, this knowledge is typically seen as being communicated by the gods and very often, though not always, concerns the future. In the Rome of Cicero and Nigidius, there was a

---

[8] On this process, and on Late Republican intellectual culture in general, see Moatti 1997. For the details, Rawson 1985 remains indispensable.

[9] See Beard *et al.* 1998: 1.149–56, who discuss this process specifically for the field of religion. Comprehensive treatments of the systematisation and 'textualisation' of Roman religion in the Late Republic are found in Rüpke 2012 and MacRae 2013a. The most prominent figure in this context is M. Terentius Varro, whose *Antiquitates rerum diuinarum* are discussed – with a twist – by Leah Kronenberg in this volume.

[10] See the fragments of his *De augurio priuato*, *De extis*, *De somniis* and *Sphaera*, as well as the brontoscopic calendar in John the Lydian, *De ostentis* 27–38 ascribed to him. These works will be discussed further in the final section.

[11] Cicero's *De auguriis*: Müller 1879: 312; writings on augury: Harries 2006: 162–9; Rawson 1985: 302–3; *disciplina Etrusca*: Rawson 1985: 303–6; Turfa 2012: 286–92.

[12] As the *TLL* articles s.vv. reveal, *diuinare* and *diuinus* are attested from the beginnning of Latin literature, but the abstract concept *diuinatio* is first found in Cicero.

[13] More strongly North 1990: 57: 'the fact that Cicero wrote a dialogue on divination should not be taken to imply that earlier generations of Romans would have recognised such a category'.

bewildering array of actors who claimed expertise in providing such knowledge and a multitude of ways of arriving at it, as well as different models of conceiving of such purported communication with the gods.[14]

On the one hand, certain forms of divination were part and parcel of Roman civic religion, the highly formalised system of practices designed to ensure divine cooperation in public undertakings.[15] Roman magistrates had the right and duty to take the auspices; the members of the College of Augurs were distinguished individuals of the ruling class; and the senate routinely received and debated reports of *prodigia* and on occasion referred them to such specialist bodies as the *haruspices* or *X(V)uiri sacris faciundis*. By virtue of being senators and having held magistracies, both Cicero and Nigidius could thus automatically claim a certain authority and expertise in divination. In addition, Cicero had been elected augur in 53 or 52 BC and was therefore privy to specialist knowledge in this particular divinatory discipline. As part of the political process, divination was something upper-class Romans engaged in as a matter of course and may not have perceived as distinct from other aspects of their public activities.

On the other hand, there was a large number of what we might call freelance diviners, who provided predictions in a variety of contexts and to a variety of audiences, using divers methods from inspired prophecy to astrology to dream interpretation. These practitioners did not enjoy official sanction, but might nevertheless come to play political roles, for example, as advisers to individual public figures or by pronouncing on matters of state. As we will see, there is reason to believe that Nigidius on occasion acted as an independent purveyor of predictions, some of which may be interpreted as interventions in politics.

This amorphous mass of divinatory practices could be subjected to intellectual enquiry in a number of different ways. Some of the studies composed in the Late Republic were probably specialised treatises, meant for other experts in the same discipline;[16] others were 'scholarly' in the widest sense, that is, works dedicated to the systematic presentation and preservation

---

[14] For the diversity of divinatory theory and practice in the Late Republic, see now Santangelo 2013.

[15] The scholarship on Roman civic or state religion is enormous and cannot be discussed here. Titles that I have found especially helpful include Beard *et al.* 1998 on Roman religion in general; Rosenberger 2007 on the religious roles of the political class; and North 1990 and Scheid 1987–89 on Roman divination. Bendlin 2000 provides a good example of the recent trend in Roman religious studies to move beyond narrow study of the state religion in order to arrive at a more comprehensive view of Roman religious beliefs and practices.

[16] This may be true of many of the works on augury, which were presumably part of a specialised debate within the College of Augurs itself. On contemporary controversies among augurs – in

of knowledge about divinatory practices both in Rome and abroad; finally, Cicero's *De diuinatione* is a philosophical discussion not of specific ways of making predictions, but of the underlying assumptions of divination itself.

We may speculate as to why this topic gained such currency in the mid-first century. While the 'antiquarian impulse' of Late Republican intellectual culture will have played a role, as will the rapid development of a Roman philosophical discourse, it is also the case that with the collapse of the Republican system, divination became a contested political issue. Political players and parties availed themselves of the traditional divinatory practices of the civic religion to further their goals and obstruct their opponents. It was especially the beleaguered optimates who again and again resorted to the time-honoured practice of *obnuntiatio*, the announcement of unfavourable auspicial signs that were supposed to render a political initiative moot;[17] most famously, Caesar's fellow-consul of 59, M. Calpurnius Bibulus, attempted to invalidate his colleague's legislation by repeatedly announcing his intention to 'observe the sky', a speech act tantamount to the registration of an inauspicious sign itself. In an example from the other side of the political spectrum, P. Clodius Pulcher's attempts to prevent Cicero from rebuilding his destroyed Palatine house involved, among other religious arguments, the use of a pronouncement by the *haruspices*, whose correct interpretation became a personal and political battleground between the two archenemies.[18]

While auspicy and haruspicy were thus instrumentalised in the conflicts of the age, non-traditional, freelance divination began to play an important political role as well. As individual political players rose to unprecedented power, they increasingly employed personal diviners, including practitioners of a newly fashionable method: astrology.[19] As we learn from Cicero (*Div.* 2.99), Crassus, Pompey and Caesar all consulted astrologers, who without fail predicted brilliant careers and long lives to their clients. We may thus imagine the Civil War as, at the same time, a battle of conflicting

---

particular between Cicero's colleagues App. Claudius Pulcher and C. Claudius Marcellus – about the nature of their discipline, see Cic. *Leg.* 2.32–3, *Div.* 2.75.

[17] On the practice of *obnuntiatio*, see esp. Linderski 1971 (and compare the more comprehensive discussion of augural law in Linderski 1986). On its political (ab)use in the Late Republic, see, e.g., Bergemann 1992: 89–113; Burckhardt 1988: 178–209; and de Libero 1992: 56–68.

[18] The affair is reflected in Cicero's speech *De haruspicum responso* of 56, on which see Lenaghan 1969 and now Beard 2012 and Corbeill 2012, as well as n. 58 below.

[19] On the rise of individual prophecy, see, e.g., Engels 2007: 786–97 and Rosenberger 2007: 300–3; specifically on astrology, see Barton 1994a: 38–44; Cramer 1954: 44–80; and Volk 2009: 127–37.

prophecies, in which each contestant claimed the support of the gods.[20] Of course, the battle was won by Cicero's and Nigidius' enemy Caesar, who in the years of his dictatorship worked to consolidate the image of himself as a divinely favoured – and, indeed, divine – figure.[21] This process continued with Caesar's successor Augustus, who not only declared his adoptive father a god, but perfected the propagandistic use of divination, especially astrology.[22]

At a time, then, when divination, in its various shapes and forms, was both a newly defined subject of scholarship and a highly contested political tool; at a time, furthermore, when their political fortunes were waning and their opponent was triumphing, at grave personal costs to themselves and (so they believed) to the *res publica*; at this time, Cicero and Nigidius turned their intellectual efforts to the topic of prophecy and prediction. Being both practitioners and theorists of divination, they brought to their subject their own expertise as well as their philosophical and political convictions. The two friends approached the question of divination in remarkably different ways; at the same time, they were both reacting to, and intervening in, the same intellectual and political context.

How to Do Things with Divination

Cicero's *De diuinatione* has caused controversy and at times even disgusted aporia, with scholars differing over the message of the work and worrying about perceived contradictions between Cicero's stance in the text and views he espouses elsewhere.[23] Traditionally, the dialogue has been read as a rationalist attack on superstition: Marcus' superior intellectual arguments carry the day against Quintus' bumbling enumeration of anecdotes.[24] Not infrequently, such interpretations see Cicero as making a principled stand against a supposed religious decline, the rise of irrationality or the abuse of religion by Caesar. However, scholars who subscribe to this reading have to contend with the fact that in the second book of *De legibus* (written in the late 50s BC), Cicero not only legislates state-sponsored predictive practices

---

[20] See Jal 1961. [21] See the classic study of Weinstock 1970. [22] See esp. Schmid 2005.
[23] For a (somewhat biased) summary, see Engels 2007: 153–64. Generally on Cicero and religion, see Goar 1972 and now Gildenhard 2011: 245–384; on his views on divination, see Guillaumont 1984 and Kowalski 1995 (specifically on augury).
[24] Versions of this interpretation are found in Blänsdorf 1991; Cramer 1954: 71–2; Goar 1968; Linderski 1982; Momigliano 1984: 204–11; Pease 1920–23; Timpanaro 1988; and Wardle 2006.

along the Roman model (20–1), but twice seems to avow a personal belief in divination: *diuinationem ... esse sentio* ('I feel that divination exists', 32) and *non uideo cur diuinationem negem* ('I don't see why I should deny divination', 32). If in *De diuinatione*, Cicero rejects divination, he must either have changed his mind or the two attitudes must somehow be shown to be reconcilable.[25] Thus, critics often maintain that in *De legibus*, Cicero speaks as a lawgiver and statesman concerned with the political uses of divination, not as a philosopher, and they point to the fact that even in *De diuinatione* the author explicitly upholds such divinatory practices as augury as part of civic religion, despite the fact that he denies them any actual predictive properties.[26]

More recently, in particular in a pair of influential 1986 papers by Mary Beard and Malcolm Schofield, *De diuinatione* has been viewed as a more balanced exploration of the pros and cons of the validity of divination.[27] On this reading, Cicero follows the practice of Academic Scepticism in presenting arguments on both sides of an issue, leaving the question open for the reader to decide. By not tying any position espoused in the dialogue to Cicero's personal views, this interpretation avoids the problem of self-contradiction (or change of mind), while also taking seriously as contributions to the debate the beliefs voiced in Quintus' speech. Some readers, however, have found it hard to swallow the idea that we are not supposed to privilege Marcus' arguments over those of his brother: not only do they seem superior,[28] but

---

[25] Scholars who posit a change of mind (or evolution of thought) on the part of Cicero include Blänsdorf 1991; Guillaumont 1984: 165–9; and Momigliano 1984: 204–11. That his views remained fundamentally the same and do not involve contradiction is maintained, among others, by Goar 1968 and 1972; Linderski 1982; Rasmussen 2000, 2003: 183–98; Timpanaro 1988; and Wardle 2006. In reconciling Cicero's stances in *De legibus* and *De diuinatione*, scholars occasionally refer to the model of the *theologia tripartita*, made famous by Varro (*Antiquitates rerum diuinarum* frr. 6–11, Cardauns 1976), according to which civic, mythological and philosophical ideas about the gods exist in separate conceptual spheres and do not interfere with one another (this is the phenomenon that Feeney 1998: 14–21, following Paul Veyne, dubs 'brain-balkanisation'). In the opinion of Moatti 1997: 181–3 and Timpanaro 1988: lxxvi–lxxxiii, the failure of thinkers like Cicero to extend their religious scepticism to a critique of political structures shows the limits of the Roman 'Enlightenment'.

[26] See esp. *Div.* 2.70–1, 74 and further below.

[27] See Beard 1986 and Schofield 1986, as well as Krostenko 2000; Scheid 1987–89; and Schultz 2009 and 2014. A sustained critique of the Beard/Schofield interpretation is found in Wardle 2006: 8–28; see also Timpanaro 1994: 257–64 and Green 2014: 78–9.

[28] The rhetorical qualities of both speeches are examined by Krostenko 2000 and Schofield 1988, both of whom point to the shortcomings of Quintus' presentation (engineered, of course, by Cicero himself). By contrast, Denyer 1985 argues that Quintus presents the Stoic theory of divination in a coherent fashion and that it is Marcus who fails to respond in an effective way; this view is attacked by Timpanaro 1994: 241–57.

the fact that the second speaker is identified with the author and gets the last word in the debate[29] is often felt to indicate that we are meant to understand that Marcus has won the day.

I believe that the questions raised by these divergent readings respond to tensions not only in Cicero's dialogue itself, but also in Cicero's actions and literary output over time and in the social practices and cultural ideas of his period. Before exploring some of these questions, though, it should be pointed out that the two readings can to some extent be reconciled if we view the dialogue as, indeed, an Academic debate – not one that ends in aporia, however, but one in which one opinion is determined to be more probable than another.[30] The closest parallel is *De finibus*, written just a few months before *De diuinatione*. There, Marcus Cicero, likewise playing the role of the Academic Sceptic, listens to and responds to the speeches of an Epicurean, a Stoic and an Antiochean, demolishing each person's arguments in turn. In each case there is a serious debate, but each time the Sceptical objections clearly show up the shortcomings of the respective dogmatic position. Given the declared identity of the Sceptical speaker, and in light of the fact that it is his contribution that concludes each dialogue, I think that it is fair to say that he more or less represents Cicero's own view.

Similarly, Marcus' speech in *De diuinatione* does not leave Quintus with a leg to stand on. This does not mean that Marcus himself, or the dialogue as a whole, has reached a position of absolute certainty. Throughout, Marcus (and, indeed, Cicero in his own persona in the prologues) stresses his doubts on the matter of divination, and the dialogue begins and ends with an affirmation of the Academic method of open-minded, non-dogmatic discussion.[31] I thus assume that we are meant to come away from *De diuinatione* with the position that the existence of divination remains unproven and that supposed manifestations of it do not deserve belief, while continuing to reserve ultimate judgement on the question – and (for what it is worth) that this was presumably Cicero's own attitude. The supposed contradiction with *De legibus* seems less grave if we consider not only that personal scepticism need not preclude a belief in the political and societal usefulness of divinatory practices, but also that Cicero's professions in favour of divination are but tentatively phrased (esp. *non uideo cur diuinationem negem*, *Leg.* 2.32) and that, furthermore, the earlier dialogue explicitly bans the Academic method of enquiry, creating a space

---

[29] To be precise, the actual last word goes to Quintus (2.150), but his short sentence is but an acceptance of his brother's invitation to similar discussions in the future.
[30] Compare Santangelo 2013: 15–23.
[31] Expression of doubts: 1.7; 2.8, 28, 48; affirmation of the Academic method: 1.7; 2.8, 150.

where dogmatic views may be aired without being liable to Sceptical attack (*Leg.* 1.39).[32]

Interpretations of *De diuinatione* such as the ones referred to have understandably focused on the central question up for debate, which can be formulated as *Estne diuinatio?*, 'Does divination exist?'[33] *Diuinatio* here is the actual *praedictio atque praesensio* (1.9) of what is otherwise unknowable, and Quintus and Marcus are discussing whether or not such a thing exists in the world, employing forms of *esse* as the *verbum existentiae*. Used in this sense, *diuinatio* is Divination with a capital D, that is, true prophecy or prophetic ability.

However, just like English 'divination', Latin *diuinatio* more often than not refers not to the ontological fact of accurate prediction, but to the diverse predictive practices (whether successful or not) found in human society, that is, what we might call divination with a lowercase d. Throughout *De diuinatione*, Cicero's usage of the term veers from one meaning to the other, beginning with the very first sentence of the prologue, where the author observes that it is a widespread opinion that *uersari quandam inter homines diuinationem, quam Graeci* μαντικὴν *appellant, id est praesensionem et scientiam rerum futurarum* ('there dwells among men a certain "divination", which the Greeks call *mantikē*, that is, the foresight and knowledge of future things', 1.1). While *quandam* introduces an element of vagueness, *uersari... inter homines* (as opposed to *esse*) gives the impression that we are talking about a social practice or custom. If the immediately following definition points us instead to (uppercase) Divination, the ensuing dialogue again and again makes it clear that (lowercase) divination is just something that people do, whether actual *diuinatio* exists or not. Thus, Quintus and Marcus talk about types (*genera*) of divination (1.3, 9, 10), especially those that are traditionally practised in their own society (*de quibus accepimus quaeque colimus*, 1.9; cf. 1.86), and they wonder whether such divination is 'true' or not (e.g. *cur esset uera diuinatio*; cf. 1.9, 37).[34] In spite of the ultimate demolition of the concept of Divination, then, *De diuinatione* leaves us with the impression of a world where divination has been and will no doubt continue to be practised in myriad ways, some of which are judged to

---

[32] Compare Görler 1995, who stresses that this momentary ban does not mean that Cicero has given up on his general Sceptical stance.

[33] The phrase does not show up as such in Cicero's text, but many similar conjunctions of *diuinatio* and the verb 'to be' do (1.1, 10, 83, 124, 125, 128; 2.8, 12, 25, 41, 74, 102, 106, 131). Some of these occur as part of the Stoic argument 'If there are gods, there is divination', which is put forward by Quintus and subsequently disproved by Marcus (1.10, 82–3; 2.41, 102, 105–6).

[34] Note that in cases like these, *esse* is used as a copula: the question is not whether Divination exists, but whether existing divination has certain properties.

be better than others. As we will see, Marcus Cicero himself turns out to be particularly adept at 'doing' divination: all the while denying the existence of *diuinatio*, he nonetheless shows himself to be an authority on and expert in its practice.

As readers of De diuinatione have often noted, Quintus throughout Book 1 attempts to enlist his brother in support of his own argument in favour of Divination.[35] He quotes from three different poetic works – Cicero's translation of Aratus (1.13–16), his poem on his consulship (1.17–22) and the *Marius* (1.106) – and refers repeatedly to De natura deorum (1.33, 93, 117); recounts a prophetic dream by Marcus (1.58–9) and reminds his brother of a true prediction he received from a rower in the fleet at Dyrrhachium during the war against Caesar (1.68–9); and does not miss any chance to bring up Marcus' status and expertise as an augur (1.24, 25, 28, 29, 30, 72, 90, 103, 105).[36] Of course, this is not a bad rhetorical strategy (and perhaps an indication that Quintus' speech is not as poor as is sometimes supposed), but we must bear in mind that it was Cicero himself who wrote the book and who must have introduced these references to his own divinatory experiences on purpose. No doubt the author relished the opportunity to quote his own poetry but this cannot explain his having Quintus harp continuously on his brother's being such an excellent *auctor* and *testis* for his own arguments (1.17), a rhetorical move that risks making Marcus – and, indeed, Cicero himself – look self-contradictory. Thus, Quintus asks triumphantly, apropos of the list of prodigies during the Catilinarian conspiracy as explicated by the Muse Calliope in *De consulatu suo* (1.22):

> *tu igitur animum poteris inducere contra ea, quae a me disputantur de diuinatione, dicere, qui et gesseris ea quae gessisti, et ea quae pronuntiaui accuratissime scripseris?* Can you really bring yourself to argue against my points about divination, you who have done what you have done and written (and very well, too) what I just quoted?.

Marcus' response to Quintus' provocations is remarkable. He proceeds to demolish, in hilariously sarcastic fashion, the predictive signs and practices his brother has cited with reference to his own words and deeds. He makes fun of the all-too-obvious 'portents' during the Catilinarian conspiracy, previously employed by himself to great effect not only in *De consulatu suo*,

---

[35] See esp. Krostenko 2000: 380–5; Santangelo 2013: 25–7; and Schofield 1986: 56–8.
[36] Note in addition Quintus' numerous references to friends and acquaintances of Marcus who were objects or recipients of true predictions, or whose belief in divination ought to be taken seriously: Deiotarus (1.26–7; cf. 2.20, 78–9), Roscius (1.79; cf. 2.66), Aesopus (1.80), Divitiacus (1.90), App. Claudius Pulcher (1.29, 132; cf. 2.75).

but also in the third Catilinarian: how clever (*scite*, 2.47) of Jupiter to strike the statue of the she-wolf and twins with lightning and thus create an easily interpretable sign of danger to Rome![37] Dreams, including Marcus' and Quintus' own, have no prophetic force, but arise from the dreamer's waking thoughts and preoccupations (2.136–42). As for the crazed prediction of the rower at Dyrrhachium, nerves were frayed at the time, and it is not surprising if fear made certain individuals descend into hysteria (2.114). If the gods had wished to communicate with the Pompeian party, would they not rather have chosen a more elevated person such as Cato, Varro, the fleet commander Coponius or indeed Cicero himself as the recipient of their message (ibid.)? Finally, augury – so the augur Marcus informs us blithely – is in this day and age little more than a scam (70–83), which is kept up *et ad opinionem uulgi et ad magnas utilitates rei publicae* ('because of the beliefs of the common people and the great benefit to the commonwealth', 70).

While Marcus is thus more than ready to disclaim any belief in supposed instances of Divination associated with his own life and work, he refuses to engage with Quintus' accusation of self-contradiction (2.46):

> '*tu igitur animum induces*' (*sic enim mecum agebas*) '*causam istam et contra facta tua et contra scripta defendere?*' *frater es, eo uereor. uerum quid tibi hic tandem nocet? resne quae talis est an ego qui uerum explicari uolo? itaque nil contra dico, a te rationem totius haruspicinae peto.*

> 'Will you really bring yourself' (for thus you were attacking me) 'to defend this cause against both your actions and your writings?' You are my brother; therefore I will let it go. But really, what are you so upset about? The subject matter, which is what it is, or I, whose objective is the truth? Therefore I won't contradict you, but instead I ask you to explain the whole system of haruspicy.

This non-response is evasive in the extreme. Marcus invokes the respect owed to his brother (*frater es; eo uereor*)[38] as an explanation for his refusal to

---

[37] Cf. *De consulatu suo* fr. 2.42–6, Soubiran 1972 (*Div.* 1.20) and *Cat.* 3.19. Marcus likewise mocks the supposed significance of lightning damage to the statue of one Pinarius Natta and to a bronze legal inscription (cf. *De consulatu suo* fr. 2.39–40 [*Div.* 1.19] and *Cat.* 3.19), and of the coincidental rededication of a previously damaged statue of Jupiter at the very moment when the Catilinarian conspiracy was exposed (*De consulatu suo* fr. 2.55–65 [*Div.* 1.20–1] and *Cat.* 3.20–1).

[38] Pease 1920–23 *ad loc.* glosses this phrase with 'you are my brother, and I am accordingly showing you all due respect'. Similarly, Timpanaro 1988 translates, 'Sei mio fratello; per questo ti devo rispettare', but notes Cicero's 'reticenza' in his commentary. Santangelo 2013: 26 suggests a more aggressive undertone: 'something along the lines of: "I am your brother and I have to respect you – but how could you possibly take that argument seriously in the first place?"'. Compare also Schofield 1986: 58.

answer Quintus' accusations, apparently implying that a potentially aggressive discussion of personal motivations would detract from the purpose of the dialogue, which aims solely at determining the truth. By first having Quintus challenge his brother's attitude and then failing to let Marcus offer an explanation or disavow his earlier statements and actions, Cicero draws attention to the fact that his persona, while denying the existence of Divination, has in the past repeatedly made use of divination, a practice for which he offers no apology whatsoever.

What is more, Marcus explicitly recommends augury, whose divinatory claims he ridicules but whose political usefulness – in particular in controlling assemblies, an objective dear to the optimate cause – he openly avows (2.70, 74, 75). When he recommends that augury be maintained and that magistrates who disregard the auspices be punished, Cicero does not refer to the recommended practice as *diuinatio*. Instead, he speaks of the *mos religio disciplina ius augurium collegi auctoritas* ('custom, religious obligation, discipline, augural law, authority of the College', 2.70) that ought to be preserved and the *religio* ('religious obligation', 2.71) and *patrius mos* ('inherited custom', ibid.) that officials need to follow. Augury is a social practice and a public institution. The fact that it is not Divination is no reason why it should not be done.

At two points, Marcus makes explicit his distinction between what I have been calling Divination (the ontological fact) and divination (the social practice). Apropos of an interpretation of portents by the haruspices, he exclaims, *quasi ego artem aliquam istorum negem! diuinationem nego* ('As though I denied that they had some method! What I deny is the existence of Divination', 2.45). Similarly, he observes apropos of augury, *quis negat augurum disciplinam esse? diuinationem nego* ('Who denies that the augurs have a system? What I deny is the existence of Divination', 2.74). Augury and haruspicy are *disciplinae* or *artes*. They are knowledge-based practices that can be systematically taught, learned and executed by their practitioners, including Marcus the augur himself.[39] This is lowercase divination, which in the minds of enlightened men like the two Ciceros need not have anything to do with uppercase Divination.

If divination can – and, in certain forms such as augury, should – be practised even if no Divination exists, it can also be helpfully employed in

---

[39] Compare Scheid 1987–89: 128: 'Pour "Cicéron", la divination est... une simple technique humaine'. Quintus, too, considers so-called 'artificial' divination (i.e. prediction that rests on established principles of matching recognised signifiers to recognised signifieds [e.g. augury] – as opposed to 'natural' divination via random prophetic signs [e.g. dreams or chance omens]) an *ars*, but believes that it rests on the empirical long-term observation of predicted outcomes (*Div.* 1.25).

other contexts. As Quintus' examples go to show, Marcus is a master of using divination for poetic, rhetorical and political purposes. By including divine signs in his epic poems, Marcus not only proves his skill at this most sublime poetic genre, in which prophecy had been a standard motif since the Homeric poems; he also elevates his own role (in *De consulatu suo*) and that of his fellow townsman (in the *Marius*) by having his epic protagonists receive a plethora of spectacular portents that, even if they momentarily foretell disaster, ultimately signal the heroes' glory and status as recipients of divine favour.[40] Similarly, the dreams of Quintus and Marcus recalled in 1.58–9, both of which prefigure Cicero's triumphant return from exile, would have served, in their telling and retelling,[41] to add a kind of supernatural lustre to the events. This is true even for the disastrously accurate prophecy at Dyrrhachium: presumed communication from the gods raises a dreary and ignominious episode to a level of epic heroism. Divination grants grandeur: significant events are foretold by significant signs, and a master rhetorician like Cicero knew how to employ this motif, both in his speeches and in his poetry, and even, we learn, in the oral or perhaps epistolary narratives of his own experiences. While Marcus dodges the question, it seems reasonable to infer that in his opinion, lack of belief in Divination is no reason why one should not use divination for one's own purposes, just as the *ars* and *disciplina* of augurs and haruspices exist independently from the question of the existence of Divination.

Quintus' potentially embarrassing examples thus ultimately support the dialogue's point that while the existence of Divination is very much in doubt, this need not preclude the practice or use of divination in a variety of contexts. Marcus may be a sceptic, but this, we learn, has not prevented him from playing the divination card in the past. Nor, we may add, does it prevent Cicero, who in this dialogue manages to have his cake and eat it: while denouncing, via Marcus, the belief in Divination, he manages, via Quintus, to cast a supernatural light on his own political achievements from the Catilinarian conspiracy to his exile to the Civil War.[42] While *De consulatu suo* was a published work, there is no reason to believe that the dreams of the two brothers and the prophecy at Dyrrhachium were in the public domain before Cicero chose to tell of them in *De diuinatione*.[43] At a time

---

[40] On *De consulatu suo*, see further Gildenhard 2011: 292–8 and Volk 2013: 101–5.
[41] In 1.59, Quintus mentions that he has heard Marcus' dream not only from his brother, but also 'repeatedly' (*saepius*) from their friend Sallustius.
[42] I thus disagree with the assertion of Krostenko 2000: 380–5 that in *De diuinatione*, Cicero attempts to 'distance himself from ... [his] poetry and its claims of personal relationships, and even direct contact, with the divine' (383).
[43] Cf. Rosenberger 2007: 302. Marcus' dream is attested also in Val. Max. 1.7.5, but his source is presumably *De diuinatione* itself.

when Julius Caesar – himself by all accounts a religious sceptic – was hard at work projecting an image of himself as a descendant of the gods, Cicero continued to burnish his own legacy, endeavouring to present his career as no less favoured by the divine. Whatever his emotional involvement may have been in the case of the supposed signs and prophecies received during his life, his intellectual stance was one of rational scepticism. However, as *De diuinatione* teaches us, one need not believe in Divination in order to do divination.

Improbable Prophet

Cicero, an innovator in many fields and endeavours, was clearly an unusually versatile user of divination, participating in its civic manifestations as a magistrate and augur, discussing it from a philosophical perspective in *De diuinatione* and employing references to it to great effect in a variety of rhetorical and literary genres. Nevertheless, his expertise in the subject was still very much in keeping with that of his fellow members of the Roman political upper class. By contrast, the activities of Nigidius Figulus, in the words of Wilhelm Kroll, 'go considerably beyond what was customary for a Roman senator at the time'.[44] Solely within the area of divination, Nigidius was evidently involved in the practices of civic religion, as well as an author of numerous works on different types of prediction making; in addition, however, he was himself apparently a freelance diviner and is associated in our sources with particularly spectacular and politically significant prophecies.

Since, unfortunately, the surviving fragments of Nigidius' treatises on augury, extispicy and dream interpretation take up less than two pages in Swoboda's 1889 edition (frr. 79–82), little can be said about the content of these works. His famous *Sphaera*, a discussion of the starry sky divided into the *sphaera Graecanica* (the Greco-Roman constellations) and the *sphaera barbarica* ('foreign' constellations, esp. Mesopotamian and Egyptian), probably had an astrological purpose, treating the constellations as *paranatellonta*, that is, as exerting influences when rising simultaneously with signs of the zodiac.[45] However, this divinatory aspect is not easily apparent from the extant fragments (frr. 84–101). By contrast, an Etruscan brontoscopic calendar preserved in John the Lydian's *De ostentis* 27–38

---

[44] Kroll 1937: 210: 'Umfang und Charakter der Studien des N. gehen über das, was damals bei einem römischen Senator üblich war, erheblich hinaus'.

[45] See Boll 1903: 350–63; cf. Della Casa 1962: 114.

(= fr. 83) is explicitly a tool of prediction: this list of thunder omens 'by the Roman Figulus from the works of Tages, translated word for word' (*Ost.* 27) correlates the occurence of thunder on any given day of the year with its significance.[46] This document and the purported role of Nigidius in its preservation and/or composition have been much debated; I return to it below.

As far as Nigidius' own divinatory activity is concerned, the evidence is what one might describe as anecdotal, if certainly intriguing. In a story related by Apuleius and drawn from Varro,[47] Nigidius assists a certain Fabius in recovering 500 lost denarii by hypnotising some boys (*ab eo pueros carmine instinctos*, Apul. *Apol.* 42.7), who under his spell reveal the whereabouts of the money. Not only is this one of the earliest attested cases of divination by child medium,[48] but the incident is remarkable for the upperclass milieu in which it takes place: Fabius is presumably a member of the patrician *gens* of that name,[49] and one of the missing coins amusingly turns out to be in the possession of none other than Cato (8).

That Nigidius may have supplied his fellow senators with the occasional customised prediction is suggested also by the more famous anecdote of his casting the horoscope of Augustus (Suet. *Aug.* 94; Cass. *Dio* 45.1.3–5).[50] When Octavius arrives late for a meeting of the senate, Nigidius asks him about the reason for his delay; having learned that his colleague's wife has just given birth to a son, Nigidius construct's the child's birth chart and reveals to the stunned father that the ruler of the world has been born (*dominum terrarum orbi natum*; δεσπότην ἡμῖν ἐγέννησας). While the story in this shape and form is clearly apocryphal, it is striking how the optimate Nigidius is credited with predicting what amounts to the downfall of the Republic and the rise of an autocratic political system.[51]

The same holds true for Nigidius' cameo appearance at the end of the first book of Lucan (1.638–72), where the savant 'whose endeavour it was to know the gods and the secrets of the heavens' (*cui cura deos secretaque*

---

[46] For an English translation of the calendar, see Turfa 2012: 86–101; an Italian translation is found in Domenici and Maderna 2007: 81–98.

[47] Apul. *Apol.* 42.5–8. It is unclear to which Varronian work Apuleius is referring (cf. Cardauns 1960: 49). On the episode, see Cardauns 1960: 45–50; Liuzzi 1983: 7–13; and Mevoli 1992.

[48] On the practice, see Johnston 2001. Mevoli 1992 argues unconvincingly that the *pueri* enchanted by Nigidius are not boys but Fabius' slaves, who have stolen the money.

[49] See Cardauns 1960: 47 on the possible identification with Q. Fabius Maximus (cos. 45).

[50] On this prediction, see Engels 2007: 622–3 and Vigourt 2001: 274–5, 351 and 400–4.

[51] Vigourt 2001: 400–4 suggests that the prediction – as well as similar ones associated with other anti-Caesarians – was invented in the Augustan period in order to rehabilitate Nigidius. See, however, further below on the tone and possible politics of the prophecy.

caeli / nosse fuit, 638–9) is the second of three purveyors of dire predictions at the outset of the Civil War, being preceded by the haruspex Arruns and followed by a crazed Roman *matrona*. Nigidius casts a horoscope, which Lucan relates in some detail, referring to or hinting at the position of six of the seven planets (651–63). The details and plausibility of the horoscope have been much debated, with scholars differing greatly on whether the described constellation of celestial bodies bears any relation to the real situation in the heavens in early 49 BC, the date of the scene in Lucan's epic.[52] Bound up with this issue is the question of whether Lucan invented Nigidius' prophecy or used as his source some account of an episode of this kind or even a work by Nigidius himself.[53]

Neither the anecdote of Augustus' horoscope nor Lucan's Nigidian horoscope can pass as a historically reliable source for any specific action or utterance on the part of Nigidius. However, the two episodes paint a similar image of a conservative Roman politician who uses his non-traditional divinatory expertise for predictions to warn of events that will lead to a sole ruler holding sway over Rome and, indeed, the world. As we have seen, Nigidius' supposed interpretation of Augustus' birth chart leads him to the conclusion that a *dominus*/δεσπότης has been born for the 'world' (Suetonius) or, more pointedly, 'for us' (ἡμῖν, Cassius Dio).[54] In the version told by Dio, this prediction so terrifies Octavius that he wants to kill his own infant son; he is stopped by Nigidius himself, who points out that fate cannot be escaped. In Lucan, too, the senator-astrologer predicts that the Civil War will inexorably lead to the rise of a *dominus* and the loss of Roman freedom (1.669–70): *et superos quid prodest poscere finem? / cum domino pax ista uenit* ('And what is the point of asking the gods for an end? That peace comes with a master'). The *dominus* in Lucan's text must likewise be Augustus, whose rise to power the historical Nigidius could hardly have prophesied.[55] However, it is perfectly imaginable that the senator in his own time used his astrological expertise to issue predictions that warned of the dangers of civil strife and in particular of the rise of another powerful

---

[52] See Domenicucci 2003; Getty 1941, 1960; Hannah 1996; Housman 1926: 325–7; Lewis 1998; and now Roche 2009: 360–75.

[53] Boll 1903: 363; Domenicucci 2003: 100; Getty 1941: 22; and Rosillo López 2009: 109–10 all consider it likely that some Nigidian material lies behind Lucan's episode, but differ on the details; Beard *et al.* 1998: 1.152, n. 106 consider the prophecy a 'brilliant parody' of the actual Nigidius; MacRae 2013a: 272–5 now suggests that by introducing Nigidius as a character, 'Lucan comments on the co-option of Roman civil theology by the Julio-Claudian house' (274).

[54] On *dominus*/δεσπότης as a 'titre exceptionnel pour des Romains', see Vigourt 2001: 274–5 (quotation at 274).

[55] See Domenicucci 2003: 101; Narducci 1974: 100; and Roche 2009 *ad* Lucan 1.670; differently Lebek 1976: 168–78.

individual and potential *dominus*, Nigidius' enemy Julius Caesar. I suggest that both the apocryphal story of Augustus' horoscope and the appearance of Nigidius in Lucan may be inspired by actual partisan prophetic activity on the part of the optimate politician – or at the very least by a tradition that Nigidius, fabled for his occult knowledge, may have engaged in such activity. Having both the motive and the means, Nigidius was the very man who could have used divination as a form of political intervention.

A tantalising further piece of evidence that the senator may have done so comes from the above-mentioned brontoscopic calendar. Scholars have long been divided over whether this fascinating document – surviving in Greek but supposedly going back to the Latin translation of an Etruscan text – reflects Mesopotamian, Etruscan, Roman or Byzantine material and concerns, with many identifying a number of different cultural strata.[56] In this context, it has been suggested that some of the predictions contained in the calendar may have specific applications to Late Republican Rome and that Nigidius may have adapted an Etruscan document to fit his own purposes.[57] Indeed, it is striking how often thunder is supposed to indicate (in addition to weather conditions, crop failures, diseases, etc.) that dissent or even civil war will arise in the commonwealth or that power will become concentrated in the hands of a single individual. Consider, for example, the following two lemmata (Lydus, *Ost.* 28 and 30):

> [14 July:] If it thunders, it signifies that power over everything will fall into the hands of one, and he will be the most unjust in public affairs.

> [25 September:] If it thunders, a tyrant will arise from the disagreement of the citizenship, and he himself will die, but the powerful will undergo unbearable suffering.

While neither the described incidents nor the dates can be matched exactly to events in the Roman sphere, the overall impression with which the calendar leaves the reader is one of an unstable political situation in which civil strife is endemic and strong individuals have the disquieting potential to rise to (or fall from) power, to the detriment of the commonwealth as a whole. Even if the calendar is indeed a faithful translation of an Etruscan document and thus reflects the political situation in early Etruria, Nigidius may well have decided to publish this gloomy set of predictions not only out of an

---

[56] See esp. Ampolo 1990–91; Piganiol 1951; and Turfa 2012.
[57] Proponents of some version of this thesis include Della Casa 1962: 128–9; Domenici and Maderna 2007: 29–30; Guittard 2003: 462–4; Kroll 1937: 207–9; Piganiol 1951; and Weinstock 1951: 140–2; differently Turfa 2012, esp. 111–13, who argues that the calendar is fundamentally an Etruscan document, reflecting conditions in the eighth and seventh centuries BC.

antiquarian interest in the *disciplina Etrusca*, but also to warn his fellow citizens of the precariousness of a political situation that could easily lead to the rise of an autocratic *dominus*. That an intention of this kind may lie behind the Latin version of the calendar is indicated by the sentence with which John the Lydian concludes his own translation of the document: ταύτην τὴν ἐφήμερον βροντοσκοπίαν ὁ Νιγίδιος οὐ καθολικὴν ἀλλὰ μόνης εἶναι τῆς Ῥώμης ἔκρινεν ('Nigidius believed that this calendar of thunder omens did not apply generally, but only to Rome', *Ost.* 38).[58]

While the evidence remains circumstantial, it seems to me that there is reason to believe that Nigidius employed his considerable knowledge both as an active diviner and as a scholar of divination as a tool of political propaganda. This non-traditional and aggressive use of divination is a far cry from Cicero's versatile but much more mainstream handling of prophetic theory and practice. On one crucial point, however, the two thinkers and political allies appear to be very much agreed: divination (whether true or false, and practised in whatever shape or form) is a matter of central importance to the Roman state and one in which they themselves, qua members of the political elite, have an important stake. Thus, Marcus observes apropos of augury (*Div.* 2.74[59]):

> *quod quidem institutum rei publicae causae est, ut comitiorum uel in iudiciis populi uel in iure legum uel in creandis magistratibus principes ciuitatis essent interpretes.*

> This practice was established for the sake of the commonwealth, so that the leaders of the state would be the arbiters of the *comitia*, whether for the purpose of criminal trials or passing laws or electing magistrates.

Members of the senatorial class act as *interpretes comitiorum*, an expression that indicates that being a diviner (*interpres* [*deum*]) and being in charge of

---

[58] We may view Nigidius' publication of the brontoscopic calendar in the context of other Late Republican Etruscan predictions with a political and/or eschatological bent (cf. Turfa 2012: 26–33), including the Prophecy of Vegoia (on which see Valvo 1988, who discusses parallels to the Calendar) and the response of the haruspices at issue in Cicero's speech of 56, which shows verbal parallels to the Calendar (see Lenaghan 1969: 157–9; Piganiol 1951: 84–5; and Turfa 2012: 31–3 and 317–18). Rawson 1978 examines Etruscan political attitudes in the Civil War, concluding tentatively that 'it is perhaps true that experts in the *disciplina Etrusca*... were likely to be sympathetic to the Roman optimates; and that is likely to have meant opposition to Caesar, at least at certain times of his life' (146). It has been suggested that Nigidius was himself Etruscan (e.g. Harris 1971: 321–2; cf. Rawson 1978: 138), but there is little evidence beyond the fact that he must have known the Etruscan language in order to be able to translate the Calendar.

[59] Cf. *Div.* 1.89, 95; 2.70, 75; *Leg.* 2.30–1.

the political process (here, the *comitia*) amount to the same thing.⁶⁰ In this ideal political world, the divinatory authority and expertise of the political elite are part and parcel of a system designed to maintain senatorial power and ensure the smooth running of the *res publica*.

In the time Cicero and Nigidius were active, this system was showing signs of severe strain and ultimately collapsed. Clinging to their position as *principes ciuitatis* and attempting to stem the tide of political dissolution, the two conservatives continued to consider divination their area of expertise and a viable sphere for their political activities. While their attempts to save the Republic were, of course, doomed – no horoscopes, prophetic dreams, signs from Jupiter or thunder claps in Etruria could prevent the rise of the threatening *dominus* – and both men ultimately paid for them with their lives, the learning, commitment, originality and occasional quirkiness with which Cicero and Nigidius handled divination contributed substantially to the creation of a new intellectual discourse and the cultural flourishing of the Late Republic.⁶¹

---

[60] The pointed phrase may have been coined either by Cicero himself or otherwise by Ti. Gracchus (cos. 177, 163), in the context of an incident alluded to by Marcus immediately following this quotation (*Div.* 2.74–5; cf. 1.33) and told more fully at *Nat. D.* 2.10–11: there, Tiberius, annoyed at a pronouncement of the haruspices that seems to call his authority into question, asks sarcastically whether *Tusci ac barbari* are supposed to be *interpretes... comitiorum* (2.11).

[61] The senatorial status and political conservatism of Cicero and Nigidius militate against the well-known thesis of Wallace-Hadrill 1997 and 2008 that the 'Roman cultural revolution' of the Late Republican and Augustan periods involved a 'shift in the control of knowledge from social leaders to academic experts' (1997: 12). See now MacRae 2013a: 148–213, who shows that the study of Roman religion in the Late Republic was largely driven by the political elite. The sociology and politics of knowledge in this period are complex and cannot be discussed in greater detail here; I hope to revisit the issue in future publications.

15 | The Public Face of Expertise: Utility, Zeal and Collaboration in Ptolemy's *Syntaxis*

JOHANNES WIETZKE*

The second-century CE polymath Claudius Ptolemy is a foundational author in the history of science. Surprisingly, despite his scientific importance, and despite the incredible volume of his work that survives in manuscript, he has often been overlooked in scholarship investigating the textual practices of ancient scientists and, in particular, the interplay of authority, tradition and expertise in their writings. More and more, however, Ptolemy is gaining due recognition as an important source for such questions,[1] and this chapter aims to involve Ptolemy more deeply in the discussion with a close analysis of his *Mathêmatikê Syntaxis*, the astronomical masterpiece perhaps better known to readers after antiquity as the *Almagest* (hereafter in this chapter: *Syntaxis*).[2] To be specific, by examining the formal features of the *Syntaxis* and other, non-mathematical texts, I will develop three claims: (1) that Ptolemy and other expert authors present authorship as a kind of euergetism; (2) that Ptolemy uses euergetic language to create a unique model of astronomical collaboration; (3) that this collaborative rhetoric, in spite of itself, ultimately underscores Ptolemy's position as the mastermind of the *Syntaxis*. In the course of my argument I hope to trace key similarities and differences, on the one hand, between the *Syntaxis* and other ancient compendia of expertise; and on the other hand, between authorship in general and other cultural practices that shaped the social worlds of antiquity.

The Literary *Euergetês*

Before engaging directly with Ptolemy and his *Syntaxis*, I begin with a general question: Why did ancient experts (be they mathematical, historical or

---

* I am deeply grateful to Maud Gleason, Geoffrey Lloyd, Reviel Netz, Josiah Ober, Nikolaos Papazarkadas, Ava Shirazi and the editors of this volume for their beneficial advice and criticism at various stages of the writing process. I owe special thanks to Markus Asper for sharing an early version of Asper 2013c, and again to the editors for sharing Beagon 2013 in advance of publication. Errors of fact and interpretation are my own.

[1] See esp. Bernard 2010; Jones 2005b; and now Tolsa 2013. For briefer treatments, see Feke 2012: 89; Mansfeld 1998: 66–75; and Netz 1999: 40, n.79.

[2] Citations of the *Syntaxis* will refer to Heiberg 1898–1907 by part and page number.

technical) publish their works? Posed so simply, the question may prompt a simple answer: to establish authority (no surprises here). But we can probe more deeply the notion of 'publish'. Authorship is the making public, in some sense, of one's expertise and creates an author's social identity. No one, it seems, claimed to drop out of society in order to become an author. On the contrary, it is remarkable how urgently authors in various prose genres express their social engagement. In what follows I briefly survey some of the different conventions that Greek prose authors of the Imperial period use to claim that engagement.[3] The survey is far from exhaustive, however, and not every surveyed author uses each of the conventions described. But we may regard each convention as a contribution to the construction of authorial social identity, and taken together, they present a general, composite image of a socially engaged author.

I focus on three conventions in particular, beginning with the impetus for writing. (1) Various writers claiming technical or historical expertise (the latter to a lesser degree) ascribe the initiative to compose their works to an external stimulus, often the prompting or demand of a patron or 'friends'.[4] (2) Once given a literary charge, authors profess to take it seriously: 'zeal' or 'care' (e.g. σπουδή, ἐπιμέλεια or cognate verbs) for the subject matter defines the attitude proper to an author, who frequently encourages the same in his readers.[5] In turn, this combination of social orientation and earnest intent underwrites (3) an author's general interest in utility. The form of a work may be designed to help the reader learn the material, but authors also trumpet the beneficial nature of a work's content, even to a wide, non-specialist readership.[6]

---

[3] Such expressions are frequently made in prefatory sections, on which see Alexander 1993: 11–101 and Mansfeld 1998. Moreover, while my present focus is on Greek authors of the Imperial period, the survey could readily be expanded to include Latin authors, as well as authors from other periods, without requiring any radical alteration of the general conclusion; cf. Formisano 2001: esp. 27–31; Janson 1964; and Santini et al. 1990–98.

[4] Discussed with many examples by Alexander 1993: 27–9, 50–63 and 73–5; and König 2009, esp. 40–4. Cf. Janson 1964: 116–24.

[5] While some authors do proclaim their own zeal (e.g. Nicomachus, *Manual of Harmonics* preface; Ael. *NA*, epilogue), the attitude appears more often to define the efforts of 'third-party' authors: these include generic predecessors, whether a named individual (e.g. [Longinus], *Subl.* 1.2) or an unspecified mass (e.g. Dion. Hal. *Ant. Rom.* 1.1.3, Josephus, *AJ* 1.1, Ael. *NA*, preface), as well as idealised authors (e.g. Strabo 1.1.19). Perhaps to speak overtly of one's own zeal was to hazard the thin ice of self-praise. Cf. Alexander 1993: 100.

[6] Examples are legion; see Alexander 1993: 97 for a non-exhaustive list. On the general orientation towards a wide readership and lack of secrecy, see Long 2001: 16–45. Of course, some authors did withhold information, e.g. those of pharmacological works: see Totelin 2009. She notes that it is less clear whether those withholdings reflect a premium on secrecy or on oral and practical instruction for transmitting knowledge (244).

It is not immediately clear, of course, that reading Strabo's *Geography* offers exactly the same 'benefit' (1.1.2: ὠφέλεια), and to the same sort of reader, as reading Philostratus' *Life of Apollonius of Tyana* (1.3), but these and other texts do broadcast claims of earnestness and utility, in the same terms, to a wide range of readers. On a formal level, then, many authors construct a public social context for themselves, sometimes even in spite of claimed intentions to the contrary.[7] Building on this point I will push the general conclusion even further: by publishing his expertise an author claims a political impact by benefitting a community (or at least literate or elite segments thereof).[8] Authorship can thus be expressed as a benefaction to a political community, and this invites us to consider such an author as the literary equivalent of a *euergetês*.[9]

A 'literary *euergetês*'? I use the term to evoke those *euergetai* who directed their beneficence towards a political body and, in turn, were publicly honoured for doing so: from the Classical period until a steep decline in the third century CE, elite individuals – often citizens, but not always; usually male, but not always – used their own resources to finance building projects, warships, festivals and distributions of food and other commodities for the good of a city – 'private liberality for public benefit', as Paul Veyne put it[10] – and for their expenditures they received public commemorations, typically in the form of honorary decrees. By the Imperial period orators and sophists were receiving honorary decrees for services rendered to cities.[11] Authors of technical treatises conceptualised their contributions in similar terms.

In both technical authorship and civic benefaction, the practitioner's authority is bolstered through similar linguistic conventions. Here, I pick up John Ma's analysis of the conventional language of honorary decrees, and in particular his claim that such language legitimates benefactors by proclaiming not only their benefactions towards a political body, but also the zeal and enthusiasm with which they perform them.[12] Idealised benefaction is thus defined by proper attitudes in addition to proper actions.

---

[7] See, e.g., König 2009 on Galen's ambivalence towards the spotlight.

[8] Perhaps as expected, a mark of historical writing (e.g. Dion. Hal. *Ant. Rom.* 1.6.3–5) and geography (e.g. Strabo 1.1.16); less intuitively of literary criticism ([Longinus], *Subl.* 1.2). Galen calls *himself* a 'common good' for the people of Rome (*De praecogn.* 4.1 = K14.619).

[9] Beagon 2013 develops a parallel claim in a Roman aristocratic context, focusing on Pliny the Elder.

[10] Veyne 1990: 10.

[11] Gauthier 1985 and Veyne 1990 remain classic treatments; see now Zuiderhoek 2009 on the eastern Roman Empire, with up-to-date bibliography. On orators and sophists, see Puech 2002.

[12] Ma 1999: 182–94, esp. 191, with references to Ma's Epigraphical Dossier. Cf. Gauthier 1985: 56.

Although Ma's specific interest lies in Seleucid monarchs and the dominated *poleis* of Hellenistic Asia Minor, his analysis can be applied to honorary inscriptions in the wider Hellenistic world as well as under the Roman Empire.[13] Thus, in second-century CE Aphrodisias we find a *xustarch* being honoured for executing his duties 'with care and great zeal' (ἐπιμελείᾳ μετὰ σπουδῆς ἀπ[ά]|σης),[14] and a first-century CE inscription recovered at Giza honours one Cn. Pompeius Sabinus for constructing the levees 'with great care' (μετὰ πάσης ἐπιμε[λείας]).[15] An animating zeal similarly defines the orators honoured in Imperial-period inscriptions.[16] Assuming thus a general consistency in such presentations, we can posit salient parallels between the respective legitimising strategies of civic benefactors and civic-minded authors: the nexus of political orientation, proper attitude and beneficent interest outlined above. While a number of social and historical differences warrant a longer study,[17] for now I maintain that an author may present himself as a literary *euergetês* in a manner that invites profitable comparison with other social practices. As Jason König and Tim Whitmarsh have recently affirmed, 'the world of knowledge... is never neutral, detached, objective'.[18]

Claudius Ptolemy: The Quiet Alexandrian?

Most ancient mathematical writing in fact does not seem to fit the euergetic typology. Here, I build on Reviel Netz's general conclusion that the production of creative, original works of mathematics[19] was an activity largely isolated from political life[20] (which is not to say that non-mathematicians

---

[13] Ma 1999: 182–3.   [14] *Corpus Inscriptionum Graecarum* 2811b.13–14.
[15] *SEG* 8.527.12.   [16] Puech 2002: 30. See, e.g., Puech 2002: 69, ll. 12 and 20.
[17] E.g. an honorary decree seems not to have been the typical result of an author's outlay – but it is not without precedent: consider P. Herennius Dexippos, whose statue was set up in Athens in the third century CE, specifically in recognition of his historical writing – though recorded as well are the many civic offices he held (Puech 2002: 220).
[18] König and Whitmarsh 2007a: 7.
[19] The terms 'creative' and 'original' here are not to be overlooked: I mean the vanguard of the exact, geometric sciences – Archimedes, Apollonius – not a Nicomachus, whose compilation shows greater interest in the philosophical aspects of numbers than in geometric proof (D'Ooge 1926: 16).
[20] Netz 1999: 292–311. Even apparent exceptions corroborate this point: in the *Sand-Reckoner*, Archimedes engages *as a mathematician* with King Gelon, treating him as one who should 'examine carefully' his proof, as if he were among those who 'lay claim to' mathematics (*Aren.* 157). The subject of the treatise, a system for naming astronomically large numbers, has no practical value, perhaps appealing to a Hellenistic interest in the grandiose (Netz 2009: 57). Moreover, when Archimedes does offer something 'of no small use' (*Method* Vol. 2, p. 73,

never appropriated mathematical discourse in making claims of proof).[21] While the conclusion may well encompass the genre of mathematical writing as a whole, it is largely based on a survey of evidence from the Classical and Hellenistic periods. What can be said about the mathematics and the exact sciences in the first few centuries of the Roman principate? Not much, as it turns out. Following Netz again, we find that our evidence suggests that creative mathematical activity, in general, dried up around the turn of the millennium.[22]

In the midst of this desert, however, in the second century CE, Ptolemy brings to flower an oasis of mathematical writing: 180,000 words of mathematical astronomy; 30,000 of harmonics; 11,000 of theoretical cartography.[23] Ptolemy affords an opportunity to compare the textual-social constructions of an Imperial-period mathematician and those of authors professing other kinds of expertise.[24] In this chapter, I offer an account of Ptolemy, the author of the *Syntaxis*. What kind of social identity does the mathematical author construct for himself in this text?

Immediately we see that the case for Ptolemy as a civic-minded or otherwise politicised mathematician is thin. In general, scholars have found Ptolemy's works to be a poor source for understanding the connection between their author and his wider social and political context.[25] This is not to deny outright that König and Whitmarsh's assessment could also

---

Netz et al. 2011), the utility is strictly mathematical; cf. Apollon. Perg. *Con.* Vol. 2, p. 4, Heiberg 1891–93. Note, however, Eratosthenes' letter to King Ptolemy, which illustrates the problem of 'doubling the cube' with reference to supersizing monumental tombs, catapults, etc. (preserved in Eutocius, *In Archim. Sph. Cyl.* 88–96, Heiberg and Stamatis 1915; authenticity defended by Knorr 1989: 131–46).

[21] Cf. Lloyd 1990, esp. 73–97, and Netz's contribution to this volume, which argues that this was quite rare.

[22] See the tables in Netz 1997: 6–10. See, however, Cuomo 2000: 9–56 on the very public profile of mathematics in general in the first centuries CE.

[23] The magnitude and self-sufficiency of Ptolemy's mathematical works, likely overshadowing any recent predecessors' efforts, may be at least partially responsible for the appearance of the foregoing dry spell (Jones 2010a: xii).

[24] Ptolemy of course authored non-mathematical works, but he himself carefully draws methodological and discursive distinctions between, for instance, theory aiming to define planetary movements geometrically and that which investigates planetary natures and their terrestrial effects, i.e. 'astrology' (*Syntaxis* 1.5–7; *Tetr.* 1.1.1–3). In isolating a discourse defined by a specific form, method and objective, I thus follow Ptolemy's own practice. On astrology as social practice, see Barton 1994b: 27–94.

[25] Pedersen 2011: 11; Jones 2010a: xi. See now, however, Tolsa 2013. Ptolemy's philosophical interests, on the other hand, have resulted in more fruitful enquiries: see Feke 2009, 2012; Taub 1993. But ethical-philosophical interests need not equate to political engagement: thus Dio, perhaps a generation before Ptolemy, complains to the Alexandrians that the city's philosophers have withdrawn from public life, having lost any hope of improving the masses (32.8).

hold true for Ptolemy.[26] To be sure, Roman power is not invisible in the works of Claudius Ptolemy, himself bearing the name of a Roman citizen.[27] Moreover, the *Tetrabiblos* opens with a defence of the broad 'usefulness' (χρησίμου) of specifically astrological prognostication, a science that, Ptolemy reports, was 'readily criticised by the many'.[28] But this is to affirm that Ptolemy makes no obvious statement that would position his *mathematical* work in front of a wider, political audience.[29] We know nothing for certain, for instance, about the extra-textual identity, social position or even existence of Syrus, the formal addressee of the *Syntaxis* and several of Ptolemy's other works,[30] nor about Ptolemy's activity in Alexandria apart from his astronomical observations. There is no indication from either Ptolemy himself or later commentators that he, in keeping with so many expert authors of the period, maintained connections to Rome.[31] One sixth-century anecdote conversely isolates him for forty years at the 'Wings of Canobus', a name that likely refers to a sparsely populated district on the outskirts of Alexandria.[32] As for the theoretical astronomy exemplified by *Syntaxis*' geometric models, papyrological evidence indicates that this itself was a niche practice in Roman Egypt.[33] All of this conforms well with Netz's general conclusion about the appearance, at least, of the mathematician's

---

[26] See n. 18 above.
[27] Probably conferred on an ancestor (Toomer 1975: 187). Moreover, Roman emperors routinely appear throughout the *Syntaxis*, but strictly in their capacity, shared with Alexandrian, Persian and Babylonian kings, as units in a chronological 'king-list' (see Toomer 1984: 11). In Ptolemy's literary *parapêgma* the *Phaseis*, 'Caesar' appears *passim* as an authority on weather signs; cf. Lehoux 2007: 492. Clarke 1999: 141–2 notes that Ptolemy's *Geography* includes occasional ethnographic commentary, itself consequent to Roman expansion; but Berggren and Jones 2000: 41 point out the lack of clear representation of Roman power on Ptolemy's maps.
[28] *Tetr.* 1.1.3. Cf. Favorinus' attack on astrology at Gell. *NA* 14.1.1–36.
[29] Ptolemy's *Geography*, for instance, while in some sense displaying the Roman Empire, does not present itself as useful to the Empire, nor is there any indication that it was received as such. It thus supports and extends the conclusions of Gehrke 1998, who denies Classical-period mathematical geography any practical value. Ptolemy does present his work as an ethical-philosophical endeavour, which includes an ethical justification for the study of mathematical astronomy: see Taub 1993: 135–153 and Feke 2012.
[30] Toomer 1984: 35, n. 5. Tolsa 2013: 207–67, however, now makes a case for identifying Syrus as M. Petronius Sura, *curator aquarum* of Rome under Hadrian.
[31] At *Syntaxis* 2.30 and 33, Ptolemy states that the astronomer Menelaus observed in Rome, who likewise appears in Plutarch's learned dialogue *De facie* (930a). While present for the conversation, Menelaus does not actually participate, remaining on the margins.
[32] Olymp. *In Phd.* 10.4, granted new credence by Jones 2005a: 61–4. Jones also considers whether Ptolemy may have been a priest at the temple of Serapis in Canobus, where he plausibly dedicated the astronomical 'Canobus Inscription' (surviving only in manuscript) to the 'saviour god' (*Inscriptio Canobi*, Heiberg 1898–1907: Vol. 2, p. 149). On the inscription, see also Hamilton *et al.* 1987 and now Tolsa 2013.
[33] Jones 1994: 37–8.

detachment from *polis*-life. What kind of social relationships does Ptolemy then construct through the *Syntaxis*?

Ptolemy's Readers – Present and Future

Like many ancient technical treatises, the *Syntaxis* sets up a didactic relationship between author and reader. It is important, however, not to overlook two qualifications that define that relationship as Ptolemy frames it in the *Syntaxis*. First, from the beginning Ptolemy presents himself as a teacher of 'specifically mathematical theories' (τῶν θεωρημάτων ... τῶν ἰδίως καλουμένων μαθηματικῶν),[34] proceeding then to define that kind of theoretical knowledge in contrast to others, namely, theology and physics.[35] So Ptolemy has established his enterprise in the *Syntaxis* as a mathematical one; the second qualification follows, namely that his readers will be mathematicians.[36] Indeed, Ptolemy states as much, proposing an account that favours conciseness over a full explanation of matters established by earlier authorities (τῶν παλαιῶν), thus intending the *Syntaxis* for readers 'who have already made even some progress' (οἱ ἤδη καὶ ἐπὶ ποσὸν προκεκοφότες).[37] Progress in what, exactly, will be clarified shortly. For now, the important point is that Ptolemy explicitly writes on a specialised topic for a specialised audience. Lacking here is the pretense of Hero, who expressly digests a mass of previous mechanical writings so that they may be understood by 'everyone' (πᾶσιν); no less lacking is the flattery of Galen, who dedicated a number of works to the consular Flavius Boethus, described as a learned amateur.[38] And Ptolemy means what he says: the mathematics prerequisite for serious study of the *Syntaxis* makes clear that it is for the already initiated and committed.[39] Indeed, Ptolemy's late-antique

---

[34] *Syntaxis* 1.5. [35] Cf. Feke 2009: 23–67.

[36] In general, the connotations of *mathêmatikos* and its cognates are not easily pinned down (Lloyd 2012). But the impression left by the *Syntaxis*' first commentaries is that through the fourth century, students approached the text specifically interested in learning its computations and geometric models, as opposed to anything of, say, astrological concern (Pingree 1994: 75–8).

[37] *Syntaxis* 1.8.

[38] Hero, *Bel.* p. 73.11, Wescher 1867 (= Marsden 1971: 18); Gal. *Lib.Prop.*, K19.13; *AA*, K2.215–16. Here again we may ask, *Who* was Syrus?

[39] See Pedersen 2011: 47–56. Likewise, Ptolemy apparently assumes that readers of his *Geography* will already be familiar with the work of his predecessors, especially Marinus (Jones 2011: 16–17).

commentators eventually resorted to supplementing the text with introductory instructions on calculation.[40]

This initial reading further corroborates notions of a rather narrow readership for advanced mathematical writings.[41] From here, we might consider more deeply the question of the roles and expectations that Ptolemy constructs for his audience. Ptolemy writes for those 'who have already made even some progress'. Progress in mathematics is surely implied here, but can we determine the use to which those readers will put the *Syntaxis*? Does Ptolemy mean for readers simply to understand and admire his collection of proofs, or does he encourage further theoretical-astronomical work? An initial answer to the latter question would seem to be negative. In his preface, Ptolemy presents his massive treatise as a 'complete' (1.8: τελείου) theory of terrestrial orientation and solar, lunar, stellar and planetary motion; it promotes no obvious, systematic program for further theoretical or observational research.[42] Moreover, in a later work, the *Planetary Hypotheses*, Ptolemy treats the mathematical theory from the *Syntaxis* as a usable foundation for work in *physical* astronomy.[43] Following these points, along with the claims of thoroughness touched on in the preceding discussion, we may well wonder whether Ptolemy intended for his mathematical readers to keep their eyes fixed downward on his proofs rather than upward on the heavens.

Several points, however, require address before we can accept that interpretation. First, Ptolemy, with ostensible modesty, characterises his achievements as the result of the progress of time, namely, 'the time that has advanced [between his predecessors and his own era]'.[44] Likewise, in the closing statements of the *Syntaxis*, Ptolemy affirms that 'the time leading up to the present contributed to the discovery and more exact correction [of data]'.[45] Ptolemy thus credits the advancement of astronomy not to his

---

[40] Pedersen 2011: 52; cf. Bernard 2010: 519.
[41] Bernard 2010: 513 seems right to infer that 'most [readers of Ptolemy's *Syntaxis*]... were probably astrologers', based on the relative popularity of 'predictive' over 'theoretical' astronomy evident from papyri. But the inverse, that most astrologers were probably readers of Ptolemy's geometric work, need not follow.
[42] Which is not to say that the exposition lacks apologies: e.g. Ptolemy admits the difficulty in obtaining accurate results for the latitudinal motion of planets with his complicated models, but he nevertheless maintains the viability of those models (2.532–4); cf. Lloyd 1987a: 306, n. 78.
[43] *Hyp.* 70.
[44] *Syntaxis* 1.8: ὁ προσγεγονὼς ἀπ' ἐκείνων χρόνος μέχρι τοῦ καθ' ἡμᾶς. Noted by Bernard 2010: 509.
[45] *Syntaxis* 2.608.

own genius, but to the advancement of time itself, during which attentive observers collected new data. It is tempting to infer from Ptolemy's statements that he was optimistic about the advancements that could be made following the passage of even more time – a progressive view apparently espoused by the Hellenistic astronomer Hipparchus,[46] whose work, we shall see, Ptolemy presents as foundational to his own – but do Ptolemy's attributions to time amount to more than modest understatement (conveyed in a perhaps 'Hipparchean' style)? He himself makes no predictions, a perhaps telling silence. Given his claim of producing a complete theory of celestial motion, there is no reason to suppose that Ptolemy expected the passage of time to offer anything other than the confirmation of his own work.

Hence a second point: Ptolemy in fact does envision that task of confirmation for astronomers-yet-to-come. Two brief passages in the *Syntaxis* offer glimpses of future observational activity: first, at *Syntaxis* 2.8 Ptolemy states that he will offer 'those [who come] after us' (τοὺς μεθ' ἡμᾶς) stellar data that will be useful to measure the apparent effects of precession; second, at *Syntaxis* 2.508 he mentions 'those [who come] later' (τοῖς ὕστερον) in regard to their needing to make adjustments to his model for the maximum elongation of the planets Venus and Mercury. In both passages, it must be stressed that the observers cannot be Ptolemy's immediate successors: in the former passage, he specifies that the measurements of precession be made after an interval longer than the 260 years that separate him from his predecessor Hipparchus; in the latter, he affirms, more vaguely, that the aforementioned adjustments will not be needed for an 'exceedingly long time' (ἐπὶ πλεῖστον χρόνον). In both cases, then, Ptolemy conceives of astronomers in a far future still conducting observations. But observations for what purpose? The key point here is that those future astronomers are imagined as making observations not to develop a new theory, but to confirm the durability of Ptolemy's. Ptolemy thus imagines a distant future in which mathematical astronomers still adhere to, still *read* the *Syntaxis* (which, to his credit, turned out to be far from science fiction).

We may yet wonder, however, what role the text assigns to readers who are closer to Ptolemy in time. In order to find itself in the hands of those future observers, of course, the *Syntaxis* required a textual tradition, but this need not have been more than a tradition of readers (of some mathematical sophistication). For observers who might have directly followed Ptolemy, on

---

[46] Pliny, *HN* 2.95. Seneca promotes a similar vision of future astronomical discovery at *QNat.* 7.25.4. Thus, we touch on notions and narratives of scientific progress, on which see Asper 2013c: esp. 414–18 on 'progress as a story of growth', which describes well Hipparchus' sentiment.

the other hand, the text makes no explicit provision, and upon initial consideration his introductory claims of completion would seem to preclude their participation in observations that contribute to the development, and not just the confirmation, of theory.

Yet the image of development that emerges from the *Syntaxis* is itself not a simple one, but entails an iterative process of 'correction' (διόρθωσις), which consists of further observations that, in turn, necessitate adjustments to the initial theory.[47] We may well suppose that this process of correction may involve the participation of individuals other than Ptolemy. Indeed, part of the *Syntaxis* itself consists of Ptolemy's correction of theory received from predecessors, and Ptolemy himself offers some general remarks that suggest it is an inclusive activity. After describing his own corrections to Hipparchus' lunar theory, Ptolemy asserts that he himself has corrected his own theories concerning the movements of the planets Saturn and Mercury using 'more secure (ἀδιστακτοτέραις) observations' (*Syntaxis* 1.328).[48] The assertion motivates some general remarks about the importance of correction:

> For it behooves those who approach this science with a real love for truth and inquiry (φιλαλήθως καὶ ζητητικῶς) to use any new methods they discover that give a more secure result, to correct not merely the ancient theories, but their own, too, if they need it. They should not think it disgraceful (αἰσχρόν), when the goal they profess to pursue is so great and divine, even if their theories are corrected and made more accurate by others (ὑπ' ἄλλων) and not only by themselves.[49]

Here, Ptolemy claims that the divine nature of the celestial bodies makes the accurate definition of their movements trump any personal concern for prestige. Thus, for one devoted to the truth, there is no shame in having one's theory corrected, if the correction nudges the theory closer to the truth. By this statement Ptolemy upholds the prestige of Hipparchus, whose faults in developing a theory of lunar position he had just laid bare.[50] We may reasonably infer, moreover, that the generality of the statement invites Ptolemy's readers to correct *his* theory. At the very least, we may regard such an expression of modesty as a rhetorical device that builds trust between

---

[47] Or so Ptolemy claims. The controversy over whether Ptolemy himself strictly adhered to his empirical principles or fabricated observational data need not be addressed here, however, as I am interested simply in the claims and constructions of the text.
[48] Cf. *Syntaxis* 2.208–10 on the challenges of planetary observation.
[49] *Syntaxis* 1.328, adapted from Toomer 1984: 206.
[50] See below for more on Ptolemy's presentation of Hipparchus.

Ptolemy and his readers, and likewise encourages them to devote themselves more deeply to his text. But did Ptolemy actually expect 'others' to correct his work? One way to approach an answer to this question is to investigate whom he credits with corrections, and in doing so we find that only Ptolemy corrects Ptolemy.[51] Moreover, his command over correcting the *Syntaxis* extends beyond its publication. In his later work on physical astronomy, *Planetary Hypotheses*, he lists a number of revisions made to the theoretical models of the *Syntaxis* since that text appeared, but once again he claims sole credit for them.[52] Whatever the sincerity of his modest and encouraging statements, then, Ptolemy consistently presents himself as the only one of his generation, at least, who in practice has developed and refined astronomical theory.

Netz's claims about the Imperial downswing in mathematical activity notwithstanding, Ptolemy's claimed solitude need not, of course, indicate real isolation. On the contrary, by the early third century, and possibly before, one Artemidorus was already criticising (though misguidedly) Ptolemy's lunar theory.[53] Even more compelling, we have strong papyrological evidence of other celestial theoretical models – other *syntaxeis* – that seem to have been devised near in time to Ptolemy's.[54] Out of the rubbish dumps of Egypt, then, a picture of theoretical astronomy is developing that suggests more diversity of contemporary activity and debate than Ptolemy's own texts imply. If those suggestions actually point to the reality of Ptolemy's intellectual environment, then all the more reason to regard Ptolemy's self-correction as competitive exclusivity, which reinforces Ptolemy's dominant position in the minds of his present and future readers.[55]

Ptolemy and Hipparchus: Astronomical *Euergetai*?

To be sure, Ptolemy is not a lone authority in the *Syntaxis*, but to find those close to par with him we must look to his predecessors. While others will be treated in the next section, here I focus on Hipparchus, the most important of Ptolemy's predecessors and most prominent in the text. We will examine how Ptolemy characterises Hipparchus and himself with the same language

---

[51] The above example is the only instance where Ptolemy describes a correction to *theory*. Throughout the *Syntaxis*, he refers to adjustments to *data*.
[52] *Hyp.* 72.   [53] Jones 1990: esp. 3–4 (on Artemidorus' date) and 10–12.   [54] Tihon 2010.
[55] Both Ptolemy's silence about other contributions and Artemidorus' criticism of other theories are hallmarks of Hellenistic mathematical discourse: see Netz 2004: 60–3.

of utility and zeal that we saw in inscriptions honouring civic *euergetai*; Ptolemy's language valorises his and Hipparchus' respective efforts aimed at a shared goal – one that is astronomical, however, rather than political. In brief, I argue that intentions and attitudes matter as much for estimating what we may consider scientific benefactions as they do for civic ones, before probing further in the next section the dynamics of Ptolemy and Hipparchus' constructed relationship.

Let me begin by probing why Hipparchus is so elevated in Ptolemy's account. First, simply, there is no reason to doubt that the historical Hipparchus was any less an astronomical authority than the Hipparchus presented in the *Syntaxis*. Second, it is important to recognise as well that Hipparchus was already esteemed in wider culture: he was not only a useful predecessor, but a famous one – although 'infamous' might better describe certain of Hipparchus' rhetorical manoeuvres, even non-mathematical authors preceding and contemporary with Ptolemy revere him as a scientific authority, and his profile emblazoned Nicaean coins between the second and third centuries CE.[56] In addition to his esoteric scientific worth, then, Hipparchus' name was valued as currency in the wider economy of prestige.

Ptolemy's own relationship to Hipparchus was complex. At several points, Ptolemy appears to subvert Hipparchus' authority for the sake of his own, distorting not only his predecessor's observational results, but also the methods by which he obtained them.[57] Such distortions are subtle, however, and Ptolemy's general and more obvious stance is to defend and praise Hipparchus' abilities as an astronomer, even when he presents him as being in error, and Ptolemy only once criticises Hipparchus outright for his errors.[58]

More frequently, when Ptolemy describes Hipparchus himself or his habits of observation, he does so with expressions that overwhelmingly demonstrate his confidence in the man. At different turns Hipparchus is 'truth loving' (φιλαλήθης) or 'toil loving' (φιλόπονος), and he conducts his work 'with devotion to enquiry' (ζητητικῶς), 'in most accurate fashion'

---

[56] Non-mathematical authors: e.g. Strabo 12.4.9; Plin. *HN* 2.95; Ael. *NA* 7.8. Coins: see Neugebauer 1975: Vol. 1, 275, n. 8 and Toomer 1978: 222. Strabo (2.1.21–2) criticises Hipparchus for employing seemingly underhanded argumentative tactics; cf. Dueck 2000: 58–9.

[57] See, e.g., Jones 2005b: 18–27, concerning Ptolemy's calculation of the length of the year with respect to Hipparchus', which draws the charge of 'sharp practice' from Jones. See also n. 76 below.

[58] *Syntaxis* 1.450: for calculating parallax 'in an exceedingly neglectful and irrational manner'.

(ἀκριβέστατα), 'with the utmost care in observation' (παρατηρητικώτατα) and 'with great zeal' (μετὰ πάσης σπουδῆς).[59] All attributes of a good man, perhaps, but why does Ptolemy take such pains to frame his predecessor as upstanding in intention and effort? Such praise for a predecessor is exceptional in the history of Greek mathematics: although Ptolemy was not the only one ever to apply a soft touch, elite mathematicians were generally more prone to ignore the accomplishments of predecessors, if not challenge or even attack them.[60] Note that, without exception, Ptolemy's expressions specifically validate the integrity of Hipparchus' work and underscore his trustworthiness as an observer. On the one hand, Ptolemy may be prescribing a model for his readers to follow.[61] But there is more at stake: unlike the various endeavours of Archimedes or Apollonius in purer mathematics, Ptolemy's astronomical work, especially that on precession and planetary motions, has an essential, historical component and entails comparing observations from different times and checking their conformity with models. Here, then, I stress the functional aspect of Hipparchus' work: Ptolemy so continuously makes positive claims about the reputation of Hipparchus and, more precisely, the validity of Hipparchus' data, because he needs them for the validation of his own theories and, ultimately, his own reputation. With specific regard to precession and planetary motions, Ptolemy purportedly founds his own achievements on his predecessor's work, using, for the former, Hipparchus' fixed-star observations which were handed down 'thoroughly worked out' (*Syntaxis* 2.3: μετὰ πάσης ἐξεργασίας) and for the latter, 'a great stock of accurate [planetary] observations from earlier times' (*Syntaxis* 2.210: τοσαύτας ἄνωθεν ἀφορμὰς ἀκριβῶν τηρήσεων).[62] But even while promoting his own accomplishments as superseding those of Hipparchus, Ptolemy does not denigrate the character

---

[59] φιλαλήθης: 1.191, 1.200, 2.210, 2.211; φιλόπονος: 1.191; ζητητικῶς: 1.8 (Hipparchus surely to be understood among τῶν παλαιῶν); ἀκριβέστατα: 1.195, 1.204 (twice); παρατηρητικώτατα: 1.276; μετὰ πάσης σπουδῆς: 1.233 (following Toomer 1984: 153, n. 45).

[60] Netz 2004: 60–3. Consider, however, the kindnesses with which Archimedes eulogises Conon in the prefaces to *Sphere and Cylinder I*, *On Spiral Lines*, and *On the Quadrature of the Parabola*. In his *Sphaerica*, Menelaus alludes to work done well by predecessors (Krause 1936: 118), though without naming names.

[61] Bernard 2010: 511.

[62] How much Ptolemy actually used historical observations to develop his models is an open question. We find only indirect reference to Hipparchus' planetary data (*Syntaxis* 2.267; cf. Toomer 1984: 452, n. 66), and Ptolemy largely depends on his own, and relatively few, dated observations in his presentation (Goldstein 2007: 274–5). But Ptolemy does 'test' each planetary model with at least one observation from the third century BCE (see the appendix of observations at Pedersen 2011: 411–12).

of his predecessor. Indeed, he apologises for him, stating that the very lack of similarly dependable historical data prevented Hipparchus from developing a better theory of precession or any planetary theory of his own. In the latter case, moreover, Ptolemy claims that Hipparchus, aware of the insufficiencies and imperfections of his data, prepared the way for an eventual theory by 'making a more useful arrangement [of them]' (*Syntaxis* 2.210: ἐπὶ τὸ χρησιμώτερον συντάξαι).

This last point is crucial for understanding how closely Ptolemy views himself and his project in relation to Hipparchus. Following his assertions about Hipparchus' intentional preparation of observations for some future successor who would devise a planetary theory, Ptolemy first dismisses the non-geometric and generally faulty efforts of those who used 'Aeon-tables'[63] for this purpose, then he speculates on what Hipparchus foresaw as necessary for success:

> [Hipparchus] reasoned (ἐλογίσατο) that one who had advanced to such a degree of accuracy (ἀκριβείας) and love of truth (φιλαληθείας) throughout the mathematical sciences would not be satisfied (οὐκ ἀπαρκέσει) to rest with such great [imperfection], as the others did for whom it made no difference (καθάπερ τοῖς ἄλλοις οὐ διήνεγκεν); but for one intending to persuade himself and his readers, [Hipparchus reasoned] it would be necessary to demonstrate the size and periods of the two anomalies through manifest and acceptable phenomena, and having combined them, to discover the position and order of the circles by which the anomalies occur and the manner of their movement, and finally, to fit practically all the phenomena to the specific form of the circles in his theory. And this, I think, clearly appeared difficult even to him. I say these things not for the sake of boasting [of my own achievement].[64]

Ptolemy's concluding statement shows that Hipparchus' outline for a successful planetary theory summarises Ptolemy's own accomplishment. As Bernard has noted, Ptolemy thus frames the planetary theory of the *Syntaxis* as the project of Hipparchus carried through to fruition.[65] It remains, however, to draw out the degree to which Ptolemy here reinforces a direct connection to his predecessor. That connection was previously established by claims of sharing the goal of modelling the celestial

---

[63] On which see Neugebauer 1975: Vol. 2, 789; and Toomer 1984: 422, n. 12.
[64] *Syntaxis* 2.211, adapted from Toomer 1984: 422. See Bernard 2010: 509–11 on the methodological and philosophical issues raised by the passage.
[65] Bernard 2010: 513.

phenomena with uniform, circular motions.⁶⁶ Moreover, in his presentation here, Ptolemy indirectly praises both Hipparchus and himself: Hipparchus had the logical foresight to know what a theory needed to succeed, while Ptolemy achieved what appeared too difficult *even* for Hipparchus. But notice, too, that Ptolemy imagines Hipparchus detailing not only the requirements of a successful theory, but prior to this, the personal characteristics necessary to devise it: a high regard for accuracy, a love of truth and neither satisfaction with nor indifference to imperfection (i.e. to conduct the work with great zeal). Using Hipparchus as a mouthpiece, then, Ptolemy defines his own character in the same conceptual terms in which he defined Hipparchus'. This shared characterisation is strengthened by Ptolemy's own prefatory and concluding claims for the 'usefulness' of his text.⁶⁷ Ptolemy's account, then, unifies predecessor and successor through its affirmations of a common project and reciprocated definition of character.

The shared characterisation of Ptolemy and Hipparchus is continuous with the euergetic image of benefactors and authors developed at the beginning of this chapter. In particular, Ptolemy draws attention to the interest in utility that motivates his and Hipparchus' respective authorial activities, and he also emphasises the care and zeal with which they undertake them, using the same terms and similar phrasing to what we find in the stock language of public inscriptions.⁶⁸ To generalise from our discussion of the *Syntaxis*, we can say that an Imperial-period author in the exact sciences, a genre so distinctive in its form and content, follows essentially the same practice of creating an authoritative persona as is found in other, contemporary expository and even epigraphic genres. That practice entails the production of a beneficent social identity, founded on overt claims about one's utility and zeal for what might be seen as a collective enterprise. However, unlike a civic benefactor or, say, a compilatory author such as Pliny, Ptolemy makes no claim about his work serving the well-being of a political community.⁶⁹

---

⁶⁶ Ptolemy states that the general goal of the 'mathematician' (μαθηματικῷ) is to demonstrate that heavenly phenomena may be represented through 'uniform circular movements' (*Syntaxis* 1.208), a goal identical to the one he states Hipparchus accomplished 'as far as he was able' (*Syntaxis* 2.210). For commentary, see Goldstein 2007: 272.

⁶⁷ For its limited readership, of course. *Syntaxis* 1.8 and 2.608.

⁶⁸ Ptolemy makes observations 'with care' (1.314: ἐπιμελέστατα and 1.339: ἐπιμελῶς), while Hipparchus solves problems 'with great zeal' (*Syntaxis* 1.233: μετὰ πάσης σπουδῆς). Compare the latter to, for instance, the roughly contemporary inscription from Aphrodisias honouring a *xystarch* for likewise conducting his office 'with great zeal' (*Corpus Inscriptionum Graecarum* 2811b.13–14: μετὰ σπουδῆς ἀπ[ά]|σης).

⁶⁹ Cf. Ma 1999: 179–243, esp. 226–8, who argues that even acts of domination by kings were portrayed as individual efforts directed towards the collective benefit of a polis. On compilatory authors, see Beagon 2013: esp. 92–3.

Instead, he directs his beneficence towards 'saving the phenomena', a goal common, at least in Ptolemy's account, to Ptolemy, Hipparchus and mathematicians in general.[70] But Ptolemy is not unique in applying euergetic rhetoric in a disciplinary context. Other authors, who may otherwise take pains to establish their work as political discourse, similarly use the language of utility and zeal to define scholarly activity;[71] indeed, we may expect such a linguistic transfer in authors who invest themselves in both political and learned endeavours.[72] What is noteworthy is that Ptolemy, seemingly so removed from Imperial-period culture in other respects,[73] adopts the same rhetorical tactics that feature in diverse kinds of literary activity. In this way, at least, Ptolemy's mathematics are not so specialised.

Ptolemy's Diachronic Collaboration

Ptolemy is like other authors in using euergetic rhetoric, but the purpose to which he applies it in his text emerges as unique, because with it he constructs expressly collaborative relationships to support the project of the *Syntaxis*. These collaborations are not 'synchronic', entailing the participation of contemporaries, consideration of which I will take up in the next section; rather, the *Syntaxis* presents a model of 'diachronic collaboration', in which Ptolemy's collaborators are predecessors and successors, extending across generations.

The very term 'collaborator' might seem ill-fitting in a diachronic model, since the processes of time and mortality restrict diachronic collaborators' actual ability to work together. As we have seen, however, the *Syntaxis* entails what Ptolemy frames as the intentional participation of individuals in the past and future: he sets tasks for successive readers to accomplish and presents Hipparchus as consciously furnishing planetary data for his own future reader. Moreover, the text presents these different participants as united by a common goal, namely, the demonstration of uniform, circular movements by the heavenly bodies, which is explicitly shared by Hipparchus and Ptolemy and which the *Syntaxis* in turn broadcasts to future readers. The *Syntaxis* thus constructs its overall project as a collaboration, but one that can only be realised diachronically. It is important to note, however, that the different participants do not collaborate equally: according to

---

[70] See n. 66 above. To be sure, 'saving the phenomena' is hardly a uniform notion among ancient authorities: see Lloyd 1987a: 293–319.
[71] E.g. Dion. Hal. *Ant. Rom.* 1.1.2; Josephus, *BJ* 1.6; [Longin.], *Subl.* 1.2; Arr. *Cyn.* 1.4–5.
[72] Beagon 2013: 87.   [73] See above discussion; cf. Jones 2010a: xi.

Ptolemy, the development of a coherent theory of planetary motion requires a differentiation of roles. With no reliable historical data on which to build his own theory, Hipparchus' role could only be to compile observations for a future synthesist; Ptolemy, claiming a sufficiently advantageous position in time, both develops a coherent theory using Hipparchus' and his own observations and in turn refines it; and delivered a refined theory in the *Syntaxis*, future readers still contribute, not through any creative theorising of their own, but by confirming the lasting validity of Ptolemy's epicycles, equants and eccentrics.

This model of diachronic collaboration may appear incomplete: where are the other astronomical authorities whom Ptolemy names in the course of the *Syntaxis*? We consider, first, the case of those who are Hipparchus' predecessors. Certainly, Ptolemy cites the observations of Timocharis, Meton, Euctemon and others in support of his models, but his treatment of these individuals is always brisk and purely instrumental. He provides only their names, observational data and, occasionally, observing locations, with nothing of the ethical apparatus that undergirds his connection with Hipparchus. Indeed, rather than uphold the observational practice of these predecessors, Ptolemy refers to Hipparchus' assessment that Meton, Euctemon and Aristarchus observed solstices 'rather imprecisely' (ὁλοσχερέστερον),[74] and, following Hipparchus again, he declares Timocharis' observations of fixed-star positions 'not trustworthy' (μήτε ... ἀξιοπίστους) on account of their having been made 'very imprecisely' (πάνυ ὁλοσχερῶς).[75] Pointing to imprecision in Hipparchus' predecessors helps Ptolemy make the case that Hipparchus was at a loss (and *knew* he was at a loss) to make, for instance, an accurate measure of precession, which Ptolemy himself claims to have achieved – thanks, of course, to Hipparchus' diligent efforts.[76] It also establishes a contrast, however, between the keen accuracy that defines Hipparchus' (and Ptolemy's) practice and the crudeness that came before. The key point is that Hipparchus is different from, i.e. superior to, his predecessors. Moreover, Ptolemy makes no comment about whether Hipparchus' predecessors share his and Hipparchus' goal of demonstrating the celestial bodies' uniform, circular motions. To be sure, we have some idea of Aristarchus' interest in geometric models from his treatise *On the Sizes and Distances of the Sun and Moon* (though he offers no commentary on his aims), but Meton, Euctemon and Timocharis may well have had other

---

[74] *Syntaxis* 1.203. [75] *Syntaxis* 2.18.
[76] To be sure, approximation was a part of Ptolemy's own observational practice, and the degree to which his claims are made in good faith has excited controversy. Lloyd 1987a: 236–41 reviews the issues.

interests.⁷⁷ To sum up: while observations made by Hipparchus' predecessors are sometimes adduced to support Ptolemy's theories, Ptolemy ranges from neutral to negative in evaluating them. Moreover, he is silent about any interest the earlier astronomers may have had in the goal that Ptolemy claims to share with Hipparchus. On balance, then, Hipparchus' predecessors maintain only dubious footing in Ptolemy's model of diachronic collaboration.

Does this satisfy? Before drawing final conclusions about this collaborative model, we need to consider those astronomers situated in the roughly 300 years separating Hipparchus and Ptolemy. Remarkably, the *Syntaxis* features only two: Menelaus and Theon. The minimal number may correlate to a general decline in interest in mathematics; again, this is the period Netz describes as 'a wilderness between two deserts'.⁷⁸ But as with the astronomers preceding Hipparchus, Ptolemy shows no interest in praising their virtues or defining their goals as observers, only specifying that Menelaus was a 'geometer' who observed at Rome (2.30) and that Theon was a 'mathematician' (2.296). Moreover, Ptolemy makes little use of the observations of either: only two citations of Menelaus and four of Theon.⁷⁹ There is not much else to say about them; about any theoretical interests either may have held in, say, precession or planetary motions, Ptolemy is silent. Like Hipparchus' predecessors, then, Menelaus and Theon are relegated to the margins of Ptolemy's collaborative scheme.

Ptolemy thus excludes these other astronomical authorities from his model of diachronic collaboration. Whatever the injustice done by Ptolemy's criticisms of and silences about other astronomers, they reinforce and highlight the positive connections between himself and Hipparchus. The diachronic model of collaboration implicit in the *Syntaxis* thus does not illustrate a history of gradual progress of mathematical astronomy, but instead encapsulates a rhetorical strategy with a manifold effect: it designates Ptolemy as an essentially unmediated successor to Hipparchus; it defines Ptolemy as the type of individual worthy to continue Hipparchus' theoretical project; it establishes that Ptolemy succeeded in accomplishing that project; it broadcasts this success to future readers and assigns them a role in its final confirmation. At its core, then, this is a story of progress, but a complex one: Ptolemy presents himself as using data accumulated since Hipparchus to accomplish the goal of Hipparchus, all while

---

⁷⁷ On Meton: Bowen and Goldstein 1988; Euctemon: Hannah 2002; Timocharis: Goldstein and Bowen 1991.
⁷⁸ Netz 1999: 284. Cf. n. 23 above.
⁷⁹ Menelaus: *Syntaxis* 2.30 and 33; Theon: 2.275, 296, 297, 299.

embodying the virtues of Hipparchus – it is thus a mash-up of sorts of the three progress-narratives of growth, ending and return recently analysed by Markus Asper.[80]

The details of Ptolemy's diachronic collaboration likewise distinguish it from the manner in which other Imperial-period authors uphold predecessors imbued with a certain canonical *mana*. Strabo, for one, elevates Homer as the ideal historian-geographer: while acknowledging certain holes in the poet's account, he still makes the case that Homer's geographical knowledge is supreme and extends to the ends of the earth.[81] Galen, for another, maintains the persisting dominance of Hippocrates' authority, aligning himself with the master as '*Hippocrates alter*' and at times even ascribing more recently devised doctrines to him.[82] Both Strabo and Galen thus treat these esteemed predecessors in a mode that returns to and reclaims an original, essentially untarnished authority; neither author claims to be completing a project that Homer or Hippocrates had begun and knowingly left unfinished. In contrast, Ptolemy presents himself not as a 'second coming' of Hipparchus, but as the one for whom Hipparchus has intentionally prepared the way.

Yet not all predecessors are appropriated as models for emulation. More and more we regard ancient scientific practices as a venue for self-assertion, and these contests for authority entailed a temporal dimension, realised through critical and often vivid, textual engagement with past authors.[83] Asper has described such activity as 'diachronic competition',[84] and on the surface, Ptolemy's collaborative rhetoric is starkly opposed to, for instance, the direct, textual attacks that Galen slings at the Hellenistic Erasistratus in support of his own claims to physiological mastery.[85] As discussed in the previous section, however, Ptolemy's own treatment of Hipparchus is somewhat complicated by subtle distortions and even open criticism of his predecessor's methods. To be sure, others had criticised Hipparchus even more vociferously,[86] but the key point is that Ptolemy is still competing with

---

[80] Asper 2013c, which concludes with a different account of mathematics as a 'diachronic group-effort', one that 'favors a group ideology that downplays individual competition' (427). This conclusion is largely drawn from later authors, namely Pappus and Proclus, but Asper notes that earlier mathematicians seem more interested in competition and self-promotion.
[81] Kim 2010: 47–84.
[82] Asper 2013c: 423–4. But Hippocrates' standing was not absolute, even for Galen: Lloyd 1991a.
[83] E.g. Asper 2007: 351–6; Barton 1994b; Lloyd 1987a: 50–108; Netz 2000, 2009. On the vividness of these encounters, see König 2012: 41–52.
[84] Asper 2007: 352.
[85] E.g. in *Ven.Sect.Er* (K11.147–86), on which see Smith 1979: 79–82. Cf. Lonie 1964.
[86] E.g. Strabo accuses Hipparchus of reducing Eratosthenes' geographical arguments to straw 'for his own gratification' (2.1.22: ἑαυτῷ κεχαρισμένως) and otherwise 'treating [Eratosthenes]

Hipparchus. Why, then, the collaborative rhetoric? On the one hand, genre has to be considered: the Greek exact sciences did not afford the same discursive space to strident personal attack as we see, for instance, in Galen's medical writings.[87] But even if Ptolemy wanted to shatter the generic mould and Hipparchus with it, he still needed some predecessor's observations to develop historically valid models of celestial motion. Hence, Ptolemy's approach: if you can't beat Hipparchus (directly), join Hipparchus – then subordinate Hipparchus and finally outdo Hipparchus. In the *Syntaxis*, then, diachronic competition takes the form of diachronic collaboration. While Ptolemy's handling of Hipparchus looks like a genteel embrace, it feels like a bear hug.

Synchronic Collaboration in the *Syntaxis*?

Ultimately, then, Ptolemy's diachronic collaboration is a device that reinforces his individual authority. Tailored to the historical nature of his astronomical project, it is apparently unique to the *Syntaxis*. But do we ever find ancient productions of synchronic, scientific collaboration? To clarify: by 'synchronic collaboration' I do not mean teacher-student relationships, nor a 'school' whose affiliates share a doctrinal orientation. Rather, I define synchronic collaboration as the activity of two or more contemporary authorities who are jointly recognised for the product of their work. In the conclusion we will see that synchronic collaboration was a practical reality among civic benefactors, who, I have argued, were otherwise characterised with the same utilitarian interest and zeal as expert authors. Did those similarities extend to the organisation of their efforts? Our intuitions about ancient authorial convention strongly suggest the negative. Authorship in antiquity was a one-man show,[88] allowing for a few qualifications: on the one hand, the Hippocratic corpus presents a case of *composite authorship*, in which one larger body of work is a later collection of texts written by multiple individuals.[89] On the other hand, we can be sure that some authors, including Ptolemy, received help from *uncredited contributors*.[90] But with

---

unfairly' (2.1.23: ἀγνωμονεῖν), tactics Strabo dismisses as 'childish' (2.1.29: παιδικά). But Strabo's account of Hipparchus is ultimately balanced; cf. Dueck 2000: 58–9.

[87] Cf. Asper 2011: 98–9 and Cuomo 2000: 196–99, who discuss how the late-antique Pappus of Alexandria (674.20–682.23) even promotes civility and community among mathematicians through a moral comparison of Euclid and Apollonius.

[88] Cf. Beagon 2013: 84.   [89] Lloyd 1975; cf. Smith 1979.

[90] On Ptolemy, see Pedersen 2011: 55–6. Cf. Quint. *Inst.* 10.1.128 on Seneca's use of 'research assistants'.

few exceptions, the final products of these collective endeavours (however intentional their 'collective' aspect) are attributed to a single authority.[91]

Norms of authorship notwithstanding, the history of Greek astronomy appears exceptional. While there is no question that Ptolemy takes full credit for the *Syntaxis*, there is a lingering possibility that he, seemingly so lonely in the second century, looks back to earlier days of synchronic collaboration. However, two ambiguities in the text suggest otherwise. The first concerns the construction 'οἱ περί + [proper name in the accusative]', which occurs nine times in the *Syntaxis*.[92] It is well known that the expression may either denote a group (literally, 'those around so-and-so') or simply be taken as a periphrasis for the named individual,[93] and commentators on both the *Syntaxis* and other astronomical texts generally leave open the question as to how it should be understood.[94] In Ptolemy, the construction is better understood to denote an individual. Thus, it can be considered parallel to first-person, plural expressions that denote a semantically singular subject, which of course feature ubiquitously in Greek literature (including Ptolemy).[95] First, comparative evidences make clear that the οἱ περί construction, even when governing more than one name (e.g. *Syntaxis* 1.203:

---

[91] The exceptions are also vague, both on how any supposed collaboration operated and, sometimes, on the conceptual nature of the text itself. Four cases: (1) Lloyd 1975: 181 discusses the so-called *Cnidian Sentences*, referred to in the Hippocratic *On Regimen in Acute Diseases* ch. 1 (Littré 1839–61: Vol. 2, 225). As Lloyd points out, however, the work is attributed to authors whose individual identities are, at least in that context, inconsequential (simply, οἱ ξυγγράψαντες), and we may well question just what 'the sentences called Cnidian' (τὰς Κνιδίας καλεομένας γνώμας) refers to, whether to one 'work' or multiple. To be sure, by Galen's time the title appears to refer to a defined work, some version of which, at least, was attributed to a single author (Gal. *Hipp.Epid.*, K17a.886). (2) The fragmentary historians Agias and Dercylus are cited together in most testimonia and fragments (*BNJ* 305). But we can surmise little about the two beyond their Argive associations and dates (not necessarily concurrent) before Callimachus; scholars have assumed, however, a successive rather than collaborative relationship: see Engels 2013. (3) The Septuagint raises questions about collective *translation*, but ancient accounts of its production, for example, from the *Letter of Aristeas*, offer little detail about a collaborative process. The high number of translators, whether 72 or 70, conveys more symbolic value (Honigman 2003: 56–8) and, perhaps, Ptolemy II Philadelphus' characteristic appetite for grand-scale productions, than it reflects any actual, practical requirements for the undertaking. Honigman 2003: 45–8 and 119–43 also suggests parallels between the collective effort behind the Septuagint and Alexandrian textual criticism. (4) Pairs of architects share credit for construction treatises at Vitruv. *De arch.* 7.pr16, and multiple architects may likewise share credit for monumental constructions themselves: e.g. Ictinus and Callicrates for the Parthenon at Plut. *Per.* 13.4.

[92] Meton and Euctemon: 1.203, 205, 207; Euctemon (alone): 1.206; Aristarchus: 1.203, 206; Timocharis: 2.18 (twice), 19.

[93] Cf. Schwyzer 1968: Vol. 2, 416–17.

[94] Toomer 1984: 137, n. 19; Evans and Berggren 2006: 183, n. 23.

[95] The 'authorial we' apparently never refers to joint-authors, nor do Hine 2009 or von Staden 1994 consider it as a possibility.

τῶν περὶ Μέτωνα καὶ Εὐκτήμονα), may denote discrete individuals who are not even contemporaries.[96] Moreover, in his other works, Ptolemy never denotes doctrine-sharing groups ('schools') with the οἱ περί construction, but only with a denominative form.[97] Finally, twice in the *Syntaxis*, Ptolemy shifts his attribution of the same accomplishment between an individual and a οἱ περί construction governing that individual within a short space of text and with no other distinguishing commentary.[98] Consequently, if a οἱ περί construction indeed signifies a group or 'school', then the individual's name alone is able to denote the same. In Ptolemy, the expressions are semantically identical.

In addition to these linguistic arguments, we should consider a cultural one. Almost exclusively in the *Syntaxis*, οἱ περί constructions indicate those who observed solar and stellar positions, and it is likely that Ptolemy and others availed themselves of some kind of (uncredited) assistance in these activities.[99] In one instance, however, the οἱ περί construction refers to those who *calculated* the length of the year, an actual scientific achievement (1.207). Given our developing picture of Greek science, in general, and Greek mathematics, in particular, as venues for competition and individual assertion,[100] it follows that an intellectual feat such as precisely calculating year-length is an accomplishment over which an individual would surely claim ownership, if not be assigned it – even if Ptolemy points out its inaccuracy.

The second ambiguity pertains to this last point. Even putting aside the interpretative difficulties of the οἱ περί construction, we still find that Ptolemy ascribes observations and that year-length calculation to *two* individuals, Meton and Euctemon. Both were apparently citizens of Athens in the last third of the fifth century BCE, but beyond this we know little for certain about their individual lives or relationship to one another.[101] But does Ptolemy regard them as a collaborative pair, and should we? Most evidence prior to Ptolemy, including even references from Old Comedy, presents Meton and Euctemon as authorities independent from one

---

[96] E.g. Plu. *De Pyth. or.* 402f–3a, where the astronomical prose-writers Aristarchus, Timocharis, Aristyllus and Hipparchus (οἱ περὶ Ἀρίσταρχον καὶ Τιμόχαριν καὶ Ἀρίστυλλον καὶ Ἵππαρχον) are contrasted with Eudoxus, Hesiod and Thales (Εὐδόξου καὶ Ἡσιόδου καὶ Θαλοῦ), taken as poets. There is an obvious chronological spread between the first four, and the contrast with the three discrete names further reinforces their identification as individuals.

[97] Thus, throughout the *Harmonics* we never find 'those around' Pythagoras or Aristoxenus, but only 'the Pythagoreans' or 'the Aristoxeneans'.

[98] Describing observations by, first, τῶν περὶ Ἀρίσταρχον, then Ἀριστάρχου (1.206); similarly describing observations by Τιμοχάριδος, then τῶν περὶ τὸν Τιμόχαριν (2.17–18).

[99] See n. 90 above. [100] See n. 83 above. [101] Cf. Mendell 2008a and 2008b.

another.¹⁰² Ptolemy is in fact one of only a few authors who appear to conceive of them as a unit, but Ptolemy does not do so consistently.¹⁰³

How, then, to explain the pairing? It is difficult to explain away completely. One explanation might be that Ptolemy simply followed the convention of Hellenistic sources,¹⁰⁴ but this merely defers the question back in time. Another hypothesis is that, given Meton and Euctemon's close proximity in fifth-century Athens and the uncertainty that historical distance may induce, certain achievements of their respective careers may have been confused in the course of time.¹⁰⁵ Anecdotes concerning Meton's 'publication' of astronomical stelai, and the individual acclaim he won for doing so, likewise suggest a more competitive relationship.¹⁰⁶ Here, we must admit

---

[102] Of the two, Euctemon enjoys a more secure position in astronomical documents: for epigraphic and literary references, see Neugebauer 1975: Vol. 2, 588 and 623 (with n. 12). Dicks 1971: 460 notes that he may also be included as a geographical authority in the fourth-century CE *De ora maritima* of Avienus. Outside the *Syntaxis*, Meton's individual presence in astronomical writings amounts only to scattered references in literary *parapêgmata*: Geminus, *Calend*. 99; Ptol. *Phas*. 22 (twice), 31 (twice), 43 (twice), 44, 58; cf. n. 104 below for another possible reference in Hipparchus. On the other hand, Meton seems to have enjoyed (suffered from?) a higher profile than Euctemon in wider literary culture: lampooned as a geometer at Ar. *Av*. 992–1019 (cf. *Scholia vetera ad Ar. Av.* 997 for references in Phrynichus and Philochorus), but commemorated for the 'Metonic cycle' at Ps.-Thphr. *Sign*. 4; Diod. Sic. 12.36.2–3; Columella, *Rust*. 9.14.12; Ael. *VH* 10.7; Censorinus *DN*. 18.8. See Plut. *Alc*. 17.4–5 and *Nic*. 13.5; Ael. *VH* 13.12 on Meton's draft-dodging activity during the Peloponnesian War.

[103] Paired at *Syntaxis* 1.203, 205, 207; Meton alone at 1.207; Euctemon alone at 1.206. Pairings prior to Ptolemy: Theodosius, *De diebus et noctibus* 152; Eudemus may have paired them, though this rests on a late-antique quotation (Simpl. *in Cael*. 497). Contemporary with the *Syntaxis*: Ptol. *Phas*. 2.67 and Vett. Val. 9.12 (in the latter, however, grouped with Philippus). The papyrological record does not support dual attribution, insofar as it has been catalogued in Jones 1999.

[104] E.g. Theodosius, *De diebus et noctibus* 152. But cf. *Syntaxis* 1.207, where Ptolemy, citing Hipparchus 'word for word' (κατὰ λέξιν), credits only Meton with the value for year-length.

[105] Proper attribution of other astronomical achievements generated ancient controversy: see Evans and Berggren 2006: 86–7 for Imperial-period uncertainties about whether Cleostratus or Eudoxus developed the eight-year astronomical cycle (presumably, in this case the gap of several generations between the two prevented any joint-attribution). Furthermore, Ptol. *Phas*. 2.67 may present a conflated biography of Meton and Euctemon: Ptolemy states that the pair conducted observations in Athens, the Cyclades, Macedonia and Thrace. Is it credible that they travelled and conducted observations so extensively around the Aegean *as a pair*? For Ptolemy, apparently, it was. But consider Avienus' statement that Euctemon was an 'inhabitant of Amphipolis' (*Ora maritima* 337), into which some have read a tradition holding our Euctemon to be among the Athenians who re-colonised the site in 437 BCE (Dicks 1971: 460). The interpretation may require too much from a late and vague source, but it gives a foothold to Euctemon, at least, in the northern Aegean. Did this detail of Euctemon's history in turn become part of a shared biography?

[106] On Meton's acclaim: Diod. Sic. 12.36.2–3. Individual acclaim for astronomical writings is likewise at stake in an anecdote about Eudoxus (Diog. Laert. 8.88). For fifth-century astronomy in its wider social context, see Bowen and Goldstein 1988: 73–7; and Hannah

that a clinching argument cannot easily be made against a collaboration between Meton and Euctemon, but we must at least recognise the problems that follow the assumption.[107] In addition to its clashing with hints of competition, collaboration presents conceptual difficulties: if the relationship between Meton and Euctemon is not one of teacher and student or vice versa (for which there is no evidence), what would a collaboration to calculate the number of days in the year actually entail, and what form would its outcome have taken – a text with two, discrete, named authors?[108] Such a text would be anomalous, on the one hand, in the competitive debate culture of Meton's and Euctemon's Athens, and, on the other, in the history of ancient expertise: for whatever one's professed devotion to knowledge, to be perceived as sole author of that knowledge was at least as important.[109]

Conclusion: The Individuality of the Collaborative Author

Publication in antiquity established individual authority. This is true for authors who explicitly situate their works in front of a wide, public audience, as well as for those, like Ptolemy, who seemed to restrict themselves to a more specialised readership. A jointly edited volume on *Authority and Expertise* (a 'König and Woolf') would have hardly been imaginable. But what about the notion of collaborative euergetism (a 'Carnegie and Mellon')? I began by comparing the author to a *euergetês*, owing to his professed zeal and utilitarian purpose. Here, at the close, I can only briefly address whether the author's individuality itself might be continuous with that euergetic image: was the civic *euergetês* defined as much by individuality as by attitude and utilitarian interest?

2002: 129–32. On the nature of Meton's 'publication', see Bowen and Goldstein 1988: 53, n. 67 and n. 108 below.

[107] Explicitly assumed by Toomer 1974a: 339, n. 14; cf. Manitius 1894: 268, n. 19; Mendell 2008a, 2008b; and van der Waerden 1960. Exercising more caution: Bowen and Goldstein 1988: 43, n. 26; Lehoux 2007: 88–90; Neugebauer 1975: Vol. 2, 623, n. 12.

[108] Whether they 'published' in a literary or epigraphic medium is uncertain; cf. Hannah 2001: 148–9, who favours inscription. This need not alter the conclusion, but the degree to which authorial norms differed for inscriptional texts cannot be addressed here.

[109] Similarly curious are Timocharis and Aristyllus, who have an even more tenuous standing in the historical record than Meton and Euctemon. Both are remembered as authors of astronomical prose (see n. 96 above), but our main source for their activity is *Syntaxis* 2.17–32: Ptolemy reports observations conducted in Alexandria in the early third century BCE. But whether the two worked in collaboration or in competition is ambiguous: see Goldstein and Bowen 1991: 101–7.

We quickly find that the question is complicated by certain ambiguities specific to civic euergetism. First, material benefactions varied greatly in their extravagance, such that they could entail either an entire temple (which, indeed, only the elite of the elite could finance) or a single column.[110] Was the benefactor who gave only the latter an individual donating a column, or a collaborator contributing one part of a temple? Second, as epigraphic evidence attests, individual columns were donated not just by individuals, but, for instance, by husband-and-wife pairs (a 'Bill and Melinda Gates').[111] Did this represent a collaboration of two individuals, or perhaps the effort of an individual family unit? Ambiguities such as the foregoing, as well as the need to differentiate the contexts of benefaction in which they were more or less common, make clear that we should expect nuanced answers to the question of the civic benefactor's individuality, which cannot here be investigated thoroughly.[112]

Still worth considering, however, is the now common observation that practices of civic euergetism served to enhance and reinforce the authority and prestige of the benefactor with respect to the beneficiaries.[113] Elite interest in maintaining a collective social distance from the *dêmos* is surely compatible with aspirations of individual euergetic achievement, and a number of scholars have promoted the idea that in the Imperial period, no less than before, euergetism was an arena in which elites competed among themselves for honour and authority.[114] This must be correct up to a point, but as was noted in the previous paragraph, economic and perhaps other social forces (including family ties) sometimes resulted in the fragmentation of euergetic actions, and with it the diffusion of authority and prestige.[115] The key point is that those benefactors who did contribute were of such status as to receive due recognition for their contribution. Whether grudging or not in its realisation, then, some kind of open collaboration did occur in the sphere of civic euergetism.

Literary euergetism offers a contrasting model. Actual synchronic collaboration among expository authors seems a mirage produced by imprecise attestations, and collaborative rhetoric is itself exceedingly rare. We find the

---

[110] Zuiderhoek 2009: 28–31 and 78–109.   [111] E.g. at Aphrodisias (Reinach 1906: 220–2).
[112] Cf. Reynolds 1996: 122.
[113] E.g. Gordon 1990: 224–31; van Nijf 1997: 111–20; Veyne 1990: 110; Zuiderhoek 2009: 92–106, 113–53.
[114] Duncan-Jones 1963: 160–1; Veyne 1990: 108, 133; Zuiderhoek 2009: 106–7. See also Sickinger 2009 on the case of democratic Athens.
[115] Zuiderhoek 2009: 106–7 argues that the checks of an anxious *boulê* would aim both to regulate the ambition and to absorb some of the prestige of 'political mavericks' threatening the elite status quo.

most explicit, if not the only, sentiment for the latter in Ptolemy's *Syntaxis*, which nevertheless separates the collaborators by centuries and finally upholds Ptolemy's individual authority at the centre of the collaboration. To be sure, Hipparchus contributed his column to the edifice: Ptolemy claims that his observations prop up the model of celestial motions, and for his purported good works, zeal and utilitarian interest Ptolemy gives Hipparchus a certain due. There is no question, however, that Ptolemy remains the sole architect of the *Syntaxis*, a facade of collaboration.

16 | The Authority of Mathematical Expertise and the Question of Ancient Writing *More Geometrico*

REVIEL NETZ

An Ancient *More Geometrico*? Stating the Question

'Someone told me that each equation I included in the book would halve the sales... However, I *did* put in one equation.' A well-known quote, from Hawking's *A Brief History of Time*.[1] Sociologists of science can immediately get to work: the author descends from the 'high' position (where he resides, there are plenty of equations), investing his 'low' or at least 'lower', more engaging text with the authority of pure science; we are reminded, for sure, of Bourdieu's *cultural capital* and of the notion that, in the modern literary system, 'difficulty' (which implies lesser commercial value) is correlated with higher, cultural capital.[2] And so: the authority, the cultural capital of science; conveyed by the formal marker of the equation. This brief history of scientific authority may or may not be true of the late twentieth-century system of scientific genres,[3] but it will serve to illustrate a sociological structure. Mathematics is socially powerful because it conveys a difficult and marked expertise.

Sociology may not be all. Let us recall another, equally famous quote: 'Philosophy is written in that great book which ever is before our eyes – I mean the universe – but we cannot understand it if we do not first learn the language and grasp the symbols in which it is written. The book is written in mathematical language, and the symbols are triangles, circles and other geometrical figures, without whose help it is impossible to comprehend a single word of it; without which one wanders in vain through a dark labyrinth.' This is from Galileo,[4] of course, and is equally a creature of its time and place. The seventeenth century is an age of writing *more geometrico*;

---

[1] Hawking 1988: vi.   [2] Bourdieu 1993.
[3] I am not aware of a formal study of this perception, but consider the related study in Healy and Hoyles 2000: high school students and their teachers are offered a variety of proof techniques for the same problem, and are asked to evaluate them as the approach they personally would prefer, and the approach they think will get the higher grade. Systematically the more verbal approach is personally preferred by nearly all students as well as by many teachers, but the more symbolic one is considered to be the more likely to get a high grade. (I am grateful to Pablo Mejia-Ramos for pointing me to this study.)
[4] *The Assayer* (1623) (Drake 1957: 237).

Spinoza's *Ethics*[5] was merely the most famous book from an entire genre of works of philosophy, natural or otherwise, written in explicitly mathematical format.[6] Writing *more geometrico* was considered by many authors to possess, in and of itself, a claim to truth because the validity of mathematics was universal: it was understood to be the very language of universe, of truth. Here is another model, not (just) social but (also) epistemic: mathematics was socially powerful in the seventeenth century because of a widely shared view that its method had special access to truth. Perhaps, such epistemic origins explain the social structure of Hawking's quote.

Social or epistemic, Hawking-like or Galileo-like: did the ancient mathematical expertise carry *authority*? For tackling this question, my strategy would be to focus not on what the ancients thought, but on what they did. Had mathematics possessed authority, we would expect those 'single equations', those writings '*more geometrico*'. What do we find, of such treatises, such passages?

My main interest is not in counting, but in understanding: looking at the passages where the authority of a mathematical way of writing is implied, and seeing just what is implied by such an authority. And so, a qualitative survey of two kinds of cases where a non-mathematical text picks up the formal properties of a mathematical genre. In the first, a non-mathematical text is cast, as a whole, in a form that might emulate the formal properties of a mathematical text. The four cases I identify are: Aristotle's *Prior Analytics*; Aristoxenus' *Elements of Harmonics*, Book III; Asclepiodotus' *Tactics*; and Proclus' *Elements of Theology*. In the second, a non-mathematical text includes an inset of a mathematical discussion with some formal properties. The four cases I identify are: the discussion of the halo in Aristotle's *Meteorology*; passages from Strabo's methodological discussion in Book II of his *Geography*; Galen's treatment of the geometry of the eye in *The Use of Parts*; and Julius Africanus' discussion of surveying in his seventh *Cestus*.

I do not survey all the ancient sources, and my main project is qualitative. But it should be emphasised that this survey already gets us near the bottom of the barrel. The fully mathematised treatise is of course fairly easy to spot and as for the inset mathematical passage, this is highly marked formally.[7] Both are extremely rare. As documented in footnote 7, the only

---

[5] In the full title: *Ethica more geometrico demonstrata* or *Ethica ordine geometrico demonstrata* (both titles have textual authority and were apparently considered interchangeable: Steenbakkers 1994: 159, n. 1).

[6] For a quick survey of the seventeenth-century *more geometrico* genre, see Garrett 2003: 9–12.

[7] My two rough tools for identifying a sample of explicitly mathematical passages were (i) a TLG search of the string '_AB_' (single letter searches bring up many false positives in the form of

context where the brief mathematical passage is at all common is the philosophical commentary. The most famous case is Simplicius' passage concerning Hippocrates' *Quadrature of Lunules* (Simplicius *in Ph.* 60.22–68.32), but there are a number of other examples, for instance in the papyrus fragment of the anonymous Commentary to Plato's *Theaetetus* (29.42–31.28, Theodorus gives a series of mathematical examples ending with 17. Why? The commentator offers a rough, elementary mathematical discussion), or in Alexander of Aphrodisias' treatment of Aristotle's system of spheres (*in Metaph.* 704.22–705.6: Aristotle discusses the manner in which the spheres move the stars; Alexander amplifies the account with the aid of a geometrical figure). Such examples – perhaps about 100 altogether – form the bulk of the extant cases where mathematical passages occur within a mostly non-mathematical context. Many passages in Plato and Aristotle touch upon mathematics in a non-mathematical manner so that a commentator may expand on such passages with a more explicit mathematical statement: in this well-defined, almost unique context, mathematics was established as a source of authority. I note this fact, and shall return to it through our discussion. The task of this chapter is to look for mathematical authority, outside the limited domain of the philosophical commentary.

Examples (I): The Mathematised Treatise
Aristotle, *The Prior Analytics*

Already Einarson in 1936 pointed out the terminological similarities between Aristotle's usage in the *Organon* and the mathematical lexicon. In

numerals; of double letters, 'AB' is the most natural). This indeed brings up all the explicitly mathematical authors, typically with many dozens or even hundreds of examples (Euclid, with a large corpus and a preference for simple labels, dominates the search with 2,339 occurrences). Non-mathematical occurrences are typically spurious (most often, the TLG citation might refer to codices with the sigla AB). Chronologically, the earliest non-mathematical author picked up by this search is Aristotle (obviously a very special case), followed by (philosophical commentary underlined): Asclepiodotus (2), Sextus Empiricus (1), <u>Alexander of Aphrodisias (36)</u>, Galen (3), Julius Africanus (7), <u>Themistius (8)</u>, <u>Ammonius (44)</u>, <u>Proclus (excluding *in Eucl., Hyp.*: 107)</u>, <u>Simplicius (147)</u>, <u>Philoponus (119)</u>, <u>Asclepius (2)</u>, <u>Olympiodorus (16)</u> (97 per cent philosophical commentary). (ii) A related search is for '@3', TLG's code for a diagram. Non-mathematical authors later than Aristotle picked up by this search (excluding magic and astrology) are: Asclepiodotus (13), Plutarch (6), <u>Anon., *in Tht.* (2)</u>, Galen (5), <u>Alexander of Aphrodisias (10)</u>, Julius Africanus (4), Iamblichus (1), <u>Themistius (1)</u>, <u>Ammonius (1)</u>, <u>Proclus (as above: 39)</u>, <u>Simplicius (30)</u>, <u>Philoponus (8)</u>, <u>Asclepius (2)</u>, <u>Olympiodorus (14)</u> (82 per cent philosophical commentary). That the two surveys agree on the whole seems to confirm their validity and the impression is that we have a grasp of the totality of mathematics in non-mathematical contexts: infrequent as a whole and, outside the philosophical commentary, extremely rare.

particular, in the *Prior Analytics*, Aristotle refers to three terms – a 'greater', a 'middle' and a 'smaller'. Such terms may be 'filled up' and, most strikingly, a couple of them may form an 'interval'. No doubt the analogy of music theory – whose object is the filling up of interval relations – is at least among the influences on Aristotle's presentation. It was left to Smith 1978 to point out that the similarity is closer as we consider the reliance upon labelled terms such as A, B and Γ; with those terms in place, it becomes possible for Aristotle to follow the usual structure of a mathematical proposition, a general, letter-less enunciation followed by a translation into lettered terms and then a discussion (or, indeed, proof). Smith follows Rose 1968 in assuming that such labels referred to original diagrams.[8]

The mathematisation is remarkable; it is incomplete in many ways. First of all, the diagrammatic base of the argument is not the only one; Aristotle very often invokes an alternative mode of argumentation, based not on diagrammatic labels, but on concrete and so more vivid stock examples.[9] Thus, instead of the mathematical dual structure of general statements and concrete diagrams, Aristotle works with a triple structure: general statements, concrete diagrams and vivid examples. Scholars of Aristotle's logic are used to the frustrations of trying to fit together Aristotle's use of *ekthesis* or setting out, with its mathematical counterpart: essentially, because Aristotle's basic structure of generalities and particularities differs from that of a mathematical text.[10] The triple structure of general statements, diagrammatic labels and vivid examples implies that Aristotle's text is not entirely predictable: while there are good reasons for Aristotle to use this or that format for a particular discussion, the text never assumes the rigidity of a mathematical text. Indeed, it maintains the normal discursiveness of an Aristotelian text, such as it is. While the text is clearly broken into very small segments, they are not separately headed and they are connected by the particles

---

[8] Rose 1968: 16–24, 133–6. Why then are the diagrams absent from the manuscript tradition? But the problem runs much deeper: there are many cases in the Aristotelian corpus where diagrams are called for and yet they were lost from the manuscript tradition, in some cases – most notably, in the *Mechanics* – restored by Byzantine readers (Van Leeuwen 2012).

[9] Most typically, 'man', 'horse', 'animal'. In general for such examples, see Ierodiakonou 2002, who explains (137) that there are in fact good reasons for employing concrete examples and not just diagrammatic labels: in the example we actually see how a result holds, or fails to hold (the concrete example fulfils somewhat the function of a Venn diagram). In the domain of geometrical configurations, labelled diagrams allow the same visual verification of relations; not so, however, with the diagrams of number theory, proportion theory or music – Aristotle's proximate inspiration (in this case, commentators indeed typically resort to the examples of concrete numerical values: see, e.g., Porphyry, *in Harm.* 117.2–21, an entire discussion where concrete numerals stand for abstract labels).

[10] Smith 1983; Thom 1993; Ierodiakonou 2002: 150–1 offers once again a sound interpretation: Aristotle means by *ekthesis* simply a proof by example, taken in a very general sense which therefore does not fit precisely any particular mathematical procedure.

(οὖν, μὲν οὖν, δέ) of an Aristotelian discourse. The overall sense is of a division not so much into propositions as into – normal – Aristotelian paragraphs.[11] An Aristotelian paragraph is characterised by a thesis (or a problem) followed by a brief discussion that provides an account of the thesis, and the structure of the *Prior Analytics* is within this practice. While the discussions corroborate the main claims, they are not structured as fully spelled-out demonstrations; γάρ vastly outnumbers ἄρα.[12] The deductive structure is very shallow. Aristotle occasionally shows the validity of a syllogism by reducing it to another, already proven, but there is no further iterative structure where stronger and more general results are gradually built: as is indeed typical of the Aristotelian paragraph elsewhere, each paragraph tends to be fairly isolated and the paragraphs as a whole are not so much arranged by deductive, as by thematic, structures. Above all, Aristotle engages – typically to him – in a project of classification;[13] the survey of syllogisms in the *Prior Analytics* is not different in kind from the survey of dialectical arguments in the *Topics*.

An Aristotelian treatise typically has an introductory passage setting out the goal and some of the key terms of the discussion, and the *Prior Analytics* is no exception (24a10–25a13). This introductory passage is thoroughly discursive and is nothing like the clusters of definitions, sometimes amplified by postulates, at the beginning of most extant Greek mathematical texts.[14] In particular, no claim, laid out in advance in this introductory passage, is ever required for the various arguments made later on in the treatise. Thus, not only is the deductive structure extremely shallow; it is, effectively, not an axiomatic treatise at all.[15]

---

[11] For the overall Aristotelian practice of segmenting his (extant) works by paragraphs, see Netz 2001.

[12] 765 occurrences of γάρ as against 31 of ἄρα. The ratio in Euclid's elements is 3,778 ἄρα, 964 γάρ. Aristotle resorts so often to γάρ because his main emphasis is on explicating a thesis, not proving a point.

[13] This, however, may also have been typical of the mathematics of the time, engaged with such questions as the classification of irrationals (by Theaetetus, underlying Euclid's book 10?), that of numbers into 'prime', 'square', 'cube', 'odd', 'even' (the key terms of Euclid, *Elements* 7–9). Closest to Aristotle's *Prior Analytics* could have been Archytas' classification of means. I argue further for the role of non-geometrical classification in the mathematics of the fourth century in future work.

[14] See Mueller 1991 for a description of the Greek mathematical starting point in practice.

[15] It is not clear that Stoic authors ever structured their discussions of arguments in an axiomatic form, and yet, if anything, they seem to have put more emphasis on the relation of derivation between arguments, delimiting five forms of argument as 'indemonstrables' (e.g. 'if it is day, it is light; it is day; therefore it is light') and arguing, then, that other valid arguments might be deduced from such indemonstrables. Demonstrables, and their relation to indemonstrables, form the main theme of Bobzien 1996.

The overall impression, then, is that Aristotle imported, for the purposes of his syllogistic, the tool of the lettered diagram, in particular in the form it took in non-geometrical contexts such as proportion and music. A natural choice, given that he wanted to refer to his terms by single-letter, and not many-letter, labels: hence he was thinking through the diagrams depicting the ratios of the terms Α, Β and Γ, not through the diagrams of intersecting lines and figures. The diagrams and labels served to present claims in simple, succinct form. Beyond that, however, Aristotle kept to his standard style.

Aristoxenus, *Elements of Harmonics*, Book III

Aristoxenus, a critical younger colleague of Aristotle, shared his acute scientific mind, but not his breadth of scholarly interest: he seems to have dedicated himself above all to the scientific study of music, in works now mostly lost. Even the extant *Elements of Harmonics* – likely, his major oeuvre – is transmitted not in the form intended by its author. Barker's view is that books II to III represent the beginning of Aristoxenus' original major work (whose continuation – we can only speculate how long – is now lost), and that Book I is a separate and less demanding treatment of the same subject.

Book 2 is very close in character to the more polished of Aristotle's extant treatises, such as the *Ethics*. It states the subject matter of a theory of music, offers careful distinctions and introduces the key terms and observations such as the meaning of 'tone' and the relationship between the octave and the tetrachord. So far, then, ordinary philosophical prose. In Book III we have the beginning of a detailed study of the structure of the tetrachord (Barker believes it could have served, in the lost part of the treatise, as a basis for a study of modulation between 'keys'). And we also move to a different style of writing. The extant text consists of twenty-eight brief statements followed by arguments. With few exceptions, the text consists of statements and arguments alone, no further interventions made by the voice of the author. Indeed, some of the statements are presented in the form of a task: 'it is to be proved that…' (ὅτι… δείκτεον), a formula familiar (if in a different word order) from mathematical texts. In general, compared to all other ancient non-mathematical texts, this Book III stands out primarily by the precision of the claims made at each argument. Furthermore, the proofs are indeed in principle valid (though readers still need to do some work so as to clarify some missing assumptions and definitions: Barker of course discusses such possible logical completions). Even though

Aristoxenus' approach to pitch is non-quantitative, he has clear constraints of a combinatorial kind. So, in what is known as 'conjunct' tetrachords, the structure is *defined* to be that of a *pyknon* followed by a ditone, repeated; so that it follows directly that a ditone is bound by two *pykna*, its lower note being the higher note of a *pyknon*, its higher note being the lower note of a *pyknon*. Such is proposition 6 of the treatise:[16] obviously a proof, then, albeit a simple one. Nearly all the proofs in the book are of this order of conceptual opacity and yet logical simplicity.

The sense of a 'proof' is clear; no less is the sense of a deductive order. The structure is still mostly thematic and 'shallow' (indeed, with such a simple structure to be studied, it is no surprise that there is little need for complex demonstrative iteration). As Barker notes, results on equal intervals are followed by results on unequal intervals; but then, the discussion moves on to the characterisation of 'routes' (continuous sequences of notes, whether ascending or descending), and such results are directly built on the more basic ones concerning intervals, equal or unequal. It is likely, as Barker points out, that even more deductive complexity would have followed: the entire sequence of results in Book III provided the tools, then, for characterising, say, possible modulations. If so, the original book must have displayed a clearer 'axiomatic' structure though, once again, in the absence of an axiomatic introduction. Instead, the whole of (what is now) Book II served as a very detailed conceptual preparation in advance of the proofs.

The entire sequence of books I to III is titled, in the manuscripts, as 'Elements of Harmonics'. This would make it the earliest extant work so titled. The title is likely authorial. Porphyry seems to have known (what is now) Books II to III under the same title;[17] Aristoxenus refers, in the so-called Book I, to *another* book as 'Elements', and then mentions, in Book II, that his readers are *about* to embark on the study of the Elements. The implication is that perhaps the entirety of Books II to III and their continuation were circulated, by Aristoxenus himself, under the heading 'Elements', and that he identified those 'Elements' primarily in the sequence of proofs in Books III and following. Now, Proclus made Hippocrates of Chios – almost a century before Aristoxenus – the author of the first Elements of geometry, followed by later fourth-century mathematical authors. It is unclear if this is to be trusted and then, even if it is, it is unclear how many other

---

[16] The manuscripts do not number the propositions; it is not necessary, however, that proposition numbering was used in the mathematics of Aristoxenus' time (Netz *et al.* 2011: 279–80).

[17] See discussion in Barker 2007: 134–5.

'Elements', in whichever disciplines, might have preceded Aristoxenus. Choose one account – where the title 'Elements' is associated in the fourth century primarily with the discipline of geometry – and Aristoxenus' title becomes, in and of itself, *more geometrico*. Choose other accounts, and it is not.

That Aristoxenus himself conceived of his project as in some sense involving a 'mathematisation' may be suggested by his own words. If indeed Books II to III represent a separate book, then we find, at its start, the famous anecdote of Plato's lecture on the Good. Plato promises to give a lecture on the Good; it is a complete failure, because no one expects such a purely mathematical account; hence, concludes Aristoxenus, better say in advance what you plan to do. This anecdote may be taken to motivate Aristoxenus' choice to begin Book II with a statement of the scope of the study of music; it can also be taken to reflect on the position of Book II, as a whole. As it were, Aristoxenus might have contemplated the option of circulating a book consisting of Books III and following alone (perhaps with a compressed, mathematical style, axiomatic-definitional introduction), and opted against it, appending, instead, a discursive Book II at the beginning and thus avoiding the fate of Plato's lecture.

The case that Aristoxenus' *Elements of Harmonics* Book III was deliberately intended *more geometrico* is, I think, even clearer than it is for Aristotle's *Prior Analytics*. But let us note that Aristoxenus' treatise is, arguably, less 'mathematical'. This of course has to do with the choice not to employ diagrams. Now, Aristotle's A, B and Γ, in his syllogistic, are essentially just convenient pegs – however useful – extraneous to his actual argument. Aristoxenus' subject matter, by contrast, cries out for visual support. He engages, essentially, in the combinatorics of discrete elements arranged along a linear sequence and it would therefore be extremely helpful to have a figure where lines are punctuated by marks of their elements. Modern discussions of Aristoxenus' subject matter almost invariably include such an illustration and it is very puzzling that an author, deliberately choosing to write *more geometrico*, and having so much to gain from an illustration, nevertheless draws no such figures.

The absence of figures means the absence of labels or, in general, tokens standing in for the terms discussed. Aristoxenus has no shorthand terms referring to particular instances of 'ditone' or *pyknon* and so his discussion is conducted, entirely, at the level of the general concepts themselves. It thus does not differ at all from any other philosophical, conceptual discussion and in generic terms it is very much the same as the normal Aristotelian paragraph. The entire *Elements of Harmonics* contains not a single instance of ἄρα. Of its 225 instances of γάρ, ninety-two are in Book 2 and fifty-two are in the substantially briefer Book 3.[18] The key point is that the entire architecture of the mathematical proposition depends on the diagram. Avoid it, and what you have left no longer has the feel of mathematics. Instead, the text of *Elements of Harmonics*, Book III, can best be characterised as the normal Peripatetic sequence of paragraphs, in a distilled form which helps make the claims of each paragraph, and so their mutual logical relations, clearer than usual. What we can say, then, is that in this case Aristoxenus saw, in the mathematical genre, a model for the clear statement of claims and of their overall logic. It is curious that Aristotle, in the *Prior Analytics*, and Aristoxenus, in the elements of *Harmonics* Book III, took out of mathematics complementary features. The one took the diagrammatic label, but not the setting out in discrete propositions; the other took the setting out in discrete propositions, but not the diagrammatic label. Perhaps, emblematic for the relationship between these two authors.

Asclepiodotus (Posidonius?), *Tactics*

Aristotle's and Aristoxenus' motives can be contextualised through our understanding of their overall character. For Asclepiodotus, we have nothing. A mere name from a Florence manuscript collecting ancient tactical works;[19] perhaps to be identified with a student of Posidonius cited by Seneca for information in natural philosophy. Posidonius, you say? One's curiosity is piqued. There was a treatise on tactics by Posidonius himself, now lost;[20] Asclepiodotus' treatise is in fact not unique, but is instead merely the first of an entire series of Roman-era Greek tactical treatises, all so closely related as to force the postulate of a common source.[21] It seems quite

---

[18] Book II takes twenty-one TLG 'screenshots', Book III takes thirteen; ninety-two and fifty-two uses of γάρ, respectively, are then almost exactly the same ratio.

[19] Laur. 55.4 132v–42v, an early minuscule Byzantine manuscript later copied by several Renaissance hands.

[20] Kidd 1988: 333–5.

[21] The other extant treatises are by Aelian (who is dated by the addressee of his treatise, Trajan), by Arrian, the second-century author, as well as an anonymous *glossarium militare*. In what follows, I will concentrate on Asclepiodotus as (presumably) the earliest extant author, always meaning the presumed lost source. The various treatises are nearly identical, that of Asclepiodotus being however more compressed; all appear, therefore, like different recensions of a single source (so Stadter 1978: 117–18). The major difficulty is that Arrian refers to Posidonius as among the past authors of tactical arts, and then goes on to disparage all past authors as not useful (1.1–2: because they are written for those who are already experts). It is meaningful, however, that both Aelian (1.2) and Arrian (1.1) conclude their list of past authors

likely, then, that the mathematical touches in Asclepiodotus' treatise could represent an innovation by an original that could well be by Posidonius – yet another mathematically aware philosopher, comparable in this sense to Aristotle and Aristoxenus.

Technical writing relating to war, in the Hellenistic world, could concentrate on war engines. In this case, such writings would generically belong to the mechanics and so expected to be 'mathematical': we have in fact extant Hellenistic works on war engines (by Philo and by Biton) whose style clearly includes mathematical dimensions (alongside other, more discursive forms of presentation).[22] There are therefore two vectors in which we can understand the mathematical touches in Asclepiodotus (or in his Posidonian model): as a direct reflection of the mathematical genre as a whole; or as an extension of the mathematisation of war, from war engines to war tactics. As it were, Asclepiodotus' (Posidonius'?) innovation could have been to conceive of the phalanx as a kind of war engine.

Let us glance briefly at this curious text. It turns away from history: not a single anecdote is recounted. The only sense that war is made by historical people, not by abstract tokens, is the occasional reference to ethnics – the central type under discussion is 'Macedonian'; certain formations are 'Scythian' or 'Thracian' (7.3), certain movements are 'Cretan' or 'Persian' (10.15). This is no more concrete than the musical *tonos* 'Lydian' or the architectural order 'Corinthian'. And the discussion as a whole could equally have come from a musical or architectural manual, so abstract it is and, indeed, so quantitative. The treatise is structured by an overarching

---

of tactical arts with Posidonius. It seems to me perfectly possible that Arrian added Posidonius to his list of previous authors and yet quoted verbatim Posidonius' own complaint concerning his, Posidonius', predecessors, inadvertently turning the complainer's complaint upon the complainer himself. As for the identity of the source: Eramo 2010: 150–1, n. 80 suggests the lost source could be by Posidonius or by Polybius: but we can be certain such would not be Polybius' style. The discussion in Kidd 1988: 32–3, 333–5 seems not to be aware of the very tight interrelation between the various ancient tactical treatises – the tightness which forces us to see them all as derived very closely from a single source. Thus, Kidd may have failed to acknowledge a curious result: the work by Posidonius which we come closest to recover, in its original form, could very well be his tactical art! Finally, it is striking that Aelian refers explicitly to his use of diagrams as an original contribution, a statement which once again cannot be literally true and must go back to his source (ch. 18); Arrian, however, omits the diagrams (to be avoided by a more literary author?). We are led to believe that Posidonius invented a new kind of military manual, based on the use of diagrams and a generalised 'mathematical' approach. Of lasting influence: both Asclepiodotus and Aelian had a significant afterlife in the Renaissance, and established then the role of the diagram in military writing (see Hale 1988).

[22] See Roby 2016: 71–2 for the mathematical dimension in Philo or Biton; chapter 1 surveys the generic range of influences on ancient mechanical writing, widely understood.

conceptual taxonomy: the arrangements for war are landed or naval (we are to discuss just the landed), which are then either footed, or mounted, each further subdivided. Once the conceptual analysis reaches the military units themselves, they are discussed in a similar taxonomic, terminological vein (hence the use of ethnics to characterise the different *taxa*).[23] Within each basic term, finally, the discussion is always framed in the most quantitative and mathematically seeming terms. This takes two main forms. For the hoplite phalanx, the author is engaged primarily with the numbers to be deployed within given units, especially concerned to make sure that the numbers will be divisible by two: hence, everything is governed by the powers of 2 and the ideal phalanx is declared to have $16,384 = 2^{14}$ soldiers. Chapter 2 is almost entirely given to the various divisions arising out of this number. In chapter 3, the parts of the phalanx are arranged by military valour and we are told that the rightmost fourth of the line should be composed of the best soldiers, the leftmost and second leftmost should be composed of those who follow, and the worst should stand as the second rightmost, as in this way the right and the left are balanced ('the geometers say'): clearly a reference to a:b::c:d → ad = bc.[24] Further numerical discussions involve the gaps between the soldiers in the phalanx as well as the lengths of the spears they carry – all indeed directly comparable to the numerical discussions in architecture (chapters 4 to 5). Later follows a very abstruse discussion (chapter 7) of the geometry of cavalry formations, which are of course more complex than the purely rectangular phalanx. It is in this context that diagrams are added in great profusion, typically lettered (each letter, in this case, standing for an individual horseman). Once again, the author touches briefly on terminology and then, the most substantial discussion in the treatise (chapter 10) involves motions – typically rotations, either of each individual separately, or of formations as a whole. Once again, this discussion is carried through extensive reference to diagrams. This brings in a type of geometry which is not typically discussed in explicit terms in Greek mathematics and it makes apposite, if trivial observations: for instance, that following three forward quarter-rotations it is more convenient to return to the original position via one further forward quarter-rotation than by three

---

[23] One wonders if this terminological emphasis may not be the point of the complaint, that previous military arts took for granted that the readers are familiar with their terminology? Indeed, we may perhaps reconcile the apparent military uselessness of such treatises, together with the apparent zeal with which so many versions were made and transmitted, in that their function was not so much military as literary: tools for the reading of *other* writings about war.

[24] This connection being made between the idea of human value and that of a geometrical progression is reminiscent of the ancient piece of ideology of 'geometrical equality' that justifies inequality: de Ste. Croix 1981 413–14.

backward quarter-rotations (10.9). Chapter 11 combines, in a sense, the discussions of formation and of motion, to discuss formations in march, once again of course considered via lettered diagrams.

The language is throughout spare, and very reminiscent of mathematical terminology: τὰς δὲ τάξεις αὐτῶν κατὰ σχῆμα οἱ μὲν τετράγωνον πεποίηνται, οἱ δὲ ἑτερόμηκες (7.2), τόπου δὲ γίνεται διπλασιασμὸς κατὰ μῆκος μέν... κατὰ βάθος δὲ... (10.18–19). This, however, recalls throughout not the language of a Greek mathematical proof, but, if anything, of a Greek mathematical definition. Indeed, the text lacks any argument and is instead a series of statements, mostly of a terminological or conceptual character (obviously, there is no ἄρα; even the frequency of γάρ – sixty-two occurrences – is unremarkable; this is about half Aristoxenus' frequency. Even for a non-mathematical text, this work shows little argumentative structure of any kind). While the vocabulary is often mathematical, the overall use of language could equally fit in a philosophical or even a grammatical context. The diagrams, too, do not serve in an argumentative, but in an explanatory role. For this reason, one does not manipulate the labels within the text, but merely points to them in the diagram. For this reason, the very logic of the diagrams is distinct from the mathematical one: most obviously, one often has the very same letter repeated over and over again (standing, in this case, for a horseman or a hoplite). This, once again, is paralleled by some mechanical diagrams.[25]

There is nothing original about pointing out the mathematical character of Aristotle's *Prior Analytics* or of Aristoxenus' *Elements of Harmonics*, Book 3; indeed, I started out by citing scholars who remarked on that character. I am not aware of similar claims made for Asclepiodotus (Posidonius?). This may be because such tactical manuals were studied much less often; but the main reason appears to be, simply, that those treatises are not obviously mathematical. There are several features that may or may not be intentionally mathematical-like – diagrams, numerical and configurational comments, a spare, mathematical terminology, a concern with definitions – but there is no trace of the key feature of the mathematical style, namely the proposition organised around a diagram. Aristotle's *Prior Analytics* has a distinct variety of that, where he emphasises the diagram as organising principle; Aristoxenus has a complementary variety, where he emphasises the proposition and avoids the diagram. In Asclepiodotus there are no

---

[25] In a mechanical diagram, the labels may sometimes refer to a functional element which may be repeated: so, for instance, the two places for the spring in the machine are labelled L in a figure in Philo's *Belopoiica* 65.9, or the two sides of a beam may both be labelled G so that the beam as a whole becomes GG, in Biton's *Construction of War Engines* 45.5.

propositions and so no writing *more geometrico*. In this case, mathematisation is obviously just that – a *move* in the generic space that nods towards mathematics without actually invoking it.

What is the purpose of this move? In fact, we are puzzled by the purpose of such treatises as a whole. Asclepiodotus (or Posidonius) is surely not a practical manual. I suggested it could have some function for the reading of other military texts, perhaps of a historical character. It is indeed strangely behind the times, referring as it does to elephant warfare and taking the Macedonian phalanx as its primary reference. Remarkably (and the sure sign this treatise does not go back specifically to Polybius), Romans are not in evidence at all. The entire cluster of military *technai* we introduce through Asclepiodotus – reaching forward to Aelian, Arrian and the *Glossarium militare*, perhaps reaching back to Posidonius – was all a Greek genre, written in a world dominated by Roman military might. That it invokes a past of Greek military greatness, and that it does so successfully (in this case, after all, we consider a work which is revised and re-issued again by many authors), suggests a specific attunement to a cultural moment. One is reminded of Whitmarsh's comments on the 'post-colonial', as it were, politics of the Second Sophistic – that of Greek authors invoking a lost military Greek grandeur while acknowledging Rome's ascendancy.[26] If we understand such military *technai* in this context, then, perhaps, the mathematical components of the style might be seen as yet another element of specific Hellenism. Mathematics is an exotic genre, resonant of Greek-ness; it comes naturally in this evocation of the lost, exotic Macedonian phalanx. (Arguably, the mathematical components of the style could play a similar role in Vitruvius' *Architecture*, another work which, under the guise of a *technê*, performs for its readers the drama of a cultural transfer.)

Proclus, *Elements of Theology*

The mathematical character of Asclepiodotus is problematic; that of Proclus' works of *Elements* is self-evident. Let us start by noting that Proclus is also the author of an *Elements of Physics*,[27] a brief, explicitly geometrical re-telling, with labelled diagrams, introductory definitions and clearly

---

[26] See, e.g., Whitmarsh 2005: 69–71 for the figure of Alexander – that is, implicitly, the Macedonian phalanx – in such fantasies.

[27] The title transmitted is not στοιχεῖα, but στοιχείωσις; I am not sure this should carry much meaning, as Proclus could also refer (e.g. *in Euc.* 74.11) to Euclid's *Elements* (as well as to Aristoxenus'! – *in Ti.* 2.169.17–18) as στοιχείωσις.

defined propositions, of Aristotle's basic claims in the *Physics* concerning locomotion (e.g. proposition 1.7: if motion is partless, its time is partless as well). As we noted above, Aristotle occasionally employs in the *Physics* the system of referring to diagrammatic tokens such as A, B, Γ; Proclus, then, merely amplifies and systematises a practice present already in his source. Such is a commentator's practice and as such it belongs, perhaps, to the general tendency of philosophical commentators to engage in the occasional piece of mathematics. The treatise is indeed brief, especially by Proclean standards. It differs, however, from the standard fare of commentator's mathematics in its extended scope, in its conceptual extension of the field of mathematics and so, as a consequence, its independence from Euclid's *Elements*.[28]

The *Elements of Theology* (in the Neoplatonist tradition, the title implies a much wider scope than just our 'theology') is considerably more ambitious. With 211 propositions, it is much longer than any normal mathematical 'book' and has more the character of an entire theory. Proclus, as noted in footnote 27 above, was familiar with Aristoxenus' *Elements of Harmonics* and he of course wrote, himself, a commentary on Euclid. His emulation of the mathematical model was certainly intentional. That he avoided the labelled diagram is perhaps a mark of the subject matter. Indeed, Proclus' approach does not lend itself to a mathematical-like treatment. I quote the relatively brief proposition 32, in Dodds' translation,[29] introducing my own numerals:

> All reversion is accomplished through a likeness of the reverting terms to the goal of reversion.
>
> (1) For that which reverts endeavours to be conjoined in every part with every part of its cause, and desires to have communion in it and be bound to it. (2) But all things are bound together by likeness, (3) as by unlikeness they are distinguished and severed. (4) If, then, reversion is a communion and conjunction, (5) and all communion and conjunction is through likeness, (6) it follows that all reversion must be accomplished through likeness.

'Reversion' is the process through which a lower entity becomes more like a higher one; this then is a theorem concerning the Platonic ascent. Thus, statement (1) resonates with Platonic doctrine and in this way perhaps it

---

[28] Most evocations of mathematics within philosophical commentary involve a display of the commentator's geometrical knowledge, typically that of Euclid's *Elements*; but Proclus' *Elements of Physics* relies, instead, on Aristotelian reasoning.
[29] Dodds 1963: 37.

calls, as far as Proclus is concerned, for no further argument. Statement (2) perhaps relies upon (3), or perhaps is clarified by it (ὥσπερ: an ambiguous connector); at any rate, once again, it seems to spell out a fairly self-evident claim. Step 4 re-states what, looking back, we recognise was a *presupposition* of Step 1 (but one which, however, Proclus seems happy to postulate), while Step 5, finally, might be seen as a special case of Step 2. Step 6, finally, indeed follows from Steps 4 to 5. We remain with a clear and meaningful logical structure: Steps 1 and 2 to 3, each providing some material for the final derivation from Steps 4 to 5 to 6. It does remain perplexing, and typical, that we do not even know for sure the function of the ὥσπερ; most fundamentally, this does differ substantially from an axiomatic derivation, in that one is allowed to bring in at will extra assumptions, for the sake of each individual argument, as long as those seem reasonable within Proclus' overall Platonism.

The obvious alternative would have been for Proclus to step back and try and spell out certain key assumptions, and then state those as explicit axioms for his system, trying to deduce the rest from that: the standard move of a seventeenth-century treatise *more geometrico*. In fact, the *Elements* contain no axiomatic introduction. Proclus proves the first twenty-four proposition without any reliance on previous propositions, many later propositions rely on little or no previous results, and I am not sure even a single proposition can be seen to unfold purely from proved statements. Axiomatic reduction to first principles simply does not seem to have mattered to Proclus.

What did matter to him? Why produce a στοιχείωσις in the first place? We do not need to speculate as Proclus – the prolific author that he was – reflected upon this very question, admittedly in the context of mathematics. In his commentary to Euclid's *Elements*, he reflects on the goals of writing '*Elements*' as such, and to sum up a somewhat involved discussion (prologue 2.7), his main thrust seems to be that an *Elements* is judged by its clarity (σαφήνεια) and concision (συντομία), in turn valuable because of their contribution to knowledge (ἐπιστήμη). As ever in Proclus, the pedagogic dimension is not far from sight and the point appears to be that an *Elements* helps the student (or even the advanced scholar) gain a better understanding of its field by having its major claims surveyed in clear and concise form. Seen in such a view, Proclus' project – a perfectly Euclidean presentation of the key elements of his metaphysics, which however lacks in any axiomatic ambition – suddenly makes sense. The Euclidean format is useful not as a tool for uncovering axiomatic starting points, but as a tool, a mnemonic almost, for systematising one's doctrine.

The Mathematised Treatise: General Observations

A generalisation? None is immediately forthcoming. A handful of mathematised treatises – and yet none alike. Which in a sense is not surprising: a tiny set of isolated experiments, this never could coalesce into a 'genre'. Different mathematisations focused on different aspects of the mathematical practice. In Asclepiodotus (Posidonius?), we have mostly the mathematical lexicon, together with the use of diagrams; in Aristotle's logic we have the diagram in its most rudimentary form, together with some of the features of the proposition which the diagram allows; in both Aristoxenus and Proclus, we see the features of the proposition that can be reproduced even without the diagram, together with the overall structuring of the treatise by its propositions. What we do not see is any mathematical-like introduction where definitions and postulates are explicitly spelled out, which implies that never do we see the effort to use the mathematical model so as to analyse a theory into its constituent suppositions. (We are reminded that the ancient axiomatic introduction was far less stabilised than one would tend to imagine, based on Euclid, *Elements* Book I alone.)[30] The variety spells not merely the lack of a clear, unified sense of 'what is mathematics', but also the fact that mathematics is used not for its own sake, but for the sake of a goal interior to the mathematised treatise itself. It is not as if Aristotle, Aristoxenus, Asclepiodotus or Proclus strive to transform their material so as to make it mathematics-like; rather, they identify a feature of the mathematical style which they find useful for their own purposes and which, as it happens, is distinct in each case: the definite reference of the lettered diagram is helpful to Aristotle in his logic; the explicit statement of individual claims in each individual enunciation is useful to Aristoxenus, in one

---

[30] As emphasised by Mueller 1991. And when did Euclid attain his dominant position in the mathematical field? Clearly, judging by the evidence of the papyri, the *Elements* penetrated to some extent the mathematical education (Netz *et al*. 2011: 254–5). Hero – whose work is perhaps to be understood as an attempt to elevate school mathematics into elite literary status – is the first extant author to cite Euclid; later references are not common and their Euclid remains the mathematical pedagogue (perhaps the earliest non-mathematical reference to Euclid, other than a citation of his *Elements*, is in Aelian's *Histories of Animals* – a contemporary of Africanus, for which see below – who points out in VI.57 that the spiders have their web in the form of a circle 'without need of a Euclid' – i.e. without the benefit of mathematical education). It is only with Pappus that Euclid becomes a fully imagined historical figure (Cuomo 2000: 71–2, 196–201). Even then, Euclid certainly never developed the kind of persona of Hippocrates (Jouanna 1999: 348–57) or of Archimedes (Jaeger 2008). What we now take as the supreme model of axiomatisation was, for antiquity, the modest achievement of elementary education.

presentation of his harmonic theory; the ability to overview the entire set of enunciations, apparently, is the one prized by Proclus. As for Asclepiodotus, his mathematical echoes may be prompted by cultural, rather than cognitive, motivations: they are a way of invoking a genre specific to the Hellenic tradition. Macedonians have phalanxes; and Greeks have diagrams.

The Greeks have diagrams, that is: a particular generic feature carries certain cultural connotations. Surely, mathematics was recognised as a genre, but, we now notice, none of the mathematisations surveyed here ends up producing a work in the mathematical genre. Early in the fourth century, the mathematical genre was elaborated and certain fields associated with it; following that moment of inception, there is no well-attested example of the same genre being introduced to any new fields.[31] The mathematisations observed are actually no more than gestures *towards* a mathematisation, employing some features of the genre and yet clearly written within non-mathematical generic boundaries: an Aristotelian esoteric treatise, an Aristoxenic philosophical science, a *technê*, a commentator's summary.[32]

What we do not see is any trace of the seventeenth century: any sense that mathematics is considered to have any special access to truth. To the extent that mathematics has epistemic value that may be exported across generic boundaries, this is not in that it finds truth otherwise hidden, but in that it makes claims, once found, more manifest. The various mathematisations always pick up a property of the mathematical genre that serve, not in the finding of truth, but in its exposition.

Not the seventeenth century. And yet: was there an authority to be gained by the sheer flaunting of one's mathematical skills? To see this, let us look for another, more localised form of writing *more geometrico*: the mathematical-like passage.

---

[31] The most likely exception could be mechanics, where we cannot prove that the Peripatetic treatise may not have been the first, in which case here we see a mathematisation late in the fourth century or even early in the third century (so Berryman 2009 – however, perhaps too pessimistically for theoretical mechanics in the fourth century?) Since Burnyeat 2005, mathematical optics is seen to go back possibly to Archytas himself, as mathematical music surely does; mathematical astronomy surely went back to Eudoxus.

[32] That the mathematisation of non-mathematical fields was rarely undertaken in antiquity might be taken together with a piece of Aristotelian doctrine: the prohibition of domain-crossing. The precise extent of such a prohibition is far from clear (the best recent survey of this problem is Hankinson 2005b – who ends up arguing how little restrictions Aristotle's approach imposed in practice, given the structure of ancient science envisaged by Aristotle himself). Nor is it clear why any non-Aristotelian should have subscribed to this epistemological tenet. It is always preferable to anchor general tendencies in Greek cultural life not to any particular doctrine, but to the shared conditions of cultural practice: we seem to look, in this case, at the cultural significance of generic boundaries.

Examples (II): The Mathematical-Like Passage Aristotle's Meteorological Phenomena: *Meteorology* 373A4–19; 375B16–7A11

In the third book of his *Meteorology*, Aristotle discusses phenomena such as the halo and the rainbow. The discussion is admirable in its rigour and insight: Aristotle correctly identifies such phenomena as optical and atmospheric.[33] For both the halo and the rainbow, following his qualitative, discursive account, Aristotle appends a mathematical passage based on a diagram, quite brief in the case of the halo, very extensive – as it stands in our manuscripts – in the case of the rainbow. The overall result is to mark *Meteorology* 3 as nearly a hybrid of philosophical and mathematical prose, a text that could equally have been written by Galileo or Descartes (and, indeed, could well have influenced the latter).[34]

Our impression of a highly mathematical text may well, however, be misleading. Vitrac 2002, following the proposal of Tannery 1929 (originally published 1886), suggests that the bulk of the mathematical rainbow discussion could have been introduced through several waves of scholastic interpolation. Indeed, as it stands in the manuscripts, the rainbow discussion is perhaps unique in the Aristotelian corpus in presenting a mathematical argument in an attempted precision of detail that goes well beyond the requirements of the passage at hand.[35] What we have if we follow Vitrac, then, is a fundamentally standard, discursive text, twice expanded with a brief diagrammatic account. Those expansions still count as the most substantial pieces of evidently mathematical discussions in the Aristotelian corpus, excluding the *Mechanics*.[36] I will look first at the halo passage, and then return to the textual problem regarding the rainbow passage.

---

[33] Johnson 2009.   [34] Johnson 2009: 354–6.

[35] Knorr 1986: 107–8 was puzzled as well and considered the entire proof an addition to the corpus by a later Aristotelian disciple (because of archaism in the language, he avoided seeing it all as a much later interpolation, and he did not consider Vitrac's option of a brief text, later expanded).

[36] For a full survey, see Vitrac 2002: 248–55. As Vitrac noted, other than the mere allusion to mathematical practice as seen most clearly in the *Prior Analytics*, arguments in the mathematical style are nearly confined to the triad of works *Physics* 6–8 – *De Caelo* – *Meteorology*. While mathematical in style, the discussions in the *Physics* and the *De Caelo* typically involve a conceptual, philosophical discussion based on a mathematical diagram (which is what inspired Proclus' *Elements of Physics*). The mathematical passages of *Meteorology* III remain, then, as the most explicitly mathematical in a non-mathematical context.

Aristotle accounts for halos by a refraction[37] in the medium – a cloud. This, together with the assumption that the refraction is similar on all sides (373a3: πάντοθεν... ὁμοίως ἀνακλωμένης) yields the result that the halo must be circular, which Aristotle immediately moves on to discuss in mathematical terms. The next sentence makes a general claim in the enunciation style; this is followed by a proof based on a labelled diagram. I discussed this proof in an earlier publication,[38] as an example of an atypical proof structure (it is atypical in that it is backwards-looking, blends construction and proof, and has many second-order explications), resulting from its atypical setting within a more discursive context, a point taken and extended by Liba Taub.[39] Most probably, as she points out, this text comes, in some sense, from a lecturing context. Johnson 2009 is a very thorough study of Aristotle's account of the halo and his main interest is to highlight Aristotle's skill. He was offended by my rather dismissive notes on the mathematics of the passage (admittedly, I called it 'hardly worthy of being called mathematics') and so he set out, in pages 350 to 351, to defend the passage.[40] If I understand him correctly, however, his point is that this is not a mathematical passage at all (351): 'Aristotle is not trying to *prove* that the halo is a circle, but to *explain* why it always appears as such ... the diagram is still drawn for the same reason in modern textbook' (emphasis in the original). That is, we should not look for a mathematical proof: instead, Aristotle illustrates a physical point by aid of a diagram in the mathematical style.

I think this is correct and so captures the ambiguity of the passage, but it should also be noted that the surface appearance suggests an attempted geometrical proof. The general enunciation is purely geometrical: 'for, from the same point to another, the equal <lines> shall bend[41] always on the line of a circle'. This opaque statement is clarified by the claims Aristotle takes for granted through his proof: that, in the figure,[42] the lines ΑΓ, ΑΖ ΑΔ are mutually equal, as are the lines ΓΒ, ΖΒ, ΔΒ. We learn, therefore, with some surprise, that Aristotle's notion of 'being equally bent' means not an equal

---

[37] An anachronistic term, of course; Aristotle uses the verb ἀνακλάω, 'bend back', 'reflect'.
[38] Netz 1999: 210–12.   [39] Taub 2003: 113–14.
[40] He also said some nice things about my book, which I wholeheartedly reciprocate about his article, and pointed out a silly error (I referred to this as a discussion of the 'rainbow').
[41] As Aristotle moves from a strictly physical, to a mathematical treatment, ἀνακλάω – 'reflected' – becomes κλάω – 'bent'.
[42] The diagram, as is the general rule in Aristotle, is omitted from the manuscripts; my reconstruction from Netz 1999: 211 tries to be faithful to the practices of the ancient diagram (a 'flat' representation of a three dimensional arrangement, a generally horizontal structure with labels moving left to right, top to bottom, and a maximally symmetrical structure).

angle, but equal *distances* from both origin and target, at the point at which the bending occurs; it is also being taken for granted (perhaps, as a supposedly innocuous simplification? Or is it meant to be implicitly proved?) that the bending occurs at a plane orthogonal to the straight line joining the origin and the target. The argument then is straightforward: we now have triangles AΓB, AΔB, AZB, in which all the sides are equal – so, obviously, congruent; finally, the unspoken presupposition that the bending occurs at a single plane, orthogonal to the line AB, yields directly the equality of ΓE, ΔE, EZ, hence the circularity of ΓΔZ. I must say I still cannot see this as anything but an attempted mathematical proof and, as such, I still find it lame. The understanding of 'being equally bent' in terms of equal line segments would be very unhelpful within the terms of mathematical optics;[43] all the derivational work is done through the implicit, unmotivated claim that the bending occurs at a plane passing perpendicularly to the line AB.

I agree with Johnson that none of this should be construed precisely as a criticism of Aristotle. Aristotle does not set out to produce a mathematical proof as such; he uses, rather, the form of a mathematical proof as a tool within a philosophical explanation, essentially to help us visualise how the halo could appear circular. I would still count it as a *scientific* problem that the assumption of the perpendicular plane of refraction is never discussed, but it is true that Aristotle's passage is not exactly a failure (which is where I agree with Johnson) because it is not exactly a piece of mathematics (which is where Johnson agrees with me). The key point, then, is that we do not have here Aristotle striving to display his mathematical skills, and failing. The motivation for the passage is not a desire to display one's mathematical skills; rather, Aristotle imports a mathematical tool for the sake of a more evident visualisation.

Is the same true of the rainbow passage? Everything depends on our reconstuction of the text. If we keep the text in its full form, we have a very detailed locus problem[44] whose solution is not free of blemishes (textual corruptions, perhaps), but is still quite impressive as a piece of mathematics. If we reduce it according to Vitrac's suggestion, we have once again no more than a mathematical-looking illustration where the key claim – that

---

[43] Once 'being equally bent' is understood as 'of equal angle', we get a rather different result: the locus of points at which the line from A to B bends at a given angle is formed by an arc on the line AB as a chord, rotated around the line AB; a sphere when the angle is right, a pumpkin when it is acute, a courgette when it is obtuse; the intersection with a plane passing perpendicularly to the line AB would remain a circle.

[44] It is: given two points A and B, to find the locus of points C such that AC:AB is a given ratio. This results in a circle (which, cut by the horizon and seen obliquely, forms the semi-elliptic shape of the rainbow).

the points where the ray falls to form a rainbow form a circle – is taken for granted. In this case, one wonders if the assumption could not be taken over from the previous discussion of the halo. Then, later readers, sensing that Aristotle's claim is valid but cannot, in fact, be reproduced from the discussion of the halo, interpolated the mathematical proof.

We run into circular reasoning. Tannery, looking for non-circular grounds for excision, identified certain 'archaic' expressions or registers in some, but not all, portions of the text (the archaic, then, is Aristotelian!); but as Vitrac points out, the text employs not two but at least three mathematical registers; what's more, commentators can easily pick up and employ an archaic register.[45] The best approach is to consider this backwards: assuming the text missed a part – could its introduction be a likely interpolation? Vitrac's main point is that the central mathematical discussion can indeed be motivated as a (series) of such interpolations, whereas it is little motivated as a single, extended discussion. We end up with a more consistent Aristotle. Most usefully, we have removed this passage from the (otherwise unattested) field of *unmotivated mathematical proofs in Aristotle* and reincorporated it into the (fairly widespread) field of *mathematical elaborations in philosophical commentary*. Circular as the argument is, it provides a much more plausible context for the current text of the *Meteorology* so that I prefer to follow Vitrac. And so, we have excised away Aristotle's one best claim for the inset mathematical passage. 'Aristotle flaunting his mathematics' disappears; and, once again, we see the same Aristotle from our previous discussion – the one conscious of the *expository* advantages of mathematics.

Mathematical Geography in Strabo: *Geography* II.1–4

Bad enough that we need to rely on philological reconstructions. In general, I would prefer to analyse extant examples and not speculate on those we have lost. So it should be admitted: we have not a single clear-cut extant example of an inset mathematical proof within a non-mathematical geographical treatise. For, in general, among ancient genres, geography is more lost than most.[46]

---

[45] Vitrac 2002: 245–8.
[46] Many works – many of which, anonymous – survive from the genres of the *periplous* or the itinerary; typically, the barest compilations of land or sea distances along a route. They form not so much geographical treatises as geographical datasets. Only two fully-fledged geographical treatises survive: by Strabo, and by Ptolemy (of course, there are many geographical passages surviving within historical works).

In very broad terms, ancient geography and music form, as it were, the inverse mirror images of each other. Both fused a descriptive element together with a mathematical one. Mathematics in music arises from the division of the tetrachord; mathematics in geography arises from the calculation of positions. Now, most extant musical treatises give a central place to the mathematical division of the tetrachord; Aristoxenus is the exception in his insistence that tuning is fundamentally a qualitative exercise. For this reason, the inset mathematical passage is ruled out in most of ancient music: it was all too mathematical, to begin with.

In geography, the reverse. The great bulk of ancient geographical treatises seems to have been almost exclusively dedicated to qualitative ethnographic and physical descriptions (with some bare data on distances of a non-mathematical character), close in genre not to mathematics, but to history.[47] The inverse mirror image of Aristoxenus, within the field of geography, was Ptolemy, whose *Geography* is dedicated entirely to the mathematical treatment of the problem of locating positions in a world map – and whose main original contribution is a discussion of the even more specialised problem of plane projection. It is not clear how many prior works of geography in antiquity were so conceived as primarily mathematical treatises.

So, in geography we should have fertile ground for the inset mathematical passage. Did any Greek geographer insert, within the context of a discursive, non-mathematical discussion, insets produced in the mathematical genre?

The answer appears to be: not exactly. Different authors employed more or less of a mathematical terminology, and put more or less emphasis on explicit calculations of a rudimentary geometrical character. But it does not appear as if any of those authors ever produced a passage, within his geographical writing, with an explicit mathematical proposition employing a diagram. This, in a sense, is true even of Ptolemy. He does produce diagram-based propositions in his treatment of the plane projection of a spherical map (Ptolemy, *Geography* 1.24), but other than this his treatment is entirely discursive, if of course heavily numerical. His treatment of positions is structured, more specifically, as a *polemical* discourse. He takes his near predecessor Marinus as his main source and then dedicates the bulk of the first book of the geography to critiques of Marinus, some extended (having to do with the entire length of the *oikoumenê*), some very succinct. An example of the

---

[47] So Clarke's well-chosen title for her study of the Hellenistic genre of Geography, *Between Geography and History* (Clarke 1999) – treating some of the *more* mathematical of ancient geographers, including Posidonius!

latter sort will give a sense of the ancient debates concerning geographical positions (Ptolemy, *Geography* 1.15):[48]

> [Marinus says] that Pachymus is opposite Leptis Magna, and Himera opposite Thena, yet the distance from Pachymus to Himera amounts to about 400 stades, while that from Leptis Magna to Thena amounts to over 1,500, according to what Timosthenes records.

Pachymus and Himera are in Sicily, Leptis Magna and Thena are in the African shores. 'Opposite' here means 'lying on the same meridian' and it is tacitly assumed that the two Sicilian localities, as well as the two African localities, are each on the same latitude (ancient geographers could never entirely shake off the convention of taking the north shores of Africa as lying nearly on a single straight line). Obviously, the two opposite sides of a rectangle should not be so far apart and we conclude that Marinus was wrong. But notice also that this discussion does not include a diagram, and does not even bring in such terms as a 'rectangle'. This is a descriptive, non-mathematical discussion. The sphericity of the earth is ignored;[49] there is no systematic approach, even, so that the data still fail to determine any resolution of Marinus' alleged inconsistency. We merely learn that Marinus was wrong. Ptolemy's correction, implied by his tables and maps (he pushes Leptis Magna to the East), appears to be a matter of Ptolemy's intuition, no more.

So much from Ptolemy. Strabo himself was no admirer of mathematics. But we can recover a fair bit of Hellenistic geography from this one author – forthcoming about his sources, explicit about their methodologies. In particular, Strabo, *Geography* 2.1–4 is a critical discussion of past treatments of the overall shape and structure of the known world and, in this context, Strabo provides us with a very full sense of the methodologies of his key predecessors, the chief of which appear to have been – based on Strabo's account – Eratosthenes, Hipparchus, Polybius and Posidonius. It must be emphasised: these are the chief geographical authors whose work even contended with such broad theoretical issues.

---

[48] Berggren and Jones 2000: 77.

[49] In general, Greek geographical data are either on a small scale, relative to the size of the Earth (as in this case), where the assumption of a flat Earth makes little difference; or, in the large scale, they are extremely vague and error-prone. In *Geography* 1.13, Ptolemy is aware of the fact that estimates of the length of the *oikoumenê* based on land travel (i.e. through the latitudes of Central Asia) should be shorter than those based on sea travel (i.e. through the latitudes of the Indian Ocean). This, however, once again, is merely implied in the discussion and is not set out in explicit geometrical terms.

Strabo is acutely aware of the position of authors within a virtual scale, from the 'mathematical' to the 'geographical' (Eratosthenes, we hear – Strabo, *Geography* 1.41 – was 'in a way, a mathematician among geographers, a geographer among mathematicians'). But what does being 'mathematical' consist of? We are reminded of Asclepiodotus' (Posidonius'?) tactical art: the more mathematical authors rely more on a geometrical terminology for shapes (thus, the various parts of Asia in Eratosthenes' treatment – Strabo, *Geography* 1.22 – 'rhomboidal', 'parallelogram'). Explicit mathematical discussions, such as there are, seem to be confined largely to Ptolemy's numerical (but only implicitly geometrical) critique. Strabo himself, in *Geography* 1.37, makes a general geometrical point: the error involved in taking a diagonal, passing through a parallelogram, to stand for its length, becomes the more pronounced as the parallelogram becomes wider. This claim is stated discursively, no diagram attached and, indeed, no geometry adduced beyond plain visual intuition. But then again, Strabo may not be an ideal example, as he so clearly prefers the more geographical end of the mathematics/geography spectrum. Posidonius? Strabo's very extensive treatment of Posidonius' *On the Ocean* – Strabo, *Geography* 2.2 – provides the impression of a piece of natural history, written with a great deal of attention to astronomical knowledge. Whether or not it included any inset mathematical proofs, we cannot tell. When Cleomedes recounts Posidonius' measurement of the size of the Earth, based on the visibility of Canobus, Cleomedes' own account[50] – as everywhere in his treatise – is discursive: a stylistic feature Cleomedes could have imposed on his source, or borrowed from it.

Polybius provides the closest we ever find to the mathematical inset within a non-mathematical geographical discussion. What we have is Strabo's critique of Polybius; which is framed as a series of responses to Polybius' own critiques of Pytheas, Dicaearchus and Eratosthenes. The critique of Dicaearchus is cited in detail (Strabo, *Geography* 2.4.2.23–55) and perhaps even verbatim.[51] Dicaearchus' claim was that the length of the western Mediterranean, from the strait of Sicily (i.e. what we mean by the strait of Messina) to the Pillars, was 7,000 stadia[52] (to which 3,000 stadia are added

---

[50] *The Heavens*, 1.7.7–47.
[51] Strabo bookends the key discussion by the expression φησι (lines 29, 43, 47, 54); the geometrical expressions are not in Strabo's own style and the φησι, among other things, serves to distance Strabo from a mode of argument he finds repellent.
[52] As usual, correcting one's predecessors was not merely a matter of mathematical geography: Polybius fashioned himself as the – first – Greek authority on the Roman Mediterranean. It is significant that he finds a *bigger* Western Mediterranean.

from the strait of Messina to the Peloponnese). Polybius criticises this as follows:

> The coast-line [of the northwest Mediterranean] is most like an obtuse angle, standing on both: the strait and the pillars; having Narbo as its vertex. So that a triangle is set up, having the line through the sea as a base, and, as sides, those making the said angle, of which the one from the strait to Narbo is more than 11,200 stadia, while the remaining is a little less than 8,000. [A slightly less technical result with the following claim:] And let the depth of the gulf at Narbo is 2,000 stadia, as a perpendicular from the vertex to the base of the obtuse-angled <triangle>. Now, it is clear... from a schoolchild's measurement, that... if we add [the 3,000 stadia] it would be more than twice what Dicaearchus said [i.e. more than twice 10,000].

The geometrical reference is obvious – the shores are turned into a geometrical structure,[53] and a reference is made to a schoolchild's measurement, παιδική μέτρησις; the reference is clearly to Pythagoras' theorem. Significantly, an explicit calculation is not made and the impression is that the patronising reference was original and implied, already in the original, that the actual geometrical operation was elided; while we cannot rule out a diagram in Polybius, the implication is that no diagrammatic labels were used.

Could we find more explicitly geometrical passages through Strabo's reports of Hipparchus' geography? But Hipparchus could very well be Ptolemy's own model, here as elsewhere: critical of a single predecessor – in an elementary, non-technical way. Strabo, indeed, presents him strictly as an (unfair) critic of Eratosthenes. The details provided by Strabo closely remind us of Hipparchus' extant 'commentary' to Aratus' *Phaenomena*, where Hipparchus takes Aratus' poetic language and submits it to a critical mathematical treatment (which indirectly indicts Aratus' prose source, Eudoxus). So here: Strabo's main point (Strabo, *Geography* 2.1.21–36) is that Hipparchus took data points from Eratosthenes and, disregarding

---

[53] When commenting upon this passage in Netz 2002: 210–12, I was especially shocked by the cavalier failure to distinguish the measurements of latitudes and longitudes: clearly, a scientific approach would have been based on an estimate of Narbo's latitude (Polybius nearly halves his perpendicular!). But no less a difficulty arises from the over-geometrical approach: in the zeal to find triangles in his geography, Polybius takes two quasi-arcs, one convex (from Messina to Narbo), one concave (from Narbo to the Pillars), and turns them into straight lines; his sides are therefore overestimated (the underestimate of the perpendicular, together with the overestimate of the sides, ends up with an overall overestimate, roughly as bad as Dicaerchus' underestimate; the true value is roughly 10,000–12,000 stadia, depending on the units chosen, i.e. about 50 per cent above Dicaearchus and 50 per cent below Polybius).

Eratosthenes' own disclaimers and admissions of uncertainty, subjected such data points to an analysis appropriate only for numbers which are taken to be certain and precise. Take for example 2.1.28–9: Eratosthenes, according to Hipparchus (1) represented the line joining (i) the Caspian Gates and (ii) the Carmanian/Persian border by a meridian, (2) asserted that the two lines running perpendicularly west of this line, from Thapsacus and from Babylon, are 10,000 and 9,000 stadia respectively, from which it easily follows that (3) Thapsacus is 1,000 stadia east of Babylon, which is directly contradicted (1.36) by Eratosthenes who asserted that (4) Thapsacus is more than 2,000 stadia east of Babylon. Strabo responds that (1) Eratosthenes takes the line joining (i) the Caspian Gates and (ii) the Carmanian/Persian border as a convenient border of a geographical section without making it a meridian, and that (2) the stated numbers of distance to that line from Thapsacus and Babylon are rough estimates, nor are they taken to be measured along a parallel. This seems like a fair critique and the key point appears to be that the deductive geometrical habit of clean verdicts of truth and falsehoods was simply irrelevant to the problem at hand.

How can one deal with numerical data which are subject to uncertainty? Ancient mathematicians were not unfamiliar with this particular difficulty, which is of course at the heart of mathematical astronomy. Their solutions sometimes involved a more explicit awareness of the problem of error. Lloyd 1982 provides, as it were, the upper boundary, Hon 1989 the lower boundary on the ancient theories of observational error. The ancients had a clear grasp of the challenge presented by observational error (Lloyd 1982), and yet never developed any sophisticated, systematic way of dealing with it (Hon 1989). The most remarkable solution adopted by the Greeks was to present the calculations based on empirical data in terms of, well, boundaries: the distance of the Earth and the moon presented as more than a minimum, less than a maximum. Aristarchus' treatment of this problem has the boundaries emerge purely from the boundaries on a trigonometric calculation (the observations themselves are taken to be precise), but Hipparchus himself bracketed the observations, themselves, within boundaries and, incidentally, hit upon an extraordinarily precise estimate of the distance of the moon.[54] In short, a more sophisticated approach to the problem of error in geography was within the grasp of ancient mathematical geographers. However, no meaningful application of such an approach could be mounted: what one needed, in truth, was a fully-fledged statistical treatment of *many* data, one that did not emerge, not even in

---

[54] For this remarkable episode, see Toomer 1974b.

rudimentary form, prior to the eighteenth century. Which is why Ptolemy – and all ancient geographers – had to rely on no more than intuition: as always, the tool to replace a statistical model.

Ancient geography was descriptive for a reason. The sphericity of the Earth mattered – but not given the roughness of the data; and, for the truly relevant task at hand – reducing the many, rough data into usable form – there was as yet no mathematical solution. The tools available made it possible to refute, yet not to substantiate, claims of geographical position. Hence the polemical tone of ancient mathematical geography, turning even Ptolemy – elsewhere, not a polemicist – into Marinus' critique.

Optics in Galen: *On the Usefulness of the Parts*, 10.12–15

This is the closest we get, in antiquity, to Hawking's single equation. In his Book 10 of *On the Usefulness of the Parts*, Galen discusses the eye. Following a long discussion of the providence of the eye's parts, Galen says (K3.812.7–13.3) that he has nearly covered everything; however there is only one matter, which he planned to leave out – so that it will not repel the many (ὅπως μὴ δυσχεραίνοιτο τοῖς πολλοῖς). The dangers, you see, are the lack of clarity of the discourse (ἀσάφεια τῶν λόγων) and – you can't make that up – length of the treatment (μῆκος τῆς πραγματείας). Why? Because the matter involves 'geometry' (θεωρία γραμμική – 'line-based contemplation' – is this a reference to proofs? To diagrams?).[55] Then a dream intervened; Galen was admonished by the deity that it would be blasphemous to leave out this account of the divine arrangement of the eyes. The ensuing discussion is of course long – thirty Kühn pages, to K3.841. It is Galen's explicit foray into mathematics – the one passage in his writing where he does not merely exhort his readers to learn mathematics,[56] but becomes a mathematical teacher himself.

This passage is Galenic: a mosaic of episodes. Three mathematical sections stand out. One is a passage of an introductory/definitional character: a discursive exposition of the key geometrical terms (K3.815.5–18.6). Second – the centrepiece – is an account of why the optical nerves meet chiastically. (The geometrical portion of this discussion is K3.818.7–31.9). Finally,

---

[55] Towards the conclusion of this discussion (K3.838.12–18), Galen returns to the same point and presents mathematical knowledge, this time, not so much as a cognitive but as a social liability: as soon as fellow-physicians realise he was trained also in mathematics, they avoid his company.

[56] For a survey of such passages, see Lloyd 2005.

Galen provides a much briefer geometrical account for why the lens is a flattened, not a perfect, sphere (K3.839.1–41.8).

The introductory/definitional passage has two main functions: to persuade the readers that it is useful to think of sight in terms of straight visual rays; and to introduce the notion of the 'cone' of vision and especially the key term of the axis of the cone. Galen relies on natural language and natural examples (the line is likened, say, to a thin hair) and on a discursive, explicitly pedagogic style: the student, indeed, is addressed in the second person singular (remarkably, the passage concludes by asking the reader to re-read it, in case the terms are not yet clear!). It seems that Galen's concern regarding the 'lack of clarity of the discourse' may have had to do primarily with this, almost lexical level.

The second, and key, mathematical discussion runs as follows. The basic idea of parallax – that different things are seen from different positions – is explained in the terms of a diagram based on the account of vision as drawn by straight lines and cones. Galen, however, does not rely solely on this geometrical argument, offering instead also a set of observations that display the same phenomenon without any need for geometry. He argues next (based, if anything, on physical rather than mathematical arguments) that double vision would result if the two axes of the visual cones happen not to lie on the same plane: this is why the notion of the axis of the cone of vision was so important for Galen. At that point, he shows that two intersecting lines must lie in the same plane, for which he quotes the *authority* of Euclid, *Elements* 11.2 (but without an explicit geometrical treatment). This, then, explains the chiasm of the optic nerves: they are made to meet at a point to make sure the two lines are on a plane.

As for the flattened spherical shape of the lens, this arises from the observation that the lines from the pupil to the edges of the crystalline body will not be entirely contained within it: a less spherical shape involves, therefore, less waste.[57] Now, as Galen hastens to remind us, we do want, to begin with, a spherical shape for the eye. The mathematical argument accounts, then, for a *compromise*. The same, let us now notice, is true of the previous argument: the chiasm of the optic nerves could, after all, strike one as a strange waste. Geometry, we find, is being brought to bear as a tool accounting for

---

[57] The figure is, I believe, forced on us by the mathematical argument employed by Galen, but it is inconsistent with his anatomy of the eye, as the pupil is positioned, strangely, behind the vitreous body; could Galen have considered – in error – that the precise position of the pupil is immaterial to the argument, so that the diagram could serve as a geometrical approximation? This may be related to what Lloyd emphasises (2005: 126) as a key weakness of Galen's discussion: the approximation of the optical nerves – thin, curving cylinders – by straight lines.

what, otherwise, might appear improvident: it shows a necessity that constrains the perfection of the body. Sure, one would wish to have parallel and not crossing optic nerves. But then we might have double vision. Sure, one would wish to have a perfectly spherical lens. But then more of the vitreous body would be wasted.

I emphasised the mosaic of many styles. There is an expository, definitional passage; there is the appeal to Euclid's authority; there are diagrams presented mostly for the sake of illustrations; but also at least one passage intended as fully-fledged proposition. The claim of the parallax between the two eyes is stated in general terms (K3.821.4–9), followed by a γάρ-statement where the two pupils and the perceived object are identified in the diagram (K3.821.9–2.2), a construction of visual rays passing through the pupils and the edges of the perceived object (K3.822.2–5), an argument (K3.822.6–11) and a general conclusion (K3.822.11–13). Here is an author who genuinely knew his Euclid, and knew how to imitate him. At last, a mathematical inset within a non-mathematical corpus. But let us be clear about the size of the inset, the size of the corpus: an extant corpus, whose word count is roughly 3 million words – and, in it, a tiny proposition of some 150 words. And even that, the god had to extract through the power of a dream!

Surveying in Julius Africanus: *Cesti* 7.15.

Julius Africanus' two known works – the *Chronography* and the *Cesti* – straddle the two worlds of his third century, the Christian and the pagan. The first aligns sacred history with the dates of Greek myth and history; the second is a huge encyclopedia (in the Homeric panoply of twenty-four books) of all kinds of useful knowledge, with not a trace of the author's Christianity. Circulating widely in Byzantium, the *Cesti* were anthologised in collections dedicated to specific *technai*, so that Africanus' *Embroideries*, originally emphasising the variety of knowledge, were taken apart and put each in their separate compartments. The Seventh Book put somewhat more emphasis on the applicability of knowledge to war and so ended up being included in its entirety within a Byzantine military anthology; we thus get a sense of the original form. The variety is striking and the components of the book are as follows: '(1) [introduction], (2) Medical Knowledge, (3) Hippiatric section, (4) Knowledge from various other domains, (5) Geoponic section, (6) [conclusion]'.[58]

[58] Wallraff et al. 2012: xlviii.

Let us look at some of this 'knowledge from other domains'. In section 14, Africanus recommends that the army should do some of its training by hunting lions. Advice on lion hunting ensues. In section 16, advice is given for how to hear well from a distance. In between, in section 15, is advice on how to find the width of a river and the height of a wall. Africanus' introductory words are: 'Those in possession of a middling general education have, in all likelihood, dealt with the *Elements* of Euclid to some extent. With the aid of his first book, it is not at all difficult to solve these problems, too ... '. Following this, the text presents a geometrical theorem: if a side of a right-angled triangle is bisected and from the point of bisection a parallel is drawn to the other side and, from the point at which it cuts the hypotenuse, another parallel is drawn to the original side, all three sides are bisected. This may serve, as Africanus explains, to find the width of a river when one cannot access the other bank. Following that, Africanus moves to the more general solution based on a dioptra and that assumes any ratio (and not just bisection). The mention of the first book of Euclid's *Elements* is significant, as indeed the first theorem, concerning bisection, does not require proportion theory and so can indeed be derived from Euclid, *Elements* Book 1 alone; the other theorems are essentially applications of *Elements* 6.4. Thus, the first theorem is not as silly as it looks: it shows how the application of the dioptra can be approximated even with the most rudimentary knowledge of Euclid's *Elements*. If so, the structure of the text suggests a serious mathematical understanding from the very introductory words themselves, and the likeliest account is that the entire passage, from beginning to end, was lifted by Africanus practically verbatim from some unknown source (perhaps comparable to Hero's *Dioptra*).[59]

This mathematical passage is, for once, unequivocally mathematical. The theorems are in the geometrical style and are accompanied by diagrams; that Africanus' text circulated with a set of diagrams in this section, but not in others, is made less surprising when we consider that in the several magical passages of Africanus' collection, magical diagrams were employed, one of which is even papyrologically attested.[60] In short, Africanus' project engaged with diversity and such diversity encompassed the appropriation of materials from various sources, ranging even into mathematics.

---

[59] It is difficult to argue that Africanus' basic modus operandi was plagiarism – if this is indeed the right term – since almost all our evidence is poorly attributed: perhaps a case of seeming plagiarising *by* Africanus is really a case of misattribution *to* Africanus? See Wallraff *et al.* 2012: lvi–lx. For the close relationship of Africanus' dioptra passage to other Greek and Arabic sources, see Lewis 2001: 62–6 (based on Africanus' willingness to use the letter I in his diagram, Lewis goes on to argue that the cited source must be early, a claim I find plausible; a further conclusion is that Africanus lifted his *diagram*, too, letter-for-letter).

[60] Wallraff *et al.* 2012: xxx–xxxi.

Once again, note can be made of lost cases of the inset mathematical text, in this case closely parallel to Africanus, and likely of the same era. Sporus' *Keria*, 'Honeycombs', are almost entirely lost; the title itself suggests a similar encyclopedic collection, and since it is given alternatively as 'Aristotelian *Keria*' one presumes the contents were at least somewhat philosophical;[61] the very few mentions we do have[62] have all to do with mathematics or with astronomy. The most substantial fragment is a solution to the problem of finding four lines in continuous proportion, reported in Eutocius' commentary to Archimedes' *Sphere and Cylinder* 2 (76–8), a much more substantial piece of mathematics than Africanus'. The impression is that these *Honeycombs* were primarily a work in the theoretical sciences, though once again having the character of a free-wheeling anthology. The little we know of Sporus derives not just from Eutocius, but also from Pappus (who provides Sporus' terminus *ante quem*). Seen in the light of both Sporus and Africanus, can we not see in Pappus' *Collection*, itself, yet another instance of '*Embroideries*'? The technical miscellany emerges as a viable genre in the third century AD, and the passage in Africanus stands out in that, within the context of a very wide-ranging epitome of technical knowledge, a little mathematics is thrown in as well.

The Mathematical-Like Passage: General Observations

For the mathematised treatise, I emphasised variety. For the inset mathematical passage, one must emphasise absolute scarcity. Africanus does not so much inset mathematics within a non-mathematical context as he displays a mosaic with distinct tesserae made more brilliant by their juxtaposition. He serves, if anything, to underline the *difference* of mathematics. Both the Aristotelian and the geographic passages have their mathematics reduced to a minimum: Aristotle uses the convention of the diagram so as to illustrate his meteorological point, Polybius evokes the vocabulary and techniques of mathematics without engaging in any mathematics directly. We are left – once again, ignoring the somewhat more common examples among the philosophical commentators – with a single context where a fully mathematical passage is set in a non-mathematical context, Galen's

---

[61] The most extensive discussion is Knorr 1989: 87–93; he tends to put Sporus (93) 'closer to the 2nd than to the 4th century' on what are, however, very slight grounds (essentially, Pappus' failure to cite Sporus when effectively repeating his solution becomes less surprising to Knorr if we put the two far apart).

[62] Summarised in *EANS*, s.v. Sporos of Nikaia.

stereometric account of the optical nerve. This, from the most prolific author, the author shrillest in his trumpeting of mathematics!

Not only rare, but also lacking in conceptual ambition. What we do not find is the brief, inset act of mathematisation, where a non-traditional subject for mathematics is ambitiously treated as if it were mathematical. Instead, the rare mathematical insets arise more naturally, as the text touches upon a subject already established as properly mathematical: optics, in the examples from Aristotle and Galen; geometry as applied to concrete land measurements, in the examples from Polybius and Africanus.

The absolute scarcity of the inset piece of mathematics could be as banal as the absolute scarcity of the visual. Illustrations were very rare in the ancient literary context[63] and it is telling that Africanus, who does include a piece of geometry within his enormously wide-ranging *Embroideries*, does so in the presence of other, *magical* illustrations. Put in less banal terms, we see here, once again, a generic constraint. To engage in mathematics, after all, is to bring in an alien genre: a particular lexicon (a deep concern for Galen, as we recall), a particular syntax. Could the inset mathematical passage have been avoided for just this reason?

Reflections

A banal avoidance of the mixing of genres; speaking, however, to a fundamental feature of the ancient literary system. We have seen two things: that the mathematical inset passage could be introduced – as long as one already did mathematics, that is, say, as one did optics vel sim.; and that the mathematised treatise took, out of mathematics, not its claim of validity, but its claim for clarity of exposition. Taken together, we see that mathematics is valid – for its own domain; this Antaeus is irrefutable – but only as long as it is moored to its ground of the genres established, already, as mathematical.

Why did the ancients not assume, as people did as a matter of course in the seventeenth century, that the validity of mathematics is generally exportable to other fields? Such questions need to be unpacked carefully. It is not profitable to produce negative accounts that explain why a modern practice was not available in antiquity (usually, we find out that this was because the ancients were not modern). Instead, let us begin from a

---

[63] I argue for this – and insist upon the marked nature of the visual in Greek mathematics – in future work.

positive account of the practice we do see in antiquity. Why did (a handful of) ancient writers use mathematics, in non-mathematical contexts, to the extent they did? Starting from that, we may perhaps say something about the scope, and limits, of the ancient authority of the mathematical genre.

First, the handful of authors in our survey tend to be prolific authors. Aristotle, Galen and Africanus wrote on a remarkably wide variety of subjects (Eratosthenes, who may be key to the mathematisation of geography, such as it was, belongs to this group). Aristoxenus, Proclus and Galen, again, dealt intensively with their central topic, seeking as it were to saturate it with writing; the same may be said for Polybius' historical/geographical project. (We do not know enough of Posidonius to be able to say if he was an extensively, or intensively, prolific author; either way, if Asclepiodotus' military *technê* does represent Posidonius' influence, it emerges, once again, from an extremely rich dossier of writing.) When does one do mathematisation? When one does everything, anyway.

Second, the handful of authors in our survey share a familiarity with philosophy in the fourth-century tradition of Plato and Aristotle. So Aristotle himself, of course, and Aristoxenus; but also Eratosthenes (the author of the *Platonikos*), Posidonius (the 'Aristoteliser'), Galen and Proclus. This indeed is the context for the only generic setting in which the inset mathematical passage, while rare, is somewhat 'normalised' – the philosophical commentary. In late antiquity, this familiarity is commonplace, but in the Hellenistic world – the high tide of the Greek specialised genres – such a familiarity could not be taken for granted and authors such as Eratosthenes and Posidonius were, in fact, isolated exceptions. So, even the rare mathematisations we saw were made possible at all only against the background of a particular philosophical position where mathematics has an extraordinarily elevated position.

The upshot of the scarcity of ancient writing *more geometrico* was not that ancient mathematics did not carry authority, but that such authority was delimited: backed up by a particular philosophical tradition; restricted to a particular domain; outside of this domain, perceived as a vehicle of clarity, not of validity.

Let us step back and reflect, briefly, on the structure of authority based on difficulty. As pointed out right at the beginning, this structure is familiar from Bourdieu's studies of cultural capital: we can understand Proust against the Vaudeville, marking different status, carrying different authority. And we can easily map this into Hawking's scientific publications against his popular science. Difficulty, on the one side; commercial success, on the other. Now, not all boring and difficult texts and performances need carry an authority. Sometimes, boring is just boring: it makes you yawn without making you genuflect. Galen, indeed, suggests just as much for the ancient audience for mathematics. So, what makes one not yawn, but genuflect, at Proust? (Aside, that is, from the fact that Proust was such a genius – you don't take me for a yokel, right?) What seems key to this particular type of the construction of cultural capital is the convertibility across a continuum. Bourdieu maps an entire set of social settings and practices, overlapping and gradually forming a spectrum from the 'low' to the 'high'. The character of cultural capital in modernity is created by, well, capitalism: with economic growth we have an expanded, many-niched cultural system, and its economic underpinnings are such that all is made immersed in the language of money.

Now, this continuum of convertibility, we find, is missing from the ancient system of genres. It is not entirely absent – after all, when Strabo notes that Eratosthenes was 'a mathematician among geographers, a geographer among mathematicians', he envisages a kind of spectrum of geographical writings. But as we noted, geography is the closest to an ancient spectrum between the genres – and one where, ultimately, mathematics remains isolated. Eratosthenes, of course, was a creature of the Hellenistic experiment with genres, one where we can find, indeed, the mixing of genres within, and with, science.[64] Aratus' *Phaenomena* was a wild success in antiquity: such was the ancient equivalent to Hawking. 'A Brief History of Eudoxus', as it were. For Aratus' poem was essentially (in the part that concerns us most) a conversion, into poetic form, of Eudoxus' astronomy. Here, then, the convertibility of science to other genres; the making of genres that, somehow, overlap.

But, crucially, Aratus' *Phaenomena* is entirely bound within the generic constraints of the didactic epic: its subtlety consists precisely in translating the contents of Eudoxus' sphere into poetic language without allowing any generic overlap and, in particular, without allowing any intrusion of the mathematical genre into its discourse. The Hellenistic generic experimentation was not an experiment in flouting genre, but, to the contrary, an experiment that made genres more marked through their clever juxtaposition. Hellenistic epic – this is the key observation of Fantuzzi and Hunter 2004 – was *more* generically regimented than its archaic model. Archimedes did likely write an epigram presenting the puzzle of the number of the cattle of the Sun. What matters to me here is that this epigram did not evoke the

---

[64] This is the topic of Netz 2009: 115–229.

mathematical formulaic language and of course did not include a diagram. The generic properties of mathematics were banned, outside of it, even for an experimental author such as Archimedes. Mathematics – apart.

And so, no ancient genre of writing *more geometrico*. There is no simple single explanation, as we note: this absence is over-determined. The seventeenth century's fascination with mathematics would be different, for many reasons as well. It surely matters that, in the seventeenth century, everyone was quasi-Platonist. It may well also be that the cognitive and cultural values of mathematics – its clarity, its relative value-neutrality, its Greek origins – meant more in the specific context of post-reformation, post-Renaissance Europe, than they did in a Hellenistic or Roman world. Of course, early modern authors did worry about the Aristotelian prohibition on domain-crossing (Aristotelian epistemology mattered, in early modern Europe, much more than it ever did in antiquity).[65] But in early modern Europe, the ancient genres were revived – apart from their ancient systematic arrangement. The role of the visual was becoming more normalised in a print culture; the difference *between* Greek genres mattered less: they were all Greek; the entire structure of the literary field was shaken up, in a multiple universe of vernaculars and prose forms. Galileo was Cervantes' contemporary. And then – Antaeus unbound: mathematics could be written everywhere, overlapping across genres, and so its room of application was universal. Its validity was no longer seen as constrained by a genre; convertible, it could set up the cultural capital, the authority of mathematical expertise.

---

[65] See, e.g., Dear 1995, chapters 32–62, a survey primarily of the Jesuits' grappling with the problem of the authority of the Aristotelian mixed sciences; an anxiety, however, not by any means confined to the Jesuits.

17 | Authority and Expertise: Some Cross-Cultural Comparisons

G. E. R. LLOYD

The collection of articles brought together here by the editors already presents a great diversity. This relates first to the subject matter in which expertise or authority is claimed, for that ranges from agriculture, architecture and warfare, to historiography, law, philosophy, religion, mathematics, the study of the heavens and medicine, howsoever each of those may be construed. Then there is the variety in the bases on which such claims are made, whether that be personal experience, training, understanding of the past, mastery of the accepted canons, or the correct methodology. Finally, there are important differences in the matter of the audience that has to be persuaded that the claims made are valid. That may be some patron or the addressee to whom the text is dedicated if there is one, in the first instance, or a peer group, or a more general public.

These are all important and difficult questions so far as Greek and Roman writers are concerned. But I conceive my task here to be to complicate the issues even further, by considering just some of the material from other cultures to investigate where there are commonalities, where there are culturally specific differences, and what light either of these may throw on our understanding of the strategic problems raised by authority and expertise, as discussed by the editors in their introduction. I shall concentrate on the Chinese data, but must begin with some remarks about two other ancient civilisations, Mesopotamia (the conventional term used to cover the successive empires that controlled the Tigris-Euphrates valley, the Babylonian and Assyrian ones especially) and Ancient India.

As soon as we start to compare societies, the difficulty of saying what any given learned discipline stands for – which may already be contested within any particular society – becomes acute. In *Disciplines in the Making*,[1] I reviewed some of the evidence for the different ways in which different societies or groups within society at different periods have understood law, or historiography, or religion, or philosophy, or mathematics or medicine. Of course, at a very general level we may say that 'mathematics', wherever and whenever it is engaged in, deals with numbers and shapes, medicine

---

[1] Lloyd 2009.

with healing or at least with the treatment of the sick, and so on. But just how mathematics should be pursued, and what it is useful for, are questions on which very different answers have been given. More strikingly still, ideas about health and disease, about what counts as a successful cure or at least a treatment that does some good, have varied very considerably.

Such points have immediate bearing on the type of authority claims that were made. Sometimes what is at stake is control of a discipline itself, sometimes its standing vis-à-vis other disciplines, the pecking order, we may say, between them. So we must distinguish intra- from inter-disciplinary rivalries, as well as the various modes of self-justification and sources of prestige we may identify.

Given the recurrent phenomenon of both types of rivalry, it will be more economical to structure my discussion by taking each ancient civilisation in turn, cross-referencing where appropriate, rather than attempt a thematic organisation to the material. So first I shall collect some relevant points from each civilisation, before attempting a summary of conclusions at the end.

In Mesopotamia, the extensive extant cuneiform texts enable us to identify several different, if overlapping, groups of experts that all fall under the overarching category of 'able scholar', *ummâni lē'ûti*. Simo Parpola[2] gave a far from exhaustive list of these, as follows: (1) *Ṭupšarru*, 'astrologer/scribe', (2) *Barû*, 'haruspex diviner', (3) *Āšipu*, 'exorcist', (4) *Asû*, 'Physician' and (5) *Kalû*, 'Lamentation chanter', to which we can add, for example, *šā'ilu*, 'dream interpreter'. In many cases, these crafts passed down from father to son. But one well-documented case, that of Marduk-šāpik-zēri, shows that one could practise more than one.[3] He followed in his father's footsteps, as *kalû*, but also claimed expertise in healing, in the study of the heavens and in physiognomy.

When talking about his skill as a healer, Marduk-šāpik-zēri says that he has 'examined healthy and sick flesh'. But in the case of most of his other claims he refers to the texts he has mastered. This is a crucial point. Sometimes the scribes, *ṭupšarru*, are identified by the texts on which they are expert, as most notably the *ṭupšarru Enūma Anu Enlil*.[4] This was put together some time between 1500 and 1200 BCE, but incorporates even older material, and it covered celestial phenomena of all kinds. On the one hand, mastery of this text was the key requirement to join the ranks of the *ṭupšarru EAE*, and on the other, the existence of that elite group of interpreters of celestial omens ensured the continued status of that text.

[2] Parpola 1993: xiii.   [3] Parpola 1993: 122.   [4] Rochberg 2000.

Yet book learning was evidently not enough. The scribes in question explain how conscientiously they studied the heavens – as well as how thoroughly they looked up what *EAE* and other texts had to say about particular ominous phenomena. On the one hand, these scribes were obviously on intimate terms with the kings to whom they communicated their results and advice – sometimes even reprimanding the king for what they considered inappropriate behaviour. On the other hand, the position of any given scribe was far from permanently secure. The tablets provide vivid evidence of the disputes that arose, not only on what the correct interpretation of a sign was, but even on what had actually been observed in the heavens.

Thus, on one occasion, there was a dispute over whether the planet Venus was visible.[5] One scribe had reported that it was, but he was contradicted in vitriolic terms by another who accused him of being a fool, an ignoramus, a cheat. In that instance, there was a problem of saying whether a particular sighting was of Venus or of Mercury: they are easy enough to distinguish if you are fully equipped with tables of periodicities, built up over the years, but otherwise, if you don't have such tables, the task can be quite tricky. When as was already happening in the seventh century BCE, the scribes were making predictions about planetary visibilities before and after periods of invisibility, as well as about lunar and even solar eclipses, then some of those predictions were, one might think, straightforwardly falsifiable.[6] Except that generally excuses were available to get round the apparent failures (in China, the non-occurrence of a predicted eclipse was sometimes put down to the special virtue of the emperor! That scored two points in one, for the emperor would no doubt be pleased and at the same time the move saved the astronomer's reputation). But for now the essential point about the cuneiform data is that the authority of the scribes depended *both* on their mastery of the canons *and* on their careful observation of the skies. The canons enabled you to make predictions about what to expect and how to interpret the omens (for even regular phenomena were still considered ominous), but at the same time you still had to keep watch.

Of course, there were many types of expertise, in Mesopotamian societies as elsewhere, where scholarship and book learning were not the foundations of a person's reputation. Some of those we hear of with official positions in the palace or at court need not necessarily have had to acquire and exhibit special skills – what would those of 'cup-bearers' have been? – but others were no doubt judged by their results, from generals to cooks or tailors or

---

[5] Parpola 1993: 37–8.   [6] Cf. Brown 2000; Rochberg 2004.

butchers.[7] Here, too, there was room for dispute about just what counted as success.

In medicine, in particular, where as Markham J. Geller has shown,[8] there were several different traditions of healers, overlapping and competing with one another, the questions of whether a cure had been achieved thanks to a particular treatment, or whether indeed the treatment had done any good whatsoever, could be disputed. But as usual with the medicine practised in ancient literate societies, authority was often built up by a combination of factors, by learning in the canons, by your perceived or imagined or claimed successes, or just by belonging to a family of doctors.

Our evidence for Mesopotamia comes overwhelmingly from highly literate sources. We do not find them doubting the very possibility of their profession. They certainly admit, even underline, that not everyone is a successful or even a competent practitioner. We have seen that individual reports and interpretations may be recognised to be wide of the mark. But the scribes are united in the assumption that there are signs out there, in the heavens, and in terrestrial omens, that do not determine what will happen, but give indications about what may.[9] You need all sorts of qualities, including moral ones, to succeed: that is the challenge of the profession. But there is no generalised scepticism that the whole enterprise is founded on an error. And after all there were plenty of predictions that turned out to be correct. One might always quarrel with an inference from the heavens to events on earth. But from the seventh century BCE onwards, predictions of events in the heavens themselves, planetary visibilities and even eclipses, were being made which were, as I noted, clearly falsifiable, but where the outcomes showed them to be correct. So while some claims to expertise depended on mastery of a learned tradition, others were, on occasion, bolstered by empirical success.

Several similar points apply also to my next ancient civilisation, India, where we have abundant evidence for gurus, teachers, wise persons (mostly but not all men) of different types, with complex relations with patrons of various kinds. Once again, the importance of scriptures is clear, ranging from the *Vedas* and the *Upaniṣads* all the way to technical treatises in medicine and mathematics, not to mention the arts of the bedchamber. The situation in India is complicated first because of the existence – from at least the eighth or seventh century – of quite well-established rival philosophical or religious schools, Brahmin, Jain, Buddhist and eventually many others,

---

[7] All in Fales and Postgate 1992.   [8] Geller 2010.   [9] Rochberg 2004.

each with their own distinct models of sagehood, moral teaching and range of technical interests.

A considerable portion of our evidence takes the form of debates, both inter- and intra-sect. In the *Upaniṣads*, for example, we find gurus locked in dispute, often on the most abstruse issues, the nature of Brahman, or of atman, for instance.[10] Some of the rules of the game are clear: you should not ask a question to which you do not know the answer yourself. But quite why some answers are accepted, others not, is obscure. The guru who literally has the last word is the winner, and the consequences for the loser are sometimes dire: he becomes the winner's servant reduced to eating his leftovers.[11] In other instances, the defeated party is represented as undergoing a religious conversion, to the beliefs that the victor represented, as in the famous case of the *Milindapañha* where Milinda (the Indian version of the Greek name Menander) first wins a series of contests with Indian gurus, but is then defeated by the last, Nagasena, and so converts to Buddhism.[12]

But why a particular guru wins is often difficult, or rather impossible, for an outsider to fathom. In such cases, one of the main messages would seem to be the very esoteric nature of the wisdom in question. Ordinary folk are not expected to have a clue. Even kings do not have the range of learning of the gurus, although we know that some of these debates were staged in courts or palaces for the entertainment, as much as the edification, of the ruler, while the ruler for his part was an important source of patronage and prestige for those whom he invited to perform. The superiority of one guru to another was a matter of who could keep going longest, but the only persons who could question their authority were other gurus, who would have to mount a challenge and then themselves be faced with possible defeat.

The intensity of the rivalry exhibited when the highest wisdom is at stake is mirrored in other more technical fields as well. As in Mesopotamia, Indian medicine was pluralist, but at the learned end of the spectrum two sets of texts, the *Caraka Samhitā* and the *Suśruta Samhitā*, came to acquire canonical status, even though (like the Hippocratic Corpus) each was an amalgam of sometimes inconsistent writings. It was up to the Ayurvedic doctor to use these texts to guide and justify his diagnoses and treatments, though the opacity of the texts left plenty of room for manoeuvre on both

---

[10] Lloyd 2014: 30–1, cites material from the *Bṛhadāraṇyaka Upaniṣad*; cf. Olivelle 1996: 40–7.
[11] Bronkhorst 2002.   [12] Lloyd 2014: 43.

counts. However, like the gurus of the *Upaniṣads*, he could expect to be challenged.

An extensive section in the *Caraka Samhitā* actually offers advice about how to deal with this (III, 8).[13] You should assess the ability of your potential opponent. You are advised not to take on someone superior to yourself, but faced with inferior opponents you are encouraged to use all sorts of tricks designed to embarrass him, frighten him, reprehend him. If he seems rather unlearned, then overwhelm him with long citations from the *sūtras*, if he is 'unable to retain sentences', then fire off a 'continuous series of sentences composed of long-strung *sūtras*' and so on. Superior medical skill, however that was judged, was not necessarily what counted: rather superior rhetorical skill did, in ways that are strongly reminiscent of the claim that Gorgias made in Plato's dialogue (*Gorgias* 456b-c), that his skill as an orator meant he was better able to persuade his brother's patients to take their medicine than his brother – himself a doctor – was. While learned doctors, in India as everywhere else, were keen to show off their learning and to distance themselves from those who had not mastered the canons, they were themselves vulnerable to attack from rivals. The best or at least most honest ones, conscious of their own limitations, desisted from extravagant claims, even though that could play into the hands of those attackers.

Now what about China? Many of the features of authority and expertise that we have already met recur, though sometimes in distinctive manifestations. Take first the role of canonical texts, knowledge of which, in China, was essential for anyone to consider themselves a 'gentleman' (*junzi*) or member of the literate elite. (The term for that was *ru*, often translated as 'Confucian', though the connections with Confucius or with the texts and teaching that passed as his were often tenuous.) Indeed, the mastery of particular texts was the key to a career in the increasingly important imperial civil service. The official appointment that a person held was, naturally, a major source of that person's prestige.

That process begins already in Han times, where the Emperor Wu Di founded the Imperial Academy and appointed official teachers in what became the so-called Confucian classics. These were the *Shi* (poems or odes), *Shu* (documents), *Li* (Rites), *Chunqiu* (Spring and Autumn Annals) and *Yi* (the book of Changes). Entry to the Academy was selective from the start, and later came to be by examination on those texts. Those who graduated successfully (that too came to be by examination) had an assured

---

[13] Cf. Matilal 1998: 31–59; Prets 2000: 369–71.

career with official appointments in one or other department of the imperial bureaucracy.

This certainly contributed to a certain hierarchisation in intellectual disciplines in China, though the pecking order between them was rather different from what we find in the Greco-Roman world. It is nonsense to say that the ancient Chinese had no 'philosophy', only 'wisdom', for it is absurd to consider their extensive and sophisticated discussions of ethics as anything but philosophical.[14] Yet the term that only much later came to be used for 'philosophy', namely *zhexue*, was not the name of any discipline, let alone of the most prestigious one, in ancient times. Again, mathematics was prized in China for its ability to solve all sorts of problems: but it was not valued, as it was in the Greco-Roman world, for its exactness and as a model for demonstration. In China, the high-prestige texts taught you how to attain the *Dao*. The claims made for more technical disciplines, mathematics, the study of the heavens, medicine, warfare, were that they too contributed to that eventual goal, for the *Dao* had many manifestations. But they ranked as inferior to the classics that were held to give instruction in the *Dao* as such, and this perception was duly reflected in the curriculum of the Imperial Academy and eventually in the subjects set for examination.

Yet what we think of as the Chinese 'classics' were originally anything but the stable well-authenticated, standardised writings some might expect from our use of that term, and the same applies to other more technical canons as we shall see in a moment. As I have already intimated, Confucius' authorship of the works attributed to him, even in the case of the *Analects*, is doubtful, for that is clearly an amalgam of texts composed by different authors at different times.[15] In many instances, the 'classical' texts existed in quite different versions, where this was not just a question of a few variant readings or possible interpolations. The *Dao De Jing* was not one of the five or six canonical texts taught in the Academy, but associated rather with the legendary Lao Zi. But the evidence of versions of this and other texts that is coming to light from excavated tombs dating to the second century BCE shows a very great diversity. Thus, one version puts the *De* section before the *Dao*, so *De Dao Jing*. It is clear that different groups of scholars, lineages, or *jia*, handed on and taught their own editions of these texts. A concern to establish *the* original version only emerges much later: in Han times there was no such concern, even though different lineages all shared the view of the importance and the prestige of the texts in question.

---

[14] Cf. Lloyd 2009: 5–27.   [15] Brooks and Brooks 1998.

The respect for the authority of the past is clear, even when we might say that what that past stood for was largely a construct of the imagination. This is not just seen in the reverence with which the key texts were held – and their importance, as noted, in the education of the literati. The texts themselves sometimes cite even earlier figures in support of their own authority. This is best seen in the more technical literature in medicine or in mathematics. The medical classic that came to form the basis for the education of elite doctors in China down the ages was the *Inner Canon of the Yellow Emperor, Huangdi neijing*, though this was not a single text, for the material we associate with it has come down to us in three different recensions (the *Lingshu, Suwen* and *Taisu*). The main format in which medical lore is transmitted is by way of a series of dialogues between the Yellow Emperor himself and his various advisers. We know of a series of other medical texts that circulated among different groups,[16] and the eventual canonisation of the *Huangdi neijing* was some time in the making. But its tactics, to claim authority, were to invoke an implied endorsement of its contents from a purely legendary figure of the distant past.

In the case of mathematics, too, we have evidence of similar processes at work. Here, the book that acquired canonical status and was the core of the mathematical curriculum for many generations was the *Nine Chapters on Mathematical Procedures, Jiuzhang suanshu*, dated to around the turn of the millennium. But it was preceded by other less systematic texts, notably the *Suanshushu* (*Writings on Reckoning*), again found in an excavated tomb. But when Liu Hui composed what is the earliest extant commentary on the *Nine Chapters*, in the third century CE, he chose to trace the origin of mathematics back to Baozi, the legendary inventor of the trigrams on which the hexagrams of the *Book of Changes* were based, and to the Yellow Emperor himself.

Liu Hui complains that the subject is neglected in his own day and the fact that he feels he has to apologise to his readers for the more technical bits of his discussion puts one in mind of Galen's observation that his contemporaries, and not just his fellow doctors, were turned off by mathematics. Yet an earlier Chinese text, the *Zhoubi suanjing*, shows that in China, as in Greece, mathematics staked a claim for its general importance, though not on the grounds Greek mathematicians used, namely its exactness and role as a model for demonstration, but rather for its ability to resolve problems in astronomy.[17] Chinese mathematics made no attempt at axiomatisation and more particularly did not engage in proving theorems by way of the

---

[16] Harper 1998.   [17] Cf. Lloyd 2002: 44–68.

construction and manipulation of the lettered diagram (as Reviel Netz showed for Greece:[18] the points and shapes in Chinese diagrams were in any case of course not identified by letters, but either by colours or by the areas being given numbers, the first, the second, and so on). Particular diagrams are referred to in our texts, in connection, for instance, with the investigation of the properties of right-angled triangles, though – as in Greece – the ones that appear in the manuscripts and later editions are of dubious authenticity. But more importantly such diagrams did not form the basis of Euclidean axiomatic-deductive proofs of what the Greeks called Pythagoras' theorem and of other results. Rather, they were more generally used in the manipulation of shapes in a type of scissors and paste technique that was known as the *churu xiangbu*, or Out-in Complementary Principle.

Nevertheless, in the *Zhoubi*, we are told how to calculate the size and distance of the sun by taking gnomon sightings (assuming a flat earth) and in the astronomical treatises that were regularly included in the great Chinese dynastic histories, starting with the *Shiji* around 86 BCE, mathematical calculations figure prominently, naturally enough, in the discussion of the calendar, of planetary regularities and of eclipse cycles. Yet, unlike in Greece, where astronomy was often used to support an argument to the effect that the cosmos is under divine guidance, the heavens were studied in China in the first instance for the sake of the information they provided that would be useful to the emperor. This was not just a matter of his needing to make sure that the calendar kept in step with the seasons. Any untoward events in the heavens of which he was unaware (and even some he knew all about) might be taken as a sign that his Mandate was failing. While Greek mathematical astronomy sometimes contributed to the claim of philosophy to be the supreme discipline, the Chinese study of the heavens acquired its – lower – status for its usefulness to government.

Many Chinese examples show the importance of past authority in building up the prestige of the text and of those who taught it – not that mastery of the text by itself turned a person into a good practitioner, that is a good doctor or mathematician, let alone someone fit to govern 'all under heaven'. Confucius himself is represented as having said that he did not create, he only transmitted: the idea was that the truth had been there all along and just had to be recovered. That did not stop innovation in just about every department of knowledge: but it did mean a certain mode of presentation of new ideas, aligning them with tradition even when they tended to depart from it.

---

[18] Netz 1999.

But while the past, accessed through canonical texts, was undeniably one of the chief sources of authority and the basis of the justification for many claims to expertise in China, that picture needs to be qualified in relation to three types of evidence in particular, first where the past is explicitly said *not* to be a guide for the present, second where the authority of the book must be supplemented by personal experience, and third where the type of expertise in question has nothing to do with book learning.

Against the argument mounted by the traditionalists, that the past was the repository of all the wisdom you needed, there were eloquent advocates of a rival view. The tactic the third-century-BCE writer Han Fei uses consists in casting doubt on just what the lessons of the past were (*Hanfeizi* 50). Nobody could be sure what the iconic teachers, Confucius and Mozi, stood for, and going back to even earlier Sage Kings was, a fortiori, even more dubious. Another third-century-BCE text, the *Lüshi chunqiu*, has a chapter ('Scrutinising the present', 15/8) that states that the standards or principles laid down by former kings may have been appropriate for their times, but since times have changed, they could no longer be applied straightforwardly to the present day. The lesson in both cases was that you always have to take the particular circumstances of the present situation into account: the present is of overriding importance.

The point can be extended. While rivalry between living contemporaries is common enough in China (as we have already seen in Mesopotamia and in India), we sometimes find in China generalised criticism that extends to the basis of widely accepted beliefs and practices. Thus, attempts to forecast the future were as common in China as anywhere in the world, and two favourite methods for doing so were based on the interpretation of the cracks made by firing turtle shells and on using the *Book of Changes* as a guide to understanding the configurations obtained by sorting sticks of milfoil. However, in the first century CE, Wang Chong expresses his doubts about the validity of the whole exercise (*Lun Heng* 71). What is so special about turtles or milfoil, he asks, that they can yield predictions? Yet while Wang Chong distances himself from many existing practices, he still casts himself, traditionally enough, in the role of adviser. Although he was never to hold high office himself, his book was composed to stake his claim to be taken seriously in just that role.

The second type of text I mentioned may be illustrated by the doctor Chunyi Yi, whose biography is contained in the first great dynastic history, the *Shiji* (ch. 105). Called to give an account of himself, he explains who taught him and lists the many books they handed on for his instruction, some of them described as 'secret' learning. But in his account of his individual case histories, he repeatedly tells us that he was able to diagnose what the patient was suffering from thanks to what he learned from taking their pulse. Two of the books he said he studied related to pulse lore, one ascribed to the legendary Yellow Emperor, the other to a doctor named Bian Que (whom we shall shortly meet again). But of course his correct diagnoses depended on what he felt when he took each patient's pulse. His own classification of pulses is a good deal more complex than any that can be found in earlier medical texts (such as those excavated from tombs at Mawangdui), so it possible, even probable, than he was responsible for elaborating their classification.[19] But whatever he may have learned from the books he studied, or from the teachers with whom he studied them, everything depended on his being able to interpret the pulse movements of each patient he encountered when he encountered them.

His exchange with another doctor named Sui about the authority of Bian Que is also significant. In a dispute over what that doctor himself was suffering from, that doctor cited this authoritative figure of the past to contradict Chunyu Yi's opinion. But to that Chunyu Yi replies, not by saying that the doctor had misquoted Bian Que, but by claiming that he had not properly understood how to apply that advice. There was no question, here, of rejecting a past authority, but Chunyu Yi exploited the room for manoeuvre that always existed in interpretation, where he evidently based his own view on the case in hand partly, at least, on his own first-hand experience.

The other types of expertise, third, that we have to take into account, in China as elsewhere, are those where book learning was strictly irrelevant. Recall how Aristotle, the son of a doctor, emphasised the limitations of such learning in medicine (*Nicomachean Ethics* 1081b2). The *Zhuangzi* compilation is particularly rich in examples from other fields, though many of these can be found in other Warring States and Western Han texts, the *Lüshi chunqiu* and the *Huainanzi* especially, as well.

Thus, the story of cook Ding appears in all three of those texts (*Zhuangzi* 3, *Lüshi chunqiu* 9/5, *Huainanzi* XI 15a). He acquired his skill as a butcher through long experience. For three whole years he did not even see a live ox. But eventually his mastery was such that after nineteen years his knife did not need sharpening, such was his understanding of how to use the interstices in the ox's body to do his work. Charioteering was another highly prized skill, where a character called Zao Fu excelled. As soon as he mounted his chariot and seized the reins, his horses were off, galloping or trotting in perfect harmony (*Huainanzi* VI 8b). Yet even Zao Fu was no match, as a

---

[19] Hsu 2010.

judge of horses, for Bole (*Huainanzi* II 9a), and even as a charioteer he was surpassed by Qinfu and Dabing, who were able to do without bit or reins.

Zhuangzi 13 has the story of the wheelwright Bian who was chipping away at a wheel at the bottom of the hall of Duke Huan while the duke was reading a book at the top of the hall. Bian puts aside his work and asks the duke what he is reading. When told it is the words of a sage, Bian asks whether he was alive or dead, and when told he was dead, concludes that the duke is reading the dregs of men of old. Duke Huan is naturally affronted and commands Bian to explain himself, which he promptly does by explaining what it feels like to move the chisel correctly, neither too fast, nor too slow, a feeling that cannot be put into words: it cannot be transmitted to anyone else nor learned from anyone else.[20]

The *Dao* has, we said, many manifestations, but the aim was to embody it, not to be able to give an account of it. In such texts, the emphasis is on self-cultivation, sometimes in deliberate contrast with the book learning prized by others. The irony was, to be sure, that some of these authors and texts, *Zhuangzi* especially, were later to be turned into models to be followed and thus to stand as authorities for the views they represented, even though one of their main lessons was to challenge the generally accepted views of authority.

So let me now turn back briefly to the Greco-Roman situation to compare the various tendencies that are revealed to be at work in the effort to secure authority there. When we compare our Mesopotamian, Indian and Chinese data, three contrasts in particular stand out. From the fourth century BCE at least, thanks in large part to the iconic figure of Socrates, and how he was interpreted, 'philosophy' came to stand, for many, as the supreme intellectual discipline. If it did not guarantee 'happiness', it was nevertheless, so the philosophers claimed, a necessary condition for such, with its three components, logic providing you with method, ethics teaching you about morality, and physics supplying what knowledge you needed about the external world – even though how those three were interpreted differed enormously from one school to another. Of course, philosophers were also figures of fun, well known for not living up to the ideals they professed.[21] But the hold that that ideal came to have is clear. For Galen, the best doctor is also a philosopher. For Vitruvius, the best architect also is. Even Hero of Alexandria stakes his claim for his mechanics in philosophical terms. High-prestige intellectual leaders in our other three civilisations were not expected to qualify as such in logic and physics, whatever may be said about ethics.

---

[20] Graham 1989: 187.   [21] Cf. Trapp in this volume.

Second, there is, as discussed, the position of mathematics, or rather one image of that subject, that is peculiar to the Greco-Roman world. We have seen that large claims were made for mathematics in China especially, but that was as a tool to resolve particular problems in, for example, the study of the heavens. In Greece alone, of the ancient societies we have mentioned, mathematics came to stand for providing a method that would yield incontrovertible proof. As I have written elsewhere, it is important to get that ideal into perspective, for first it was in practice unattainable in many of the fields to which the Greeks applied it (theology, for instance), and second even in mathematics itself it came to distract attention from heuristics. Nor was it an ideal for the whole of what we call Greco-Roman mathematics.[22] On the one hand, the competitiveness in constructing such an ideal may be compared with what we find elsewhere, where rival individuals and groups battle for prestige and authority. On the other, I would see the distinctive features of that particular Greek ideal as reflecting the particular circumstances in which those battles played out in the Greek world, where it was as much the philosophers as the mathematicians themselves who saw or imagined they saw the need for arguments that would silence the opposition for good.

My third point is a generalisation of the last. Many Greek experts, not just philosophers, but also, for example, medical writers, often back up their first-order claims to knowledge by second-order justifications that they had arrived at their conclusions by using the correct method. The author of the Hippocratic treatise *On Ancient Medicine* contrasts his – sound – understanding of medical matters, based on experience, with the misguided ideas of his opponents who founded their – speculative – theories on arbitrary, unverifiable postulates or 'hypotheses'. Conversely, the philosopher Parmenides dismisses the opinions of ordinary folk en bloc with the argument that they are misled by trusting sense-perception rather than reason.

There are some signs of classical Chinese authors being interested in the sources of knowledge: the Mohists, for instance, identified three such, 'report', 'explanation' and 'observation'. But in general the Chinese were less concerned than their Greek counterparts to use a methodological or epistemological argument to show that their opponents *must* be mistaken. In the hard-hitting debates that characterise much Greek intellectual argument in the Classical period and later, many had recourse to what they hoped would be a knock-down refutation of the opposition. They must be wrong, since they adopted entirely the wrong methodology. The Chinese were more likely to refer to the consequences of following their rivals' ideas, generally

---

[22] Cuomo 2001.

construed as leading to disorder and chaotic government, although whether that would indeed be the consequence was of course always a matter of opinion.

My survey has been drastically selective, but what general remarks, not to say conclusions, does it prompt? Some patterns certainly recur in all our ancient societies. Where, given the different political, economic, social and institutional circumstances of those I have considered, one might have expected some radical differences in the way in which efforts to claim authority were played out, the differences we have found tend to be matters of degree, with variations not just across societies and periods, but also across disciplines. Let me consider briefly, in turn, the role of canons, appeals to the past, the role of patrons and the argument from personal experience.

In all of the societies under discussion, certain texts came to be given the status of canons, that is to be considered the authoritative repositories of the essential knowledge relevant to a particular enquiry. Once in place, such a corpus normally forms the basis for the training of the next generation of aspiring practitioners, though the extent to which they may be expected to go beyond it may vary, as also may the degree to which dissent from its teaching is permissible. Clearly open disagreement is not necessary, when the original texts are sufficiently open-ended, not to say ambiguous, in their interpretation. Inspecting how different scholars used the *Enūma Anu Enlil*, the *Upaniṣads*, the *Caruka Samhitā*, the so-called Confucian 'classics' and the various medical and mathematical canons in China and the Greco-Roman world can tell us a good deal about the self-confidence of those elites and the ways they sought to control the manner in which the corresponding discipline was interpreted and practised.

The past was mostly the subject of respect, even of awe, but it was of course subject to creative reinterpretation and manipulation. In China and, as is well known, also in Greece, there were explicit statements qualifying or criticising the usefulness of past models and suggesting the need for new ones to replace them. Strident claims to innovate may be a distinctive feature of certain Greek writers of the fifth and fourth centuries BCE especially,[23] but Mesopotamian astronomers, Indian doctors and Chinese ones could and did push their subjects forward, even when they chose to do so under the banner of recovery rather than of discovery.

In all our societies, at different times, patrons played a major role, both as sponsors of research and as guarantors of its respectability. Dedicating a

---

[23] Lloyd 1987a: 109–71.

work to a ruler or other powerful person could add to its claim to be authoritative, though the price that always had to be paid was not to include anything that would offend or displease the individual concerned. Obviously, books so dedicated rarely, if ever, contain materials that are radically subversive of existing ideas and practices.

The point can be extended to include other audiences – and there were always other potential readers even of books commissioned for the private use of the ruler. Clearly, the tactics to be used in persuading an emperor of the value of your work differed rather from those appropriate to a peer group or to a more general public. As Reviel Netz's chapter in this volume illustrates, Archimedes sometimes chose to address his treatises not to a king, but to whoever he believed to be competent enough to appreciate them. At the point where those at the forefront of a discipline are reporting work at its cutting edge (as we say), tradition is left behind and readers are expected to appreciate just how original these authors are. Archimedes is, to be sure, exceptional from many points of view. But claimed improvements based on the authors' own expertise can be documented from all our ancient societies, even while the extent to which those authors trumpet their own originality varies appreciably.

Evidently, the rhetoric of presentation is one thing: the reality of actual innovation another. The actual pace of change in a given discipline at a given time may contrast with the protestations of its leading practitioners. But if it is apparent that many claims for authority and expertise are little more than bluff, that does not mean that all were. There is a delicate balance between what the fashions of presentation dictated and the underlying substance of those claims, and from the point of view of a modern investigator, it is clearly crucial to bear that contrast in mind.

Bibliography

Agosta, G. (2009) *Ricerche sui Cynegetica di Oppiano*, Amsterdam.
Albrecht, M. von (2008a) 'Seneca's language and style, I', *Hyperboreus* 14: 68–90.
  (2008b) 'Seneca's language and style, II: linguistic differences and connections between Seneca's philosophical works and his tragedies', *Hyperboreus* 14: 124–50.
Alexander, L. (1993) *The Preface to Luke's Gospel: Literary Convention and Social Context in Luke 1.1–4 and Acts 1.1*, Cambridge.
Algra, K. (2009) 'Stoic philosophical theology and Graeco-Roman religion', in R. Salles (ed.), *God and Cosmos in Stoicism*, Oxford: 224–51.
Allmand, C. (2009) 'A Roman text on war: The *Strategemata* of Frontinus in the Middle Ages', in P. Coss and C. Tyerman (eds), *Soldiers, Nobles and Gentlemen: Essays in Honour of Maurice Keen*, Woodbridge: 153–68.
  (2011) *The De Re Militari of Vegetius: The Reception, Transmission and Legacy of a Roman Text in the Middle Ages*, Cambridge.
Amato, E. (2003) 'Per la cronologia di Dionisio il Periegeta', *Revue de philologie* 77: 7–16.
  (2005) *Dionisio di Alessandria: Descrizione della terra abitata. Prefazione, introduzione, traduzione, note e apparati*, Milan.
Ambaglio, D. (1981) 'Il trattato *Sul Comandante* di Onasandro', *Athenaeum* 59: 353–77.
Ampolo, C. (1990–91) 'Lotte sociali in Italia centrale. Un documento controverso: il calendario brontoscopico attribuito a Nigidio Figulo', *Opus* 9–10: 185–97.
Anderson, G. (1986) *Philostratus: Biography and Belles Lettres in the Third Century AD*, London.
Ando, C. (ed.) (2003) *Roman Religion*, Edinburgh.
  (2008) *The Matter of the Gods: Religion and the Roman Empire*, Berkeley, CA.
Andrén, A. (1960) 'Dionysius of Halicarnassus on Roman monuments', in *Hommages à Léon Herrmann* (Collection Latomus 44), Brussels: 88–104.
Anglo, S. (2002) 'Vegetius' *De re militari*: The triumph of mediocrity', *Antiquaries Journal* 82: 247–67.
  (2005) *Machiavelli: The First Century*, Oxford.
Arnar, A. S. (1990) *Encyclopedism from Pliny to Borges*, Chicago, IL.
Asheri, A., Lloyd, A. and Corcella, A. (2007), *A Commentary on Herodotus, Books I–IV*, Oxford.

Asper, M. (1997) *Onomata allotria: Zur Genese, Struktur und Funktion poetologischer Metaphern bei Kallimachos*, Stuttgart.

(2007) *Griechische Wissenschaftstexte: Formen, Funktionen, Differenzierungsgeschichten*, Stuttgart.

(2011) '"Frame tales" in Ancient Greek science writing', in K.-H. Pohl and G. Wöhrle (eds), *Form und Gehalt in Texten der Griechischen und Chinesischen Philosophie*, Stuttgart: 91–112.

(ed.) (2013a) *Writing Science: Mathematical and Medical Authorship in Ancient Greece*, Berlin.

(2013b) 'Introduction', in Asper 2013a: 1–13.

(2013c) 'Making up progress – in Ancient Greek science writing', in Asper 2013a: 411–30.

Astbury, R. (ed.) (2002) *M. Terentius Varro: Saturarum Menippearum Fragmenta* (2nd edn), Munich.

Astin, A. E. (1978) *Cato the Censor*, Oxford.

Atherton, C. (ed.) (1998) *Form and Content in Didactic Poetry*, Bari.

Aubert, J.-J. (1994) *Business Managers in Ancient Rome: A Social and Economic Study of Institores, 200 BC–AD 250*, Leiden.

Aubert, J.-J. and Sirks, B. (eds) (2002) *Speculum Iuris: Roman Law as a Reflection of Social and Economic Life in Antiquity*, Ann Arbor, MI.

Baier, T. (1997) *Werk und Wirkung Varros im Spiegel seiner Zeitgenossen*, Stuttgart.

Baldwin, B. (1963) 'Columella's sources and how he used them', *Latomus* 22: 785–91.

Baltussen, H. (ed.) (2013) *Greek and Roman Consolations: Eight Studies of a Tradition and Its Afterlife*, Swansea.

Baragwanath, E. (2008) *Motivation and Narrative in Herodotus*, Oxford.

Baragwanath, E. and de Bakker, M. (2012a) 'Introduction: Myth, truth, and narrative in Herodotus' *Histories*', in Baragwanath and de Bakker 2012b: 1–56.

(eds) (2012b) *Myth, Truth, and Narrative in Herodotus*, Oxford.

Baraz, Y. (2012) *A Written Republic: Cicero's Philosophical Politics*, Princeton, NJ.

Barker, A. (2007) *The Science of Harmonics of Classical Greece*, Cambridge.

Barnes, J. (1993) 'Galen and the utility of logic', in J. Kollesch and D. Nickel (eds), *Galen und das hellenistische Erbe*, Stuttgart: 33–92.

(2003) 'Proofs and syllogisms in Galen', in Barnes and Jouanna 2003: 1–24.

Barnes, J. and Jouanna, J. (eds) (2003) *Galien et la philosophie*, Vandoeuvres-Geneva.

Baron, C. A. (2013) *Timaeus of Tauromenium and Hellenistic Historiography*, Cambridge.

Barthes, R. (1987) *Writer Sollers* (tr. P. Tody), Minneapolis, MN.

Bartley, A. N. (2003) *Stories from the Mountains, Stories from the Sea: The Digressions and Similes of Oppian's Halieutica and the Cynegetica*, Göttingen.

Barton, T. (1994a) *Ancient Astrology*, London.

(1994b) *Power and Knowledge: Astrology, Physiognomics, and Medicine under the Roman Empire*, Ann Arbor, MI.

Bartsch, S. (2006) *The Mirror of the Self: Sexuality, Self-Knowledge and the Gaze in the Early Roman Empire*, Chicago, IL.

Bauman, R. A. (2000) *Human Rights in Ancient Rome*, London.

Beagon, M. (1992) *Roman Nature: The Thought of Pliny the Elder*, Oxford.

(2013) '*Labores pro bono publico*: The burdensome mission of Pliny's *Natural History*', in König and Woolf 2013: 84–107.

Beard, M. (1986) 'Cicero and divination: The formation of a Latin discourse', *Journal of Roman Studies* 76: 33–46.

(2012) 'Cicero's "Response of the haruspices" and the voice of the gods', *Journal of Roman Studies* 102: 20–39.

Beard, M. and North, J. (eds) (1990) *Pagan Priests: Religion and Power in the Ancient World*, London.

Beard, M., North, J. and Price, S. (1998) *Religions of Rome* (2 volumes), Cambridge.

Bekker-Nielsen, T. (2002) 'Fish in the ancient economy', in K. Ascani, V. Gabrielsen, K. Kvist and A. H. Rasmussen (eds), *Ancient History Matters: Studies Presented to Jens Erik Skydsgaard on His Seventieth Birthday*, Rome: 29–37.

(2006) 'The technology and productivity of ancient sea-fishing', in T. Bekker-Nielsen (ed.), *Ancient Fishing and Fish Processing in the Black Sea Region*, Aarhus: 83–95.

Bellemore, J. (1992) 'The dating of Seneca's *Ad Marciam De Consolatione*', *Classical Quarterly* 42: 219–34.

Bendlin, A. (2000) 'Looking beyond the civic compromise: Religious pluralism in late Republican Rome', in E. Bispham and C. Smith (eds), *Religion in Archaic and Republican Rome: Evidence and Experience*, Edinburgh: 115–35, 167–71.

Bendz, G. (1938) *Die Echtheitsfrage des vierten Buches der Frontinischen Strategemata*, Lund.

Bergemann, C. (1992) *Politik und Religion im spätrepublikanischen Rom*, Stuttgart.

Berggren, J. L. and Jones, A. (2000) *Ptolemy's Geography: An Annotated Translation of the Theoretical Chapters*, Princeton, NJ.

Bernard, A. (2010) 'The significance of Ptolemy's *Almagest* for its early readers', *Revue de Synthèse* 131: 495–521.

Berryman, S. (2009) *The Mechanical Hypothesis in Ancient Greek Natural Philosophy*, Cambridge.

Bettalli, M. (2010) 'Il militare', in G. Zecchini (ed.), *Lo storico antico: mestieri e figure sociali*, Bari: 215–29.

Bing, P. (1988) *The Well-Read Muse: Present and Past in Callimachus and the Hellenistic Poets*, Göttingen.

Blank, D. L. (1998) *Sextus Empiricus: Against the Grammarians*, Oxford.

(2012) 'Varro and Antiochus', in D. Sedley (ed.), *The Philosophy of Antiochus*, Cambridge: 250–89.

Blänsdorf, J. (1991) '"Augurenlächeln": Ciceros Kritik an der römischen Mantik', in H. Wißmann (ed.), *Zur Erschließung von Zukunft in den Religionen:*

Zukunftserwartung und Gegenwartsbewältigung in der Religionsgeschichte, Würzburg: 45–65.

Bobzien, S. (1996) 'Stoic syllogistic', *Oxford Studies in Ancient Philosophy* 14: 133–92.

Boll, F. (1903) *Sphaera: Neue griechische Texte und Untersuchungen zur Geschichte der Sternbilder*, Leipzig.

Bömer, F. (1969–86) *P. Ovidius Naso: Metamorphosen, I–VI*, Heidelberg.

Boshnakov, K. (2004) *Pseudo-Skymnos (Semos von Delos?)*, Stuttgart.

Bosman, P. R. (2012) 'Lucian among the Cynics: The *Zeus Refuted* and Cynic tradition', *Classical Quarterly* 62: 785–95.

Bosworth, A. B. (1993) 'Arrian and Rome: The minor works', *Aufstieg und Niedergang der römischen Welt* II, 34(1): 253–64.

Boudon-Millot, V. (2009) 'Galen's *bios* and *methodos*: From ways of life to paths of knowledge', in Gill, Wilkins and Whitmarsh 2009: 175–89.

Boudon-Millot, V. and Jouanna, J. (2010) *Galien: Ne pas se chagriner*, Paris.

Boudon-Millot, V. and Pietrobelli, A. (2005) 'Galien ressuscité: edition "princeps" du texte grec du *De propriis placitis*', *Revue des études grecques* 118: 168–213.

Bourdieu, P. (1993) *The Field of Cultural Production*, Cambridge.

Bowen, A. C. and Goldstein, B. R. (1988) 'Meton of Athens and astronomy in the late fifth century B.C.', in E. Leichty, M. de J. Ellis and P. Gerardi (eds), *A Scientific Humanist: Studies in Memory of Abraham Sachs*, Philadelphia, PA: 39–81.

Bowersock, G. W. (1965) *Augustus and the Greek World*, Oxford.

Bowie, E. (1990) 'Greek poetry in the Antonine age', in D. A. Russell (ed.), *Antonine Literature*, Oxford: 53–90.

(2004) 'Denys d'Alexandrie: un poète grec dans l'empire romain', *Revue des études antiques* 106: 177–85.

(2006) 'Portrait of the sophist as a young man', in B. McGing and J. Mossman (eds), *The Limits of Ancient Biography*, Swansea: 141–53.

(2013) 'Libraries for the Caesars', in König, Oikonomopoulou and Woolf 2013: 237–60.

Boylan, M. (1984) 'The Galenic and Hippocratic challenges to Aristotle's conception theory', *Journal of the History of Biology* 17: 83–112.

Boys-Stones, G. (2013) 'The *Consolatio ad Apollonium*: Therapy for the dead', in Baltussen 2013: 123–37.

Brancacci, A. (2000) 'Dio, Socrates and Cynicism', in Swain 2000: 240–60.

Branham, R. B. (1989) *Unruly Eloquence: Lucian and the Comedy of Traditions*, Harvard, MA.

Braund, S. (2009) *Seneca De Clementia: Edited with Translation and Commentary*, Oxford.

Bremer, J. M. (1981) 'Greek hymns', in H. S. Versnel (ed.), *Faith, Hope and Worship: Aspects of Religious Mentality in the Ancient World*, Leiden: 193–215.

Briscoe, J. (2008) *A Commentary on Livy: Books 38–40*, Oxford.

Brisson, L., Congourdeau, M.-H. and Solère, J.-L. (eds) (2008) *L'Embryon: formation et animation. Antiquité grecque et latine, traditions hébraïque, chrétienne et islamique*, Paris.

Brodersen, K. (1994) *Dionysios von Alexandria: Das Lied von der Welt*, Hildesheim.

Bronkhorst, J. (2002) 'Discipliné par le débat', in L. Bansat-Boudon and J. Scheid (eds), *Le Disciple et ses maîtres*, Paris: 207–25.

Brooks, E. B. and Brooks, A. T. (1998) *The Original Analects*, New York.

Broughton, T. R. S. (1951–86) *The Magistrates of the Roman Republic* (3 volumes), New York.

Brown, D. (2000) *Mesopotamian Planetary Astronomy-Astrology*, Groningen.

Brown, P. (1978) *The Making of Late Antiquity*, Cambridge, MA.

Brunt, P. A. (1972) 'Cn. Tremelius Scrofa the agronomist', *Classical Review* 86: 304–8.

 (1989) 'Philosophy and religion in the late Republic', in Griffin and Barnes 1989: 174–98.

Bryan, J. (2013) 'Neronian philosophy', in Buckley and Dinter 2013: 134–48.

Buckley, E. and Dinter, M. T. (eds) (2013) *A Companion to the Neronian Age*, Chichester.

Burckhardt, L. A. (1988) *Politische Strategien der Optimaten in der späten römischen Republik*, Stuttgart.

Burnyeat, M. F. (2005) 'Archytas and optics', *Science in Context* 18: 35–53.

Cam, M.-T., Liou, B. and Zuinghedau, M. (eds) (1995) *Vitruvius: De l'architecture: Livre VII*, Paris.

Cameron, A. (2004) *Greek Mythography in the Roman World*, New York.

 (2011) *The Last Pagans of Rome*, New York.

Campanile, D. (2005) 'Il sofista allo specchio: Filostrato nelle *Vitae Sophistarum*', *Studi Ellenistici* 16: 275–88.

Campbell, B. (1987) 'Teach yourself how to be a general', *Journal of Roman Studies* 77: 13–29.

Campbell, B. and Purcell, N. (1996) 'Iulius Frontinus, Sextus', in S. Hornblower and A. Spawforth (eds), *The Oxford Classical Dictionary* (3rd edition), Oxford: 785.

Cardauns, B. (ed.) (1960) *Logistoricus über die Götterverehrung (Curio de cultu deorum)* (with translation and commentary), Würzburg.

 (ed.) (1976) *M. Terentius Varro, Antiquitates Rerum Divinarum: I Die Fragmente; II Kommentar*, Wiesbaden.

 (2001) *Marcus Terentius Varro: Einführung in sein Werk*, Heidelberg.

Carey, S. (2003) *Pliny's Catalogue of Culture: Art and Empire in the Natural History*, Oxford.

Carlsen, J. (1995) *Vilici and Roman Estate Managers until AD 284*, Rome.

Carvounis, K. and Hunter, R. (eds) (2008) 'Signs of Life? Studies in Later Greek Poetry' *Ramus* 37(1–2).

Cavavero, A. (2002) 'The envied Muse: Plato versus Homer', in E. Spentzou and D. Fowler 2002: 47–67.

Cèbe, J.-P. (ed.) (1985) *Varron, Satires ménippées*, Vol. 7 (with translation and commentary), Rome.
  (ed.) (1999) *Varron, Satires ménippées*, Vol. 13 (with translation and commentary), Rome.
Chandler, C. (2006) *Philodemus: On Rhetoric, Books 1 and 2*, New York.
Charles, M. (2007) *Vegetius in Context: Establishing the Date of the Epitoma Rei Militaris*, Stuttgart.
Chilton, H. W. (1971) *Diogenes of Oenoanda: The Fragments*, London.
Civiletti, M. (2002) *Filostrato: Vite dei Sofisti*, Milan.
Clark, H. (1990) *The Fictional Encyclopedia: Joyce, Pound, Sollers*, New York.
  (1992) 'Encyclopedic discourse', *SubStance* 67: 95–110.
Clarke, K. (1997) 'In search of the author of Strabo's *Geography*', *Journal of Roman Studies* 87: 92–110.
  (1999) *Between Geography and History: Hellenistic Constructions of the Roman World*, Oxford.
Clauss, J. J. (2006) '*Theriaca*: Nicander's poem of the earth', *Studi italiani di filologia classica* 4: 160–82.
Clay, D. (1992) 'Lucian of Samosata: Four philosophical lives (Nigrinus, Demonax, Peregrinus, Alexander Pseudomantis)', *Aufstieg und Niedergang der römischen Welt* II, 36(5): 3406–50.
Cole, T. (1990) *Democritus and the Sources of Greek Anthropology*, Atlanta, GA.
Conte, G. B. (1994) *Genres and Readers: Lucretius, Love Elegy, Pliny's Encyclopedia* (tr. G. Most), Baltimore.
Corbeill, A. (2012) 'Cicero and the Etruscan haruspices', *Papers of the Langford Latin Seminar* 15: 243–66.
  (2013) 'Cicero and the intellectual milieu of the late Republic', in C. Steel (ed.), *The Cambridge Companion to Cicero*, Cambridge: 9–24.
Cornell, T. (ed.) (2013) *The Fragments of the Roman Historians* (3 volumes), Oxford.
Costanza, S. (1991) 'Motivi callimachei nel proemio dei *Cynegetica* di Oppiano d'Apamea', in *Studi di filologia classica in onore di Giusto Monaco, I: Letteratura greca*, Palermo: 479–89.
Courrént, M. (2011) *De architecti scientia: Idée de nature et théorie de l'art dans le De architectura de Vitruve*, Caen.
Cramer, F. H. (1954) *Astrology in Roman Law and Politics*, Philadelphia, PA.
Cribiore, R. (2001) *Gymnastics of the Mind: Greek Education in Hellenistic and Roman Egypt*, Princeton, NJ.
Crook, J. (1995) *Legal Advocacy in the Roman World*, London.
Crowther, N. B. (1992) 'Second-place finishes and lower in Greek athletics (including the pentathlon)', *Zeitschrift für Papyrologie und Epigraphik* 90: 97–102 (reprinted in Crowther 2004).
  (2000) 'Resolving an impasse: Draws, dead heats, and similar decisions in Greek athletics', *Nikephoros* 13: 125–40 (reprinted in Crowther 2004).
  (2004) *Athletika: Studies on the Olympic Games and Greek Athletics*, Hildesheim.

Cuomo, S. (2000) *Pappus of Alexandria and the Mathematics of Late Antiquity*, Cambridge.
  (2001) *Ancient Mathematics*, London.
  (2007a) 'Measures for an emperor: Volusius Maecianus' monetary pamphlet for Marcus Aurelius', in König and Whitmarsh 2007b: 206–28.
  (2007b) *Technology and Culture in Greek and Roman Antiquity*, Cambridge.
  (2011) 'Skills and virtues in Vitruvius' Book 10', in Formisano and Böhme 2011: 309–32.
Cuypers, M. (2004) 'Prince and principle: The philosophy of Callimachus' *Hymn to Zeus*', in M. A. Harder, R. F. Regtuit and G. C. Wakker (eds), *Callimachus II (Hellenistica Groningana 7)*, Leuven: 95–115.
Dalle Vedove, E. (2009) 'Aspetti della presenza del dedicatario nel *De beneficiis* di Seneca e raffronto con le prefazioni di Seneca Padre', in G. Picone, L. Beltrami and L. Ricottilli (eds), *Benefattori e beneficati: La relazione asimmetrica nel De beneficiis di Seneca*, Palermo: 97–120.
Dalzell, A. (1996) *The Criticism of Didactic Poetry*, Toronto.
d'Assigny, M. (1686) *The Stratagems of War, or, A Collection of the Most Celebrated Practices and Wise Sayings of the Great Generals in Former Ages Written by Sextus Julius Frontinus, One of the Roman Consuls; Now English'd*, London.
De George, R. T. (1985) *The Nature and Limits of Authority*, Lawrence, KS.
De Jong, I. (2001) *A Narratological Commentary on the Odyssey*, Cambridge.
De Jong, I., Nünlist, R. and Bowie, A. (eds) (2004) *Narrators, Narratees, and Narratives in Ancient Greek Literature*, Leiden.
De Jonge, C. (2005a) 'Dionysius of Halicarnassus and the method of metathesis', *Classical Quarterly* 55: 1–18.
  (2005b) 'Dionysius of Halicarnassus as a historian of linguistics: The history of the "parts of speech" in *De compositione verborum 2*', *Henry Sweet Bulletin* 44: 5–18.
  (2008) *Between Grammar and Rhetoric: Dionysius of Halicarnassus on Language, Linguistics and Literature*, Leiden.
De Lacy, P. (1972) 'Galen's Platonism', *American Journal of Philology* 93: 27–39.
  (ed.) (1992) *Galen: On Semen* (*CMG* 5.3.1), Berlin.
  (ed.) (2005) *Galen: On the Doctrines of Hippocrates and Plato* (3 volumes) (3rd edn) (*CMG* 5.4.1.2), Berlin.
De Libero, L. (1992) *Obstruktion: Politische Praktiken im Senat und in der Volksversammlung der ausgehenden römischen Republik (70–49 v.Chr.)*, Stuttgart.
De Pizan, C. (1999) *The Book of Deeds of Arms and of Chivalry* (tr. S. Willard, ed. C. C. Willard), Philadelphia, PA.
De Ste. Croix, G. E. M. (1981) *The Class Struggle in the Ancient Greek World*, Ithaca, NY.
Dean-Jones, L. (2003) 'Written texts and the rise of the charlatan in ancient Greek medicine', in H. Yunis (ed.), *Writing into Culture: Written Text and Cultural Practice in Ancient Greece*, Cambridge: 97–121.

Dear, P. (1995) *Discipline and Experience: The Mathematical Way in the Scientific Revolution*, Chicago, IL.
Debru, A. (1991) 'Expérience, plausibilité et certitude chez Galien', in López Férez 1991: 31–40.
Delbrück, H. (1975–85) *History of the Art of War within the Framework of Political History* (4 volumes) (translated from the German by Walter J. Renfroe, Jr.), Westport, CT.
Delcourt, A. (2005) *Lecture des Antiquités romaines de Denys d'Halicarnasse: Un historien entre deux mondes*, Brussels.
Della Casa, A. (1962) *Nigidio Figulo*, Rome.
Denyer, N. (1985) 'The case against divination: An examination of Cicero's *De divinatione*', *Proceedings of the Cambridge Philological Society* 31: 1–10.
Detienne, M. (1996) *The Masters of Truth in Archaic Greece*, New York.
Devillers, O. and Krings, V. (1996) 'Autour de l'agronome Magon', in M. Khanoussi, P. Ruggeri and C. Vismara (eds), *L'Africa romana: Atti dell' XI Convegno di Studio, Cartagine, 15-18 dicembre 1994*, Ozieri: 489–516.
Devine, A. M. (1989) 'Aelian's manual of Hellenistic military tactics: A new translation from the Greek with an introduction', *The Ancient World* 19: 31–9.
Dicks, D. R. (1971) 'Euctemon', in Gillespie 1970–90: 4.459–60.
Diederich, S. (2007) *Römische Agrarhandbücher zwischen Fachwissenschaft, Literatur und Ideologie*, Berlin.
Dihle, A. (1996) 'Die Theologia tripertita bei Augustin', in H. Cancik, H. Lichtenberger and P. Schäfer (eds), *Geschichte-Tradition-Reflexion: Festschrift für Martin Hengel zum 70. Geburtstag*, Tübingen: 183–202.
Dillon, J. (1993) *Alcinous: The Handbook of Platonism*, Oxford.
Dillon, J. N. (2012) *The Justice of Constantine: Law, Communication and Control*, Ann Arbor, MI.
Dobbin, R. (1998) *Epictetus: Discourses Book 1*, Oxford.
Dodds, E. R. (1963) *Proclus: The Elements of Theology*, Oxford.
Domenici, I. and Maderna, E. (2007) *Giovanni Lido: Sui segni celesti*, Milan.
Domenicucci, P. (2003) 'La previsione astrologica attribuita a Nigidio Figulo in Luc. I, 639-70', *Schol(ia)* 5(3): 85–106.
Donini, P. (2008), 'Galen's Psychology', in Hankinson 2008b: 184–209.
Doody, A. (2007) 'Virgil the farmer? Critiques of the *Georgics* in Columella and Pliny', *Classical Philology* 102: 180–97.
Doody, A., Föllinger, S. and Taub, L. (2012) 'Structures and strategies in ancient Greek and Roman technical writing', *Studies in History and Philosophy of Science*, Special Issue 43(2): 233–6.
D'Ooge, M. L. (1926) *Nicomachus of Gerasa: Introduction to Arithmetic* (tr. with studies in Greek arithmetic by F. E. Robbins and L. C. Karpinski), New York.
Dougherty, C. (2001) *The Raft of Odysseus: The Ethnographic Imagination of Homer's Odyssey*, Oxford.

Drake, S. (1957) *Discoveries and Opinions of Galileo*, New York.
Du Plessis, P. J. (ed.) (2013) *New Frontiers: Law and Society in the Roman World*, Edinburgh.
Ducatillon, J. (1977) *Polemique dans la collection hippocratique*, Paris.
Dueck, D. (2000) *Strabo of Amaseia: A Greek Man of Letters in Augustan Rome*, London.
Dumont, J. C. (1986) 'Quelques aspects de l'esclavage et de l'économie agraire chez Pline', *Helmantica* 37: 293–306.
Duncan-Jones, R. P. (1963) 'Wealth and munificence in Roman Africa', *Papers of the British School at Rome* 31: 159–77.
Dyck, A. R. (2004) *A Commentary on Cicero, De Legibus*, Ann Arbor, MI.
Eck, W. (1982) 'Die Gestalt Frontinus in ihrer politischen und sozialen Umwelt', in Frontinus-Gesellschaft (ed.), *Wasserversorgung im Antiken Rom*, Vol. I: *Sextus Julius Frontinus, Curator Aquarum*, Munich.
Edwards, C. (1996) *Writing Rome: Textual Approaches to the City*, Cambridge.
   (1997) 'Self-scrutiny and self-transformation in Seneca's *Letters*', *Greece and Rome* 44: 23–38 (reprinted in Fitch 2008: 84–101).
Effe, B. (1977) *Dichtung und Lehre: Untersuchungen zur Typologie des antiken Lehrgedichts*, Munich.
Einarson, B. (1936) 'On certain mathematical terms in Aristotle's logic', *American Journal of Philology* 57: 33–54, 151–72.
Elderkin, G. (1935) 'Two mosaics representing the Seven Wise Men', *American Journal of Archaeology* 39: 92–111.
Elm, D. (2003) 'Die Kontroverse über die "Sondergötter": ein Beitrag zur Rezeptions- und Wirkungsgeschichte des Handbuches Religion und Kultus der Römer von Georg Wissowa', *Archiv für Religionsgeschichte* 5: 67–79.
Engels, D. (2007) *Das römische Vorzeichenwesen (753–27 v.Chr.): Quellen, Terminologie, Kommentar, historische Entwicklung*, Stuttgart.
Engels, J. (2013) 'Agias and Derkylos (305)', in *Brill's New Jacoby* (Brill Online).
Eramo, I. (2010) *Discorsi di guerra*, Bari.
Erdkamp, P. (1999) 'Agriculture, underemployment and the cost of rural labour in the Roman world', *Classical Quarterly* 49: 556–72.
Erren, M. (1958) 'ΑΣΤΕΡΕΣ ΑΝѠΝΥΜΟΙ (Zu Arat 367–385)', *Hermes* 86: 240–3.
Eshleman, K. (2012) *The Social World of Intellectuals in the Roman Empire: Sophists, Philosophers and Christians*, Cambridge.
Evans, J. and Berggren, J. L. (2006) *Geminos's Introduction to the Phenomena: A Translation and Study of a Hellenistic Survey of Astronomy*, Princeton, NJ.
Fajen, F. (1999) *Oppianus Halieutica: Der Fischfang*, Stuttgart.
Fales, F. M. and Postgate, J. N. (1992) *Imperial Administrative Records, Part I: Palace and Temple Administration (State Archives of Assyria 7)*, Helsinki.
Fantar, M. H. (1998) 'De l'agriculture à Carthage', in M. Khanoussi, P. Ruggeri and C. Vismara (eds), *L'Africa romana: Atti del XII Convegno di Studio, Olbia, 12–15 dicembre 1996*, Sassari: 113–21.

Fantham, E. (2002) 'Ovid's *Fasti*: Politics, history, and religion', in B. W. Boyd (ed.), *Brill's Companion to Ovid*, Leiden: 197–233.
  (2004) *The Roman World of Cicero's De oratore*, Oxford.
Fantuzzi, M. and Hunter, R. (2004) *Tradition and Innovation in Hellenistic Poetry*, Cambridge.
Feeney, D. (1998) *Literature and Religion at Rome: Cultures, Contexts, and Beliefs*, Cambridge.
Feke, J. (2009) 'Ptolemy in Philosophical Context: A Study of the Relationships between Physics, Mathematics, and Theology' (PhD diss., University of Toronto).
  (2012) 'Ptolemy's defense of theoretical philosophy', *Apeiron* 45: 61–90.
Fitch, J. G. (ed.) (2008) *Seneca*, Oxford.
Flach, D. (ed.) (2002) *Marcus Terentius Varro, Gespräche über die Landwirtschaft, Buch 3*, Darmstadt.
Fögen, T. (ed.) (2005) *Antike Fachtexte: Ancient Technical Texts*, Berlin.
  (2009) *Wissen, Kommunikation und Selbstdarstellung: Zur Struktur und Charakteristik römischer Fachtexte der frühen Kaiserzeit*, Munich.
Ford, A. (1992) *Homer: The Poetry of the Past*, Ithaca, NY.
  (2002) *The Origins of Criticism: Literary Culture and Poetic Theory in Classical Greece*, Princeton, NJ.
Forhan, K. (2002) *The Political Theory of Christine de Pizan*, Burlington, VT.
Formisano, M. (2001) *Tecnica e scrittura: Le letterature tecnico-scientifiche nello spazio letterario tardolatino*, Rome.
  (2002) 'Strategie da manual. L'arte della guerra, Vegezio e Machiavelli', *Quaderni di Storia* 55: 99–127.
  (2009) 'The Renaissance tradition of the ancient art of war', in G. Beltramini (ed.), *Andrea Palladio and the Architecture of Battle*, Venice: 226–39.
  (2011) 'The *Strategikós* of Onasander: Taking military texts seriously', *Technai* 2: 39–52.
  (2012) 'Review of Allmand 2011', *Bryn Mawr Classical Review* 2012.11.59.
  (2014) 'Kriegskunst', in M. Landfester (ed.), *Renaissance-Humanismus: Lexikon zur Antikerezeption* (*Der Neue Pauly*, Supplemente 9), Stuttgart.
Formisano, M. and Böhme, H. (eds) (2011) *War in Words: Transformations of War from Antiquity to Clausewitz*, Berlin.
Formisano, M. and van der Eijk, P. (eds) (2017) *Knowledge, Text and Practice in Ancient Technical Writing*, Cambridge.
Fowler, D. P. (2000) 'The didactic plot', in M. Depew and D. Obbink (eds), *Matrices of Genre: Authors, Canons, and Society*, Cambridge, MA: 205–19.
Fox, M. (1993) 'History and rhetoric in Dionysius of Halicarnassus', *Journal of Roman Studies* 83: 31–47.
  (1996) *Roman Historical Myths: The Regal Period in Augustan Literature*, Oxford.
  (2001) 'Dionysius, Lucian, and the prejudice against rhetoric in history', *Journal of Roman Studies* 91: 76–93.

Foxhall, L. (1990) 'The dependent tenant: Land leasing and labour in Italy and Greece', *Journal of Roman Studies* 80: 97–114.

Frakes, R. M. (2011) *Compiling the Collatio Legum Mosaicarum et Romanarum in Late Antiquity* (Oxford Studies in Roman Law and Society 2), Oxford.

Frede, M. (2003) 'Galen's theology', in Barnes and Jouanna 2003: 73–106.

Frischer, B. (1982) *The Sculpted Word: Epicureanism and Philosophical Recruitment in Ancient Greece*, Berkeley, CA.

(1991) *Shifting Paradigms: New Approaches to Horace's Ars poetica*, Atlanta, GA.

Fromentin, V. (1988) 'L'attitude critique de Denys d'Halicarnasse face aux mythes', *Bulletin de l'Association Guillaume Budé* 1988: 318–26.

(1993) 'La définition de l'histoire comme mélange dans le prologue des *Antiquités Romaines* de Denys d'Halicarnasse (I, 8 3)', in *Les Actes du IIe colloque sur Denys d'Halicarnasse historien* (Pallas 39): 177–92.

(1998) *Denys d'Halicarnasse: Antiquités Romaines, Livre I* (with introduction and commentary), Paris.

(2006) 'Denys d'Halicarnasse et Hérodote III, 80–82 ou comment choisir la meilleure constitution pour Rome?', in *Le Monde et les mots: Mélanges Germaine Aujac* (Pallas 72): 229–42.

(2010) 'Les Moi de l'historien: récit et discours chez Denys d'Halicarnasse', in M.-R. Guelfucci (ed.), *Jeux et enjeux de la mise en forme de l'histoire: recherches sur le genre historique en Grèce et à Rome* (Dialogues d'histoire ancienne Suppl. 4), Besançon: 261–77.

Fuhrmann, M. (1960) *Das systematische Lehrbuch: Ein Beitrag zur Geschichte der Wissenschaften in der Antike*, Göttingen.

Gabba, E. (1984) 'Dionigi, Varrone e la religione senza miti', *Rivista storica italiana* 96: 855–70.

(1991) *Dionysius and the History of Archaic Rome*, Berkeley, CA.

Galimberti, A. (2002) 'Lo *Strategikós* di Onasandro', in M. Sordi (ed.), *Guerra e diritto nel mondo greco e romano*, Milan: 141–54.

Gallia, A. B. (2012) *Remembering the Roman Republic: Culture, Politics, and History under the Principate*, Cambridge.

Garnett, R. (1895) 'On the date of the Ἀποτελεσματικά of Manetho', *Journal of Philology* 23: 238–40.

Garrett, A. V. (2003) *Meaning in Spinoza's Method*, Cambridge.

Gauthier, P. (1985) *Les cités grecques et leur bienfaiteurs*, Paris.

Geertman, H. and De Jong, J. J. (eds) (1989) *Munus non ingratum: Proceedings of the International Symposium on Vitruvius' De architectura and the Hellenistic and Republican Architecture, Leiden 20–23 January 1987*, Leiden.

Gehrke, H.-J. (1998) 'Die Geburt der Erdkunde aus dem Geiste der Geometrie: Überlegungen zur Entstehung und zur Frühgeschichte der wissenschaftlichen Geographie bei den Griechen', in Kullmann, Althoff and Asper 1998: 163–92.

Geller, M. (2010) *Ancient Babylonian Medicine*, Chichester.

Gerson, J. (1960) *Oeuvres complètes*, Vol. 2 (ed. P. Glorieux), Paris.

Getty, R. J. (1941) 'The astrology of P. Nigidius Figulus (Lucan I, 649–65)', *Classical Quarterly* 35: 17–22.
  (1960) 'Neopythagoreanism and mathematical symmetry in Lucan, De bello civili 1', *Transactions of the American Philological Association* 91: 310–23.
Gibson, R. (1997) 'Didactic poetry as "popular" form', in C. Atherton (ed.), *Form and Content in Didactic Poetry*, Bari: 67–98.
Gildenhard, I. (2011) *Creative Eloquence: The Construction of Reality in Cicero's Speeches*, Oxford.
Gill, C. (2010) *Naturalistic Psychology in Galen and Stoicism*, Oxford.
Gill, C., Whitmarsh, T. and Wilkins, J. (eds) (2009) *Galen and the World of Knowledge*, Cambridge.
Gillespie, C. C. (ed.) (1970–90) *Dictionary of Scientific Biography* (18 volumes), New York.
Gilliver, C. (2007) 'Battle', in Sabin, Van Wees and Whitby 2007: 122–57.
Gilmartin, K. (1975) 'A rhetorical figure in Latin historical style: The imaginary second person singular', *Transactions of the American Philological Association* 104: 99–121.
Giuffré, V. (1974) *La letteratura 'de re militari': appunti per una storia degli ordinamenti militari*, Naples.
Gleason, M. (1995) *Making Men: Sophists and Self-Presentation in Ancient Rome*, Princeton, NJ.
Glinister, F. (2007) 'Constructing the past', in Glinister and Woods 2007: 1–9.
Glinister, F. and Woods, C. (eds) (2007) *Verrius, Festus and Paul: Lexicography, Scholarship, and Society*, London.
Goar, R. J. (1968) 'The purpose of De divinatione', *Transactions of the American Philological Association* 99: 241–8.
  (1972) *Cicero and the State Religion*, Amsterdam.
Goldhill, S. (2002) *The Invention of Prose*, Oxford.
  (2006) 'Artemis and cultural identity in empire culture: How to think about polytheism, now?', in D. Konstan and S. Saïd (eds), *Greeks on Greekness: Viewing the Greek Past under the Roman Empire*, Cambridge: 112–61.
Goldstein, B. R. (2007) 'What's new in Ptolemy's Almagest?', *Nuncius* 22: 261–85.
Goldstein, B. R. and Bowen, A. C. (1991) 'The introduction of dated observations and precise measurement in Greek astronomy', *Archive for History of Exact Sciences* 43: 93–132.
Goodchild, H. and Witcher, R. E. (2010) 'Modelling the agricultural landscapes of Republican Italy', in J. Carlsen and E. Lo Cascio (eds), *Agricoltura e scambi nell'Italia tardo-repubblicana*, Rome: 187–220.
Goodyear, F. R. D. (1982) 'Technical writing', in E. J. Kenney (ed.), *The Cambridge History of Classical Literature*, Vol. 2: *Latin Literature*, Cambridge: 667–73.
Gordley, M. E. (2011) *Teaching through Song in Antiquity*, Tübingen.
Gordon, R. (1990) 'The veil of power: Emperors, sacrificers and benefactors', in Beard and North 1990: 199–231.

Görler, W. (1995) 'Silencing the troublemaker: *De Legibus* I. 39 and the continuity of Cicero's scepticism', in J. G. F. Powell (ed.), *Cicero the Philosopher: Twelve Papers*, Oxford: 85–113.

Grafton, A. and Williams, M. (2006) *Christianity and the Transformation of the Book: Origen, Eusebius and the Library of Caesarea*, Cambridge, MA.

Graham, A. C. (1989) *Disputers of the Tao*, La Salle, IL.

Gray, V. (2012) 'Herodotus on Melampus', in Baragwanath and de Bakker 2012b: 167–91.

Graziosi, B. (2001) 'Competition in wisdom', in F. Budelmann and P. Michelakis (eds), *Homer, Tragedy and Beyond: Essays in Honour of P. E. Easterling*, London: 57–74.

Green, C. M. C. (1997) 'Free as a bird: Varro *De re rustica* 3', *American Journal of Philology* 118: 427–48.

(2002) 'Varro's three theologies and their influence on the *Fasti*', in G. Herbert-Brown (ed.), *Ovid's Fasti: Historical Readings at its Bimillennium*, Oxford: 71–99.

Green, P. (2006) *Diodorus Siculus: Books 11–12.37.1*, Austin, TX.

Green, S. J. (2014) *Disclosure and Discretion in Roman Astrology: Manilius and his Augustan Contemporaries*, Oxford.

Greene, J. and Kehoe, D. P. (1995) 'Mago the Carthaginian', in M. Fantar and M. Ghakhi (eds), *Actes du IIIe Congrès International des Études Phéniciennes et Puniques*, Tunis: 110–17.

Greene, K. (2000) 'Technological innovation and economic progress in the ancient world: M. I. Finley re-considered', *Economic History Review* 53: 29–59.

Gregory, A. F. and Tuckett, C. M. (eds) (2005) *Trajectories through the New Testament and the Apostolic Fathers*, Oxford.

Griffin, M. T. (1976) *Seneca: A Philosopher in Politics*, Oxford.

(1989) 'Philosophy, politics and politicians at Rome', in Griffin and Barnes 1989: 1–37.

(1995) 'Philosophical badinage in Cicero's letters to his friends', in J. G. F. Powell (ed.), *Cicero the Philosopher*, Oxford: 325–46.

(2007) 'Seneca's pedagogic strategy: Letters and *De beneficiis*', in R. Sorabji and R. Sharples (eds), *Greek and Roman Philosophy 100 BC–200 AD*, Vol. 1, London: 89–113.

(2013) *Seneca on Society: A Guide to De Beneficiis*, Oxford.

Griffin, M. T. and Barnes, J. (eds) (1989) *Philosophia Togata: Essays on Philosophy and Roman Society*, Oxford.

Grimal, P. (1965–66) 'Encyclopédies antiques', *Cahiers d'histoire mondiale* 9: 459–82.

Gros, P. (1989) 'L'auctoritas chez Vitruve: Contribution a l'étude de la sémantique des ordres dans le *De Architectura*', in Geertman and De Jong 1989: 126–33.

(1994) 'Munus non ingratum: le traité vitruvien et la notion de service', in P. Gros, *Le projet de Vitruve: objet, déstinataires et reception du De architectura*, Rome: 75–90.

Gross, A. (1996) *The Rhetoric of Science*, Cambridge, MA.
   (2006) *Starring the Text: The Place of Rhetoric in Science Studies*, Carbondale, IL.
Guillaumont, F. (1984) *Philosophe et augure: Recherches sur la théorie cicéronienne de la divination*, Brussels.
Guittard, C. (2003) 'Les calendriers brontoscopiques dans le monde étrusco-romain', in C. Cusset (ed.), *La Météorologie dans l'antiquité: Entre science et croyance*, Saint-Étienne: 455–66.
Gunderson, E. (ed.) (2009) *The Cambridge Companion to Ancient Rhetoric*, Cambridge.
Habinek, T. and Schiesaro, A. (eds) (1997) *The Roman Cultural Revolution*, Cambridge.
Hagendahl, H. (1967) *Augustine and the Latin Classics*, Göteborg.
Hahn, J. (1991) 'Plinius und die griechischen Ärzte in Rom: Naturkonzeption und Medizinkritik in der *Naturalis Historia*', *Sudhoffs Archiv* 75: 209–39.
Hale, J. R. (1988) 'A humanistic visual aid: The military diagram in the Renaissance', *Renaissance Studies* 2: 280–98.
Hamilton, N. T., Swerdlow, N. M. and Toomer, G. J. (1987) 'The "Canobic Inscription": Ptolemy's earliest work', in J. L. Berggren and B. R. Goldstein (eds), *From Ancient Omens to Statistical Mechanics*, Copenhagen: 55–73.
Hankinson, R. J. (1991a) *On the Therapeutic Method: Books One and Two*, Oxford.
   (1991b) 'Galen on the foundations of science', in López Férez 1991: 15–29.
   (1994) 'Galen's concept of scientific progress', *Aufstieg und Niedergang der römischen Welt* II, 37(2): 1775–89.
   (1998) *Galen: On Antecedent Causes*, Cambridge.
   (2003) 'Causation in Galen', in Barnes and Jouanna 2003: 31–72.
   (2005a) 'Prédiction, prophétie, prognostic: la gnoséologie de l'avenir dans la divination et la medicine antique', in R. Kany-Turpin (ed.), *Signes et prédiction dans l'antiquité*, Saint-Etienne: 147–62.
   (2005b) 'Aristotle on kind-crossing', in Sharples 2005: 23–54.
   (2008a) 'Galen's epistemology', in Hankinson 2008b: 157–83.
   (ed.) (2008b) *The Cambridge Companion to Galen*. Cambridge.
   (2009) 'Galen on the limitations of knowledge', in Gill, Whitmarsh and Wilkins 2009: 206–42.
Hannah, R. (1996) 'Lucan Bellum civile 1.649–65: The astrology of P. Nigidius Figulus revisited', *Papers of the Leeds International Latin Seminar* 9: 175–90.
   (2001) 'From orality to literacy? The case of the parapegmata', in J. Watson (ed.), *Speaking Volumes: Orality and Literacy in the Greek and Roman World*, Leiden: 139–59.
   (2002) 'Euctemon's parapêgma', in Tuplin and Rihll 2002: 112–32.
Harder, M. A., Regtuit, R. F. and Wakker, G. C. (eds) (2006) *Beyond the Canon* (*Hellenistica Groningana* 11), Leuven.
Hardie, P. R. (1992) 'Plutarch and the interpretation of myth', *Aufstieg und Niedergang der römischen Welt* II, 33(6): 4743–87.
Hardie, P. (2015) *Ovidio: Metamorfosi*, Vol. 6, Milan.

Harkness, D. (2007) *The Jewel House: Victorian London and the Scientific Revolution*, New Haven, CT.

Harper, D. (1998) *Early Chinese Medical Literature: The Mawangdui Manuscripts*, London.

Harries, J. (2006) *Cicero and the Jurists: From Citizens' Law to the Lawful State*, London.

(2013) 'Encyclopaedias and autocracy: Justinian's encyclopaedia of Roman law', in König and Woolf 2013: 178–96.

Harris, W. V. (1971) *Rome in Etruria and Umbria*, Oxford.

(1989) *Ancient Literacy*, Cambridge, MA.

Harris-McCoy, D. (2008a) 'Varieties of Encyclopedism in the Early Roman Empire: Vitruvius, Pliny the Elder, Artemidorus' (PhD Diss., University of Pennsylvania).

(2008b) 'Jason König and Tim Whitmarsh, *Ordering Knowledge in the Roman Empire*', *Bryn Mawr Classical Review* 2008.10.39.

(2012) *Artemidorus' Oneirocritica: Text, Translation, and Commentary*, Oxford.

(2013) 'Artemidorus' *Oneirocritica* as fragmentary encyclopedia', in König and Woolf 2013: 154–77.

Harrison, S. J. (2000) *Apuleius: A Latin Sophist*, Oxford.

Harrison, S. J., Hilton, J. and Hunink, V. (2001) *Apuleius: Rhetorical Works*, Oxford.

Hawking, S. (1988) *A Brief History of Time*, New York.

Healy, L. and Hoyles, C. (2000) 'A study of proof conceptions in algebra', *Journal for Research in Mathematics Education* 31: 396–428.

Heath, M. (2004) *Menander: A Rhetor in Context*, Oxford.

Heiberg, J. L. (1891–93) *Apollonii Pergaei quae Graece exstant opera*, Leipzig.

(1898–1907) *Claudii Ptolemaei opera quae exstant omnia*, Vol. 1: *Syntaxis mathematica*; Vol. 2: *Opera astronomica minora*, Leipzig.

Heiberg, J. L. and Stamatis, E. (1915) *Commentarii in libros de sphaera et cylindro*, Vol. 3: *Archimedis opera omnia cum commentariis Eutocii*, Leipzig.

Heinimann, F. (1961) 'Eine vorplatonischen Theorie der Techne', *Museum Helveticum* 18: 105–30.

Helm, R. (1956) *Die Chronik des Hieronymus = Hieronymi Chronicon (GCS 47)*, Berlin.

Henderson, J. (2002) 'Columella's living hedge: The Roman gardening book', *Journal of Roman Studies* 92: 110–33.

(2004) *Morals and Villas in Seneca's Letters: Places to Dwell*, Cambridge.

Hense, O. (1905) *C. Musonii Rufi reliquiae*, Leipzig.

Hesk, J. (2007) 'Combative capping in Aristophanic comedy', *Proceedings of the Cambridge Philological Society* 53: 124–60.

Heurgon, J. (1976) 'L'agronome carthaginois Magon et ses traducteurs en latin et en grec', *Comptes rendus de l'Académie des Inscriptions et Belles-Lettres* 120: 441–56.

(1978) *Varron: Économie rurale, Livre 1*, Paris.

Heuser, B. (2007) 'Introduction', in C. von Clausewitz, *On War*, Oxford: vii–xxxviii.
  (2010) *The Evolution of Strategy: Thinking War from Antiquity to the Present*, Cambridge.
Hidber, T. (ed.) (1996) *Das klassizistische Manifest des Dionys von Halikarnass: Die praefatio zu De oratoribus veteribus* (with translation and commentary), Stuttgart.
  (2011) 'Impacts of writing in Rome: Greek perceptions of Roman literature in the first century BCE', in Schmitz and Wiater 2011: 115–23.
Highet, G. (1962) *The Anatomy of Satire*, Princeton, NJ.
Hinds, S. (1998) *Allusion and Intertext: Dynamics of Appropriation in Roman Poetry*, Cambridge.
Hine, H. M. (2006) 'Rome, the cosmos, and the emperor in Seneca's *Natural Questions*', *Journal of Roman Studies* 96: 42–72.
  (2009) 'Subjectivity and objectivity in Latin scientific and technical literature', in Taub and Doody 2009: 13–30.
  (2010) 'Form and function of speech in the prose works of the younger Seneca', in D. H. Berry and A. Erskine (eds), *Form and Function in Roman Oratory*, Cambridge: 208–24.
  (2016) '*Philosophy* and *philosophi* from Cicero to Apuleius', in Williams and Volk 2016: 13–29.
Hirzel, R. (1895) *Der Dialog: Ein literarhistorischer Versuch* (2 volumes), Leipzig.
Hobden, F. and Tuplin, C. (eds) (2012) *Xenophon: Ethical Principles and Historical Enquiry*, Leiden.
Hon, G. (1989) 'Is there a concept of experimental error in Greek astronomy?', *British Journal for the History of Science* 22: 129–50.
Honigman, S. (2003) *The Septuagint and Homeric Scholarship in Alexandria: A Study in the Narrative of the Letter of Aristeas*, London.
Honoré, T. (1978) *Tribonian*, London.
  (1994) *Emperors and Lawyers* (2nd edn), Oxford.
  (2002) *Ulpian: Pioneer of Human Rights* (2nd edn), Oxford.
  (2010) *Justinian's Digest: Character and Compilation*, Oxford.
Hopkinson, N. (1994) *Greek Poetry of the Imperial Period: An Anthology*, Cambridge.
Horster, M. (2008) 'Some notes on grammarians in Plutarch', in A. G. Nikolaidis (ed.), *The Unity of Plutarch's Work: 'Moralia' Themes in the 'Lives', Features of the 'Lives' in the 'Moralia'*, Berlin: 611–24.
Horster, M. and Reitz, C. (eds) (2003) *Antike Fachschriftsteller: Literarischer Diskurs und sozialer Kontext*, Stuttgart.
Housman, A. E. (1926) *M. Annaei Lucani Belli Civilis libri decem*, Oxford.
Howe, N. (1985) 'In defense of the encyclopedic mode: On Pliny's preface to the *Natural History*', *Latomus* 44: 561–76.
Howley, J. (2013) 'Why read the jurists? Aulus Gellius on reading across disciplines', in Du Plessis 2013: 9–30.

Hsu, E. (2010) *Pulse Diagnosis in Early Chinese Medicine: The Telling Touch*, Cambridge.

Humfress, C. (2005) 'Law and legal practice in the age of Justinian', in M. Maas (ed.), *The Cambridge Companion to Justinian*, Cambridge: 161–84.

Humphreys, S. C. (1975) 'Transcendence and intellectual roles: The ancient Greek case', *Daedalus* 104: 91–118.

Hunter, R. (1995) 'Written in the stars: Poetry and philosophy in the *Phainomena* of Aratus', *Arachnion* 2: 1–34.

(2004a) 'Epic in a minor key', in Fantuzzi and Hunter 2004: 191–245.

(2004b) 'The *Periegesis* of Dionysius and the traditions of Hellenistic poetry', *Revue des études antiques* 106: 217–31.

(2006) 'The Prologue of the *Periodos to Nicomedes* ("Pseudo-Scymnus")', in Harder, Regtuit and Wakker 2006: 123–40.

(2012) *Plato and the Traditions of Ancient Literature: The Silent Stream*, Cambridge.

(2014) *Hesiodic Voices: Studies in the Ancient Reception of Hesiod's Works and Days*, Cambridge.

Hutchinson, G. O. (2009) 'Read the instructions: Didactic poetry and didactic prose', *Classical Quarterly* 59: 196–211.

Ierodiakonou, K. (2002) 'Aristotle's use of examples in the *Prior Analytics*', *Phronesis* 47: 127–52.

Inwood, B. (1995) 'Seneca in his philosophical milieu', *Harvard Studies in Classical Philology* 97: 63–76 (reprinted in B. Inwood (2005) *Reading Seneca: Stoic Philosophy at Rome*, Oxford: 7–22).

Jacob, C. (1981) 'L'œil et la mémoire: sur la *Périégèse de la Terre habitée* de Denys', in C. Jacob and F. Lestringant (eds), *Arts et légendes d'espaces: figures du voyage et rhétoriques du monde*, Paris: 21–97.

(1990) *La Description de la terre habitée de Denys d'Alexandrie ou la leçon de géographie*, Paris.

(1991) 'Θεὸς Ἑρμῆς ἐπὶ Ἀδριανοῦ: la mise en scène du pouvoir impérial dans la Description de la terre habitée de Denys d'Alexandrie', *Cahiers du Centre Gustave Glotz* 2: 43–53.

Jacques, J.-M. (2007) *Nicandre, Œuvres*, Vol. 3: *Les Alexipharmaques*, Paris.

Jaeger, M. (2008) *Archimedes and the Roman Imagination*, Ann Arbor, MI.

Jal, P. (1961) 'La propagande religieuse à Rome au cours des guerres civiles de la fin de la république', *Antiquité classique* 30: 395–414.

Janniard, S. (2008) 'Végèce et les transformations de l'art de la guerre aux IVe et Ve siècles après J.-C.', *Antiquité tardive* 16: 19–36.

Janson, T. (1964) *Latin Prose Prefaces: Studies in Literary Conventions*, Stockholm.

Jocelyn, H. D. (1982) 'Varro's *Antiquitates rerum divinarum* and religious affairs in the late Roman Republic', *Bulletin of the John Rylands University Library of Manchester* 65: 148–205.

Johnson, M. R. (2009) 'The Aristotelian explanation of the halo', *Apeiron* 42: 325–58.

Johnson, W. A. (2013) 'Libraries and reading culture in the high empire', in König, Oikonomopoulou and Woolf 2013: 347–63.

Johnson, W. A. and Parker, H. N. (eds) (2009) *Ancient Literacies: The Culture of Reading in Greece and Rome*, Oxford.

Johnston, I. and Horsley, G. H. R. (2011) *Galen, Method of Medicine*, Cambridge, MA.

Johnston, S. I. (2001) 'Charming children: The use of the child in ancient divination', *Arethusa* 34: 97–117.

(2008) *Ancient Greek Divination*, Malden, MA.

Jones, A. (1990) *Ptolemy's First Commentator (Transactions of the American Philosophical Society 80.7)*, Philadelphia, PA.

(1994) 'The place of astronomy in Roman Egypt', *Apeiron* 27: 25–51.

(1999) *Astronomical Papyri from Oxyrhynchus (P. Oxy. 4133–4300a)*, Philadelphia, PA.

(2005a) 'Ptolemy's "Canobic Inscription" and Heliodorus's observation reports', *SCIAMVS* 6: 53–97.

(2005b) 'In order that we should not ourselves appear to be adjusting our estimates... to make them fit some predetermined amount', in J. Z. Buchwald and A. Franklin (eds), *Wrong for the Right Reasons*, Dordrecht: 17–39.

(2010a) 'Introduction', in Jones 2010b: xi–xv.

(ed.) (2010b) *Ptolemy in Perspective: Use and Criticism of his Work from Antiquity to the Nineteenth Century*, Dordrecht.

(2011) 'Ptolemy's Geography: a reform that failed', in Z. Shalev and C. Burnett (eds), *Ptolemy's Geography in the Renaissance*, London: 15–29.

Jones, C. (1978) *The Roman World of Dio Chrysostom*, Cambridge, MA.

Jouanna, J. (1999) *Hippocrates* (tr. M. B. DeBevoise; first published in French 1992), Baltimore, MD.

Keegan, J. (1983) *The Face of Battle: A Study of Agincourt, Waterloo, and the Somme*, London.

Keen, R. (1985) 'Lucretius and his reader', *Apeiron* 19: 1–10.

Kennedy, D. F. (2011) 'Sums in verse or a mathematical aesthetic?', in S. J. Green and K. Volk (eds), *Forgotten Stars: Rediscovering Manilius' Astronomica*, Oxford: 165–87.

Ker, J. (2006) 'Seneca, man of many genres', in K. Volk and G. D. Williams (eds), *Seeing Seneca Whole: Perspectives on Philosophy, Poetry and Politics*, Leiden: 19–41.

(2009) 'Outside and inside: Senecan strategies', in W. J. Dominik, J. Garthwaite and P. A. Roche (eds), *Writing Politics in Imperial Rome*, Leiden: 249–71.

Keulen, W. (2009) *Gellius the Satirist: Roman Cultural Authority in Attic Nights*, Leiden.

Khan, Y. (2004) 'Denys lecteur des *Phénomènes* d'Aratos', *Revue des études antiques* 106: 233–46.

Kidd, D. (1997) *Aratus, Phaenomena: Edited with Introduction, Translation and Commentary*, Cambridge.
Kidd, I. G. (1988) *Posidonius*, Vol. 2: *The Commentary*, Cambridge.
Kim, L. (2010) *Homer between History and Fiction in Imperial Greek Literature*, Cambridge.
Kindt, J. (2006) 'Delphic oracle stories and the beginning of historiography: Herodotus' Croesus Logos', *Classical Philology* 101: 34–51.
King, C. (2003) 'The organization of Roman religous beliefs', *Classical Antiquity* 22: 275–312.
Klotz, F. (2007) 'Portraits of the philosopher: Plutarch's self-presentation in the *Quaestiones Convivales*', *Classical Quarterly* 57: 650–67.
  (2011) 'Imagining the past: Plutarch's play with time', in Klotz and Oikonomopoulou 2011: 161–78.
Klotz, F. and Oikonomopoulou, K. (eds) (2011) *The Philosopher's Banquet: Plutarch's Table Talk in the Intellectual Culture of the Roman Empire*, Oxford.
Kneebone, E. (2008) 'ΤΟΣΣ' ΕΔΑΗΝ: The poetics of knowledge in Oppian's *Halieutica*', in Carvounis and Hunter 2008: 32–59.
Knorr, W. R. (1986) *The Ancient Tradition of Geometric Problems*, Boston, MA.
  (1989) *Textual Studies in Ancient and Medieval Geometry*, Boston, MA.
Koch, C. (2003) 'Roman state religion in the mirror of Augustan and late Republican apologetics' (tr. C. Barnes), in Ando 2003: 296–329.
König, A. (2007) 'Knowledge and power in Frontinus' *On Aqueducts*', in König and Whitmarsh 2007b: 177–205.
  (2009) 'From architect to imperator: Vitruvius and his addressee in the *De architectura*', in Taub and Doody 2009: 31–52.
König, J. (2005) *Athletics and Literature in the Roman Empire*, Cambridge.
  (2007) 'Fragmentation and coherence in Plutarch's *Sympotic Questions*', in König and Whitmarsh 2007b: 43–68.
  (2009) 'Conventions of prefatory self-presentation in Galen's *On the Order of My Own Books*', in Gill, Whitmarsh and Wilkins 2009: 35–58.
  (2010) 'Competitiveness and anti-competitiveness in Philostratus' *Lives of the Sophists*', in N. Fisher and H. van Wees (eds), *Competition in the Ancient World*, Swansea: 279–300.
  (2011) 'Self-promotion and self-effacement in Plutarch's *Sympotic Questions*', in Klotz and Oikonomopoulou 2011: 179–203.
  (2012) *Saints and Symposiasts: The Literature of Food and the Symposium in Greco-Roman and Early Christian Culture*, Cambridge.
  (2014) 'Images of elite community in Philostratus' *Lives of the Sophists*', in J. Madsen and R. Rees (eds), *Roman Rule in Greek and Latin Writing: Double Vision*, Leiden: 246–70.
  (2016) 'Re-reading Pollux: Encyclopaedic structure and athletic culture in *Onomasticon* Book 3', *Classical Quarterly* 66: 298–315.

König, J. and Whitmarsh, T. (2007a) 'Ordering knowledge', in König and Whitmarsh 2007b: 3–39.

(eds) (2007b) *Ordering Knowledge in the Roman Empire*, Cambridge.

König, J., Oikonomopoulou, K. and Woolf, G. (eds) (2013) *Ancient Libraries*, Cambridge.

König, J. and Woolf, G. (eds) (2013) *Encyclopaedism from Antiquity to the Renaissance*, Cambridge.

Korenjak, M. (2003) *Die Welt-Rundreise eines anonymen griechischen Autors ('pseudo-Skymnos'): Einleitung, Text, Übersetzung und Kommentar*, Hildesheim.

Koster, S. (1970) *Antike Epostheorien*, Wiesbaden.

Kowalski, H. (1995) 'Cicero, Augur und Politiker: Die Theorie und Praxis der Auspizien in Ciceros Tätigkeit', *Acta classica Universitatis Scientiarum Debreceniensis* 31: 125–39.

Krause, M. (1936) *Die Sphärik von Menelaos aus Alexandrien in der Verbasserung von Abū Nasr Mansūr b. ʿAlī b. ʿIrāq*, Berlin.

Krenkel, W. A. (ed.) (2002) *Marcus Terentius Varro, Saturae menippeae* (with translation and commentary) (4 volumes), St Katharinen.

Kroll, W. (1937) 'P. Nigidius Figulus', *Pauly-Wissowa RE* 17: 200–12.

Kron, J. G. (2008) 'The much maligned peasant: Comparative perspectives on the productivity of the small farmer in classical antiquity', in L. de Ligt and S. J. Northwood (eds), *People, Land and Politics: Demographic Developments and the Transformation of Roman Italy, 300 BC–AD 14*, Leiden: 71–119.

Kronenberg, L. (2009) *Allegories of Farming from Greece and Rome: Philosophical Satire in Xenophon, Varro, and Virgil*, Cambridge.

Krostenko, B. A. (2000) 'Beyond (dis)belief: Rhetorical form and religious symbol in Cicero's *de Divinatione*', *Transactions of the American Philological Association* 130: 353–91.

Krueger, D. (2004) *Writing and Holiness: The Practice of Authorship in the Early Christian East*, Philadelphia, PA.

Kuhn, T. S. (1962) 'The structure of scientific revolutions', *International Encyclopedia of Unified Science: Foundations of the Unity of Science*, Vol. 2.2, Chicago, IL.

Kühner, R. and Gerth, B. (1904) *Ausführliche Grammatik der griechischen Sprache* (3rd edn), Hanover.

Kullmann, W., Althoff, J. and Asper, M. (eds) (1998) *Gattungen wissenschaftlicher Literatur in der Antike*, Tübingen.

Kurth, T. (1994) *Senecas Trostschrift an Polybius, Dialog 11: Ein Kommentar* (Beiträge zur Altertumskunde 59), Stuttgart.

Lammert, F. (1931) *Die römische Taktik zu Beginn der Kaiserzeit und die Geschichtsschreibung*, Leipzig.

Langlands, R. (2008) '"Reading for the moral" in Valerius Maximus: The case of *severitas*', *Cambridge Classical Journal* 54: 160–87.

(2011) 'Roman exempla and situation ethics: Valerius Maximus and Cicero, *De officiis*', *Journal of Roman Studies* 101: 1–23.

Lateiner, D. (1989) *The Historical Method of Herodotus*, Toronto.

Latour, B. (1987) *Science in Action: How to Follow Scientists and Engineers through Society*, Cambridge, MA.

Le Bohec, Y. (1998) 'Que voulait Onasandros?', in Y. Burnand, Y. Le Bohec and J.-P. Martin (eds), *Claude de Lyon empereur romain*, Paris: 169–79.

le Saux, F. (2004) 'War and knighthood in Christine de Pizan's *Livre des fait d'armes et de chevallerie*', in C. Saunders, F. le Saux and N. Thomas (eds), *Writing War: Medieval Literary Responses to Warfare*, Cambridge: 93–105.

Lebek, W. D. (1976) *Lucans Pharsalia: Dichtungsstruktur und Zeitbezug*, Göttingen.

Leeman A., Pinkster J. and Wisse J. (1981–96) *M. Tullius Cicero, De oratore, Libri III* (4 volumes), Heidelberg.

Lehmann, Y. (1997) *Varron théologien et philosophe romain*, Brussels.

Lehoux, D. (2007) *Astronomy, Weather, and Calendars in the Ancient World: Parapegmata and Related Texts in Classical and Near-Eastern Societies*, Cambridge.

(2012) *What Did the Romans Know? An Inquiry into Science and Worldmaking*, Chicago, IL.

(2013) 'Seeing and unseeing, seen and unseen', in D. Lehoux, A. D. Morrison and A. Sharrock (eds), *Lucretius: Poetry, Philosophy, Science*, Oxford: 131–52.

Lenaghan, J. O. (1969) *A Commentary on Cicero's Oration De haruspicum responso*, The Hague.

Lenoir, M. (1996) 'La littérature *De re militari*', in C. Nicolet (ed.), *Les Littératures téchniques dans l'antiquité romaine: Statut, public et destination, tradition (Fondation Hardt. Entretiens sur l'antiquité classique 42)*, Geneva: 77–115.

Levick, B. (2003) 'Seneca and money', in A. De Vivo and E. Lo Cascio (eds), *Seneca uomo politico e l'età di Claudio e di Nerone: Atti del Convegno Internazionale (Capri 25–27 marzo 1999)*, Bari: 211–28.

Lewis, A.-M. (1998) 'What dreadful purpose do you have? A new explanation of the astrological prophecy of Nigidius Figulus in Lucan's *Pharsalia* I, 658–63', *Studies in Latin Literature and Roman History* 9: 379–400.

Lewis, M. J. T. (2001) *Surveying Instruments of Greece and Rome*, Cambridge.

Lhommé, M.-K. (2007) 'Varron et Verrius au 2ème siècle après Jésus-Christ', in Glinister and Woods 2007: 11–47.

Lieberg, G. (1973) 'Die "theologia tripartita" in Forschung und Bezeugung', *Aufstieg und Niedergang der römischen Welt* I, 4: 63–115.

Liebs, D. (1976) 'Rechtsschulen und Rechtsunterricht im Prinzipat', *Aufstieg und Niedergang der römischen Welt* II, 15: 197–216.

Lightfoot, J. L. (2008) 'Catalogue technique in Dionysius Periegetes', in Carvounis and Hunter 2008: 11–31.

(2014) *Dionysius Periegetes, Description of the Known World, with Introduction, Text, Translation, and Commentary*, Oxford.

Lim, R. (1995) *Public Disputation, Power, and Social Order in Late Antiquity*, Berkeley, CA.
Lindemann, A. (2005) 'Paul's influence on "Clement" and Ignatius', in Gregory and Tuckett 2005: 9-24.
Linderski, J. (1971) 'Römischer Staat und Götterzeichen: Zum Problem der obnuntiatio', *Jahrbuch der Universität Düsseldorf 1969-1970*: 309-22 (= Linderski 1995: 444-57).
  (1982) 'Cicero and Roman divination', *La parola del passato* 37: 12-38 (= Linderski 1995: 458-84).
  (1985) 'The libri reconditi', *Harvard Studies in Classical Philology* 89: 207-34.
  (1986) 'The augural law', *Aufstieg und Niedergang der römischen Welt* II, 16: 2146-312.
  (1995) *Roman Questions: Selected Papers*, Stuttgart.
Ling, R. (1991) 'Brading, Brantingham and York: A new look at some fourth-century mosaics', *Britannia* 22: 148-53.
Littlewood, R. J. (2006) *A Commentary on Ovid, Fasti Book VI*, Oxford.
Littré, E. (1839-61) *Oeuvres complètes d'Hippocrate* (10 volumes), Paris.
Liuzzi, D. (1983) *Nigidio Figulo 'astrologo e mago': Testimonianze e frammenti*, Lecce.
Lloyd, G. E. R. (1975) 'The Hippocratic question', *Classical Quarterly* 25: 171-92.
  (1979) *Magic, Reason, and Experience: Studies in the Origins and Development of Greek Science*, Cambridge.
  (1982) 'Observational error in later Greek science', in J. Barnes and J. Brunschwig (eds), *Science and Speculation: Studies in Hellenistic Theory and Practice*, Cambridge: 128-64.
  (1983) *Science, Folklore and Ideology: Studies in the Life Sciences in Ancient Greece*, Cambridge.
  (1987a) *The Revolutions of Wisdom: Studies in the Claims and Practice of Ancient Greek Science*, Berkeley, CA.
  (1987b) 'Dogmatism and uncertainty in early Greek speculative thought', in M. Detienne (ed.), *Poikilia: Études offertes à J.-P. Vernant*, Paris: 297-312.
  (1988) 'Scholarship, authority and argument in Galen's *Quod animi mores*', in P. Manuli and M. Vegetti (eds), *Le Opere psicologiche di Galeno*, Naples: 11-42.
  (1990) *Demystifying Mentalities*, Cambridge.
  (1991a) 'Galen on Hellenistics and Hippocrateans: Contemporary battles and past authorities', in Lloyd 1991b: 398-416.
  (1991b) *Methods and Problems in Greek Science: Selected Papers*, Cambridge.
  (1996a) 'Theories and practices of demonstration in Galen', in M. Frede and S. Striker (eds), *Rationality in Greek Thought*, Oxford: 255-77.
  (1996b) *Adversaries and Authorities: Investigations into Ancient Greek and Chinese Science*, Cambridge.
  (2002) *The Ambitions of Curiosity*, Cambridge.
  (2005) 'Mathematics as a model of method in Galen', in Sharples 2005: 110-30.
  (2009) *Disciplines in the Making*, Oxford.

(2012) 'The pluralism of Greek "mathematics"', in K. Chemla (ed.), *The History of Mathematical Proof in Ancient Traditions*, Cambridge: 294–310.

(2014) *The Ideals of Inquiry*, Oxford.

Lloyd, G. E. R. and Sivin, N. (2002) *The Way and the Word: Science and Medicine in Early China and Greece*, New Haven, CT.

Long, A. (2002) *Epictetus: A Stoic and Socratic Guide to Life*, Oxford.

Long, P. O. (2001) *Openness, Secrecy, Authorship: Technical Arts and the Culture of Knowledge from Antiquity to the Renaissance*, Baltimore, MD.

Lonie, I. M. (1964) 'Erasistratus, the Erasistrateans and Aristotle', *Bulletin of the History of Medicine* 38: 426–43.

López Férez, J. A. (ed.) (1991) *Galeno: Obra, pensamiento y influencia*, Madrid.

Loraux, N. (1986) 'Thucydide a écrit *La Guerre du Péloponnèse*', *Métis* 1: 139–61.

Lorenz, T. (1965) *Galerien von griechischen Philosophen- und Dichterbildnissen bei den Römern*, Mainz.

Luraghi, N. (2003) 'Dionysios von Halikarnassos zwischen Griechen und Römern', in U. Eigler, U. Gotter, N. Luraghi and U. Walter (eds), *Formen römischer Geschichtsschreibung von den Anfängen bis Livius: Gattungen – Autoren – Kontexte*, Darmstadt: 268–86.

Lynch, C. (ed.) (2003) *Niccolò Macchiavelli, Art of War* (translated, edited, and with a commentary), Chicago.

Lyons, J. D. (1989) *Exemplum: The Rhetoric of Example in Early Modern France and Italy*, Princeton, NJ.

Ma, J. (1999) *Antiochus III and the Cities of Western Asia Minor*, Oxford.

Machiavelli, N. (2003) *Art of War: Translated, Edited and with a Commentary by Christopher Lynch*, Chicago, IL.

Mackey, J. L. (2009) 'Rethinking Roman Religion: Action, Practice, and Belief' (PhD Diss., Princeton).

MacMullen, R. (1966) *Enemies of the Roman Order*, Cambridge, MA.

MacRae, D. (2013a) 'The Books of Numa: Writing, Intellectuals and the Making of Roman Religion' (PhD Diss., Harvard).

(2013b) 'Review of Rüpke 2012', *American Journal of Philology* 134: 510–14.

Magnelli, E. (2005) 'Esiodo "epico" ed Esiodo didattico: il doppio epilogo di Dionisio Periegeta', *Appunti romani di filologia* 7: 105–8.

(2006) 'Altre fonti e imitazioni del poema di Dionisio Periegeta', *Studi italiani di filologia classica* 4: 241–51.

Mahaffy, J. P. (1889) 'The work of Mago on agriculture', *Hermathena* 7: 29–35.

Malloch, S. J. V. (2015) 'Frontinus and Domitian: The politics of the *Strategemata*', *Chiron* 45: 77–100.

Manitius, K. (ed.) (1894) *Gemini elementa astronomiae* (with translation and commentary), Leipzig (reprinted Stuttgart, 1974).

Mann, J. E. (2008) 'Prediction, precision, and practical experience: The Hippocratics on technē', *Apeiron* 4: 89–122.

Manning, C. E. (1981) *On Seneca's 'Ad Marciam'* (*Mnemosyne* Suppl. 69), Leiden.

Mansfeld, J. (1998) *Prolegomena mathematica: From Apollonius of Perga to Late Neoplatonism, with an Appendix on Pappus and the History of Platonism*, Leiden.
Marcotte, D. (2000) *Géographes grecs*, Vol. 1: *Introduction générale*, Paris.
Marincola, J. (1997) *Authority and Tradition in Ancient Historiography*, Cambridge.
Maróti, E. (1976) *The Vilicus and the Villa-System in Ancient Italy*, Budapest.
Marrou, H. I. (1938) *Saint Augustin et la fin de la culture antique*, Paris.
Marsden, E. W. (1971) *Greek and Roman Artillery: Technical Treatises*, Oxford.
Martin, J. (1956) *Histoire du texte des Phénomènes d'Aratos*, Paris.
  (1997) 'John of Salisbury's manuscripts of Frontinus and Gellius', *Journal of the Warburg and Courtauld Institutes* 40: 1–26.
  (1998) *Aratos, Phénomènes* (2 volumes), Paris.
Martin, R. (1971) *Recherches sur les agronomes latins et leurs conceptions économiques et sociales*, Paris.
  (1972) '"Familia rustica": les ésclaves chez les agronomes latins', in *Actes du colloque 1972 sur l'ésclavage*, Paris: 267–97.
  (1985) 'État présent des études sur Columelle', *Aufstieg und Niedergang der römischen Welt* II, 32(3): 1959–79.
Martínez, V. M. and Senseney, M. F. (2013) 'The professional and his books: Special libraries in the ancient world', in König, Oikonomopoulou and Woolf (2013): 401–17.
Matilal, B. K. (1998) *The Character of Logic in India* (ed. J. Ganeri and H. Tiwari), Albany, NY.
Mattern, S. P. (2008) *Galen and the Rhetoric of Healing*, Baltimore, MD.
  (2013) *The Prince of Medicine*, Oxford.
May, J. and Wisse, J. (2001) *Cicero: On the Ideal Orator*, New York.
Mazour-Matusevich, Y. and Bejczy, I. P. (2007) 'Jean Gerson on virtues and princely education', in I. P. Bejczy and C. J. Nederman (eds), *Princely Virtues in the Middle Ages 1200–1500*, Turnhout.
McEwen, I. K. (2003) *Vitruvius: Writing the Body of Architecture*, Cambridge, MA.
McGill, S. (2012) *Plagiarism in Latin Literature*, Cambridge.
McLean, D. R. (2007) 'The Socratic corpus: Socrates and physiognomy', in M. Trapp (ed.), *Socrates from Antiquity to the Enlightenment*, Aldershot: 65–88.
Meissner, B. (1999) *Die technologische Fachliteratur der Antike: Struktur, Überlieferung und Wirkung technischen Wissens in der Antike (ca. 400 v. Chr. – ca. 500 n. Chr.)*, Berlin.
Mendell, H. (2008a) 'Euktêmôn of Athens (440–410 BCE)', in *EANS*: 317.
  (2008b) 'Metôn of Athens (440–410 BCE)', in *EANS*: 551–2.
Mendelson, E. (1976) 'Encyclopedic narrative: From Dante to Pynchon', *Modern Language Notes* 91: 1267–75.
Merkel, R. (ed.) (1841) *P. Ovidii Nasonis: Fastorum libri sex* (with commentary), Berlin.
Mertz, E. (2007) *The Language of Law School: Learning to 'Think Like a Lawyer'*, Oxford.

Mevoli, D. (1992) 'Una "magia" di Nigidio (Apul. apol. 42)', *Studi di filologia e letteratura* 2: 115–25.
Millar, F. (1986) 'A new approach to the Roman jurists', *Journal of Roman Studies* 76: 272–80.
  (1999) 'The Greek east and Roman law: The dossier of M. Cn. Licinius Rufus', *Journal of Roman Studies* 89: 90–108.
Millis, W. (1961) *Military History*, Washington, DC.
Milner, N. P. (1996) *Vegetius: Epitome of Military Science*, Liverpool.
Mitchell, M. W. (2006) 'In the footsteps of Paul: Scriptural and apostolic authority in Ignatius of Antioch', *Journal of Early Christian Studies* 14: 27–45.
Moatti, C. (1997) *La raison de Rome: Naissance de l'esprit critique à la fin de la République (IIe–Ier siècle avant Jésus-Christ)*, Paris.
Moles, J. (1983) 'The date and purpose of the *Fourth Kingship Oration* of Dio Chrysostom', *Classical Antiquity* 2: 251–78.
  (1990) 'The *Kingship Orations* of Dio Chrysostom', *Papers of the Leeds Latin Seminar* 6: 297–375.
Momigliano, A. (1984) 'The theological efforts of the Roman upper classes in the first century B.C.', *Classical Philology* 79: 199–211 (reprinted in Ando 2003: 147–63).
Mora, F. (1995) *Il pensiero storico-religioso antico: Autori greci e Roma*, Vol. 1: *Dionigi d'Alicarnasso*, Rome.
Moraux, P. (1981) 'Galien comme philosophe: la philosophie de la nature', in Nutton 1981: 87–116.
Morgan, T. (2013) 'Encyclopaedias of virtue? Collections of sayings and stories about wise men in Greek', in König and Woolf 2013: 108–28.
Morson, G. S. (1989) 'Parody, history, and metaparody', in G. S. Morson and C. Emerson (eds), *Rethinking Bakhtin: Extensions and Challenges*, Evanston, IL: 63–86.
Morysine, R. (1539) *The Strategemes, Sleyghtes, and Policies of Warre, gathered together, by S. Julius Frontinus, and translated into Englyshe, by Rycharde Morysine*, London.
Mosca, B. (1937) 'Satira filosofica e politica nelle Menippee di Varrone', *Annali della Scuola Normale Superiore di Pisa* 6: 41–77.
Most, G. (1999) 'The poetics of early Greek philosophy', in A. Long (ed.), *The Cambridge Companion to Early Greek Philosophy*, Cambridge: 332–62.
  (2011) 'War and justice in Hesiod', in Formisano and Böhme 2011: 13–21.
Mras, K. (1914) 'Varros Menippeische Satiren und die Philosophie', *Neues Jahrbuch für Philologie* 33: 390–420.
Mueller, I. (1991) 'On the notion of a mathematical starting point in Plato, Aristotle, and Euclid', in A. Bowen (ed.), *Science and Philosophy in Classical Greece*, London: 59–97.
Müller, C. (1855–61) *Geographi Graeci minores* (2 volumes), Paris.
Müller, C. F. W. (1879) *M. Tulli Ciceronis scripta quae manserunt omnia*, Part 4, Vol. 3, Leipzig.

Murray, P. (2004) 'The Muses and their arts', in P. Murray and P. Wilson (eds), *Music and the Muses: The Culture of Mousike in the Classical Athenian City*, Oxford: 365-89.

Musial, D. (2001) 'Sodalicium Nigidiani: les Pythagoriciens à Rome à la fin de la République', *Revue de l'histoire des religions* 218: 339-67.

Narducci, E. (1974) 'Sconvolgimenti naturali e profezia delle guerre civili: Phars. I 522-695', *Maia* 26: 97-110.

Nederman, C. (ed.) (1990) *John of Salisbury: Policraticus: On the Frivolities of Courtiers and the Footprints of Philosophers*, Cambridge.

Nelsestuen, G. (2011) 'Polishing Scrofa's agronomical *eloquentia*: Representation and critique in Varro's *De re rustica*', *Phoenix* 65: 315-51.

Nesselrath, H.-G. (1985) *Lukians Parasitendialog: Untersuchungen und Kommentar*, Berlin.

Netz, R. (1997) 'Classical mathematics in the classical Mediterranean', *Mediterranean Historical Review* 12: 1-24.

  (1999) *The Shaping of Deduction in Greek Mathematics: A Study in Cognitive History*, Cambridge.

  (2000) 'Why did Greek mathematicians publish their analyses?', in P. Suppes, J. M. Moravcsik and H. Mendell (eds), *Ancient and Medieval Traditions in the Exact Sciences: Essays in Memory of Wilbur Knorr*, Stanford, CA: 139-57.

  (2001) 'On the Aristotelian paragraph', *Proceedings of the Cambridge Philological Society* 47: 211-32.

  (2002) 'Greek mathematicians: A group picture', in Tuplin and Rihll 2002: 196-216.

  (2004) *The Transformation of Mathematics in the Early Mediterranean World: From Problems to Equations*, Cambridge.

  (2009) *Ludic Proof: Greek Mathematics and the Alexandrian Aesthetic*, Cambridge.

Netz, R., Noel, W., Tchernetska, N. and Wilson, N. (eds) (2011) *The Archimedes Palimpsest* (2 volumes), Cambridge.

Neudecker, R. (2013) 'Archives, books and sacred space in Rome', in König, Oikonomopoulou and Woolf (2013): 312-31.

Neugebauer, O. (1975) *A History of Ancient Mathematical Astronomy* (3 volumes), Berlin.

Neugebauer, O. and Van Hoesen, H. B. (1959) *Greek Horoscopes*, Philadelphia, PA.

Newlands, C. E. (1995) *Playing with Time: Ovid and the Fasti*, Ithaca, NY.

Nickel, D. (1989) *Untersuchungen zur Embryologie Galens*, Berlin.

  (2001) *Galen: Über die Ausformung der Keimlinge* (CMG 5.3.3), Berlin.

Noè, E. (1977) 'L'agronomo Cneo Tremellio Scrofa', *Numismatica e Antichità Classiche* 6: 119-33.

  (2001) 'La memoria dell'antico in Columella: continuità, distanza, conoscenza', *Athenaeum* 89: 319-43.

North, J. (1986) 'Review article: Religion and politics, from Republic to Principate', *Journal of Roman Studies* 76: 251-8.

(1990) 'Diviners and divination at Rome', in Beard and North 1990: 49-71.

Novara, A. (2005) *Auctor in bibliotheca: Essai sur les textes préfaciels de Vitruve et une philosophie latine du livre*, Louvain.

Nutton, V. (1979) *Galen, On Prognosis* (CMG 5.8.1), Berlin.

(ed.) (1981) *Galen: Problems and Prospects*, London.

(1999) *Galen: On My Own Opinions: Text, Translation and Commentary* (CMG 5.3.2), Berlin.

(2009) 'Galen's authorial voice: A preliminary enquiry', in Taub and Doody 2009: 53-62.

(2013) *Ancient Medicine* (2nd edn; 1st edn published in 2004), Abingdon.

O'Daly, G. (1999) *Augustine's City of God: A Reader's Guide*, Oxford.

Oldfather, W. (1928) 'Onasander, Strategikós', in Illinois Greek Club, *Aeneas Tacticus, Asclepiodotus, Onasander (Loeb Classical Library)*, Cambridge, MA.

(1933-67) *Diodorus of Sicily (Loeb Classical Library)*, Cambridge, MA.

Olivelle, P. (1996) *Upaniṣads*, Oxford.

Olson, D. and Sens, A. (1999) *Matro of Pitane and the Tradition of Epic Parody in the Fourth Century BCE*, Atlanta, GA.

Önnerfors, A. (ed.) (1995) *Vegetius. Epitoma rei militaris*, Stuttgart.

Papathomopoulos, M. (2003) *Oppianus Apameensis: Cynegetica. Eutecnius, Sophistes: Paraphrasis metro soluta*, Munich.

Paret, P. (1992) *Understanding War: Essays on Clausewitz and the History of Military Power*, Princeton, NJ.

Parpola, S. (1993) *Letters from Assyrian and Babylonian Scholars (State Archives of Assyria 10)*, Helsinki.

Parry, R. (2007) '*Episteme* and *techne*', *Stanford Encyclopedia of Philosophy*, http://plato.stanford.edu/entries/episteme-techne/ (last consulted 10 August 15) (revised version; first published online 2003).

Paschalis, M. (2000) 'Generic affiliations in Roman and Greek Cynegetica', in G. M. Sifakis (ed.), *Κτερίσματα: φιλολογικά μελετήματα αφιερωμένα στον Ιω. Καμπίτση (1938-1990)*, Herakleion: 201-32.

Peachin, M. (2002) 'Introduction', in Aubert and Sirks (2002): 1-14.

Pease, A. S. (1920-23) *M. Tulli Ciceronis De divinatione* (2 volumes), Urbana, IL.

Pedersen, O. (2011) *A Survey of the Almagest* (with annotation and new commentary by A. Jones), New York.

Peglau, M. (2003) 'Varro: ein Antiquar zwischen Tradition und Aufklärung', in A. Haltenhoff, A. Heil and F.-H. Mutschler (eds), *O Tempora, O Mores!: Römische Werte und römische Literatur in den letzten Jahrzehnten der Republik*, Munich: 137-64.

Pendergraft, M. (1990) 'On the nature of the constellations: Aratus, *Ph.* 367-85', *Eranos* 88: 99-106.

Pera, M. (1994) *The Discourses of Science*, Chicago, IL.

Pera, M. and Shea, W. (eds) (1991) *Persuading Science: The Art of Scientific Rhetoric*, Canton, MA.

Perfigli, M. (2004) *Indigitamenta: Divinità funzionali e funzionalità divina nella religione romana*, Pisa.

Petrocelli, C. (2008) *Onasandro, Il generale: Manuale per l'esercizio del comando*, Bari.

Phillips, A. A. and Willcock, M. M. (1999) *Xenophon and Arrian, On Hunting*. Warminster.

Piganiol, A. (1951) 'Sur le calendrier brontoscopique de Nigidius Figulus', in P. R. Coleman-Norton (ed.), *Studies in Roman Economic and Social History in Honor of Allan Chester Johnson*, Princeton, NJ: 79–87.

Pigeaud, A. and Pigeaud, J. (eds) (2000) *Les Textes médicaux latins comme literature: Actes du VIe Colloque International sur les Textes Médicaux Latins du 1er au 3 septembre 1998 à Nantes*, Nantes.

Pingree, D. (1994) 'The teaching of the Almagest in late antiquity', *Apeiron* 27: 75–98.

Porter, J. I. (2006a) 'Introduction: What is "classical" about classical antiquity?', in Porter 2006c: 1–65.

(2006b) 'Feeling classical: Classicism and ancient literary criticism', in Porter 2006c: 301–52.

(ed.) (2006c) *Classical Pasts: The Classical Traditions of Greece and Rome*, Princeton, NJ.

Potter, D. S. (2004) *The Roman Empire at Bay, AD 180–395*, London.

Powell, J. G. F. (1996) 'Response by J. G. F. Powell', in Sommerstein 1996: 59–64.

Pownall, F. S. (1998) 'Condemnation of the impious in Xenophon's "Hellenica"', *Harvard Theological Review* 91: 251–77.

(2004) *Lessons from the Past. The Moral Use of History in Fourth-Century Prose*, Ann Arbor, MI.

Prets, E. (2000) 'Theories of debate, proof and counter-proof in the early Indian dialectical tradition', in P. Balcerowitz and M. Mejor (eds), *On the Understanding of Other Cultures*, Warsaw: 369–81.

Puech, B. (2002) *Orateurs et sophistes grecs dans les inscriptions d'époque imperial*, Paris.

Purves, A. C. (2010) *Space and Time in Ancient Greek Narrative*, Cambridge.

Rabier, C. (2007a) *Fields of Expertise: A Comparative History of Expert Procedures in Paris and London, 1600 to Present*, Newcastle.

(2007b) 'Introduction: Expertise in historical perspectives', in Rabier (2007a): 1–34.

Rasmussen, S. W. (2000) 'Cicero's stand on prodigies: A non-existent dilemma?', in R. L. Wildfang and J. Isager (eds), *Divination and Portents in the Roman World*, Odense: 9–24.

(2003) *Public Portents in Republican Rome*, Rome.

Rasmussen, T. (2001) 'Roman agriculture and rural settlement in central Italy', in P. Herz and G. Waldherr (eds), *Landwirtschaft im Imperium Romanum*, St Katharinen: 221–34.

Rathbone, D. (2008) 'Poor peasants and silent sherds', in L. de Ligt and S. J. Northwood (eds), *People, Land and Politics: Demographic Developments and the Transformation of Roman Italy 300 BC–AD 14*, Leiden: 305–32.

Rawson, E. (1978) 'Caesar, Etruria and the *Disciplina Etrusca*', *Journal of Roman Studies* 68: 132–52 (= Rawson, E. (1991) *Roman Culture and Society: Collected Papers*, Oxford: 289–323).

(1985) *Intellectual Life in the Late Roman Republic*, London.

Reay, B. (2005) 'Agriculture, writing, and Cato's aristocratic self-fashioning', *Classical Antiquity* 24: 331–61.

Rebuffat, E. (2001) ΠΟΙΗΤΗΣ ΕΠΕΩΝ: *Tecniche di composizione poetica negli Halieutica di Oppiano*, Florence.

Reeve, M. D. (2004) *Vegetius, Epitoma Rei Militaris*, Oxford.

Reinach, T. (1906) 'Inscriptions d'Aphrodisias', *Revue des Études Grecques* 19: 79–150, 205–98.

Reinhardt, T. (2003) *Cicero's Topica: Edited with an Introduction, Translation and Commentary*, Oxford.

Reis, D. M. (2005) 'Following in Paul's footsteps: Mimesis and power in Ignatius of Antioch', in Gregory and Tuckett 2005: 287–305.

Relihan, J. C. (1993) *Ancient Menippean Satire*, Baltimore, MD.

Rengakos, A. (2011a), 'Herodot', in Zimmermann 2011: 338–80.

(2011b) 'Thukydides', in Zimmermann 2011: 381–417.

Reynolds, J. M. (1996) 'Honouring benefactors at Aphrodisias: A new inscription', in C. Roueché and R. R. R. Smith (eds), *Aphrodisias Papers 3: The Setting and Quarries, Mythological and Other Sculptural Decoration, Architectural Development, Portico of Tiberius, and Tetrapylon* (Journal of Roman Archaeology Suppl. 20), Ann Arbor, MI: 121–6.

Richardot, P. (1997) *Végèce et la culture militaire au Moyen-Âge*, Paris.

Ridings, D. (1995) *The Attic Moses: The Dependency Theme in Some Early Christian Writers*, Göteborg.

Riess, W. (2008) 'Apuleius Socrates Africanus? Apuleius' defensive play', in W. Riess (ed.), *Paideia at Play: Learning and Wit in Apuleius*, Groningen: 51–73.

Riggsby, A. (2007) 'Guides to the wor(l)d', in König and Whitmarsh 2007b: 88–107.

Rihll, T. E. (1999) *Greek Science*, Oxford.

Rimell, V. (2007) 'Petronius' lessons in learning – the hard way', in König and Whitmarsh 2007b: 108–32.

Rives, J. B. (2007) *Religion in the Roman Empire*, Malden, MA.

(2010) 'Graeco-Roman religion in the Roman Empire: Old assumptions and new approaches', *Currents in Biblical Research* 8: 240–99.

Roby, C. (2016) *Technical Ekphrasis in Ancient Science: The Written Machine between Alexandria and Rome*, Cambridge.

Rochberg, F. (2000) 'Scribes and scholars: The *tupšar Enūma Anu Enlil*', in J. K. Marzahn and H. Neumann (eds), *Assyriologica et Semitica (Alter Orient und Altes Testament, 252)*, Münster: 359–76.

(2004) *The Heavenly Writing: Divination, Horoscopy, and Astronomy in Mesopotamian Culture*, Cambridge.

Roche, P. (2009) *Lucan: De Bello Ciuili Book I*, Oxford.

Rodgers, R. H. (2004) *Frontinus, De Aquaeductu Urbis Romae*, Cambridge.

Romano, E. (1987) *La capanna e il tempio: Vitruvio o dell' architectura*, Palermo.

Roochnik, D. (1996) *Of Art and Wisdom: Plato's Understanding of Techne*, University Park, PA.

Rosafio, P. (1994) 'Slaves and *coloni* in the villa system', in J. Carlsen, P. Ørsted and J. E. Skydsgaard (eds), *Landuse in the Roman Empire*, Rome: 145–58.

Rose, L. E. (1968) *Aristotle's Syllogistic*, Springfield, MO.

Rosen, R. (1990) 'Poetry and sailing in Hesiod's *Works and Days*', *Classical Antiquity* 9: 99–113.

(2008) 'Socratism in Galen's psychological works', in C. Brockmann, W. Brunschön and O. Overwien (eds), *Antike Medizin im Schnittpunkt von Geistes- und Naturwissenschaften (Beiträge zur Altertumskunde)*, Berlin: 155–71.

(2010) 'Galen, satire, and the compulsion to instruct', in H. M. F. Horstmanshoff (ed.), *Hippocrates and Medical Education*, Leiden: 325–42.

Rosenberger, V. (2003) 'Die verschwundene Leiche: Überlegungen zur Auffindung des Sarkophags Numas im Jahre 181 v. Chr.', in B. Kranemann and J. Rüpke (eds), *Das Gedächtnis des Gedächtnisses: Zur Präsenz von Ritualen in beschreibenden und reflektierenden Texten*, Marburg: 39–59.

(2007) 'Republican *nobiles*: Controlling the *res publica*', in J. Rüpke (ed.), *A Companion to Roman Religion*, Malden, MA: 292–303.

Rosenblitt, J. A. (2011) 'The "devotio" of Sallust's Cotta', *American Journal of Philology* 132: 397–427.

Rosillo López, C. (2009) 'La guerra civil de las letras: religión, panfletarios y lucha política (49–44 a.c.)', *Klio* 91: 104–14.

Ross, D. O. (1975) *Backgrounds to Augustan Poetry: Gallus, Elegy and Rome*, Cambridge.

Rossiter, J. J. (2007) 'Wine-making after Pliny: Viticulture and farming technology in late antique Italy', in L. Lavan, E. Zanini and A. C. Sarantis (eds), *Technology in Transition: A.D. 300–650*, Leiden: 93–118.

Roth, U. (2004) 'Inscribed meaning: The *vilica* and the villa economy', *Papers of the British School at Rome* 72: 101–24.

Rüpke, J. (2012) *Religion in Republican Rome: Rationalization and Ritual Change*, Philadelphia, PA.

Russell, D. (1973) *Plutarch*, London.

Sabin, P., Van Wees, H. and Whitby, M. (eds) (2007) *Cambridge History of Greek and Roman Warfare*, Vol. 2, Cambridge.

Santangelo, F. (2013) *Divination, Prediction and the End of the Roman Republic*, Cambridge.

Santini, C. (1992) 'Il prologo degli *Strategemata* di Sesto Giulio Frontino', in Santini, Scivoletto and Zurli 1990–98, Vol. 2: 983–90.

Santini, C., Scivoletto, M. and Zurli, L. (1990–98) *Prefazioni, prologhi, proemi di opere tecnico-scientifiche latine* (3 volumes), Rome.

Sanudo Torsello, M. (2011) *The Book of the Secrets of the Faithful of the Cross (Liber Secretorum Fidelium Crucis)* (tr. Peter Lock), Farnham.

Sautel, J.-H. (2010) 'Un récit de théophanie chez Denys d'Halicarnasse: L'apparition des Dioscures à la bataille du Lac Régille (*Antiquités Romaines*, VI, 13). Etude rhétorique', *Revue des études antiques* 112: 375–90.

Savage, S. (1945) 'Remotum a notitia vulgari', *Transactions of the American Philological Association* 76: 157–65.

Scheid, J. (1987–89) 'La parole des dieux: l'originalité du dialogue des romains avec leur dieux', *Opus* 6–8: 125–36.

    (1996) 'Indigetes or –ites, indigitamenta', in S. Hornblower and A. Spawforth (eds), *The Oxford Classical Dictionary* (3rd edn), Oxford: 755.

Schenkl, H. (1916) *Epicteti dissertationes ab Arriano digestae*, Leipzig.

Schiesaro, A. (1997) 'The boundaries of knowledge in Virgil's *Georgics*', in Habinek and Schiesaro (1997): 63–89.

Schiesaro, A., Mitsis, P. and Clay, J. S. (eds) (1993) *Mega nepios: Il destinatario nell'epos didascalico* (*Materiali e discussioni* 31), Pisa.

Schindel, U. (2000) 'Apuleius: Africanus Socrates? Beobachtungen zu den Verteidigungsreden des Apuleius und des platonischen Sokrates', *Hermes* 128: 443–56.

Schlapbach, K. (2010) 'The *logoi* of philosophers in Lucian of Samosata', *Classical Antiquity* 29: 250–77.

Schmid, A. (2005) *Augustus und die Macht der Sterne: Antike Astrologie und die Etablierung der Monarchie in Rom*, Cologne.

Schmitt, W. (1969) 'Kommentar zum ersten Buch von Pseudo-Oppians *Kynegetica*' (PhD Diss., Westfälische Wilhelms-Universität zu Münster).

Schmitz, T. (1997) *Bildung und Macht: Zur sozialen und politischen Funktion der zweiten Sophistik in der griechischen Welt der Kaiserzeit*, Munich.

    (2009) 'Narrator and audience in Philostratus' *Lives of the Sophists*', in E. Bowie and J. Elsner (eds), *Philostratus*, Cambridge: 49–68.

    (2010) 'A sophists's drama: Lucian and classical tragedy', in I. Gildenhard and M. Revermann (eds), *Beyond the Fifth Century: Interactions with Greek Tragedy from the Fourth Century BCE to the Middle Ages*, Berlin: 289–311.

Schmitz, T. A. and Wiater, N. (eds) (2011) *The Struggle for Identity: Greeks and Their Past in the First Century BCE*, Stuttgart.

Schofield, M. (1986) 'Cicero for and against divination', *Journal of Roman Studies* 76: 47–65.

Schultz, C. E. (2009) 'Argument and anecdote in Cicero's *De divinatione*', in P. B. Harvey, Jr and C. Conybeare (eds), *Maxima debetur magistro reverentia: Essays on Rome and the Roman Tradition in Honor of Russell T. Scott*, Como: 193–206.

    (2014) *A Commentary on Cicero, De divinatione I*, Ann Arbor, MI.

Schultz, F. (1946) *History of Roman Legal Science*, Oxford.

Schultze, C. (1986) 'Dionysius of Halicarnassus and his audience', in I. S. Moxon, J. D. Smart and A. J. Woodman (eds), *Past Perspectives: Studies in Greek and Roman Historical Writing*, Cambridge: 121–41.

(1995) 'Dionysius of Halicarnassus and Roman chronology', *Proceedings of the Cambridge Philological Society* 41: 192–214.

(2000) 'Authority, originality and competence in the *Roman Archaeology* of Dionysius of Halicarnassus', *Histos* 4: 6–49.

(2012) 'Negotiating the plupast: Dionysius of Halicarnassus and Roman self-definition', in J. Grethlein and C. B. Krebs (eds), *Time and Narrative in Ancient Historiography: The 'Plupast' from Herodotus to Appian*, Cambridge: 113–38.

Schwager, T. (2012) *Militärtheorie im Späthumanismus: Kulturtransfer taktischer und strategischer Theorien in den Niederlanden und Frankreich (1590–1660)*, Berlin.

Schwyzer, E. (1968) *Griechische Grammatik* (4th edn), Munich.

Scodel, R. (2008) *Epic Facework: Self-Presentation and Social Interaction in Homer*, Swansea.

Sedley, D. (1997) 'Plato's *auctoritas* and the rebirth of the commentary tradition', in J. Barnes and M. Griffin (eds), *Philosophia Togata II*, Oxford: 110–29.

(2007) *Creationism and its Critics in Antiquity*, Berkeley, CA.

(2013) 'Cicero and the *Timaeus*', in M. Schofield (ed.), *Aristotle, Plato and Pythagoreanism in the First Century BC: New Directions for Philosophy*, Cambridge: 187–205.

Semanoff, M. (2006) 'Undermining authority: Pedagogy in Aratus' *Phaenomena*', in Harder, Regtuit and Wakker 2006: 303–18.

Serbat, G. (2003) *Celse, De la médecine, Livres I et II*, Paris.

Setaioli, A. (2000) *Facundus Seneca: Aspetti della lingua e dell'ideologia senecana*, Bologna.

Shackleton Bailey, D. R. (1977) *Cicero: Epistulae ad Familiares* (2 volumes), Cambridge.

Shapin, S. (1994) *A Social History of Truth: Civility and Science in Seventeenth-Century England*, Chicago, IL.

Sharples, R. W. (ed.) (2005) *Philosophy and the Sciences in Antiquity*, Aldershot.

Sickinger, J. P. (2009) 'Nothing to do with democracy: "Formulae of disclosure" and the Athenian epigraphic habit', in L. G. Mitchell and L. Rubinstein (eds), *Greek History and Epigraphy: Essays in Honour of P. J. Rhodes*, Swansea: 87–102.

Sidebottom, H. (2009) 'Philostratus and the symbolic roles of the sophist and the philosopher', in E. Bowie and J. Elsner (eds), *Philostratus*, Cambridge: 69–99.

Sigsbef, D. L. (1976) 'The *paradoxa Stoicorum* in Varro's *Menippeans*', *Classical Philology* 71: 244–8.

Singer, P. N. (1997) *Galen: Selected Works*, Oxford.

(2013) *Galen: Psychological Writings*, Cambridge.

Skutsch, O. (1985) *The Annals of Q. Ennius, Edited with an Introduction and Commentary*, Oxford.

Skydsgaard, J. E. (1968) *Varro the Scholar: Studies in the First Book of Varro's De Re Rustica*, Munksgaard.
Sluiter, I. (2013) 'The violent scholiast: Power issues in ancient commentaries', in Asper 2013a: 191–214.
Smith, C. (1998) 'Onasander on how to be a general', in M. Austin, J. Harries and C. Smith (eds), *Modus Operandi: Essays in Honour of Geoffrey Rickman*, London: 151–66.
Smith, C. B. (2011) 'Ministry, martyrdom and other mysteries: Pauline influence on Ignatius of Antioch', in M. F. Bird and J. R. Dodson (eds), *Paul and the Second Century*, London: 37–56.
Smith, R. (1978) 'The mathematical origins of Aristotle's syllogistic', *Archive for History of Exact Sciences* 19: 201–9.
  (1983) 'What is Aristotelian ecthesis?', *History and Philosophy of Logic* 3: 113–27.
Smith, W. D. (1979) *The Hippocratic Tradition*, Ithaca, NY.
Sommerstein, A. H. (ed.) (1996) *Religion and Superstition in Latin Literature*, Bari.
Sorabji, R. (ed.) (1990) *Aristotle Transformed: The Ancient Commentators and Their Influence*, London.
  (2000) *Emotion and Peace of Mind: From Stoic Agitation to Christian Temptation*, Oxford.
Soubiran, J. (1972) *Cicéron: Aratea, Fragments poétiques*, Paris.
Spentzou, E. and Fowler, D. (eds) (2002) *Cultivating the Muse: Struggles for Power and Inspiration in Classical Literature*, Oxford.
Spurr, M. S. (1986) 'Agriculture and the *Georgics*', *Greece and Rome* 33: 164–87.
Stadter, P. A. (1976) 'Xenophon in Arrian's *Cynegeticus*', *Greek, Roman and Byzantine Studies* 17: 157–67.
  (1978), 'The *Ars tactica* of Arrian: Tradition and originality', *Classical Philology* 73: 117–28.
  (1980) *Arrian of Nicomedia*, Chapel Hill, NC.
Stanton, G. (1973) 'Sophists and philosophers: Problems of classification', *American Journal of Philology* 94: 351–64.
Star, C. (2012) *The Empire of the Self: Self-Command and Political Speech in Seneca and Petronius*, Baltimore, MD.
  (2009) 'Divisions of speech', in Gunderson 2009: 77–91.
Steenbakkers, P. (1994) *Spinoza's Ethica from Manuscript to Print*, Utrecht.
Steiner, D. (2005) 'Nautical matters: Hesiod's *Nautilia* and Ibycus fragment 282 PMG', *Classical Philology* 100: 347–55.
Stephens, S. A. (2003) *Seeing Double: Intercultural Poetics in Ptolemaic Alexandria*, Berkeley, CA.
Stoll, O. (2005) 'Wenig imponierend? Einige Gedanken zum Stand der römischen Technik und Agrartechnik', *Relationes Budvicenses* 4: 5–17.
Stump, E. (ed.) (1988) *Boethius In Ciceronis Topica: An Annotated Translation of a Medieval Dialectical Text*, Ithaca.

Swain, S. (1996) *Hellenism and Empire: Language, Classicism, and Power in the Greek World, AD 50–250*, Oxford.

Swain, S. (ed.) (2000) *Dio Chrysostom: Politics, Letters and Philosophy*, Oxford.

Swigger, R. (1975) 'Fictional encyclopedism and the cognitive value of literature', *Comparative Literature Studies* 12: 351–66.

Swoboda, A. (1889) *P. Nigidii Figuli reliquiae*, Vienna.

Tannery, P. (1929) 'Aristóte, *Météorologie*, livre III, ch. V', in J. L. Heiberg and H. G. Zeuthen (eds), *Mémoires Scientifiques*, Vol. 9, Paris: 51–61.

Tarrant, H. (1985) 'Alcinous, Albinus, Nigrinus', *Antichthon* 19: 87–95.

(2003) 'Athletics, competition and the intellectual', in D. Phillips and D. Pritchard (eds), *Sport and Festival in the Ancient Greek World*, Swansea: 351–63.

Tarrant, R. (1995) 'Ovid and the failure of rhetoric', in D. Innes, H. Hine and C. Pelling (eds), *Ethics and Rhetoric: Classical Essays for Donald Russell on his Seventy-Fifth Birthday*, Oxford: 63–74.

Tarver, T. (1996) 'Varro, Caesar, and the Roman calendar: a study in Late Republican religion', in Sommerstein 1996: 39–57.

Tatum, W. J. (1999) 'Roman religion: Fragments and further questions', in S. N. Byrne and E. P. Cueva (eds), *Veritatis amicitiaeque causa: Essays in Honor of Anna Lydia Motto and John R. Clark*, Wauconda, IL: 273–91.

Taub, L. C. (1993) *Ptolemy's Universe: The Natural Philosophical and Ethical Foundations of Ptolemy's Astronomy*, Chicago, IL.

(2003) *Ancient Meteorology*, London.

Taub, L. C. and Doody, A. (eds) (2009) *Authorial Voices in Greco-Roman Technical Writing*, Trier.

Tellegen-Couperus, O. and Tellegen, J. W. (2013) '*Artes urbanae:* Roman law and rhetoric', in Du Plessis (2013): 31–50.

Thom, P. (1993) 'Apodeictic ecthesis', *Notre Dame Journal of Formal Logic* 34: 193–208.

Thomas, A. (1930) *Jean Gerson et l'éducation des Dauphins de France: Étude critique*, Paris.

Thomas, R. (2000) *Herodotus in Context: Ethnography, Science, and the Art of Persuasion*, Cambridge.

Thomas, R. F. (1987) 'Prose into poetry: Tradition and meaning in Virgil's *Georgics*', *Harvard Studies in Classical Philology* 91: 229–60.

(1988) *Virgil, Georgics*, Vol. 1: *Books I–II*, Cambridge.

Tieleman, T. L. (1996) *Galen and Chrysippus on the Soul: Argument and Refutation in the De placitis Books II–III*, Leiden.

Tihon, A. (2010) 'An unpublished astronomical papyrus contemporary with Ptolemy', in Jones 2010b: 1–10.

Timpanaro, S. (1988) *Marco Tullio Cicerone: Della divinazione*, Milan.

(1994) 'Alcuni fraintendimenti del *De divinatione*', in S. Timpanaro, *Nuovi contributi di filologia e storia della lingua latina*, Bologna: 241–64.

Tolsa, C. (2013) 'Claudius Ptolemy and Self-Promotion: A Study on Ptolemy's Intellectual Milieu in Roman Alexandria' (PhD diss., University of Barcelona).

Toomer, G. J. (1974a) 'Meton', in Gillespie 1970–90: 9.337–40.

(1974b) 'Hipparchus on the distances of the sun and moon', *Archive for History of Exact Sciences* 14: 126–42.

(1975) 'Ptolemy', in Gillespie 1970–90: 9.186–206.

(1978) 'Hipparchus', in Gillespie 1970–90: 15.207–224.

(1984) *Ptolemy's Almagest*, London.

Totelin, L. (2009) *Hippocratic Recipes: Oral and Written Transmission of Pharmacological Knowledge in Fifth- and Fourth-Century Greece*, Leiden.

(2012) 'And to end on a poetic note: Galen's authorial strategies in the pharmacological books', *Studies in History and Philosophy of Science* 43: 307–15.

Traina, A. (1995) *Lo stile 'drammatico' del filosofo Seneca* (4th edn), Bologna.

Trapp, M. (1997) *Maximus of Tyre: The Philosophical Orations*, Oxford.

(2000) 'Plato in Dio', in Swain (2000): 213–39.

(2003) *Greek and Latin Letters: An Anthology with Translation*, Cambridge.

(2007) *Philosophy in the Roman Empire: Ethics, Politics and Society*, Aldershot.

(2012) 'Dio Chrysostom and the value of prestige', in G. Roskam, M. De Pourcq and L. van der Stockt (eds), *The Lash of Ambition: Plutarch, Imperial Greek Literature and the Dynamics of Philotimia*, Louvain: 119–41.

(2014) 'Philosophia between Greek and Latin culture: Naturalized immigrant or eternal stranger?', in F. Mestre and P. Gómez (eds), *Three Centuries of Greek Culture under the Roman Empire: Homo Romanus Graeca oratione*, Barcelona: 2–48.

(forthcoming) 'Visibly different? Looking at philosophoi in the Roman Imperial period', in P. Vesperini (ed.), *Philosophari: Usages romains des savoirs grecs sous la République et sous l'Empire. Actes des Colloques de l'École Française de Rome*, Paris.

Tuplin, C. J. and Rihll, T. E. (eds) (2002) *Science and Mathematics in Ancient Greek Culture*, Oxford.

Turfa, J. M. (2012) *Divining the Etruscan World: The Brontoscopic Calendar and Religious Practice*, New York.

Turner, A. (2007) 'Frontinus and Domitian: *Laus principis* in the *Strategemata*', *Harvard Studies in Classical Philology* 103: 423–49.

Usener, H. (1896) *Götternamen: Versuch einer Lehre von der Religiösen Begriffsbildung*, Bonn.

Valvo, A. (1988) *La 'Profezia di Vegoia': Proprietà fondiaria e aruspicina in Etruria nel I secolo a.c.*, Rome.

Van der Eijk, P. (1997) 'Towards a grammar of ancient scientific discourse', in E. Bakker (ed.), *Grammar as Interpretation: Greek Literature in its Linguistic Contexts*, Leiden: 77–129.

(2005) 'Introduction', in *Medicine and Philosophy in Classical Antiquity: Doctors and Philosophers on Nature, Soul, Health and Disease*, Cambridge: 1–42.

(2013) 'Galen and the scientific treatise: A case study of mixtures', in Asper 2013a: 145–76.
Van der Waerden, B. L. (1960) 'Greek astronomical calendars and their relation to the Athenian civil calendar', *Journal of Hellenic Studies* 80: 168–80.
Van Hoof, L. (2010) *Plutarch's Practical Ethics: The Social Dynamics of Philosophy*, Oxford.
Van Leeuwen, J. (2012) 'The Tradition of the Aristotelian "Mechanics": Text and Diagrams' (PhD Diss., Humboldt-Universität zu Berlin).
Van Nijf, O. M. (1997) *The Civic World of Professional Associations in the Roman East*, Amsterdam.
Van Nuffelen, P. (2010) 'Varro's *Divine Antiquities:* Roman religion as an image of truth', *Classical Philology* 105: 162–88.
  (2014) 'Galen, divination, and the status of medicine,' *Classical Quarterly* 64: 337–52.
Versnel, H. S. (2011) *Coping with the Gods: Wayward Readings in Greek Theology*, Leiden.
Vessey, M. (2014) 'Fashions for Varro in late antiquity and Christian ways with books', in C. Harrison, C. Humfress and I. Sandwell (eds), *Being Christian in Late Antiquity: A Festschrift for Gillian Clark*, Oxford: 253–77.
Veyne, P. (1983) *Les Grecs ont-ils cru à leurs mythes? Essai sur l'imagination constituante*, Paris.
  (1990) *Bread and Circuses: Historical Sociology and Political Pluralism* (tr. B. Pearce; first published in French in 1976), London.
Vigourt, A. (2001) *Les Présages impériaux d'Auguste à Domitien*, Paris.
Vitrac, B. (2002) 'Note textuelle sur un (problème de) lieu géométrique dans les Météorologiques d'Aristote (III. 5, 375 b 16–376 b 22)', *Archive for History of Exact Sciences* 56: 239–83.
Vogt S. (2005) '"... er schrieb in Versen, und er tat recht daran": Lehrdichtung im Urteil Galens', in T. Fögen (ed.), *Antike Fachtexte: Ancient Technical Texts*, Berlin: 51–78.
Volk, K. (2002) *The Poetics of Latin Didactic: Lucretius, Vergil, Ovid, Manilius*, Oxford.
  (2009) *Manilius and his Intellectual Background*, Oxford.
  (2013) 'The genre of Cicero's *De consulatu suo*', in T. D. Papanghelis, S. J. Harrison and S. Frangoulidis (eds), *Generic Interfaces in Latin Literature: Encounters, Interactions and Transformations*, Berlin: 93–112.
  (2016) 'Roman Pythagoras', in Williams and Volk 2016: 33–49.
Von Staden, H. (1994) 'Author and authority: Celsus and the creation of a scientific self,' in M. E. Vázquez Buján (ed.), *Tradición e innovación de la medicina latina de la antigüedad y de la alta edad media*, Santiago de Compostela: 103–17.
  (1997) 'Galen and the "Second Sophistic"', in R. Sorabji (ed.), *Aristotle and After*, London: 33–54.

(1998) 'Gattung und Gedächtnis: Galen über Wahrheit und Lehrdichtung', in Kullmann, Althoff and Asper 1998: 65–94.

(2009) 'Staging the past, staging oneself: Galen on Hellenistic exegetical traditions', in Gill, Whitmarsh and Wilkins 2009: 132–56.

Vottero, D. (ed.) (1998) *Lucio Anneo Seneca: I frammenti*, Bologna.

Wachsmuth, C. (1860) 'Ueber die Unächtheit des vierten Buches der Frontinischen *Strategemata*', *Rheinisches Museum* 15: 574–83.

Walbank, F. W. (1962) 'Polemic in Polybius', *Journal of Roman Studies* 52: 1–12.

Wallace-Hadrill, A. (1997) '*Mutatio morum:* The idea of a cultural revolution', in Habinek and Schiesaro 1997: 3–22.

(2008) *Rome's Cultural Revolution*, Cambridge.

Wallraff, M., Scardino, C., Mecella, L. and Guignard, C. (2012) *Cesti: The Extant Fragments*, Berlin.

Walsh, P. G. (1998) *Cicero: On the Nature of the Gods* (translated with introduction and notes), Oxford.

Wardle, D. (2006) *Cicero: On Divination Book 1* (with translation and commentary), Oxford.

Warren, J. (2007) 'Diogenes Laertius, biographer of philosophy', in König and Whitmarsh 2007b: 133–49.

Watson, A. (1974) *Law Making in the Later Roman Republic*, Oxford.

Weinstock, S. (1951) 'Libri fulgurales', *Papers of the British School at Rome* 19: 122–53.

(1970) *Divus Julius*, Oxford.

Wenskus, O. (1998) 'Columellas Bauernkalender zwischen Mündlichkeit und Schriftlichkeit', in Kullmann, Althoff and Asper 1998: 253–62.

(2010) 'Prognosegenauigkeit und Imagepflege', in U. Tischer and A. Binternagel (eds), *Fremde Rede – Einige Rede*, Frankfurt am Main: 79–92.

Werner, S. (2009) 'Literacy studies in Classics: the last twenty years', in Johnson and Parker 2009: 333–84.

Wescher, C. (1867) *La Poliorcétique des grecs*, Paris.

West, M. L. (1978) *Hesiod, Works and Days: Edited with Prolegomena and Commentary*, Oxford.

Wheeler, E. L. (1988) *Stratagem and the Vocabulary of Military Trickery*, Leiden.

Whitaker, J. (1990) *Alcinoos: Enseignement des doctrines de Platon*, Paris.

Whitby, M. (2007) 'The *Cynegetica* attributed to Oppian', in S. Swain, S. Harrison and J. Elsner (eds), *Severan Culture*, Cambridge: 125–34.

White, K. D. (1970) *Roman Farming*, London.

(1973) 'Roman agricultural writers', *Aufstieg und Niedergang der römischen Welt* II, 1(4): 539–97.

Whitmarsh, T. (2004a) 'Aelius Aristides', in de Jong, Nünlist and Bowie 2004: 441–7.

(2004b) 'Philostratus', in de Jong, Nünlist and Bowie 2004: 423–39.

(2005) *The Second Sophistic*, Oxford.

Whitney, E. (1990) *Paradise Restored: The Mechanical Arts from Antiquity through the Thirteenth Century*, Philadelphia, PA.

Wiater, N. (2011a) 'Writing Roman history – shaping Greek identity: The ideology of historiography in Dionysius of Halicarnassus', in Schmitz and Wiater 2011: 61–91.

  (2011b) *The Ideology of Classicism: Language, History, and Identity in Dionysius of Halicarnassus*, Berlin.

  (forthcoming) 'The aesthetics of the past: Language, time, and historical consciousness in Dionysian criticism', in C. de Jonge, S. Oakley and C. Schulze (eds), *Dionysius of Halicarnassus: Criticism and History in Augustan Rome*, Cambridge.

Wilcox, A. (2012) *The Gift of Correspondence in Classical Rome: Friendship in Cicero's Ad Familiares and Seneca's Moral Epistles*, Madison, WI.

Wilkinson, C. L. (2013) *The Lyric of Ibycus*, Berlin.

Willard, C. C. (1995) 'Pilfering Vegetius? Christine de Pizan's *Fait d'Armes et de Chevalrie*', in L. Smith and J. Taylor (eds), *Women, the Book and the Worldly, (Selected Proceedings of the St Hilda's Conference, 1993)*, Vol. 2, Cambridge: 31–7.

Willi, A. (1998) 'Numa's dangerous books: The exegetic history of a Roman forgery', *Museum Helveticum* 55: 139–72.

Williams, G. (1991) 'Vocal variations and narrative complexity in Ovid's Vestalia: Fasti 6.249–468', *Ramus* 20: 183–204.

Williams, G. and Volk, K. (eds) (2016) *Roman Reflections: Studies in Latin Philosophy*, New York.

Wilson, M. (1987) 'Seneca's *Epistles* to Lucilius: A revaluation', *Ramus* 16: 102–21 (reprinted in Fitch 2008: 59–83).

  (1997) 'The subjugation of grief in Seneca's *Epistles*', in S. M. Braund and C. Gill (eds), *The Passions in Roman Thought and Literature*, Cambridge: 48–67.

  (2001) 'Seneca's *Epistles* reclassified', in S. J. Harrison (ed.), *Texts, Ideas, and the Classics: Scholarship, Theory, and Classical Literature*, Oxford: 164–87.

  (2013) 'Seneca the consoler? A new reading of his consolatory writings', in Baltussen (2013): 93–121.

Winter, B. (1997) *Paul and Philo among the Sophists*, Cambridge.

Wiseman, T. P. (2002) 'History, poetry, and annales', in D. S. Levene and D. P. Nelis (eds), *Clio and the Poets: Augustan Poetry and the Traditions of Ancient Historiography*, Leiden: 331–62.

  (2009) *Remembering the Roman People: Essays on Late-Republican Politics and Literature*, Oxford.

Wisse, J., Winterbottom, M. and Fantham, E. (2008) *M. Tullius Cicero, De oratore Libri III*, Vol. 5: Book III, 96–230, Heidelberg.

Wissowa, G. (1904) *Gesammelte Abhandlungen zur römischen Religions- und Stadtgeschichte*, Munich.

Wissowa, G. (1921) 'Die Varronischen *di certi* und *incerti*', *Hermes* 56: 113–30.
Wölfflin, E. (1875) 'Frontins Krieglisten', *Hermes* 9: 72–92.
Wood, N. (1967) 'Frontinus as a possible source for Machiavelli's method', *Journal of the History of Ideas* 28: 243–8.
Woolf, G. (2009) 'Literacy or literacies in Rome', in Johnson and Parker 2009: 46–68.
  (2011) *Tales of the Barbarians: Ethnography and Empire in the Roman West*, Malden, MA.
Wycislo, W. E. (2001) *Seneca's Epistolary Responsum: The De Ira as Parody*, Frankfurt am Main.
Zanker, P. (1996) *The Mask of Socrates: The Image of the Intellectual in Antiquity*, Berkeley, CA.
Ziman, J. (1984) *An Introduction to Science Studies: The Philosophical and Social Aspects of Science and Technology*, Cambridge.
  (2000) *Real Science: What It Is and What It Means*, Cambridge.
Zimmermann, B. (ed.) (2011) *Handbuch der griechischen Literatur der Antike*, Vol. I, Munich.
Zuiderhoek, A. (2009) *The Politics of Munificence in the Roman Empire: Citizens, Elites and Benefactors in Asia Minor*, Cambridge.

Index

Academics, 44, 252, 311, 312, 313, 315, 330, 335–6
Achilles, 18, 136–7, 213
actors, 33
addressees, 9, 13, 60–7, 120, 260–82, 409
Aelian, 130, 144, 386
Aelianus Tacticus, 155–6, 169
Aelius Aristides, 24
    Response to Plato, in Defence of Oratory, 50
Aelius Stilo, 326
Aeneas, 236–40, 314, 317
Aeneas Tacticus, 141, 166, 167
Aeschines, 54
Agathyllus, 237, 238
Agricola, 165, 179
agricultural writing, 14, 101, 130, 182–202, 325, 409
Agrippina, 79
Alcinous, 42
    Manual of Platonic Doctrine, 31, 32
Alexander of Aphrodisias, 55, 376
Alexander the Great, 38, 117, 173
Alexandria, 121, 216, 353
Alfenus Varus, 94
alphabetical order, 192
alternative explanation, 15–16, 236–47, 259, 326
Amafinius, 78
Ambaglio, D., 141, 142
Ammonius, 16, 18–19, 25
anatomy, 283–305
Anaxagoras, 37, 77, 114
Anglo, Sidney, 150
Antagoras of Rhodes, 220
anthropology, 125
Antioch, 21
Antiochus Chuzon, 90
Antiochus of Syracuse, 241
Antisthenes, 54
Antistius Labeo, 104
Antoninus Pius, 98, 100, 222
Aphrodisias, 351
Aphrodite, 240
Apollo, 21, 34, 218

Apollonius (mathematician), 360
Appian, 162
Apsines, 22
Apuleius, 42, 54
    Apology, 40–1, 48, 343
    Florida, 41
Aratus, *Phaenomena*, 207–8, 210, 220, 224, 338, 398, 407
Arcadia, 238
Arcesius, 114, 115
Archimedes, 50, 114, 360, 407, 423
    Sphere and Cylinder, 404
architecture, 3, 4, 107–28, 409
Archytas, 114, 192
Ares, 255, 257
Areus, 73
Argos, 243
Ariaethus, 237, 238
Aristarchus, 364, 399
    On the Sizes and Distances of the Sun and Moon, 364
Aristippus, 29, 55
Aristobulus, 51
Aristophanes of Byzantium, 120–2
Aristotle, 12, 17, 28, 31, 32, 50, 153, 192, 197, 198, 376, 379, 383, 404, 405, 406, 408
    Ethics, 379
    Mechanics, 378, 391
    Meteorology, 375, 391
    Nicomachean Ethics, 419
    Organon, 376
    Physics, 387
    Prior Analytics, 375, 376–9, 381, 382, 385, 389
Aristoxenus, 383, 385, 406
    Elements of Harmonics, 375, 379–82, 385, 387, 389, 395
arithmetic, 11, 110
Arrian, 130, 169, 386
    Cynegeticus, 225–6
    Discourses of Epictetus, 29, 34, 44
Artemidorus (critic of Ptolemy), 358
Artemidorus, *Oneirocritica*, 127, 210
Artemis, 209, 211, 226, 229

Index

ascent, imagery of, 34–5
Asclepiodotus, *Tactics*, 375, 382–6, 389, 390, 397, 406
Asinius Pollio, 78
Aspasius, 21
Asper, Markus, 366
astrology, 330, 331, 332, 333–4, 342, 344, 410
astronomy, 4, 11, 110, 187, 207–8, 209–10, 348–73, 397, 399, 409, 410–11, 415, 416–17, 422
Ateius Capito, 88, 104
atheism, 252–4
Athenodorus, 37
Athens, 15, 18, 29, 118, 162, 169, 170, 255, 369, 370, 371
Atherton, Catherine, 203
athletes, 33
athletic contests, 16
athletic training, 4, 11, 13, 17
Atilius, 102
atomism, 125
Attalids, 121
Attalus Philometor, 192
auctoritas, 70, 71, 87, 107, 109, 112, 140
audiences, 33, 41, 47, 63, 120, 132, 265, 267
Aufidius Bassus, 63, 64
Aufidius Namusa, 102
augury, 310, 315, 331, 335, 339, 340, 342, 346
Augustine, 153
 De civitate Dei, 306, 313, 314, 317, 319, 325, 327
Augustus, 71, 72, 73, 86, 104, 107, 109, 112, 120, 122, 125, 128, 143, 250, 334, 343, 344
Aulus Gellius, *Attic Nights*, 29–30, 34, 40, 41, 48, 51, 55, 84–5, 88–9, 96, 103, 105, 326–7
authority effect, 235, 243, 257, 258
authority in modern scientific writing, 1, 374
autopsy, 2–3, 17, 83, 203, 226–7, 228, 240, 260–82, 283, 304, 411

Babylon, 399
banishment of philosophers, 39, 54, 79, 316
Baozi, 416
Baragwanath, Emily, 244
Barker, Andrew, 379
Barthes, Roland, 127
Bartsch, Shadi, 145
Beard, Mary, 317, 331, 335
beards, 54, 56, 67
benefaction, 6, 348–73
Bernard, A., 361
Bibulus, 72

biography, 5, 81
biology, 302
Biton, 383
Boethius, 103
Bömer, Franz, 137
Book of Changes, 416, 418
books, 28, 53, 102, 134, 148, 149, 161, 165, 166, 184, 201, 270, 314, 316–17, 318, 411, 418
Bourdieu, Pierre, 374, 406, 407
Brading, Isle of Wight, 53
Britain, 179
Brunt, P. A., 195
Brutus, 102

calendars, 186, 187, 190, 345–6, 417
Callimachus, *Hymn to Zeus*, 220
Calpurnius Bibulus, M., 333
Caracalla, 228
Caraka Samhitā, 413, 414, 422
Cardauns, B., 317, 326
care, 349
Carey, Sorcha, 210
Carthage, 101, 156, 168, 169, 191, 324
cartography, 352
Cascellius, 104
Cassian School, 93
Cassius Dio, 83, 343, 344
Cassius Dionysius, 191, 192
Cassius Longinus, C., 88, 93, 94, 98, 99, 104
Cato the Elder, 99, 102, 130, 149, 154–5, 160, 161–2, 165, 173, 179, 184, 185, 190, 194, 198, 199, 200
 De agri cultura, 184, 186, 187, 188
Cato the Younger, 37, 72, 73, 78, 339
causation, 284, 289–92, 297, 305
Celsus, 130, 199
 De medicina, 101
Cephalon of Gergis, 237, 238
Cervidius Scaevola, 96
Chersiphron, 114, 115
Chinese science, 1, 409, 411, 414–23
chronology, 242
Chrysippus, 29, 47, 51, 53, 61, 75, 76
Chunyi Yi, 418–19
Cicero, 49, 55, 58, 64, 66, 72, 80, 87, 88, 91, 92, 104, 111, 131, 135, 139, 184, 196, 200, 321, 329–47
 Academica, 78, 308, 317
 De auguriis, 331
 De consulatu suo, 338, 341
 De divinatione, 8, 310, 311, 312, 315, 330, 333, 334–42, 346

De domo sua, 318, 327
De fato, 311
De finibus, 65, 78, 336
De legibus, 311, 312, 320, 325, 334, 335, 336–7
De natura deorum, 309, 311–12, 314–15, 318, 320, 322, 338
De officiis, 111
De oratore, 123, 131–3
De re publica, 72, 199
Hortensius, 78
In Pisonem, 78
Letters, 78
Lucullus, 165
Marius, 338, 341
On Pompey's Command, 164, 177
Pro Fonteio, 164
Pro Murena, 78
Pro Scauro, 318
Somnium Scipionis, 72
Timaeus, 330
Topica, 103
Tusculan Disputations, 65, 78
Cincinnatus, 184, 201
citation, 68, 69, 84, 85, 89, 94, 98, 100, 115, 116, 118, 127, 128, 194, 198, 203, 225, 230, 306, 327, 365, 414
citation, avoidance of, 114–19
Clark, Hilary, 210
Claudius, 65, 69
Clausewitz, Carl von, 139, 142
Cleanthes, 34, 61, 75
Cleomedes, 397
Clodius Pulcher, P., 333
coins, 56
collaboration, 8, 16–17, 348–73
Collatio of Roman and Mosaic Law, 103
Columella, 185, 191, 192, 195, 199, 200
  De re rustica, 66, 101, 102, 126, 187
comedy, 75, 322, 369
commentary, 52, 86, 98, 100, 141, 163, 355, 376, 387, 394, 404, 406
Commodus, 204
competitiveness, 1, 7, 10, 15–16, 18–22, 36, 194, 203, 246, 264, 358, 366–7, 372, 410, 413–14, 418, 421
  avoidance of, 7, 8, 19–22, 26
comprehensiveness, 107–8, 112–14, 118, 126, 174, 208, 209, 210–11, 222, 223, 224, 225, 226, 229, 230, 231
Confucius, 414, 415, 417, 418, 422
  Analects, 415
Constantine, 103

consuls, 19, 88, 97, 98, 102, 104, 141, 155, 162, 180, 250, 268, 333
contested authority, 12, 30, 43, 56–7
controversiae, 84, 96
Cordus, 71, 72, 74
Coriolanus, 173
Cornelius Maximus, Q., 94
Cornutus, 39
Crassus, 333
Cremutius Cordus, 65
Crete, 383
criticism of predecessors, 2, 193, 241, 242, 250, 283, 357–8, 359, 365, 366–7, 395–6, 397–9
Croton, 244
Ctesibius, 115
Cuomo, Serafina, 129, 166
Cynics, 32, 33, 41, 48, 54

d'Assigny, Marius, 153, 156–7
Dao De Jing, 415
de Bakker, Mathieu, 244
de George, R. T., 234
de Pizan, Christine, 154, 157
Dean-Jones, Lesley, 189
declamation, 22, 23, 41, 92
Delbrück, Hans, 140, 152
Democritus, 101, 114, 115, 125, 192
Descartes, 391
Diades, 114
diagrams, 53, 54, 377, 379, 381, 382, 384–6, 387, 389, 390, 391, 392, 395, 398, 400, 401, 402, 403, 408, 417
dialogues, 8, 15–19, 21, 58–82, 94, 131–3, 182–202, 330, 334, 416
Dicaearchus, 197, 397–8
dictionaries, 107
didactic verse, 13, 203–30
Diederich, Silke, 193, 201
Digest, 84, 85, 90–1, 103, 105, 106
digestion imagery, 123
Dinocrates, 117
Dio Chrysostom, 13, 39, 42, 54
  Kingship Orations, 38–9
Diodorus, 252, 254–5
Diogenes, 38, 53, 55
Diogenes of Oenoanda, 44
Dion, 37
Dionysius (sophist), 24
Dionysius (tyrant), 73
Dionysius of Halicarnassus, 4, 9
  Early Roman History, 231–59
Dionysius son of Calliphon, 224

Index

Dionysius, *Periegesis*, 204, 205, 206, 208, 211, 212–13, 214, 216–19, 221–2, 225, 226, 227, 228, 230
Diophanes of Bithynia, 191
disciplines, 409–10, 415
dissection, 267, 268, 286, 291, 293
divination, 6, 262, 310, 311, 329–47, 410
divine inspiration, 24
divine knowledge, 211–14, 219, 228
Dodds, E. R., 387
Dodona, 240
Domitian, 38, 39, 54, 79, 165, 180–1
drawing, 11
dream interpretation, 332, 339, 341, 342, 410
Drusus, 71, 73
Dyrrhachium, 341

early Christian writing, 9
ease, rhetoric of, 206
editor, self-presentation as, 108–9, 119–27
education, teaching, 13, 22–4, 25, 41, 42, 49, 63, 65, 70, 83, 91, 110, 111, 117–19, 132, 134, 233, 250, 251, 277, 278, 300–1, 325, 354, 388, 401, 414, 415, 416, 419
Egeria, 317
Egypt, 342
Einarson, B., 376
Eleusis, 22
elite status, 25, 39–41
Empedocles, 27, 101
emperors, 6, 38, 39, 64, 86–7, 96, 97, 178–81, 215, 417, 419, 423
empire, 5
encyclopaedic writing, 107, 127, 130, 210–11, 402, 404
engineering, military, 129, 163
Ennius, 102
ephebeia, 18
Ephesus, 19, 119
Ephorus, 225
epic, 139, 163, 168, 171, 203, 212, 215, 222, 223, 224, 229, 341
Epictetus, 29, 34, 44, 47–8, 51, 52, 53, 65, 66, 80
Epicureans, 32, 33, 44, 56, 63, 69, 78, 80, 81, 298, 336
Epicurus, 28, 35, 44–5, 68, 314
epistemic authority, 234–5, 244
epistolography, 58
epitomisation, 191
Erasistratus, 366
Eratosthenes, 242, 396, 397, 398–9, 406, 407
Eros, 220

ethics, 32, 67, 81, 420
ethnography, 81
Etruscans, 342, 345–6
etymology, 32, 89, 201, 307, 312, 326
Euclid, 387, 402, 417
  Elements, 388, 389, 401, 403
Euctemon, 364, 369–71
Eudoxus, 398, 407
Euphrates, 42
Eusebius, 55
Eutocius, 404
exempla, 8, 142, 149, 153–9, 166, 181
experimentation, 3, 266, 267
expertise
  avoidance of narrow versions of, 10–15, 22–5
  non-professional, 12, 41–3, 58–82

Fabianus, 42
Fantuzzi, Marco, 407
farming (*see also* agricultural writing), 17
Favorinus, 19–20, 42
Feeney, Denis, 327
festivals, 16, 18, 19
first-person usage, 2, 7, 16, 266, 280, 368
Flavius Boethus, 268, 354
Flavius, C., 98
Ford, Andrew, 211
friendship, 60–7, 68, 265
Frontinus, 130
  De aquis, 178, 180
  De re militari, 14, 157, 167, 179
  Strategemata, 9, 14, 153–81

Gaius, 90
Galen, 2–3, 4, 7, 9, 26, 50–1, 260–82, 354, 366, 404, 405, 406, 407, 416, 420
  Antecedent Causes, 275
  De Placitis Hippocratis et Platonis, 287
  De simplicium medicamentorum, 260
  Method of Medicine, 260, 264, 272–82
  On Antecedent Causes, 260–9, 273
  On my Opinions, 284
  On My Own Books, 284
  On Not Grieving, 264, 269–72, 275, 282
  On the Construction of Fetuses, 10, 283–305
  On the Usefulness of the Parts, 400–2
  Prognosis, 260–9, 273, 275, 282
  Protrepticus, 11
  Subfiguratio empirica, 260
  That the Best Doctor is also a Philosopher, 12
  The Affections and Errors of the Soul, 46
  The Use of Parts, 375

Galileo, 374, 391, 408
Gaul, 324
Geller, Markham J., 412
generalship, 3, 4, 36, 153–81
  as a metaphor for writing, 143, 145, 158, 159
genre, 81, 140, 151, 183, 200, 254, 308, 342, 367, 375, 390, 395, 405, 406, 407–8
geography, 17, 81, 206, 208, 212, 214, 216–19, 221, 224, 225, 238, 241, 366, 394–400, 407
geometry, 4, 11, 19, 28, 37, 53, 110, 300, 353, 365, 374–408
Gerson, Jean, 153–4
Gill, Christopher, 293
Gilliver, Catherine, 151
Giza, 351
Glossarium militare, 386
gods, 36, 252–7, 259, 296, 297, 298, 306–28
Gordian, 21
grammar, grammarians, 11, 17–18, 19, 32, 89, 96, 270, 329, 331, 385
Granius Flaccus, *De indigitamentis*, 310
Greek identity, 249
Green, Carin, 199
Griffin, Miriam, 75
gymnasia, 56

Hadrian, 85, 90, 96, 98, 204, 222
Hadrian of Tyre, 25
Han Fei, 418
Hannibal, 132–3, 168
Hardie, Philip, 137
Harris-McCoy, Daniel, 210
Hawking, Stephen, 374, 375, 400, 406, 407
Hecaton, 75, 76
Hegesippus, 237, 238
Hellanicus, 236, 238, 239
Heracles, 35
Heraclitus, 27
Hermes, 218
Hermogenes, 114, 115, 116
Hero of Alexandria, 354, 420
  Dioptra, 403
Herodes Atticus, 21
Herodotus, 225, 232, 243–6, 259
Hesiod, 18, 20, 26, 35, 45, 101, 102, 192, 210, 220
  Works and Days, 216–17
Hestia, 252–3
Heurgon, Jacques, 195
Hiero, 192
Himera, 396

Hipparchus, 50, 356, 357, 358–67, 396, 398–9
Hippocratic writings, 2, 46, 50–1, 101, 283–8, 292, 366, 367, 380, 413
  On Ancient Medicine, 421
  Quadrature of Lunules, 376
historiography, 3, 5, 11, 110, 129, 158–9, 163, 168, 170, 173, 231–59, 348, 395, 409
  relationship between Greek and Roman, 232, 236
history of science, 1, 3
Homer, 17, 28, 45, 51, 102, 164, 341, 366
  Catalogue of Ships (*Iliad* Book 2), 108, 211, 212, 223, 224
  Iliad, 20, 101, 237, 241
  Odyssey, 224
Hon, G., 399
Horace, *Epistles*, 324
Hortensius, 193, 198
Howley, Joe, 89
Huainanzi, 419–20
humility, 9
Hunter, Richard, 210, 407
hymns, 220, 222, 229

Iavolenus Priscus, 88
Ignatius of Antioch, 9
Illyricum, 198
immunity, 39
Imperial Academy (China), 414, 415
Indian science, 409, 412–14, 418, 420–3
  Inner Canon of the Yellow Emperor, Huangdi neijing, 416
intellectual heroism, 246–7
Ionia, 19, 21
Iphicrates, 173
irony, 30, 137, 183, 193, 201, 327
Isis, 323
ius civile, 84, 85, 91, 92, 93, 97, 98
ius honorarium, 96, 98
Iuventius Celsus, P., 88, 99

Janniard, Sylvain, 151
Jerome, 329, 330
Jocelyn, H. D., 307
John of Salisbury, 153
John the Lydian, *De ostentis*, 342, 346
Johnson, M. R., 392, 393
Johnston, S. I., 331
journeying, imagery of, 34–5
Judaism, 51
Julius Africanus, 406
  Cesti, 375, 402–4
  Chronography, 402

Julius Caesar, 72, 104, 167, 169, 195, 329, 333, 334, 338, 341, 345
Julius Paulus, 83–7, 89, 90, 96, 97, 98, 99, 100, 105
Jupiter, 339, 347
jurists, 83–106
Justinian, 84, 90, 91, 103, 106
Juvenal, 324

Keegan, John, 140
König, Jason, 351, 352
Kroll, Wilhelm, 342
Kronenberg, Leah, 196, 199

Labeo, 99
labour, 246
Lactantius, 315
Langlands, Rebecca, 177
Lao Zi, 415
Late Republic, intellectual culture of, 5, 331, 333, 347
Lateiner, Donald, 244
law, 3, 4, 11, 83–106, 110, 311, 325, 331, 409
Le Bohec, Yann, 141
legal ambiguity, 84, 91, 105
Lehoux, Daryn, 311, 312
Lenoir, Maurice, 130
Leptis Magna, 396
letters, 162, 163, 270, 271, 329
Lex Aquilia, 94
Lex Voconia, 93
libraries, 56, 99, 102, 103, 120, 121, 191
limitations of knowledge, 10, 13, 25, 163–78, 203–30, 283–305
literacy, 187, 188, 189
Liu Hui, 416
Livia, 71, 72, 73
Livy, 78, 153, 317
Lloyd, G. E. R., 1, 7, 399
logic, 11, 32, 67, 68, 377, 420
Lollianus, 23, 24
Loraux, Nicole, 236, 246–7, 259
Lucan, 343–4
Lucian, 54, 55
  Fisherman, 48
  Hermotimus, 48–9
  Nigrinus, 28–9, 30, 35, 51, 53
  On Salaried Posts, 39
  On the Parasite, 14
Lucius Verus, 86
Lucretius, *De rerum natura*, 35, 125

Lucullus, 78, 165, 193, 198
Lüshi chunqiu, 418, 419–20

Ma, John, 350
Macedonia, 170, 383, 386
Machiavelli, 150, 151, 154
Madauros, 42
magic, 40–1
Mago, 101, 186, 187, 189, 190, 191–2, 194, 199, 200, 202
Manilius, 102
Marcellus, 71
Marcellus of Side, *Iatrica*, 205, 222–4
Marcus Aurelius, 85, 86, 204, 269
Marduk-šāpik-zēri, 410
Marincola, John, 140, 231–2, 235, 243
Marinus, 395–6, 400
Marius, 134–5, 136
Mark Antony, 163
Martin, R., 195
Masurius Sabinus, 97, 98, 99
mathematics, 3, 5, 8, 10, 11, 17, 111, 348–73, 374–408, 409, 412, 415, 416–17, 421
  seventeenth-century, 374, 390, 405, 408
Mattern, Susan, 260
Maximus of Tyre, 33, 42, 43, 44, 45–6, 47, 51, 52–3, 55
measurement, 228
medical imagery, 34
medicine, 2–3, 4, 11, 12, 17, 26, 27, 34, 101, 110, 130, 222–4, 260–82, 283–305, 402, 409, 410, 412, 413–14, 415, 416, 418–19, 421, 422
  hostility to, 189
memory, 22, 24, 121, 266, 275, 294, 314, 388
Menecrates of Ephesus, 192
Menecrates of Xanthus, 237
Menelaus (astronomer), 365
Menippean satire, 307–8, 327–8
Mesopotamia, 342, 345
Mesopotamian science, 409, 410–12, 413, 418, 420–3
Metagenes, 114
metaphysics, 32, 338
Metellus, 314, 317–18
meteorology, 32, 187, 391
Methodists, 275
Meton, 364, 369–71
metre, 225
Metrodorus, 69
Milindapañha, 413
military writing, 14, 17, 129–52, 153–81, 226, 382–6, 402, 409, 415

Chinese, 129
  relationship between Greek and Roman, 141–2, 155–6, 158
  Renaissance, 129, 131, 133, 134, 135, 147, 150, 153–4, 164
Millar, Fergus, 84
Millis, Walter, 152
miscellany, 404
Mithridates, 165
Modestinus, 90
modesty, 9–10, 52, 355, 357
Mohists, 421
Momigliano, Arnaldo, 311, 312
money, 10, 11
moral virtue, 2, 4, 30, 32, 42, 49–50, 67, 68, 118, 172–3, 174, 235, 248–57, 259, 412
Morysine, Rycharde, 154, 164
mosaics, 53, 56
Moses, 51
Mouseion, 216, 228
Mozi, 418
Mucius Scaevola, Augur, 88
Mucius Scaevola, Pontifex, 88, 92, 97, 100, 103, 319, 325
Mucius Volusius, Q., 104
Mucius, P., 102
Musaeus, 45
Muses, 18, 34, 123, 204, 208, 209, 211, 216–17, 218, 219, 221, 228, 338
music, 4, 11, 16, 19, 31, 34, 110, 352, 377, 379–82, 383, 395
Musonius Rufus, 34, 39, 47, 65, 80

nature, 24, 32, 108, 124–7, 205, 287, 296, 298, 311
Nelsestuen, G., 195, 196
Neoplatonists, 387
Neratius Priscus, 99
Nero, 38, 39, 58, 64, 79, 80, 81, 104
Nerva, 155, 178–9, 180
Netz, Reviel, 351, 352, 353, 358, 365, 417
Nicagoras, 22
Nigidius Figulus, 329–47
  Sphaera, 342
Nine Chapters on Mathematical Procedures, Jiuzhang suanshu, 416
North, John, 317, 331
novel, 5
Numa, 316–17, 318, 320, 323
Numenius, 51

Octavia, 71, 72, 73
Odysseus, 164

Olympic festival, 19
Onasander, 130, 131, 135, 139, 141–6, 167, 169, 172
Oppian, *Halieutica*, 204, 205, 206, 207–8, 212, 213–14, 215–16, 217–22, 226, 227, 228, 230
optics, 393, 400–5
Orchomenus, 238
Orpheus, 45
otium, 61, 133, 182, 202
Ovid, 131, 139
  Fasti, 309, 318
  Metamorphoses, 135

painting, 4, 13
Panaetius, 37, 52
Papinian, 90
Papirius (jurist), 102
Papirius Fabianus, 78
Pappus, 404
Paret, Peter, 151–2
Parmenides, 27, 421
Parpola, Simo, 410
patronage, 6, 22, 40, 65, 409, 412, 413, 422–3
Paul, 9
Paulus, 72
Peloponnesian War, 73, 246
Penates, 253, 314, 317, 318
Pergamum, 120, 121
Pericles, 37
Peripatetics, 33, 44, 267, 330, 382
Persia, 156, 170, 254, 383
persona, 1–3, 5, 38, 178, 231–3, 258, 314, 321, 324, 326, 327
persuasion, 1, 7, 49, 80, 95, 232, 361, 401, 409, 414, 423
Petronius, *Satyrica*, 187
Phaedo, *Zopyrus*, 55
pharmacology, 270
Philip of Macedon, 38
Philo, 51, 383
Philodemus, *On Rhetoric*, 12, 22
philosophy, 3, 4, 5, 6, 11, 12, 14, 16, 17, 21, 25, 27–57, 58–82, 101, 110, 133, 163, 183, 197, 198, 234, 255, 286, 289, 296, 300–1, 304, 306–28, 333, 385, 404, 406, 409, 415, 417, 421
  as an overarching discipline, 12, 18, 26, 420
  definitions of, 31–3, 41–3
  hostility to, 77–81
  relationship between Greek and Roman, 75–8

Philostratus
  Gymnasticus, 4
  Life of Apollonius of Tyana, 38, 350
  Lives of the Sophists, 19–25
Philostratus of Lemnos, 21, 22
Phormio, 132–3
Phrygia, 237
physical appearance, 30, 53–6, 67
physics, 32, 67, 81, 354, 420
physiognomy, 55, 410
physiology, 110, 284, 287, 292, 303, 305, 366
piloting, 4
Pinnius, 182
Pisonian conspiracy, 58
Plato, 3, 12, 18, 27, 31, 32, 36, 37, 45, 46, 50, 51, 53, 54, 62, 72, 73, 75, 141, 153, 196, 303, 304, 326, 376, 381, 387, 406
  Gorgias, 50, 414
  Phaedo, 35
  Phaedrus, 35
  Republic, 35, 36, 141, 256
  Timaeus, 330
Platonists, 28, 31, 33, 41, 44, 54, 141, 146, 388
Plautus, 123, 324
Pliny the Elder, 116, 126–7, 188, 189–90, 192, 199, 200, 201, 210, 318, 362
Pliny the Younger, 93, 165
Plutarch, 25, 44, 66, 80–1
  Bravery of Women, 66
  Cato the Elder, 106
  Consolatio ad Apollonium, 69
  Consolation to his Wife, 80–1
  On Curiosity, 80
  On Exile, 80
  On Feeling Good, 80
  On Self-Praise, 9–10
  On the Delays of Divine Vengeance, 66
  On the Generation of the Soul in the Timaeus, 66
  On the Proposition that the Philosopher ought to Converse above all with Political Leaders, 37
  On Tranquillity of Mind, 66
  Precepts of Statecraft, 66
  Sympotic Questions, 15–19, 66
  The E at Delphi, 66
  To an Uneducated Ruler, 37
poetry, 4, 58, 74, 120, 129, 139, 192, 203–30, 241, 256, 341
Polemo, 19–20
political authority, 5–7, 36–9, 107
political function of scientific writing, 5–7, 333–4
political theory, 234
Pollux, 25
  Onomasticon, 11
Polyaenus, 130, 171
Polybius, 130, 139, 141, 166, 167, 232, 242–3, 386, 396, 397–8, 404, 405, 406
Polydus, 115
polymathy, 110–12
Pompeii, 53
Pompey, 72, 164, 169, 329, 333
Pomponius, 92, 97, 100, 101, 102–3, 104, 105
  Enchiridion, 100
Porphyry, 380
Poseidon, 219
Posidonius, 382, 386, 396, 406
  On the Ocean, 397
practical knowledge, 10, 12–13, 14, 24, 46, 82, 110, 118, 140, 190, 216, 217, 218, 226
  in tension with theory, 129–52, 155–6, 157–78, 182–202, 204, 278
praetor, 316
Praetor's Edict, 90, 96, 98
prefaces, 5, 6, 16, 21, 23, 60–7, 117, 119–21, 124, 142–5, 159–60, 161, 184, 199, 209, 214, 215–16, 217, 221, 317, 378
Presocratic philosophy, 27
Priapus, 323
Price, Simon, 317, 331
Proclus, 380, 406
  Elements of Physics, 386
  Elements of Theology, 375, 386–8, 389, 390
Proculus, 87, 94, 95, 98, 99
Prodicus, 35
progress, 365
proof, 274, 300, 352, 355, 377, 379–81, 385, 392, 393, 394, 400, 417, 421
Proust, Marcel, 406, 407
Ps.-Manetho, *Apotelesmatica*, 205, 209–10, 211, 212, 213
Ps.-Oppian, *Cynegetica*, 204, 205, 208–9, 211, 212, 214, 226–9, 230
Ps.-Plutarch, *On Fate*, 66
Ps.-Scymnus, 224–5
psychology, 284, 287
Ptolemies, 121
Ptolemy, 397, 398
  Geography, 395
  Planetary Hypotheses, 355, 358
  Syntaxis, 348–73
Publius Aelius, 102
Pulvillus, 72
Pyrrhonists, 44
Pyrrhus, 163

Pythagoras, 37, 51, 101, 197, 398, 417
Pythagoreans, 117, 316
Pytheas, 397
Pythius, 114, 115

Q. Cornelius Maximus, 104
qualifications, lack of in ancient scientific culture, 1, 25
Quintilian, 49–50, 58, 123, 129
Quintus Cicero, 330, 334–42
Quintus Metellus, 175
Quintus Veranius, 141

readers, 9, 161, 175, 176, 177, 188, 231–59, 355, 356, 363, 423
reading practices, 68
reason, 32, 33, 71, 88, 304, 312, 322, 328, 361, 421
Reay, Brendon, 184
recipes, 189, 194, 270
Relihan, J. C., 307–8
rescripts, 84, 85, 86, 87, 103
Rhea Silvia, 255, 257
rhetoric, 4, 5, 7, 11, 13, 16, 17, 19–25, 32, 41, 49–50, 58, 84, 92, 93, 95, 96, 130, 131, 135, 136–9, 248, 262, 331, 350, 351, 414
Roman Empire, 5, 6, 8, 130, 156–7, 181, 229, 351
Roman identity, 59, 101, 170
Rome, 19, 28, 59, 77, 78, 93, 100, 102, 103, 104, 112, 144, 148, 183, 232, 241, 242, 249, 250, 251, 254, 261, 267, 307, 329, 331, 339, 344, 346, 353
Romulus and Remus, 255, 257
Rose, L. E., 377
Rüpke, Jörg, 310, 312, 319
Rutilius, 73

Sallust, 131, 136, 139, 153
　Bellum Iugurthinum, 134–5
Salvius Julianus, 90, 98
Samos, 93
Samothrace, 240, 326
Sasernas (agricultural writers), 194, 198, 199
satire, 14–15, 48–9, 55, 195, 196, 199, 306–28, 420
Saturn, 324
saving the phenomena, 363
Sceptics, 33, 44
Schofield, Malcolm, 335
schools, legal, 104
schools, philosophical, 12, 14, 17, 43, 51, 56, 67, 104, 314, 412

Schultze, Clemence, 231–2, 235
scientia, 89, 110, 157, 158, 164, 165, 166, 167, 337
Scipio, 37, 168, 173
Scopelian, 20
sculpting, 4, 11, 13
Scythia, 225, 383
second-person usage, 263–4, 267, 268, 271, 273, 401
Seleucids, 351
self-effacement, 8–9, 16, 25, 161, 171, 172, 179–80
senatorial resolutions (*senatus consulta*), 84
senators, 89, 90, 97, 184, 201, 316, 332, 342, 343
Seneca, 12, 52, 58–82, 153, 382
　Ad Marciam, 58, 68–9, 70–4, 79
　Ad Polybium, 68–70, 79
　consolations, 65, 66, 68–75
　De beneficiis, 61, 62, 66, 75–7, 79, 81
　De breuitate vitae, 61
　De clementia, 38, 66, 79
　De constantia sapientis, 61–2, 66
　De ira, 60, 66
　De otio, 66, 79
　De providentia, 60, 66, 79
　De tranquillitate animi, 60–1, 66
　De vita beata, 61
　Exhortationes, 81
　Letters, 44–5, 52, 62–3, 64, 66, 67–9, 77, 78, 79, 81
　Natural Questions, 78
Seneca the Elder, *Controversiae*, 318
Septimius Severus, 83, 85, 86
Serapis, 323
sermo humilis, 95
Sertorius, 169
Servius, 306
Servius Sulpicius Rufus, 88, 93, 94, 102, 103, 104
Seven Sages, 117
Sextus Aelius, 102
　Tripertita, 102
Sextus Empiricus, 44
　Against the Grammarians, 12
Shiji, 417, 418
Sicily, 72, 397
Simplicius, 376
Skydsgaard, J. E., 195
slaves, 13, 184, 185–91, 261
Smith, Christopher, 142, 146
Smith, R., 377
Smyrna, 19

Socrates, 12, 18, 27, 34, 38, 53, 54, 55, 73, 75, 77, 109, 117, 196, 327, 420
sollertia, 161, 163, 166, 167, 168, 172, 174
sophists, 11, 19–25, 27, 47, 350
Sophocles, 236, 239
sorcery, 262
soul, 283–305
Sparta, 162, 169, 170
Spinoza, 375
Sporus, *Keria*, 404
statutes (*leges*), 84
Stoics, 17, 29, 32, 33, 36, 37, 40, 41, 44–5, 47, 48, 51, 52, 55, 61, 63, 66, 68, 70, 74, 75, 80, 286, 312–13, 315, 317, 319, 321–3, 324, 325, 330, 336
Strabo, 252, 350, 366, 375, 394–400, 407
stream imagery, 122–3
Suanshushu (*Writings on Reckoning*), 416
subdivisions, 3, 22, 110, 126, 194, 325, 378, 384
subversion of authority, 39, 91, 203, 307, 359
succession literature, 104
Suda, 141
Suetonius, 79, 344
  Augustus, 343
Sulla, 72, 169
Suśruta Samhitā, 413
Swoboda, A., 342
Sybaris, 244
symposium, 15–19
Syracuse, 72, 73

Tablet of Cebes, 35
Tacitus, 88, 104
  Agricola, 165
Tages, 343
Tannery, P., 391, 394
Tarquinius Priscus, 320
Tarquinius Superbus, 173
Tarrant, Richard, 139
Taub, Liba, 392
technai (arts), 11–12, 14, 18, 22–5, 26, 27, 386, 402
Thales, 197
Thapsacus, 399
Thebes, 170
Theodorus, 376
Theodosian Code, 90, 103
theology, 32, 53, 220, 288, 306–28, 354, 386–8, 409, 421
Theon (astronomer), 365
Theophrastus, 50, 192
  De causis plantarum, 198
Theopompus, 232

theoretical knowledge, 13–14, 23–5, 82, 110, 267, 279, 283, 354, 355
Thesmopolis, 39, 41
Thrace, 162, 238, 383
Thucydides, 232, 236, 243, 246–7, 259
Tiberius, 65, 72, 73, 97
Timaeus, 225, 232
Timocharis, 364
Timosthenes, 396
Titius Aristo, 93
Torsello, Marino Sanudo, 153
tradition as a source of authority, 2, 4, 17, 25, 30, 51–3, 67, 84, 98–9, 100–4, 114–19, 149–51, 200, 366, 381, 409, 410, 414, 422
Trajan, 9, 155, 180
Trebatius Testa, C., 92, 102, 103
Tremelius Scrofa, 182, 183, 185, 186, 187, 193, 194, 195–6, 199, 200
Tribonian, 90, 91, 103, 106
Tubero, Q., 102, 104
ṭupšarru Enūma Anu Enlil, 410–11, 422
Turranius Niger, 182, 196–7
Twelve Tables, 89, 97, 98, 102

Ulpian, 86, 87, 88, 90, 94, 97, 98, 99, 100, 103, 105
Upaniṣads, 412, 413, 414, 422
Urseius Ferox, 94, 95, 98, 99
utility, 11, 142, 349, 353, 359, 362

Valerius Maximus, 153, 159, 170, 171, 172, 177, 318
Van Hoof, Lieve, 80
Varro, 14, 115, 252, 256, 339, 343
  Antiquitates, 201
  Antiquitates rerum divinarum, 306–28
  Antiquitates rerum humanarum, 85
  Curio, 316, 319
  De lingua latina, 201, 307, 326
  De re rustica, 65, 66, 116, 126, 182–202, 307, 325
  Menippean Satires, 307–8, 328
Vedas, 412
Vegetius, 130, 131, 135, 139, 147–52, 153, 154–5, 156, 160, 165
Venus, 323, 324
Veranius, *De verbis pontificalibus*, 310
Vespasian, 39, 79, 80
Vesta, 314, 318
veterinary knowledge, 190
Veyne, Paul, 350
Vincent de Beauvais, *Speculum maius*, 127
Virgil, 153, 199

vision, 145–6, 207, 227, 228
Vitrac, B., 391, 393, 394
Vitruvius, 4, 6, 11, 12, 26, 107–28, 166, 167, 386, 420
Volusius Maecianus, 87

Wang Chong, 418
West, Martin, 217
White, K. D., 192
Whitmarsh, Tim, 351, 352, 386
wills, 85–6
Wilson, M., 77
Wiseman, Peter, 256
Wissowa, G., 307
witnesses, 9, 260–82

writing on request, 52, 60, 65–6, 265, 349
Wu Di, 414

Xenocrates, 28
Xenophon, 54, 141, 167, 185, 192, 196
   Cynegeticus, 225–6
   Discourse on the Command of Cavalry, 166
   Oeconomicus, 200

Zeno, 28, 29, 34, 51, 53, 61, 75, 197
Zeus, 220, 221
Zhoubi suanjing, 416, 417
Zhuangzi, 419–20
zoology, 218, 227
Zopyrus, 55

For EU product safety concerns, contact us at Calle de José Abascal, 56–1°, 28003 Madrid, Spain or eugpsr@cambridge.org.

www.ingramcontent.com/pod-product-compliance
Lightning Source LLC
LaVergne TN
LVHW080309260326
834688LV00038B/1029